$89.95UCF

Fuzzy Control

Fuzzy Control

Synthesis and Analysis

Edited by

Shehu S. Farinwata
*Ford Motor Company, Research Laboratory,
Dearborn, Michigan, USA*

Dimitar Filev
Ford Motor Company, AMTDC, Redford, Michigan, USA

Reza Langari
Texas A & M University, College Station, Texas, USA

JOHN WILEY & SONS, LTD
Chichester · New York · Weinheim · Brisbane · Singapore · Toronto

Copyright © 2000 by John Wiley & Sons, Ltd
                    Baffins Lane, Chichester,
                    West Sussex PO19 1UD, England
                    National   01243 779777
                    International (+44) 1243 779777

e-mail (for orders and customer service enquiries): cs-books@wiley.co.uk
Visit our Home Page on http://www.wiley.co.uk or http://www.wiley.com

All Rights Reserved. No part of this publication may be reproduced, stored in a retrieval, system, or transmitted, in any form or by any means, electronic, mechanical, photocopying, recording, scanning or otherwise, except under the terms of the Copyright, Designs and Patients Act 1988 or under the terms of a licence issued by the Copyright Licensing Agency, 90 Tottenham Court Road, London, W1P 9HE, UK, without the permission in writing of the Publisher, with the exception of any material supplied specifically for the purpose of being entered and executed on a computer system, for the exclusive use by the purchaser of the publication.

The Editors and contributors have asserted their right under the Copyright, Designs and Patents Act, 1988, to be identified as the editors of 2nd contibutors to this work.

Neither the author(s) nor John Wiley & Sons Ltd accept any responsibility or liability for loss or damage occasioned to any parson or property through using the material, instructions, methods or ideas contained herein, or acting or refraining from acting as a result of such use. The author(s) and Publisher expressly disclaim all implied warranties, including merchantability or fitness for any particular purpose. There will be no duty on the author(s) or Publisher to correct any errors or defects in the software.

Designations used by companies to distinguish their products are often claimed as trademarks. In all instances where John Wiley & Sons is aware of a claim, the product names appear in initial capital or all capital letters. Readers, however, should contact the appropriate companies for more complete information regarding trademarks and registration.

Other Wiley Editorial Offices

John Wiley & Sons, Inc., 605 Third Avenue,
New York, NY 10158-0012, USA

Wiley-VCH Verlag GmbH,
Pappelallee 3, D-69469 Weinheim, Germany

Jacaranda Wiley Ltd, 33 Park Road, Milton,
Queensland 4064, Australia

John Wiley & Sons (Asia) Pte Ltd, 2 Clementi Loop #02-01,
Jin Xing Distripark, Singapore 129809

John Wiley & Sons (Canada) Ltd, 22 Worcester Road,
Rexdale, Ontario M9W 1L1, Canada

Library of Congress Cataloging-in-Publication Data

British Library Cataloguing in Publication Data

A catalogue record for this book is available from the British Library

ISBN 0 471 98631 3

Typeset in 10/12pt Times by Thomson Press (India) Ltd, New Delhi
Printed and bound in Great Britain by Antony Rowe, Ltd, Chippenham, Wilts
This book is printed on acid-free paper responsibly manufactured from sustainable forestry
in which at least two trees are planted for each one used for paper production.

Contents

Editor's Preface — xi
List of Contributors — xix
About the Editors — xxi
Acknowledgments — xxiii

MODELING — 1

1 Information Granularity in the Analysis and Design of Fuzzy Controllers — 3
  1.1 Introduction — 3
  1.2 The Basic Architecture of the Fuzzy Controller and its Non-linear Relationships — 4
  1.3 Set-Based Approximation of Fuzzy Sets — 7
  1.4 Information Granularity of the Rules of the Fuzzy Controller — 10
    1.4.1 Fuzzy Sets and Information Granularity — 11
  1.5 Robustness Properties of the Fuzzy Controller — 13
  1.6 Linguistic Information as Inputs of the Fuzzy Controller — 17
  1.7 Conclusions — 21
  Acknowledgment — 21
  References — 21

2 Fuzzy Modeling for Predictive Control — 23
  2.1 Introduction — 23
  2.2 Fuzzy Modeling — 24
    2.2.1 Outline of the Modeling Approach — 24
  2.3 Extraction of an Initial Rule Base — 26
  2.4 Simplification and Reduction of the Initial Rule Base — 27
    2.4.1 Similarity Analysis — 28
    2.4.2 Simplification and Reduction — 28
  2.5 Model Predictive Control — 30
    2.5.1 Basic Principles — 30
    2.5.2 Optimization in MPC — 31
    2.5.3 The Branch-and-Bound Optimization — 32
  2.6 Modeling and Control of an HVAC Process — 34
    2.6.1 Initial Modeling of the System — 35
    2.6.2 Validating the Initial Model — 35
    2.6.3 Simplifying the HVAC Model — 39
    2.6.4 Control Results — 40
    2.6.5 Summary of Results — 41
  2.7 Concluding Remarks — 42
  Appendix A: The Gustafson–Kessel Clustering Algorithm — 43

	Appendix B: The Rule Base Simplification Algorithm	44
	References	45

3 Adaptive and Learning Schemes for Fuzzy Modeling — 47
- 3.1 Introduction — 47
- 3.2 Identification Problems of the TSK Fuzzy Models — 49
- 3.3 Criteria and Schemes for Learning and Evaluation of Fuzzy Models — 54
  - 3.3.1 The Global Learning Criterion, Q_G — 54
  - 3.3.2 The Local Learning Criterion, Q_L — 55
  - 3.3.3 Evaluation Criteria — 56
- 3.4 Algorithms for Global Learning by Fuzzy Models — 56
  - 3.4.1 Comparison of the Learning Algorithm Using a Numerical Example — 59
- 3.5 Algorithm for Local Learning by Fuzzy Models — 63
- 3.6 Reinforced Learning Algorithm — 66
- 3.7 Simulation Results for Control Applications — 67
- 3.8 Conclusions — 70
- References — 70

4 Fuzzy System Identification with General Parameter Radial Basis Function Neural Network — 73
- 4.1 Introduction — 73
- 4.2 Fuzzy Systems through Neural Networks — 75
  - 4.2.1 Radial Basis Function Neural Networks — 77
- 4.3 General Parameter Radial Basis Function Network (GP RBFN) — 78
  - 4.3.1 General Parameter Method for System Identification — 79
  - 4.3.2 GP RBFN Training Algorithm — 80
- 4.4 GP RBFN Adaptive Fuzzy Systems (AFSs) — 81
  - 4.4.1 Basic Algorithm — 81
  - 4.4.2 Unbiasedness Criterion for the GP RBFN AFS — 83
- 4.5 Simulation Results — 84
- 4.6 Conclusion — 90
- References — 91

ANALYSIS — 93

5 Lyapunov Stability Analysis of Fuzzy Dynamic Systems — 95
- 5.1 Introduction — 95
- 5.2 Mathematical Preliminaries — 96
- 5.3 Construction of Fuzzy Dynamic Models from Discrete-Time Stochastic Models — 97
  - 5.3.1 Construction of Fuzzy Dynamic Models via Fuzzy Composition — 98
  - 5.3.2 Construction of a Fuzzy Dynamic Model via the Fuzzy Extension Principle — 99
- 5.4 Stability Analysis of Fuzzy Dynamic Systems — 99
  - 5.4.1 Convergence in Fuzzy Dynamic Systems — 100
  - 5.4.2 Stability of Fuzzy Dynamic Systems — 100
  - 5.4.3 The Direct Lyapunov Method for Fuzzy Dynamic Systems — 103
- 5.5 Application—First-Order Fuzzy Dynamic System — 104
- 5.6 Concluding Remarks — 110
- References — 111

6 Passivity and Stability of Fuzzy Control Systems — 113
- 6.1 Introduction — 113
- 6.2 Fuzzy Control Systems — 114
  - 6.2.1 Mamdani Fuzzy Controllers — 114
  - 6.2.2 Takagi–Sugeno Fuzzy Control Systems — 115

6.3	Stability and Passivity of Fuzzy Controllers		117
	6.3.1	Basic Concepts	117
	6.3.2	Passivity of QPI Controllers	122
	6.3.3	Passivity of DPS Controllers	123
	6.3.4	Passivity of Polytopic Differential Inclusions	126
6.4	Stability of Feedback Control with Fuzzy Controllers		130
	6.4.1	Feedback Control with QPI Mamdani Controllers	131
	6.4.2	Feedback Control with DPS Mamdani Controllers	131
	6.4.3	Feedback Control with Linear Takagi–Sugano Controllers	133
6.5	Applications		135
	6.5.1	Control of LTI Systems by Fuzzy Controllers	136
	6.5.2	Fuzzy Control of Euler–Lagrange Systems	137
6.6	Conclusions		138
Acknowledgments			139
Appendix			139
References			142

7 Frequency Domain Analysis of MIMO Fuzzy Control Systems — 145

7.1	Introduction	145
7.2	Multiple Equilibria in MIMO Fuzzy Control Systems	146
7.3	Frequency Analysis of Limit Cycles	148
7.4	Robust Analysis of Limit Cycles using Singular Values	149
7.5	Conclusions	151
Acknowledgments		151
References		151

8 Analytical Study of Structure of a Mamdani Fuzzy Controller with Three Input Variables — 153

8.1	Introduction	153
8.2	Configuration of the Fuzzy Controller	154
8.3	Analytical Study of the Fuzzy Controller Structure	157
8.4	Conclusion	162
Acknowledgment		162
References		162

9 An Approach to the Analysis of Robust Stability of Fuzzy Control Systems — 165

9.1	Introduction		165
9.2	Perspective		166
9.3	The Nominal Fuzzy Control Problem		167
9.4	Equilibrium Points for Fuzzy Controlled Processes		168
9.5	Fuzzy Robustness Analysis		168
	9.5.1	Robustness Problem Statement	169
	9.5.2	Concepts of Sensitivity and Robustness	170
	9.5.3	Formulation of Fuzzy System Robustness	171
	9.5.4	The Main Result	173
	9.5.5	Derivation of the Main Result	173
9.6	Generalization of the Robust Stability Result		174
	9.6.1	Virtual Interactions Based on Stability	175
	9.6.2	General Result for Robust Stabilization	177
	9.6.3	Minimizing dV	177
9.7	Fuzzy Extremes of Perturbations		178
	9.7.1	A Measure of Fuzzy Robustness	179
	9.7.2	Comments	180
9.8	Application Example		182

		9.8.1 Problem Statement	185
		9.8.2 Simulation Studies and Results	186
		9.8.3 Discussion	196
	9.9	Conclusions	196
	Bibliography		197

10 Fuzzy Control Systems Stability Analysis with Application to Aircraft Systems — 203

- 10.1 Introduction — 203
  - 10.1.1 Fuzzy Control — 204
  - 10.1.2 Lyapunov Stability of Non-linear Fuzzy Control Systems — 205
  - 10.1.3 The Fuzzy Control Problem — 205
  - 10.1.4 Equilibrium Points for Fuzzy Controlled Processes — 207
  - 10.1.5 The Partitioned State Space — 208
  - 10.1.6 Dissipative Mapping and Input–Output Stability — 208
  - 10.1.7 Dissipative Mapping for the Fuzzy Control System — 210
  - 10.1.8 Stability of Linear Fuzzy Control Systems — 211
  - 10.1.9 Positive Realness and Dissipativeness — 212
  - 10.1.10 Verifying Dissipativeness — 214
- 10.2 Linear Continuous-Time Model Application — 214
  - 10.2.1 A Missile Autopilot — 214
  - 10.2.2 Analysis — 215
  - 10.2.3 Simulation Studies and Results — 219
  - 10.2.4 Conclusions — 220
- 10.3 Linear Discrete-Time Model Application — 221
  - 10.3.1 Advanced Technology Wing Aircraft Model — 221
  - 10.3.2 Introduction — 221
  - 10.3.3 The ATW Problem — 222
  - 10.3.4 Control Architecture — 223
  - 10.3.5 Control Rule Synthesis — 224
  - 10.3.6 Stability Analysis — 227
  - 10.3.7 Conclusions — 232
- 10.4 Summary — 233
- Bibliography — 233

SYNTHESIS — 237

11 Observer-Based Controller Synthesis for Model-Based Fuzzy Systems via Linear Matrix Inequalities — 239

- 11.1 Introduction — 239
- 11.2 Takagi–Sugano Models — 240
  - 11.2.1 Continuous-Time T–S Models — 240
  - 11.2.2 Continuous-Time T–S Controllers and Closed-Loop Stability — 241
  - 11.2.3 Discrete-Time T–S Controllers — 242
- 11.3 LMI Stability Conditions for T–S Fuzzy Systems — 243
  - 11.3.1 The Continuous-Time Case — 243
  - 11.3.2 The Discrete-Time Case — 243
- 11.4 Fuzzy Observers — 244
  - 11.4.1 Why Output Feedback? — 244
  - 11.4.2 Continuous-Time T–S Fuzzy Observers — 244
  - 11.4.3 Separation Property of the Observer/Controller — 246
  - 11.4.4 Discrete-Time T–S Fuzzy Observers — 247
- 11.5 Numerical Example — 249
- 11.6 Conclusion — 252
- References — 252

12 LMI-Based Fuzzy Control: Fuzzy Regulator and Fuzzy Observer Design via LMIs — 253
- 12.1 Introduction — 253
- 12.2 Takagi–Sugano Fuzzy Model — 254
- 12.3 Fuzzy Regulator Design via LMIs — 255
  - 12.3.1 Parallel Distributed Compensation — 255
  - 12.3.2 Control Performance Represented by LMIs — 256
- 12.4 Fuzzy Observer Design — 262
- 12.5 Conclusions — 263
- References — 264

13 A framework for the Synthesis of PDC-Type Takagi–Sugano Fuzzy Control Systems: An LMI Approach — 267
- 13.1 Introduction — 267
  - 13.1.1 Brief Historical Overview — 267
- 13.2 Background Materials — 268
  - 13.2.1 T–S Fuzzy Model of Non-linear Dynamic Systems and its Stability — 268
  - 13.2.2 PDC-Type T–S Fuzzy Control System and its Stability — 269
- 13.3 Stability LMIs as a Framework for the Synthesis of PDC-Type T–S Fuzzy Control Systems — 271
- 13.4 Pole Placement Constraint LMIs as Performance Specifications for the Synthesis of PDC-Type T–S Fuzzy Control Systems — 274
- 13.5 An Extension to PDC-Type T–S Fuzzy Control Systems with Parameter Uncertainties — 276
- 13.6 A Simulated Example — 279
- 13.7 Concluding Remarks — 281
- References — 282

14 On Adaptive Fuzzy Logic Control on Non-linear Systems—Synthesis and Analysis — 283
- 14.1 Introduction — 283
- 14.2 Control Objective — 284
- 14.3 DFLS Identifier — 285
- 14.4 Control Law of the System — 287
- 14.5 Adaptive Law for the Parameter Vector $\bar{Y}$ — 288
- 14.6 Adaptive Law for $\hat{g}$ — 290
- 14.7 Stability Properties of the DFLS Control Algorithm — 291
- 14.8 Illustrative Application — 292
- 14.9 Concluding Remarks — 295
- Appendix: Proof of Theorem 7.1 — 296
- References — 307

15 Stabilization of Direct Adaptive Fuzzy Control Systems: Two Approaches — 309
- 15.1 Introduction — 309
- 15.2 Integral Sliding-Mode Adaptive FLC: Approach I — 310
  - 15.2.1 Structure of an Integral Sliding-Mode Adaptive FLC — 310
  - 15.2.2 Stabilization of the Integral Sliding-mode Adaptive FLC — 311
  - 15.2.3 Properties of the Integral Sliding-Mode Adaptive FLC — 313
- 15.3 New Fuzzy Logic Based Learning Control: Approach II — 314
  - 15.3.1 Structure of the New Fuzzy Logic Based Learning Control — 314
  - 15.3.2 Stabilization of the New Fuzzy Logic Based Learning Control — 314
  - 15.3.3 Discussion of the New Fuzzy Logic Based Learning Control — 316
- 15.4 Simulation — 316

		15.4.1	Approach I	316
		15.4.2	Approach II	317
	15.5	Concluding Remarks		319
	References			320

16 Gain Scheduling Based Control of a Class of TSK Systems   321

	16.1	Introduction	321
	16.2	TSK Model as a Gain Scheduled System	322
	16.3	Stability Conditions for TSK Fuzzy Systems	324
	16.4	Synthesis of TSK Compensators	327
	16.5	Analytic Form of the Polytopic TSK Compensator	330
	16.6	Parameterization of Non-parametric TSK Compensators	333
	16.7	Conclusion	334
	References		334

17 Output Tracking Using Fuzzy Neural Networks   335

	17.1	Introduction	335
	17.2	Problem Statement—Assumptions	337
	17.3	The Structure of the Controller	339
	17.4	The Main Results	340
	17.5	The Learning Algorithm	341
	17.6	Illustrative Examples	342
	17.7	Comprehensive Results and Conclusions	346
	References		347

18 Fuzzy Life-Extending Control of Mechanical Systems   349

	18.1	Introduction		349
	18.2	Architecture of Life-Extending Control Systems		351
	18.3	Life-Extending Control of a Rocket Engine		352
		18.3.1	Inner Loop Feedback Controller for LECS-1	353
		18.3.2	Outer Loop Fuzzy Controller for LECS-1	355
		18.3.3	Results and Discussion for LECS-1	359
	18.4	Life-Extending Control of a Power Plant		362
		18.4.1	Inner Loop Feedback and Gain Scheduling	364
		18.4.2	Fuzzy Controller	367
		18.4.3	Results and Discussion	372
	18.5	Summary and Conclusions		379
		18.5.1	Control System Stability	380
	Acknowledgments			381
	Appendix A: Brief Description of the Rocket Engine			381
	Appendix B: Brief Description of the Power Plant			382
	References			382

Epilogue   385

Index   387

Editors' Preface

There are 18 chapters in this edited volume addressing various aspects of design and analysis of fuzzy control systems. Though not originally intended, but in the interest of completeness, some modeling aspects have been included. The unifying theme of this volume is *Synthesis* and *Analysis*. It is intended to look deeply into methods of analyzing, in a quantitative manner, the important areas of fuzzy control systems, namely synthesis and dynamical properties. A variety of fuzzy modeling approaches are now appearing, mostly based on the Takagi–Sugeno–Kang (TSK) or Takagi–Sugeno (TS) models, including singular perturbation models, and a host of other control-theoretic models. One of the goals of this volume is to look closely at the analytical structures of these models and bring out the characteristic systematisms in them, and to ascertain methods for studying their dynamic properties in a formal manner. On the synthesis front, efforts concentrate on sound foundational and systematic formulations and implementations of these models whose applications range from decision analysis to optimization and control. The *synthesis* and *analysis* fronts are not mutually exclusive, but some of the contributions exhibit a certain inclination towards one or the other. Many of the analysis efforts have focused on model structure and the model's suitability for use in decision and control. Thus, issues pertaining to stability, controllability, observability, robustness, membership function parameterization, clustering, universe quantization, fuzzification and defuzzification methods and inferencing, have been addressed in one form or another as they pertain to the behavior of the synthesized model within a closed-loop control setting.

Let us look at two interesting fronts in this field. On the computational front, the control algorithms for fuzzy control systems have fallen into the area termed 'soft computing'. This entails the use of a number of constituent approaches such as fuzzy logic and fuzzy sets, neural computation and genetic algorithms in an effort to solve a control and decision problem. This constitutes the major core of intelligent control. The humanistic nature of these techniques is clear and is the key to the 'intelligence' in the intelligent control of systems. Of particular interest are new and recent directions in stability and robustness analysis, controller and observer designs, most of which employ the TS fuzzy model or system. On the applications front, the areas of fuzzy, and neuro-fuzzy control in particular, have themselves witnessed the dramatic entry of application-oriented material, especially from Japan and Europe, and recently from the United States. Moreover, within the last five to ten years, analytical issues such as those addressed in this volume have been rekindled to complement the vast application entries, and to expatiate upon the excellent pioneering works of L. A. Zadeh which laid the platform for everything we are doing, Kickert, King, Mamdani and his colleagues, Braae, Rutherford, Tong, Kizka *et al.*, Ray *et al.*, de Glas, Gupta, Cumani, Kania, and Sugeno; to mention just a few or run the risk of running out of space, for there were just so many that are worthy of mention. The push for the aforementioned resurgence could be attributed to two of many sources. On the one hand, fuzzy control researchers with systems and control emphasis have felt the need to address the analytical issue in a rigorous manner as these have previously been left as open issues. On the other hand, the field of fuzzy control has witnessed relentless criticisms for the lack of a formal design methodology and rigorous performance analyses that may be comparable or at least appeal to a wider spectrum of control theorists. This, by itself, has been a formidable push.

At this juncture it would be prudent to assert that concerns over the status of fuzzy logic control, particularly in the industry, are paving the way for lively and often heated panel discussions in a variety of conferences on the progress in, and acceptance level to date of, fuzzy logic and fuzzy control. Two cases in mind being at the 1996 North American Fuzzy Information Processing Society (NAFIPS) conference at Berkeley, California, and the more recent one at the 1998 IEEE Conference on Decision and Control (CDC) in Tampa, Florida. This has provided additional impetus for writing this volume. In connection with this, there may be two important issues to look at critically:

1. Control systems' historical perspective, and
2. a perspective gained from an industrial experience.

The first perspective takes an optimistic view, using as a metric the historical evolution of certain noted advances in traditional control theory. The painful paths taken by a host of these advances, ranging from the state-space methods, adaptive control, non-linear control, and even the H-infinity control methods, may be compared with the most recent and current paths that fuzzy control seems to be taking. This may tend to create a virtual and therefore unsatisfactory comfort level to the researchers and practitioners of fuzzy logic control. It is a kind of admission that things-could-be-worse or we're-not-doing-too-badly-after-all. That there is definitely an uneasy feeling among the serious practitioners of fuzzy logic control was made abundantly clear by several comments that were made at the NAFIPS meeting. The second perspective is far from historic. It addresses the nature and nucleus of our technological being: knowledge, expertise, experiences, prejudices, attitudes and dogmas. We can probably classify the last two or three as *fears and concerns*. And perhaps much more. It addresses the fuel, the catalyst in us to make or break the growth and spread of the additional tool that we have found and are just beginning to utilize. It is about the today and tomorrow of fuzzy logic control, or soft-computing as employed in the design of control systems. In this second perspective the issues that have been raised by many researchers in fuzzy control, but more so by peers from the conventional and modern control community, are basically the analytical issues and the issue of systematic design methodology. It suffices to say right now that most of these are issues relating to the lack of systematic design methodology, and the lack of analysis tools, comparable to those found in the more traditional control systems field, such as analyses of stability, robustness, controllability, observability and so on that are now being addressed, as reported in this volume. Then there are issues pertaining to implementation, integration, hybridization, applications and cost effectiveness which are of particular appeal to the industrial sector.

On a historical note, the urge to write this volume developed right after the publication of the volume *Fuzzy Control: Implementation and Applications*, edited by M. J. Patyra and D. M. Mlynek, in which the first-named editor co-authored a chapter, and felt a strong need to complement that volume and others with a similar volume on say, '*synthesis and analysis*'. It was felt that there was just too much misunderstanding by fuzzy skeptics about the claims of fuzzy logic in systems and control, some of which were ill-conceived, and that these needed to be checked. The editors are of the view that a serious control system designer who has also labored to be knowledgeable in fuzzy control, for instance, would not, and probably does not, say that fuzzy control is a panacea for all control problems. Nor say without much rationale that fuzzy control is superior to conventional control or vice versa. Rather, a genuine effort should be made, premised by a genuine interest, to learn and understand the newer techniques and ascertain where these techniques fit, if they do, in the control designer's repertoire of design tools. The issue is whether the available tools, be they modern, conventional, unconventional, can be employed either in a complementary or even a supplementary manner, as the designer sees fit, in the best interest of a given decision and control objective. Needless to say, such an objective may be engineering in nature or may even be a matter of elegance and may include, among many things, complexity, cost, ease of implementation and integration, or even suitability for further analytical or theoretical investigation. The issue is not whether fuzzy control is better than conventional control or vice versa, or whether neural networks is a *de facto* methodology, or that because we can solve this and that difficult problem using fuzzy or neural networks, we can go ahead and scrap conventional control. The challenge is not to scrap fuzzy or conventional control if they cannot solve a problem that neural networks just solved. Unfortunately, these are some of the dogmas, prejudices and non-issues that seem to pervade a great part of the control community from academia to industry. Most of the chapters in this volume are written by people from the control community who are also using fuzzy logic or neural networks in their decision and control activities and may therefore be in a better position to talk about the capabilities and limitations of fuzzy control while viewing it truly from a control engineer's point of view only as yet another welcome design tool.

Among the many aims of this volume is to reach out to those who are seriously considering fuzzy control as a viable tool in their engineering endeavors but have concerns about systematic design and analysis issues. We would therefore expect this volume to appeal to scientists, engineers, practicing professionals, researchers, technical specialists and academicians. Furthermore, the subjects would be suitable to senior undergraduates with a good background in control theory, and graduate students in systems and control from the various traditional engineering disciplines. Good background refers to adequate preparation in conventional control theory, covering frequency domain concepts and stability, and modern control theory, covering state-space methods. Familiarity with some non-linear design and analysis tools for control systems such as the Lyapunov's stability theory would be a plus, in which case the volume may be used in a 'Special Topics' course in advanced control systems for Masters and Ph.D. level students in systems and controls with a prior exposure to fuzzy sets and fuzzy logic. It is stressed that this is not intended to be another introductory book on fuzzy control; it is intended to be a volume on quite sophisticated advanced issues in control theory as they apply to fuzzy control. The adequately prepared audience should find it a joy to acquire and read.

The chapters are written by many authors from all over the world. They are experts in their respective domains and are very active in research and development in various aspects of this exciting emerging technology. The authors deserve a special recognition for their fine contributions and for making this volume a reality.

On closing this section, there is a dilemma that faces us. Not too long ago many researchers in the field of fuzzy control, in defense of the lack of analysis in fuzzy control, argued vehemently whether a rigorous analysis comparable with the type found in modern control theory is sufficient or even necessary in fuzzy control engineering. Others have pondered, from an industrial point of view, what analyses are really needed in fuzzy control. Perhaps rather surprisingly, stability was not even considered a top priority. Experimentation and testing presided over a rigorous mathematical analysis of stability, though this view is gradually changing today. It was proclaimed that, from an industrial viewpoint, a rigorous analysis of stability may neither be necessary nor sufficient. It suffices to say, whether or not such an analysis is conducted in any aspect of control theory fuzzy or otherwise, the real process needs to be simulated, experimented upon and tested rigorously. We are aware of the fact that no real, physical process works on analytical rigor alone. From this view point it would seem, indeed, that a rigorous mathematical analysis of, say, stability of the real process is certainly not sufficient and, depending on the process, it may not even be necessary. However, there seems to be a catch here. If such a rigorous analysis is conducted strictly for what has been termed as *intellectual arrogance* ([57] in Chapter 9), then it may not be useful or even usable and may have a seemingly questionable necessity. A useful and usable analysis may be, for example, one that has been employed as a major design tool, for the sake of selecting certain viable system parameters for design and implementation. Such an analysis may be considered necessary but not sufficient. In either case, in order to uphold good engineering practice, validation through experimentation and rigorous testing cannot be compromised. The yearning from the control community for rigor and analysis in the proof of stability, and the insistence that it be necessary, especially in fuzzy control, has been proclaimed the *cult of analysis* ([57] in Chapter 9). Some effort is made not to be too guilty of that in this volume.

The chapters are organized along the theme of the volume. There are three thematic parts. The first deals with some modeling issues in fuzzy control, the second delves into the analysis issue as one of the two major thrust areas of the volume, and the third part, the other major thrust area, addresses the synthesis issue.

The first chapter in the modeling section sets the stage on both the synthesis and analysis fronts by tackling the very important issue of information granularity in the analysis and design of fuzzy controllers. It is written by W. Pedrycz from University of Alberta, Edmonton, Canada. The chapter elaborates on the role of information granularity in the development of fuzzy controllers. A rare and contrasting view is taken by discussing a general framework involving data in the form of information granules, exhibiting various levels of information granularity as opposed to numeric data which is being commonly accepted by fuzzy controllers. The chapter addresses a number of design issues, including robustness of the fuzzy controller, representation of linguistic information and quantification of its granularity, and a detailed description of non-linear characteristics of the compiled version of the fuzzy controller.

The second chapter, entitled Fuzzy Modeling for Predictive Control, is a very interesting one written by M. Setness and R. Babuska of Delft University of Technology, Delft, The Netherlands. The authors noted the increased interest in data-driven approaches to process modeling, in view of the application of model predictive control (MPC) to complex non-linear processes. The chapter describes a fuzzy modeling approach and a scheme for MPC that uses the fuzzy model in a predictive control strategy based on discretized control actions, thereby enabling the use of efficient branch-and-bound optimization. In the modeling approach, both modeling and accuracy are addressed. One of the interesting features of the approach reported here is the idea of extracting models of varying complexity from an initial model without performing additional data acquisition. The authors applied this modeling approach to a heating, air-conditioning and ventilating (HVAC) system.

The third chapter, entitled Adaptive and Learning Schemes for Fuzzy Modeling, is written by G. Vachkov from Nagoya University, Nagoya, Japan. The author looks at several learning algorithms for creating TSK types of fuzzy models, namely the global learning, local learning and adaptive (reinforced) learning of fuzzy models. The author brings out the main attractive feature in these algorithms as the ability to work and converge quickly in the presence of noisy, sparse and even contradictory data.

To close the first part, the fourth chapter is one of the exceptions in this volume that is outside the original theme of the volume and thus prompted us to consider some modeling aspects of fuzzy control systems so as to make the volume more complete. It is written by D. F. Akhmetov and Y. Dote from Muroran Institute of Technology, Muroran, Japan, and is entitled Fuzzy System Identification with General Parameter Radial Basis Function Neural Network. The paper considers a fuzzy system identification problem through artificial neural networks. The problem of reconstructing an unknown fuzzy relation is formulated as a problem of identifying the structure and parameters of a neural network. The approach taken in the chapter is based on the functional equivalence between some classes of fuzzy systems and radial basis function networks. This is what gives the chapter its unique and interesting feature. Y. Dote has been active in the area of stability analysis of fuzzy control systems.

The second part, Analysis, is kicked off by J. R. Layne of Wright Patterson Air Force Base and K.M. Passino of the Ohio State University with the chapter entitled Lyapunov Stability Analysis of Fuzzy Dynamic Systems. This chapter looks at the properties of fuzzy system as a dynamic system and goes on to provide definitions fundamental to the study of dynamic systems. Lyapunov stability, asymptotic stability and exponential stability of invariant sets are defined, pointing out the parallelism with the definitions of the usual dynamic systems and functional analysis. In this kind of discourse the issue of boundedness, convergence, bounded sets and invariance usually prevail and the authors provide various definitions of these. The chapter is rich with the essential mathematical preliminaries needed to fully enjoy the exposition. Various ways of constructing fuzzy dynamic models are discussed. A detailed application is provided on a first-order fuzzy dynamic system.

The next chapter, Passivity and Stability of Fuzzy Control Systems, is written by R. Gorez from Université Catholique de Louvain, Louvain-La-Neuve, Belgium, and G. Calcev from Motorola, USA. The authors propose a unifying approach to the stability of fuzzy control systems based on passivity theory. They show that under mild conditions, the Mamdani and Takagi–Sugeno fuzzy controllers are equivalent to passive non-linear dynamical controllers. Thus, one is able to derive stability results for some general classes of plants and the corresponding controllers, including non-linear controllers and plants with uncertainties. The authors discussed how the passivity-based approach can accommodate a broader generality of compensators, such as, for example, compensators that may be used as signal filters.

The next chapter is entitled Frequency Domain Analysis of MIMO Fuzzy Control Systems written by A. Ollero, F. Gordillo and J. Aracil from Escuela Superior de Ingenieros, Sevilla, Spain. These researchers are known for their contributions to stability analysis of fuzzy control systems. In this chapter they present a frequency response method to partly analyze the existence of limit cycles in multivariable fuzzy control systems. It follows along the traditional method of decomposing the system into a linear part and a non-linear memoryless part. They provide an interesting measure of closeness to limit cycles by the combined use of singular values and the describing function.

The chapter written by H. Ying of the University of Texas Medical Branch, entitled Analytical Study of Structure of a Mamdani Fuzzy Controller with Three Input Variables, is a departure from the popular two-input variable fuzzy controller. The author investigates a fuzzy controller structure that uses error, rate of change of error and the derivative of rate of change of error of the process output as the input variables to the fuzzy controller. The structure of this controller is theoretically proven to be the sum of a global three-dimensional multi-level relay and two local non-linear controllers, where one of the local non-linear controllers is a non-linear PID controller with a variable proportional gain. The author also discusses the role of the global and local controllers in fuzzy control action. Some limiting cases of the fuzzy control structure are discussed.

The next chapter, entitled An Approach to Analysis of Robust Stability of Fuzzy Control Systems, is written by S. S. Farinwata from the Research Laboratory of Ford Motor Company, Dearborn Michigan, USA. A systematic method is introduced and outlined for practical robust stability analysis of a model-based, closed-loop fuzzy control system with respect to parameter variation. The fuzzy controller architecture considered is along the lines of the author's previous cell-space type of fuzzy controller. The robust stability method relies heavily on formulating a performance index as an energy-like function which needs to be minimized. This expression is further manipulated to incorporate the sensitivities of the systems trajectories with respect to known parameters. The sensitivities are formulated in terms of fuzzy quantities so that bounds on the sensitivities can be extended to bounds on the fuzzy quantities. The goal of this approach is simplicity, comprehensibility and ease of application rather than complexity of exposition. The result of this endeavor is an optimum robustness set which includes a fuzzy robustness measure, allowable sensitivities and cross sensitivities, allowable perturbations and convergence rate. To illustrate the approach, a previously designed fuzzy controller for an idling automotive engine is used as a test bed.

The chapter entitled Fuzzy Control Systems Stability Analysis with Application to Aircraft systems is a collaborative effort between S. S. Farinwata from Ford Motor Company, Dearborn, Michigan, USA, and S. Chiu from Rockwell Science Center. Thousand Oaks, California, USA. The chapter contains two applications placed in two parts. The theoretical background and the first application are provided by the first author in part one of the chapter, and the second application in part two is provided by S. Chiu. The chapter exploits Lyapunov stability theory which applies to most physical systems that are inherently dissipative in nature, or can easily be made so by the proper choice of inputs and outputs. Lyapunov theory is developed, leading to a passivity theory for dissipative linear and non-linear systems. However, the applications considered are of continuous-time and discrete-time linear systems that are controlled via fuzzy control. Thus, the development of the non-linear passivity theory, though very interesting, has been made very brief so as to fit the scope of the applications. The first application is on a fuzzy control system for a missile autopilot's simplified model and the second application is on a fuzzy control system or Rockwell's advanced technology wing aircraft model.

This brings us to the third part, Synthesis. The first chapter in this part is entitled Observer-Based Controller Synthesis for Model-Based Fuzzy Systems via Linear Matrix Inequality and is written by a diverse group of researchers in control theory who are making a lot of valuable contributions to fuzzy control: A. Jadbabaie from California Institute of Technology, Pasadena, California, USA, M. Jamshidi from University of New Mexico, Albuquerque, New Mexico, USA, A. Titli from LAAS du CNRS and INSA, Tolouse, France and C. T. Abdallah from University of New Mexico, Albuquerque, New Mexico, USA. The authors extend some recent results regarding the stability of continuous-time and discrete-time Takagi–Sugeno (TS) fuzzy systems to the case where the system's state is not available for measurement. The fuzzy observer is developed and the authors show that the well-celebrated Kalman's separation principle holds for the controller–observer combination. This enables them to formulate the design problem as two separate linear matrix inequality (LMI) feasibility problems. A numerical example is provided to demonstrate the effectiveness of the approach. Some interesting future direction which includes the extension to the design of a Kalman filter for this class of system is discussed.

The chapter entitled LMI-Based Fuzzy Control: Fuzzy Regulator and Fuzzy Observer Design via LMIs is written by K. Tanaka from University of Electro-Communications, Tokyo, Japan, and H. O. Wang from Duke University, Durham, North Carolina, USA. They and their colleagues have been very innovative and timely in the development and application of control-theoretic concepts based on the TS fuzzy system. This chapter presents a model-based fuzzy control of non-linear systems. The interesting and unique feature of this chapter is the design of a fuzzy regulator and observer via LMI, incorporating in the multi-objective optimization process such control performance properties as decay rate, disturbance rejection, robust stability minimization of a quadratic performance function, and constraints in the control input and output. Furthermore, the authors present new stability conditions, a relaxation of their previous results, and the new results are also utilized in the LMI design procedure.

The next chapter, entitled A Framework for Synthesis of PDC-type Takagi–Sugeno Fuzzy Control Systems: An LMI Approach, is written by J. Joh from Changwon National University, Changwon, Kyungnam, Korea, and one of the co-editors, R. Langari, from Texas A&M University, College Station, Texas, USA. This chapter capitalizes on the framework for synthesis of the PDC-type TS fuzzy control system using the LMIs of Tanaka and his colleagues. The authors derived extended stability LMIs of continuous and discrete TS fuzzy control systems from the work of H. Wang and his colleagues. The derived stability condition treats the feedback gains as unknowns which can be used as a framework for the synthesis of the PDC-type TS fuzzy systems. An interesting feature of this chapter is the representation of the desired performance as an LMI-constrained pole placement which places the poles in the desired region of the complex plane. Thus, by simultaneously solving the stability LMIs and the pole placement LMIs, the feedback shows which guarantee global asymptotic stability and satisfy performance are determined. Furthermore, the authors extend their approach to include parameter uncertainties by incorporating an additional set of LMIs for these. The effectiveness of the approach is demonstrated on a cart–pole problem.

The next chapter is written by researchers from Canada, J. X. Lee and G. Vukovich from the Canadian Space Agency, St. Hubert, Canada. The chapter is entitled On Adaptive Fuzzy Logic Control of Nonlinear Systems—Synthesis and Analysis. It makes an interesting distinction between static fuzzy logic systems that seem to be common in the literature and dynamic fuzzy logic systems (DFLS) presented in the chapter. In this chapter the authors synthesize a more practical DFLS-based indirect adaptive control scheme which generalizes their prior effort in adaptive control and which is applicable to a large class of non-linear systems. The chapter contains extensive analysis of closed-loop system performance and stability which are stated and proved as theorems. The authors provide a simulation example to illustrate the controller design procedure.

The chapter entitled Stabilization of Direct Adaptive Fuzzy Control Systems: Two Approaches is written by H. C. Myung and Z. Z. Bien from the Korean Advanced Institute of Science and Technology, Yusong-gu, Taejon, Korea, and Y. T. Kim from Samsung Electronics Company, Ltd, Suwon, Korea. The authors, in an interesting way, mesh the sliding mode control with adaptive control methods in their design of a fuzzy control system. The idea is to minimize the very heuristic aspect that pervades the conventional fuzzy control system design which the authors explain can be potentially troublesome. A non-linear control system is stabilized using this approach in a pre-specified bounded state region. Furthermore, the authors propose a direct adaptive fuzzy control method with a novel adaptation law for an improvement of the robustness of the tracking problem.

The next chapter is written by one of the co-editors, D. Filev, from Ford Motor Company, Glendale, Michigan, USA. The chapter is entitled Gain Scheduling-Based Control of a Class of TSK Systems. The author presents the synthesis of stable Takagi–Sugeno–Kang (TSK) systems based on the concept of gain scheduling. The chapter demonstrates how the polytopic representation of the TSK system can be used for the synthesis of a TSK-type compensator. The key idea of this approach is to design a compensator that is scheduled by the firing levels of the rules. A transfer function is assigned which is invariant to the firing levels and which guarantees stability of the closed-loop system. An example is provided to illustrate the approach.

The chapter entitled On Output Tracking Using Fuzzy Neural Networks is written by D. K. Pirovolou from Anadrill-Schlumberger, Sugar Land, Texas, USA, and G. J. Vachtsevanos from the Georgia Institute of Technology, Atlanta, Georgia, USA. The chapter presents a systematic methodology for the design of a neuro-fuzzy controller for the output tracking problem of non-linear systems of unknown dynamics. The authors consider systems in companion form in which the input enters the system linearly, and which are observable and bounded-input bounded-state stable. The architecture discussed falls into the category of direct adaptive, non-model-based control. To illustrate the effectiveness of the approach, the authors provide two examples on a cart-pole system and an industrial furnace.

The final chapter, entitled Fuzzy Life Extending Control of Mechanical Systems, is written by P. Kallappa and A. Ray from the Pennsylvania State University, University Park, Philadelphia, USA. The chapter presents a methodology for the synthesis of fuzzy-logic-based life-extending control systems (LECS). The objective of such a scheme is to enhance the performance, durability and life extension of mechanical structures. In this approach, knowledge of the process dynamics and the characteristics of the material is used to synthesize a hierarchical control system with a two-tier architecture. A supervisory controller at the upper tier makes decisions based on the trade-off between performance enhancement and life extension. The lower tier consists of a feedforward control policy and a family of linear multivariable robust feedback controllers which are gain-scheduled. The authors apply the approach via simulation experiments to a rocket engine that is similar to the Space Shuttle's main engine, as well as to a fossil fuel power plant model.

Many thanks to all those who helped with the reviews of the chapters in this volume. The first-named editor wishes to thank Mr Charles W. Heil Jr., P. E. of Ford Motor Company, Dearborn, Michigan, USA, for his helpful and prompt paper reviews.

In closing, the Editors wish the reader an enjoyable time reading this volume and fruitful experiences in his or her educational and professional endeavors.

Shehu S. Farinwata
Dearborn, Michigan, USA
Dimitar Filev
Detroit, Michigan, USA
Reza Langari
College Station, Texas, USA
May 1999

List of Contributors

Witold Pedrycz
Department of Electrical and Computer Engineering
University of Alberta
Edmonton T6G 2G7, Canada

Magne Setnes and **Robert Babu ka**
Control Laboratory
Delft University of Technology
P.O. Box 5031, 2600 GA
Delft, The Netherlands

Gancho Vachkov
Department of Micro Systems Engineering
Nagoya University
Furo-cho, Chikusa-ku, 464-8603 Nagoya, Japan

Daouren F. Akhmetov and **Yasuhiko Dote**
Department of Computer Science
and Systems Engineering
Muroran Institute of Technology, Muroran, Japan
27-1, Mizumoto-cho, Muroran 050-8585, Japan

Jeffrey R. Layne
AFRL/SNAT
2185 Avionics Circle
Wright Patterson AFB, OH 45433-7301, USA

Kevin M. Passino
Department of Electrical Engineering
The Ohio State University
2015 Neil Avenue
Columbus, OH 432102, USA

Raymond Gorez
Centre of Systems Engineering
and Applied Mechanics
Universite Louvain
Louvain-La-Neuve, Belgium

George Calcev
Land Mobile Product Systems
Motorola, IL 60004, USA

Anibal Ollero, Francisco Gordillo and **Javier Aracil**
Escuela Superior de Ingenieros
Camino de los Descubrimientos s/n
Sevilla-41092, Spain

Hao Ying
Department of Physiology and Biophysics
Biomedical Engineering Center
The University of Texas Medical Branch
Galveston, TX 77555, USA

Shehu S. Farinwata
International and Systems Research Laboratory
Powertrain Control Systems
Ford Research Laboratory
Dearborn, MI 48121-2053, USA

Stephen Chiu
Rockwell International Science Center
Thousand Oaks, CA 91360, USA

Ali Jadbabaie
Control and Dynamical Systems
California Institute of Technology,
Mail Code 107-81
Pasadena, CA 91125, USA

Mohammad Jamshidi and
Chaouki T. Abadallah
Department of Electrical and
Computer Engineering
University of New Mexico
Albuquerque, NM 87131, USA

Andre Titli
LAAS du CNRS and INSA
Tolouse, France

Kazuo Tanaka
Department of Mechanical and
Control Engineering
University of Electro-Communications
1-5-1 Chofugaoka, Chofu
Tokyo 182-8585, Japan

Hua O. Wang
Department of Electrical and
Computer Engineering
Duke University
Durham, NC 27708, USA

Joongseon Joh
Department of Control and Instrumentation Engineering
Changwon National University
#9 Sarimdong, Changwon
Kyungnam, 641-773, Korea

Reza Langari
Department of Mechanical Engineering
Texas A&M University
College Station, TX 77843-3123, USA

James X. Lee and **George Vukovich**
Directorate of Spacecraft Engineering
Canadian Space Agency
St. Hubert, PQ J3Y 8Y9, Canada

Hwan-Chun Myung and **Zenn. Z. Bien**
Korean Advanced Institute of
Science and Technology
373-1 Kusung-dong, Yusong-gu
Taejon 305-701, Korea

Yong-Tae Kim
SAMSUNG Electronics Company, Ltd
Pst-Box 105, Suwon Post-Office, Suwon
Kyungki-do, Korea

Dimitar Filev
Advanced Manufacturing Technology
Development Center
Ford Motor Company
24500 Glendale Avenue, Redford,
MI 48239, USA

Dimitrios K. Pirovolou
Advanced Drilling Systems
Anadrill-Schlumberger
Sugar Land, TX 77478, USA

George J. Vachtsevanos
School of Electrical and Computer
Engineering
Georgia Institute of Technology
Atlanta, GA 30332-0250, USA

Patada Kallappa, Ashok Ray and
Michael S. Holmes
Mechanical Engineering Department
The Pennsylvania State University
University Park, PA 16802, USA

About the editors

Shehu S. Farinwata was a graduate of the Federal Government College, Sokoto, Nigeria, 1976, with a WASC (West African School Certificate) Division One—Distinction (highest honor). He received the BS (honors) and MS degrees in electrical engineering from the University of Detroit, Detroit, Michigan, USA, in 1985, and the Ph.D. degree in electrical engineering from the Georgia Institute of Technology (Georgia Tech), Atlanta, Georgia, USA, in 1993 along with Computer Integrated Manufacturing Systems (CIMS) Certificate. Before coming to the United States, he had studied briefly at the Oxford Air Training School, Kidlington, Oxford, England, UK, and later at L'institut University de Technologie (I.U.T.), Genie Electrique, Lannion, France. He is currently an Engineering Specialist at the Ford Motor Company, Research Laboratory, Dearborn, Michigan. Prior to that, he held a position as Project Engineer with the General Motors Corporation, Michigan, USA. Dr Farinwata's research interests include non-linear control, robust control, adaptive control, intelligent systems, stability and robustness analyses of fuzzy control systems, automotive applications. He has published from these areas, and is also a chapter contributor in the edited book *Fuzzy Logic: Implementation and Applications*, John Wiley & Teubner, 1996. He is a member of IEEE and the national honor societies Eta Kappa Nu and Tau Beta Pi. Dr Farinwata is a member of the Graduate faculty with the Department of Electrical and Computer Engineering, Wayne State University, Detroit, MI, USA, where he has taught as an Adjunct Faculty (part-time) the graduate courses: System Identification and Adaptive Control; Linear Systems (Controls II); Optimal Control and Mathematical Methods in Engineering. He currently serves on the Editorial Board of *IEEE Transactions on Control Systems Technology* as an Associate Editor. Dr Farinwata is a Licensed Professional Engineer (P.E.) in the State of Michigan, USA.

Dimitar P. Filev is a Senior Technical Specialist with Ford Motor Company in Michigan, USA. Prior to joining Ford, Dr Filev was Associate Professor of Information Systems at Iona College and Senior Research Associate at the Machine Intelligence Institute, New Rochelle, NY, USA. He is conducting research in control theory and applications, modeling of complex systems, fuzzy modeling and control, and he has published over 150 articles in refereed journals and conference proceedings. He is a co-author of three books including the book *Essentials of Fuzzy Modeling and Control* (with R. Yager) published by John Wiley & Sons, 1994. He is the recipient of the 1995 Award for Excellence of MCB University Press. He is Associate Editor of *IEEE Transactions on Fuzzy Systems*. He is also a recipient of Ford's Henry Ford Technology Award, 1996. Dr Filev received his Ph.D. degree in Electrical Engineering from the Czech Technical University in Prague, in 1979.

Reza Langari received his BS, MS and Ph.D. degrees in mechanical engineering from the University of California at Berkeley, USA, where he worked closely with Professor

Lotfi Zadeh (the originator of fuzzy set theory) on the analysis and synthesis of fuzzy logic control systems. He has held positions with Measurex Corporation, Insight Development Corporation, and Integrated Systems, Inc. Dr Langari is presently an Associate Professor in the Department of Mechanical Engineering at Texas A&M University and Associate Director of the Centre for Fuzzy Logic, Robotics and Intelligent Systems Research at Texas A&M University, USA. His expertise is in the area of fuzzy information processing and control, non-linear and adaptive control systems, and computing architecture for real-time control. Dr Langari has published a number of articles on analysis, design and implementation of fuzzy logic control (FLC) systems. He is a co-editor, with J. Yen and L. A. Zadeh, of the book *Industrial Applications of Fuzzy Logic and Intelligent Systems*, IEEE Press, 1995. He is the co-author with John Yen of the book *Fuzzy Logic: Intelligence, Control and Information*, Prentice Hall, 1999. Dr Langari serves as associate editor of *ASME Journal of Dynamic Systems, Measurement and Control,* as well as *AutoSoft: International Journal of Intelligent Automation and Soft Computing*.

Acknowledgments

We wish to express our sincere gratitude to the staff of John Wiley and Sons, Europe, for their enthusiasm, direction, dedication and commitment to this project. We are indeed grateful to them for their patience and understanding. On a personal level, the first-named editor would like to express tremendous appreciation especially to the Publishing Editor, Peter Mitchell, to Rhoswen Cowell editorial assistant, and to John and Celia Hall, copy-editors. The publishing of this volume would not have been possible without their professional support.

MODELING

1
Information Granularity in the Analysis and Design of Fuzzy Controllers

W. Pedrycz
University of Alberta, Edmonton, Canada and Systems Research Institute, Polish Academy of Sciences, Warsaw, Poland

1.1 INTRODUCTION

There is no doubt that a substantial body of knowledge has been already accumulated about the behavior and synthesis of fuzzy controllers [5, 6, 9, 13]. Evidently, fuzzy controllers are chiefly regarded as numeric constructs. We usually refer to them as compiled versions of fuzzy controllers. Far less attention has been paid to their analysis and synthesis completed in the presence of linguistic (non-numeric) information. The literature on this subject is relatively scarce. The reader may refer to Pedrycz [10, 11] who looked into the problem of representing uncertainty and forming relevant interfaces to the fuzzy controllers capturing the linguistic facet of the controller. Non-singleton fuzzy systems were studied by Mouzouris and Mendel [8]. Some related research ideas introduced in the realm of fuzzy neural networks were reported by Ishibuchi [3].

    The objective of this chapter is to explore a facet of the fuzzy controller that is directly linked to the aspect of representing, handling and utilizing the non-numeric character of information available to the fuzzy controller. The augmentation of the fuzzy controller along this line of development is regarded as a natural evolution of the system that enhances its implementation and expands the range of potential applications. The very nature of the approach requires that we revisit a notion of information granularity and elaborate on the critical role of levels of information granulation at which a certain fuzzy controller needs to be developed. Even though this fundamental aspect of information granulation and processing was advocated by Zadeh as early as 1979 [16] (see also [17]), this notion has never been directly exploited in the design of the fuzzy controller. Such a tendency is not overly surprising, though. Fuzzy controllers have always been regarded in their numeric (compiled) versions. This does not leave too much room for any meaningful discussions geared to their linguistically-oriented extensions of the fuzzy controllers. One may even gain

Fuzzy Logic. Edited by S. Farinwata, D. Filev and R. Langari
© 2000 John Wiley & Sons, Ltd

an impression that the development of the fuzzy controller has been stalled at the beginning of the overall cycle of its evolution by channeling all effort towards the numeric side of the control structures and not pursuing more vigorously the territories of information processing where the fuzzy controller has been naturally prepared to explore.

More concisely, the fuzzy controller in its full fledged version is a mapping from a family of fuzzy sets of state to the family of fuzzy sets of control

$$\mathcal{F}(\mathbf{X}) \to \mathcal{F}(\mathbf{U}),$$

where $\mathcal{F}(\mathbf{X})$ denotes a family of fuzzy sets of state defined in $\mathbf{X}$. Similarly, $\mathcal{F}(\mathbf{U})$ is used to express a family of fuzzy sets of control defined in the control space. The compiled version of the controller is viewed as a mapping from $\mathbf{R}$ to $\mathbf{R}$. While the concept of the fuzzy controller relies on fuzzy sets as information granules, its compiled version becomes very much limited, albeit highly practical as far as the state-control transformation involved is concerned.

The agenda of this study embraces a number of detailed and essential design issues that emerge in the overall process of the development of the fuzzy controller arising at the linguistic level. The material is arranged in seven sections. We start with setting up all necessary terminology, introducing the main topology of the controller and elaborating on the resulting non-linear characteristics of the fuzzy controller (Section 1.2). Since we will be alluding to a set-based controller being induced by the fuzzy controller, we discuss an underlying problem of approximating fuzzy sets by sets (Section 1.3). The notion of information granularity and its role in the design of fuzzy controllers is studied in Section 1.4. In what follows we discuss the essential robustness properties of the fuzzy controller and show how these are quantified in terms of the granularity of information involved in the construction of the fuzzy controller. We contrast the robustness of the fuzzy controllers and set-based controllers. In Section 1.6 we concentrate on an important application scenario where the fuzzy controller has to cope with non-numeric inputs. We discuss relevant possibility–necessity mechanisms that provide necessary representation support of uncertainty and analyze ways in which the fuzzy controller needs to interact with such a linguistic environment. Owing to the factor of uncertainty residing within input data, we elaborate on the form of the resulting control actions and ways in which the uncertainty factor is reflected there. Concluding remarks are covered in Section 1.7.

1.2 THE BASIC ARCHITECTURE OF THE FUZZY CONTROLLER AND ITS NON-LINEAR RELATIONSHIPS

In what follows, we consider a single-input–single-output fuzzy controller formed with the aid of a series of rules

If x is A_k Then u is m_k

$k = 1, 2, \ldots, K$, where A_k are fuzzy sets defined in the input space $\mathbf{X}$ and m_k are selected (prototypical) numeric values of the control actions defined in $\mathbf{U}$ [10]. More descriptively, each rule links a certain region of the input space (captured by a linguistic granule) with the corresponding numeric representative defined in the output space. The use of the numeric values in the conclusion part of the rules simplifies the analysis quite considerably.

Furthermore, their usage is also fully legitimate: while the locality of the rule is captured via some linguistic granules, the associated control action needs to be numeric to fully interact with the numeric control environment. Admitting the rules of this format, a typical way of mapping (transforming) any state of the control system (input of the controller) to the control space is to carry out an aggregation of the form

$$u = \frac{\sum_{i=1}^{K} A_i(x) m_i}{\sum_{i=1}^{K} A_i(x)}, \quad (1.1)$$

where u denotes the resulting control action. Interestingly, if the linguistic terms A_k form a partition of the input space (that is they sum up to 1 at any coordinate of the input space), then the denominator of the above expression is always equal to 1 and we end up with a linear combination of the prototypical control actions m_k. More importantly, the linguistic nature of the information granules in the input space helps us achieve a smooth transition between control modes represented by the individual rules. Technically, the above methods scale easily up to a many-input–single-output fuzzy controller.

In general, the characteristics of the fuzzy controller (1.1) are non-linear. Furthermore, by carrying out this type of transformation we confine ourselves to a so-called compiled version of the fuzzy controller [1]. More descriptively, we may regard (1.1) as a *numeric manifestation* of the fuzzy controller. Note that any changes in the aggregation mechanism may produce another compiled version of the linguistic rules. As a matter of fact, we are faced with a multiplicity of numeric manifestations of the same linguistic rules. The form of the developed non-linearity relies quite considerably upon the analytical form of the membership functions of A_k. Quite often these are defined by means of triangular membership functions with the modal values located at $a_1, a_2, \ldots, a_K$ with each two successive membership functions overlapping at the membership level of 1/2. Under these additional assumptions about the character of the membership functions we end up dealing with the piecewise linear input–output relationship. The knots of the obtained piecewise linear characteristics are fully characterized through the modal values of the linguistic labels as well as the respective prototypical control values standing in the corresponding control rules. In other words, the fuzzy controller (as discussed here) is fully characterized by a $2K$-dimensional vector of its parameters $\mathbf{p} = [a_1, a_2, \ldots, a_K, m_1, m_2, \ldots, m_K]$. Or, put another way, one may view this vector as a chromosome encoding the fuzzy controller (more precisely, its compiled version) and used afterwards in any genetic-based optimization process.

It is interesting to underline some main findings about the characteristics of the fuzzy controller *vis-à-vis* the modifiable parameters of the fuzzy controller. For more details see Figure 1.1. This type of analysis sheds light on the expected performance of the fuzzy controller.

1. The calculations of control based on (1.1) reveal that the knots of the piecewise linear characteristic of the controller are situated at the modal values of the A_k's. The slope of the control relationship in this region is computed as the ratio $(m_k - m_{k-1})/(a_k - a_{k-1})$; see Figure 1.2.

   By adjusting the position of the linguistic terms or/and modifying the prototypical control values one can easily affect the gain of the controller that applies to this specific local region. In other words, one can regard the fuzzy controller (or better to say its numeric manifestation) as a multigain switchable controller. One can eventually

6 INFORMATION GRANULARITY

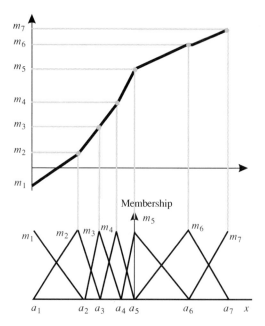

Figure 1.1 Piecewise characteristics of the fuzzy controller

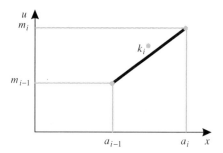

Figure 1.2 A linear segment of the control characteristics of the compiled fuzzy controller

approximate it as such by determining a single global gain coefficient k where, obviously, the values of the coefficient are situated within the interval of the gain factors $[k_{min}], k_{max}$, where k_{min} and k_{max} denote the lowest and highest gain coefficients determined for the fuzzy controller; see also Figure 1.3. For purposes of further stability analysis, it is convenient to treat the fuzzy controller as a system with an interval-valued gain coefficient. This may eventually help to investigate some stability criteria along the line of the Kharitonov theorem [4]. As a matter of fact, there have already been some research pursuits along this line.

2. If we replace fuzzy sets via their set-based counterparts, (Figure 1.4), then the resulting characteristics of the controller exhibit a series of jumps (discontinuities) which become inherent to any set-based construct. The location of these jumps (steps) is determined by the points of intersection of the adjacent linguistic terms.

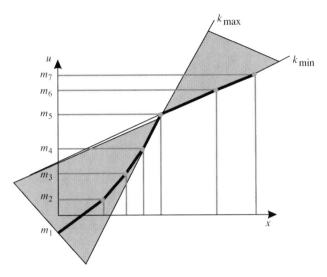

Figure 1.3 Bounds of the variable gain factor of the controller

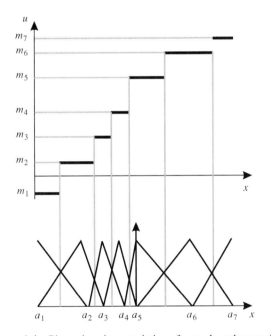

Figure 1.4 Piecewise characteristics of a set-based controller

1.3 SET-BASED APPROXIMATION OF FUZZY SETS

It is well known that fuzzy sets can be represented by families of α-cuts [18]. In the discrete case, a finite family of α-cuts fully represents any fuzzy set. In the infinite case, if we proceed with an increasing number of α-cuts, then any fuzzy set can be approximated to any required degree of accuracy. Which is the best α-cut, namely a specific value of α for which A_α approximates A to the highest possible extent? To illustrate the underlying concept, let us first confine ourselves to a unimodal fuzzy set. Then the problem can be formulated as follows: Determine such a value of α for which a total error defined as

$$V = |Q_1 + Q_2 - Q_3 - Q_4|$$

becomes minimized,

$$\min_{\alpha \in (0,1)} V(\alpha) = V(\alpha_{min}),$$

where the Q_i's stand for the areas highlighted in Figure 1.5. For pertinent details visualizing the error components refer to Figure 1.5.

The interpretation of the optimization task is straightforward: we construct such a set (α-cut) so that the deformations of the original membership function resulting from raising and lowering the corresponding membership values are balanced. In other words, one can look at the optimization criterion as having four parts representing four regions, Q_1, Q_2, Q_3 and Q_4, over which the membership functions have been modified (deformed). We expect that the two first regions counterbalance the two remaining, meaning that

$$Q_1 + Q_2 \cong Q_3 + Q_4. \tag{1.2}$$

The minimization criterion can exploit different types of distance functions. The two important examples are the Euclidean and Hamming distances defined as

$$||a - b|| = (a - b)^2$$

and

$$||a - b|| = |a - b|.$$

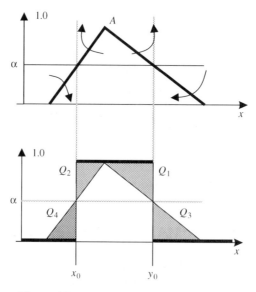

Figure 1.5 Set approximation of a fuzzy set

The calculations of the four components of the performance index are completed in the following fashion:

$$Q_4 = \int_{x_{min}}^{x_0} ||A(x) - 0|| dx,$$

$$Q_2 = \int_{x_0}^{x_{mode}} ||1 - A(x)|| dx,$$

$$Q_1 = \int_{x_{mode}}^{y_0} ||1 - A(x)|| dx,$$

$$Q_3 = \int_{y_0}^{x_{max}} ||A(x) - 0|| dx,$$

where x_{mode} denotes a point where A attains its maximum.

In general, when we encounter multimodal fuzzy sets, the optimization criterion can easily be generalized by taking into consideration all regions over which the membership values are elevated to 1 and the corresponding regions where the membership values are suppressed to 0. For subnormal fuzzy sets the integration occurs between 0 and a maximal membership value.

Summarizing, by finding an optimal threshold value, we are at the position of approximating any fuzzy set by its set representative. It should be emphasized that there we encounter inevitable losses of information and such losses are quantified by computing (1.2) which shows how much the original fuzzy set A needs to be distorted to become converted into a set. An interesting optimization problem gives rise to so-called shadowed sets [11, 12], where we introduce a concept of regions of uncertainty for which no particular membership value is assigned. In other words, by pushing fuzziness out of some regions of the universe of discourse, we allow for complete uncertainty to be assigned to some other areas to compensate for this assignment phenomenon. In this sense, a total balance of uncertainty remains constant yet the uncertainty factor is distributed differently across the universe of discourse and made more 'localized' (Figure 1.6).

In what follows we analyze the problem of set approximation in the case of triangular fuzzy sets. Furthermore, we can assume that the modal value of this fuzzy set is set at 0. The resulting areas, Figure 1.7, lead to the following expressions for the corresponding areas situated under the membership function:

$$Q_1 = \frac{\alpha^2 a}{2},$$

$$Q_2 = \frac{\alpha^2 b}{2},$$

$$Q_3 = \frac{a(1-\alpha)^2}{2},$$

$$Q_4 = \frac{b(1-\alpha)^2}{2}.$$

The equality of the right- and left-hand sides of (1.2) is accomplished by choosing a suitable value of the threshold level. Once all necessary computations have been completed, we arrive at a result that looks surprisingly familiar and intuitively sound, namely

$$\alpha = 1/2.$$

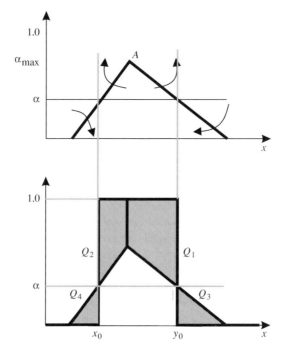

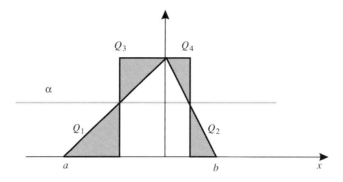

Figure 1.7 Set approximation of triangular fuzzy sets

Furthermore, what is quite interesting is that this result does not depend on the spreads of the fuzzy set as they do not show up in the final expression.

1.4 INFORMATION GRANULARITY OF THE RULES OF THE FUZZY CONTROLLER

The number of control rules induces the size of the linguistic terms being used in the condition part of the rules. Here comes an interesting design issue: How many control rules

INFORMATION GRANULARITY OF THE RULES OF THE FUZZY CONTROLLER

should one go for and what impact will this choice have on the resulting performance of the controller? We may look at this problem from two main standpoints:

1. dealing with input data of different levels of information granularity, and
2. operating in the presence of noisy inputs.

We start with the notion of information granularity that, even intuitively clear and appealing, requires a proper formalization.

1.4.1 Fuzzy Sets and Information Granularity

The granularity of fuzzy sets reflects how 'specific' they are. In other words, the term granularity deals with a size of an information granule being represented (embraced) by the concept conveyed therein. For instance, it becomes apparent that sets constitute a hierarchy; see Figure 1.8: very narrow intervals are definitely more *specific* than broader ones. In the limit, if the set covers the entire universe of discourse, then its granularity level attains a minimum.

Quantification of the granularity also seems quite straightforward: the higher the cardinality of the information granule, the lower its granularity. Let us recall that the concept of cardinality pertains to the number of elements in the information granule contained within the boundaries of this entity. In the case of sets, this reduces to a straightforward count of the elements in the set. For fuzzy sets (owing to the partial membership values of the individual elements), we end up introducing the concept of the so-called σ-count. For discrete universes of discourse $\mathbf{X}$ the granularity of A defined therein reads as

$$\sigma(A) = \sum_k A(x_k).$$

For the continuous universe $\mathbf{X}$ we consider an integral form defined as

$$\sigma(A) = \int_{\mathbf{X}} A(x) \, dx.$$

The σ-count is a meaningful measure of granularity when we are concerned with normal fuzzy sets (namely, fuzzy sets whose maximal membership grades attain 1).

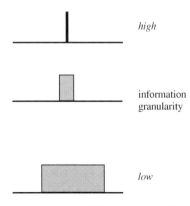

Figure 1.8 A hierarchy of sets of different levels of information granularity

12 INFORMATION GRANULARITY

It is worth emphasizing that the concept of information granularity and granulation comes hand in hand with a collection of fuzzy sets rather than a single fuzzy set. The notion of granulation becomes emphasized in this way as it tends to pertain to the entire universe of discourse (variable) providing a closer look at the way in which individual chunks of information are expressed. Nevertheless, the overall construct starts from a single information granule.

For fuzzy sets this simple definition calls for some additional refinement so that it becomes capable of dealing with elements exhibiting the phenomenon of partial membership. The definition of specificity proposed by Yager [14] seems to cope with the problem. Here we propose a slightly different version that eventually eliminates some computational drawbacks associated with the original concept of specificity. More formally, the specificity of a fuzzy set A is defined as an integral

$$\text{Sp}(A) = \int_0^{\alpha_{max}} \frac{1}{1 + \exp\left(-\frac{1}{\text{card}(A_\alpha)} - 0.5\right)} d\alpha.$$

One can easily relate some parameters of fuzzy sets with the resulting specificity. For instance, considering triangular fuzzy sets with a variable spread a, we easily derive a non-linear relationship between this parameter and the resulting specificity of the fuzzy set; see Figure 1.9. The cardinality of each α-cut, A_α, becomes a linear function of α, $A_\alpha = 2a(1 - \alpha)$. Subsequently, the specificity is expressed as a non-linear function of α.

An interesting design question concerns the choice of information granules of some specificity. How large should the information granules be that we go for? From the analysis of the characteristics of the fuzzy controller one can infer that these are affected by the granularity of the fuzzy sets standing in the condition part of the control rules. The more linguistic terms (and the higher their granularity), the more adjustment we can make to the piecewise characteristics of the fuzzy controller. There is another essential design practice of the fuzzy controller (highlighted quite early in their development) that links information granules with a level of noise (disturbances) occurring in the system under control. The intent is to make the granules meaningful enough so that they help to 'absorb' existing disturbances. It is quite apparent that very narrow membership functions (which characterize fuzzy sets of higher granularity) may not be capable of handling noise. In other words, if x is quite strongly

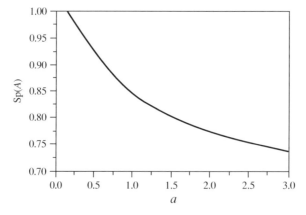

Figure 1.9 Specificity of triangular fuzzy set as a function of its spread (a)

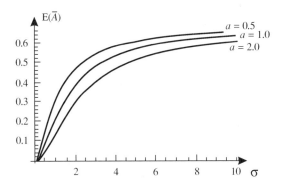

Figure 1.10 Expected value of the complement of A versus s for some selected values of the spread of the fuzzy set A

associated (with a high membership value, say 0.8 or so) with a certain linguistic term (fuzzy set), then we would expect that its noise-affected version, say $x+z$, could still be identified with the same fuzzy set. Lower granularity of the fuzzy set strongly promotes this type of binding. Again this aspect could be easily quantified by calculating the expression

$$E(\bar{A}) = \int_{-\infty}^{+\infty} \bar{A}(x)p(x)dx, \tag{1.3}$$

where $p(x)$ denotes the probability density function of noise existing in the system. More specifically, the above integral expresses the probability of an event that the noisy version of the measurement (x) falls outside the region embraced by fuzzy set A assuming that its noise-free version belongs to A with the membership value equal to 1. We may also treat the value of the integral as an expected value of the complement of A [15]. We require that the following inequality holds:

$$E(\bar{A}) \leq \gamma,$$

where γ describes a certain predefined positive threshold level situated in the unit interval and expresses the level at which we can tolerate a lack of noise absorption by the linguistic granule A. As an example, let us discuss the triangular fuzzy set A (being parameterized by its spread a), i.e.

$$A(x) = \exp(-x^2/a).$$

Assume a Gaussian model of noise N(0, σ) with the probability density function expressed as

$$p(x) = \frac{1}{\sigma\sqrt{2\pi}} \exp(-x^2/2\sigma^2)$$

that additively affects an original reading of the sensor x, and results in the sum $x+$N(0, σ) with σ standing for a standard deviation of the noise component. The resulting plots of $E(\bar{A})$ viewed as a function of σ for some selected values of the spread are included in Figure 1.10.

1.5 ROBUSTNESS PROPERTIES OF THE FUZZY CONTROLLER

There is an interesting property of the fuzzy controllers that many publications allude to but quite rarely define and places further considerations in a certain quantitative setting. This concerns the robustness of the fuzzy controller. Intuitively, it sounds quite legitimate to

14 INFORMATION GRANULARITY

expect that non-numeric information granularity can promote robustness. Yet, the term robustness needs to be defined in more detail. By robustness (in the sense introduced in this study) we mean the ability of the construct (controller) to produce meaningful results in the presence of deteriorated input information. As such, one may notice that the essence of this definition does not fully coincide with the classic definition of robustness encountered in control theory or control engineering. The reason is straightforward as we are focused on granular computing and make an attempt to anchor the definition in this setting. Taking this into account, we will not be pursuing any comparative analysis as far as these two definitions are concerned. We do believe, however, that the term robustness introduced here fully reflects the underlying feature of granular computing. In the simplest case one can envision that the controller should tolerate some deterioration in its input and still produce relevant control actions, perhaps close to the original control actions inferred for the deterioration-free scenario. The deterioration can occur for many reasons: in particular one may end up with some noisy measurements or, even more interestingly, measurements that are set-based oriented rather than arising as plain numbers [7].

It should be emphasized that fuzzy sets do not support noise filtering in any sense envisioned in statistical filtering. This is also quite apparent as neither sets nor fuzzy sets can cope with repetitive information whose usage forms the cornerstone of any averaging (i.e. filtering) effect. The robustness emerging in this case is inherently associated with the information granularity utilized in the development of the controller. Intuitively, to elaborate on this notion it is easier to start with set-based controllers. We consider the information granules in the condition part of the rules to be sets rather than fuzzy sets. All elements embraced by the same set are indistinguishable, yielding the same proper control action. The granulation helps 'absorb' disturbances (noise). The result of the sensor can be affected by some noise and still yield proper action or influence it to a negligible extent. Two examples of such deteriorated inputs are illustrated in Figure 1.11. In the first instance we are concerned with a noise factor being added to the original input. In the second case an original measurement appears at the lower level of information granularity. Owing to the set-based nature of the involved constructs, in both cases we achieve an interesting absorption effect. The most profound effect of information deterioration occurs at the edges of information granules (sets). In this sense the effect is localized yet with a substantial impact on the control actions. Fuzzy sets, on the other hand, exhibit a deterioration effect that is less radical yet distributed more evenly across the entire universe of discourse.

More formally, we can quantify the robustness of the set-based or fuzzy controller by introducing the following performance index:

$$V = \int_{x \in X} || FC(x) - FC(\Xi(x)) || dx,$$

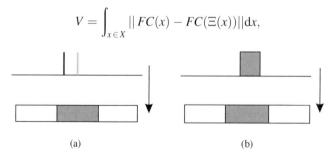

Figure 1.11 An effect of absorption realized by sets (information granules) (a) noisy input-random shift of the original measurement; (b) an effect of decreased information granularity

Table 1.1 Parameters of the fuzzy controller (fuzzy sets are defined by triangular membership functions $T(x; a, m, b)$)

Condition part	Prototype control
$T(x; -12, -10, -7)$	-6
$T(x; -10, -7, -4)$	-3
$T(x; -7, -4, -2)$	-1
$T(x; -4, -2, 0)$	-0.5
$T(x; -2, 0, 4)$	0.1
$T(x; 0, 4, 7)$	-3
$T(x; 4, 7, 10)$	6
$T(x; 7, 10, 15)$	8

where FC stands for the mapping realized by the fuzzy controller while $||\cdot||$ denotes the relevant distance function defined between the control action implied by the original input (x) and its distorted counterpart, i.e. $\Xi(x)$. Any ensuing comparative analysis of robustness can be then oriented towards contrasting fuzzy controllers and set-based controllers.

As an example we consider the fuzzy controller composed of eight rules whose parameters are outlined in Table 1.1. We assume an additive model of noise, meaning that the original measurement x is affected by noise z yielding $x+z$. The noise has a uniform distribution with amplitude d. To carry out a comparative analysis, we consider a set-based controller where each linguistic term appearing in the condition part is replaced by its set representative. As a matter of fact, the control expression (rule) reads as

If condition is $A \sim_i$ then u is u_i,

where $A \sim_i$ denotes a 1/2-cut of the respective fuzzy set. The selection of the threshold is fully justifiable in light of the findings reported in Section 1.2.

The characteristics of the controller for the original input and its noise-affected counterpart are shown in Figure 1.12. Figure 1.13 shows plots of $u(x+z)$ versus the control output derived in the presence of noise-free input. Furthermore, Figure 1.14 illustrates the relationship $u = u(x)$ for the set-based controller.

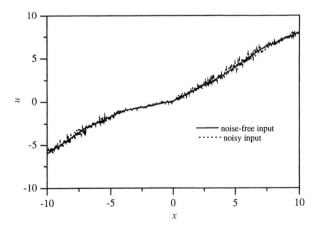

Figure 1.12 $u=u(x)$ for the fuzzy controller; $d=1.0$

16 INFORMATION GRANULARITY

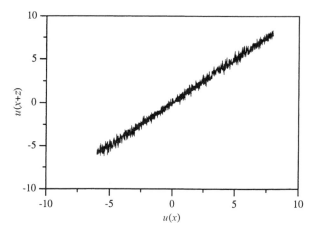

Figure 1.13 $u(x+z)$ versus $u(x)$

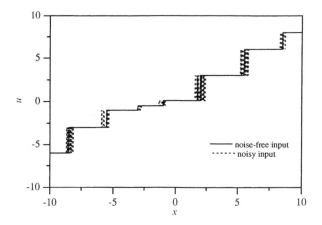

Figure 1.14 Relationship $u=u(x)$ for set-based controller; $d=1.0$

It is quite apparent from these figures that the noise effect manifests quite differently depending on whether we are talking about a fuzzy controller or its set-based counterpart. Noticeably, the set-based controller leads to a 'centralized' yet more radical impact of the noise on the performance of the controller. This impact occurs at the boundaries of the sets but becomes completely eliminated at the remaining parts of the input space.

It is also of interest to quantify the effect of noise on the performance of the fuzzy controller as well as the set-based controller. A plausible performance measure would assume the form

$$Q = \sqrt{\int_x u(x) - u(x+z)^2 dx}.$$

The values of Q are consistently lower for the fuzzy controller. The index itself expressed as a function of d for the two controllers is given in Figure 1.15.

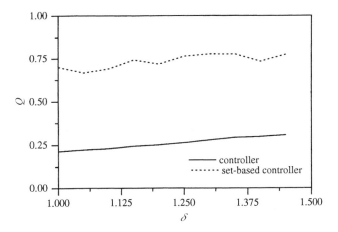

Figure 1.15 Q versus d

1.6 LINGUISTIC INFORMATION AS INPUTS OF THE FUZZY CONTROLLER

Commonly, the fuzzy controller is designed and exploited in a purely numeric format. This means that the only information provided to the fuzzy controller comes in the form of some numbers. Subsequently, fuzzy controllers arise in a so-called compiled version resulting in the form of a numeric non-linear mapping between the input and output variables. While this is a general avenue being followed, both in research and practice, there is an interesting generalization in which we admit the inputs to be non-numeric quantities, especially fuzzy sets or sets. Such imprecise quantities may arise as a result of noise being inherently associated with the sensor readings. The fuzziness could be introduced on purpose to accommodate a lack of complete knowledge about the system under control. A highly convincing and representative example arises in the case of dynamic systems with time delay. The systems of this class are modeled via the first-order differential equation

$$\frac{dx}{dt} = f(x(t), u(t - \tau)),$$

where τ denotes delay time. It becomes evident that the current control signal will exhibit some delayed effect. If the exact delay time is not known, then we need to proceed with caution: too radical changes of control may not be beneficial and we could end up with a series of oscillations of the system's output producing a highly unstable response. To alleviate this shortcoming, the input of the controller should be 'discounted' in some sense to accommodate the factor of uncertainty associated with the delay effect. This discounting effect is accomplished by treating the measurement as an interval rather than a single numerical entity. The higher the value of the delay in the system, the broader the range of the measurement interval (its granularity gets lower).

The key design question one has to address is: How do we represent uncertainty so that it becomes available to the controller's mapping capabilities? In what follows we exploit the use of the possibility and necessity measures [2]. Let X and A be two fuzzy sets defined in the same universe of discourse. The possibility measure expresses the degree of overlap between X and A while the necessity measure expresses the extent to which X is included in A:

$$\text{Poss}(X, A) = \sup_x [\min(X(x), A(x))] \quad (1.4)$$

and

$$\text{Nec}(X, A) = \inf_x [\max(1 - X(x), A(x))]. \quad (1.5)$$

These two quantities are then treated as the inputs to the fuzzy controller. Note that a single measure is not sufficient to capture the phenomenon of uncertainty. In spite of that, the possibility measure exhibits some useful properties. In particular, consider that X is *unknown*, which is modeled as a fuzzy set with the membership function equal identically to 1. Then all control rules are activated to the same extent (equal to 1) which converts the original control expression to the form

$$u = \frac{\sum_{i=1}^{K} A_i(x) m_i}{\sum_{i=1}^{K} A_i(x)} = \frac{\sum_{i=1}^{K} m_i}{K} = \bar{m},$$

which is nothing but an average of the prototypical control values. This result is highly convincing: if we are not provided with any current information about the state of the system (and what the linguistic term *unknown* stands for), then the evident control rationale would be to adhere to the mean value of the repertoire of control actions.

The minimal configuration of the model capturing the uncertainty effect involves both the possibility and necessity measures as two complementary inputs quantifying uncertainty. The intent is to infer control based upon the incoming uncertainty and to reflect this factor in the resulting control. We define an interval of control actions $U = [u_-, u_+]$ whose bounds are defined with the use of the possibility and necessity measures:

$$u \in \left[\frac{\sum_{i: m_i < 0} \text{Poss}_i m_i + \sum_{i: m_i > 0} \text{Nec}_i m_i}{\sum_{i=1}^{n} \text{Poss}_i}, \frac{\sum_{i: m_i < 0} \text{Nec}_i m_i + \sum_{i: m_i > 0} \text{Poss}_i m_i}{\sum_{i=1}^{n} \text{Poss}_i} \right],$$

where $\text{Poss}_i = \text{Poss}(X, A_i)$ and $\text{Nec}_i = \text{Nec}(X, A_i)$.

The lower and upper boundaries of this information granule of control are determined by the values of the possibility and necessity measures. The rationale behind the boundaries formed in the above manner is the following (note that the possibility measure assumes higher values than the corresponding necessity measure):

1. To deal with the lower boundary, we attempt to produce the minimal value resulting from the combination of the prototypical values (conditions of the rules) and the obtained possibilities and necessities. If the prototypical value is negative, then it is multiplied by the associated possibility measure. The necessity value enters this formula if the prototypical value is positive. This type of aggregation makes the result as small as possible.

2. The opposite aggregation occurs when determining the upper bound of the interval; now the possibility comes with positive values of the prototypes in the condition part of the rules while the negative values are visible when talking about the negative prototypes.

One can contrast this augmented way of representing uncertainty with the commonly used way of interacting with the environment by emphasizing the joint usage of the possibility and

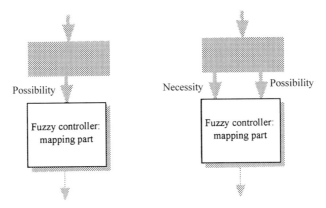

Figure 1.16 Fuzzy controller and its interface for numeric and non-numeric information

necessity measures (Figure 1.16). Since the necessity measures assume lower values than the corresponding possibility values, Nec(X, A) < Poss(X, A), the above bounds make sense.

Let us analyze some interesting scenarios highlighting the nature of the obtained set of control actions:

1. If we are concerned with the numeric input, then Poss(X, A) = Nec(X, A) and the control interval reduces to a single numeric quantity. One measure (possibility or necessity) is completely sufficient.

2. If X is *unknown*, then for all the rules of the controller we get: $\text{Poss}(X, A_i) = 1$ and $\text{Nec}(X, A_i) = 0$. The control set exhibits the boundaries developed based upon the negative and positive prototypical values of the control actions standing in the control rules:

$$u \in \left[\frac{\sum_{i:m_i<0} m_i}{K}, \frac{\sum_{i:m_i<0} m_i}{K} \right].$$

Denote by k_- and k_+ the number of negative and positive control actions, respectively. The above expression can be rewritten in an equivalent format:

$$u \in \left[\frac{k_-}{K} \bar{m}_-, \frac{k_+}{K} \bar{m}_+ \right],$$

i.e.

$$u \in [\alpha_- \bar{m}_-, \alpha_+ \bar{m}_+].$$

The last expression states that the bounds of the control interval are proportional to the averages of the ensembles of the positive and negative control actions appearing in the control rules. More specifically,

$$\bar{m}_- = \frac{1}{k_-} \sum_{i:m_i<0} m_i$$

and

$$\bar{m}_+ = \frac{1}{k_+} \sum_{i:m_i>0} m_i.$$

20 INFORMATION GRANULARITY

Continuing the example of the fuzzy controller with the rules defined in Table 1.1, we consider the non-numeric input information as a set (interval) of width d. Essentially, we end up with the input X centered around $\{x\}$ and distributed over the range $[x-d, x+d]$. The above construct leads to the interval-valued control shown in Figure 1.17. As expected, larger bounds of the input yield a broader interval of plausible control actions.

For comparative reasons, we determine control actions that rely only on the possibility measure, that is

$$u = \frac{\sum_{k=1}^{N} \text{Poss}_k \, m_k}{\sum_{k=1}^{N} \text{Poss}_k}.$$

Obviously, the resulting control actions are numeric. There is, however, an interesting effect of flattening the characteristics of the controller. With a lowering of the granularity of the inputs, the input–output characteristics of the controller exhibit lower values of the gain factor. This becomes particularly apparent at the lower and upper ranges of the inputs (Figure 1.18).

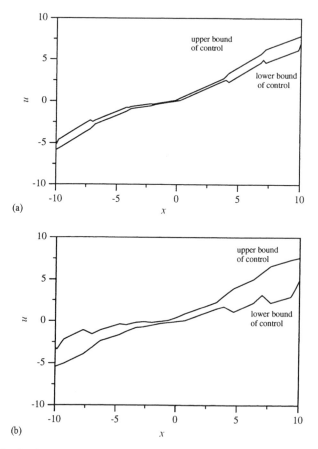

Figure 1.17 Set-based control for a non-numeric type of input: (a) $d=0.2$; (b) $d=0.7$

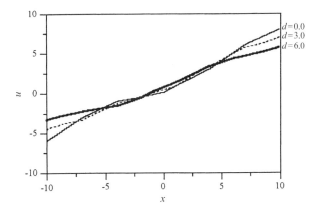

Figure 1.18 Control characteristics of the fuzzy controller derived for non-numeric information

1.7 CONCLUSIONS

We have studied a list of essential design tasks dealing with the development of the fuzzy controller in the presence of granular information. We have discussed one possible way of representing uncertainty (granularity) in the inputs of the fuzzy controller. The proposed architecture producing the possibility–necessity representation of uncertainty gives rise to a set of feasible control actions. In this sense the fuzzy controller develops a mapping of the following form:

$$\mathcal{F}(\mathbf{X}) \to \mathcal{P}(\mathbf{U}),$$

where $\mathcal{P}(\mathbf{U})$ is a family of sets of control defined in $\mathbf{U}$. An eventual augmentation of the format of the control actions can be accomplished by proceeding with more advanced methods of uncertainty representation (e.g. those exploiting compatibility measures); one should become aware of an extra computational overhead associated with them.

There is no doubt that future developments of fuzzy controllers will lie in the areas of processing and control completed in the presence of linguistic information. We have already underlined several sources of fuzziness of the inputs. They originate not only from sensors of limited resolution and sensors affected by noise; the effect of fuzziness can be a direct consequence of a lack of complete knowledge about the system under control. While this study has identified such essential issues, some of the detailed design problems have not been completed so far. They will be addressed in subsequent studies.

ACKNOWLEDGMENT

Support from the Natural Sciences and Engineering Council of Canada (NSERC) is gratefully acknowledged.

REFERENCES

1. Bonissone, P. P. and K. H. Chiang, A knowledge-based system view of fuzzy logic controllers, In: R. R. Yager and L. A. Zadeh (eds.), *Fuzzy Sets, Neural Networks, and Soft Computing*, van Nostrand Reinhold, New York, 1994, pp. 296–310.

2. Dubois, D. and H. Prade, *Fuzzy Sets and Systems: Theory and Applications*, Academic Press, New York, 1980.
3. Ishibuchi, H., Development of fuzzy neural networks, In: W. Pedrycz (ed.), *Fuzzy Modelling. Paradigms and Practice*, Kluwer Academic Publishers, Dordrecht, 1996, pp. 185–202.
4. Kharitonov, V. L., Asymptotic stability of an equilibrium position of a family of systems of linear differential equations, *Differential'nye Uraveniya*, **14**, 1978, 1483–1485.
5. Kruse, R., J. Gebhardt and F. Klawonn, *Foundations of Fuzzy Systems*, Wiley, Chichester, 1994.
6. Langari, R., Synthesis of nonlinear controllers via fuzzy logic, In: A. Kandel and G. Langholz (eds.), *Fuzzy Control Systems*, CRC Press, Boca Raton, 1994, pp. 263–274.
7. Milanese M., et al. (eds.), *Bounding Approaches to System Identification*, Plenum Press, New York, 1996.
8. Mouzouris, G. C. and J. M. Mendel, Nonsingleton fuzzy logic systems: theory and application, *IEEE Transactions on Fuzzy Systems*, **1**, 1997, 56–71.
9. Pedrycz, W., *Fuzzy Systems and Fuzzy Control*, 2nd edn, Research Studies Press, Taunton, 1993.
10. Pedrycz, W., *Fuzzy Sets Engineering*, CRC Press, Boca Raton, 1995.
11. Pedrycz, W., *Computational Intelligence: An Introduction*, CRC Press, Boca Raton, 1997.
12. Pedrycz, W., Shadowed sets, *IEEE Transactions on Systems, Man and Cybernetics, B*, **28**, 1998, 103–109.
13. Tzafestas, S. G. (ed.), *Methods and Applications of Intelligent Control*, Kluwer Academic Publishers, Dordercht, 1997.
14. Yager, R. R., Entropy and specificity in a mathematical theory of evidence, *International Journal of General Systems*, **9**, 1983, 249–260.
15. Zadeh, L. A., Probability of fuzzy events, *Journal of Mathematical Analysis and Applications*, **22**, 1968, 421–427.
16. Zadeh, L. A., Fuzzy sets and information granularity, In: M. M. Gupta, R. K. Ragade and R. R. Yager (eds.), *Advances in Fuzzy Set Theory and Applications*, North-Holland, Amsterdam, 1979, pp. 3–18.
17. Zadeh, L. A., The role of fuzzy logic in the management of uncertainty in expert systems, *Fuzzy Sets and Systems*, **11**, 1983, 1993–2227.
18. Zimmermann, H. J., *Fuzzy Set Theory and Its Applications*, Kluwer Academic Publishers, Dordrecht, 1985.

2
Fuzzy Modeling for Predictive Control

Magne Setnes and **Robert Babuška**
*Control Laboratory, Delft University of Technology,
Delft, The Netherlands*

2.1 INTRODUCTION

As model predictive control (MPC) becomes more applied in industry [19], and as computers get faster and more sampled data are becoming available, the interest in data-driven modeling increases. In order to apply MPC, a model must be available that can predict the process variables over the specified prediction horizon with sufficient accuracy. For non-linear and complex or partially known processes, deriving a mathematical (first principle) model is often not feasible, thereby hampering the application of MPC to many real-world processes. Thus, there is a need for efficient data-driven non-linear modeling tools for MPC. Another factor that restrains the application of MPC to fast, non-linear processes is that a non-linear (and non-convex) optimization problem must be solved for each sample instant. This chapter addresses both these issues by describing a fuzzy modeling approach and a control scheme using the resulting fuzzy models of the plant in a predictive control strategy based on discrete branch-and-bound optimization.

Fuzzy systems have proved to be useful in the modeling of complex non-linear processes. Like neural networks, fuzzy systems have been recognized as universal approximators [14]. Various techniques such as fuzzy clustering [1, 31, 32], neuro-fuzzy learning methods [17], and orthogonal least squares [30] have been proposed for data-driven fuzzy modeling. Most approaches, however, utilize only the function approximation capabilities of fuzzy systems, and little attention is paid to the complexity of the resulting models. This makes them less suited for applications like MPC in which the emphasis is not only on the numerical properties, but also on the computational load. Recently, methods for complexity reduction have been studied for fuzzy neural networks [6], and for fuzzy rule-based models in general [21, 23]. For MPC applications, such techniques can be useful for reducing the computational demands of the models.

The data-driven fuzzy modeling approach described in this chapter provides the user with both accurate and compact rule bases. Using fuzzy clustering, an initial rule base is

---

Fuzzy Logic. Edited by S. Farinwata, D. Filev and R. Langari
© 2000 John Wiley & Sons, Ltd

identified, focusing on numerical accuracy. Thereafter, similarity-driven rule base simplification is applied to produce various versions of the model, differing in complexity and computational demand. The user can introduce a tradeoff between model accuracy and complexity. This enables the generation of different models for varying tasks. A complex but accurate model can be useful for off-line simulations, while to understand the basic concepts of the system, a transparent and less complex rule base is needed. For the purposes of the MPC scheme presented here, the model accuracy must be sufficient, but at the same time, restrictions on computational demands are introduced by real-time requirements.

In what follows we describe the fuzzy modeling approach. Then the MPC scheme using a fuzzy process model and the discrete branch-and-bound optimization is discussed. Both the modeling and the control scheme are applied to the problem of controlling a heating, ventilation and air-conditioning (HVAC) system. From process measurements, a fuzzy model of the process is identified. The complexity of this initial model is reduced in order to better cope with the restrictions on the computation time introduced by the process. Various models of the system are obtained from the initial model, and their performance is discussed with respect to prediction accuracy, MPC performance and computational demands. Both simulation and real-time results are presented.

2.2 FUZZY MODELING

Different algorithms have been developed for the construction or tuning of fuzzy models from numerical data. Typically, expert knowledge expressed in a verbal form is translated into a collection of if–then rules. In this way, a certain model structure is created. Parameters in this structure (membership functions, weights of the rules, etc.) can be fine-tuned using input–output data.

When no prior knowledge about the system under study is available, a fuzzy model can be constructed entirely on the basis of numerical data. In what follows we address a data-driven modeling approach based on fuzzy clustering [1, 22]. An advantage of this approach is that it obviates the process of knowledge acquisition which is a well-known bottleneck for practical applications of knowledge-based systems. Instead, the expert or modeler assumes a more active role of model analysis and validation. This may lead to revealing new pieces of information, and may result in a kind of 'emergent' knowledge acquisition.

In the approach addressed here, an initial fuzzy model is acquired from sampled process data, utilizing the functional approximation capabilities of fuzzy systems. This fuzzy model is then processed in a simplification step, where the compactness and transparency are enhanced. By means of a threshold, the user can introduce a tradeoff between model accuracy and complexity.

2.2.1 Outline of the Modeling Approach

The individual steps of the modeling approach are shown in Figure 2.1. The purpose of the different steps and the related methods are outlined below, assuming that proper data collection has been performed.

FUZZY MODELING

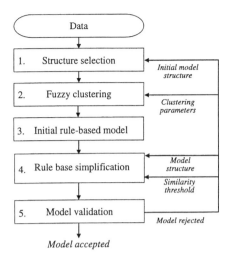

Figure 2.1 Overview of the steps in the fuzzy modeling approach

Step 1: Structure selection. The relevant input and output variables are determined with respect to the aim of the modeling exercise. For dynamic systems, structure selection allows us to translate the identification into a regression problem that can be solved in a static manner [18]. Often, a reasonable choice can be made by the user, based on the prior knowledge about the process.

Step 2: Clustering of the data. Fuzzy clustering is used to discover substructures in the product space of the available observations. Each cluster defines a fuzzy region in which the system can be approximated locally by a submodel. By applying techniques such as cluster validity measures [9] or cluster merging [13, 20], an appropriate number of clusters can be found.

Step 3: Generation of an initial fuzzy model. Fuzzy clustering partitions the space of the available data into regions in which relations exists between the inputs and the output. A rule-based fuzzy model is derived from the resulting fuzzy partition matrix and from the cluster prototypes. The rules, the membership functions and other parameters that constitute the fuzzy model are extracted in an automated way. The exact procedure depends on the type of fuzzy model required [1]. In Section 2.3 we focus on fuzzy models of the Takagi–Sugeno type.

Step 4: Rule base simplification. The initial rule base obtained from data is often redundant and unnecessarily complex as it is based on numerical optimization. The complexity is reduced in a process where similarity analysis is used to identify fuzzy sets representing compatible or redundant concepts. By merging compatible fuzzy sets and removing redundant ones, a simplified, more compact model is obtained. A tradeoff can be introduced between numerical accuracy and complexity, making it possible to generate models with varying degrees of complexity. This procedure is discussed in Section 2.4.

Step 5: Model validation. By means of validation, the final model is either accepted as appropriate for the given purpose, or it is rejected. In addition to numerical validation by means of simulation, inspection of fuzzy models plays an important role in the validation step. The coverage of the input space by the rules can be analyzed, and, for an incomplete rule base, additional rules can be provided based on prior knowledge, local linearization, or models based on physical laws. Additional information about the model can be gained, leading to important changes. Inspection of the model is made easier by the simplification of the rule base performed in the previous step.

2.3 EXTRACTION OF AN INITIAL RULE BASE

Assume that data from an unknown system $y = f(x)$ is observed. The aim is to use this data to construct a deterministic function $y = F(x)$ that can serve as a reasonable approximation of $f(x)$. In fuzzy modeling, the function F is represented as a collection of fuzzy if–then rules. Depending on the form of the propositions and on the structure of the rule base, different types of rule-based fuzzy models can be distinguished. In the following we consider rule-based models of the Takagi–Sugeno (TS) type [28]. It consist of a set of fuzzy rules, each of which describes a local input–output relation, typically in linear form:

$$R_i : w_i(\text{if } x_1 \text{ is } A_{i1} \text{ and} \ldots \text{and } x_n \text{ is } A_{in} \text{ then } y_i = a_i x + b_i), \quad i = 1, 2, \ldots, K. \quad (2.1)$$

Here R_i is the ith rule, $x = [x_1, \ldots, x_n]^T$ is the input (antecedent) variable, $A_{i1}, \ldots, A_{in}$ are fuzzy sets defined in the antecedent space, y_i is the rule output variable, and w_i is the rule weight. Typically, $w_i = 1, \forall i$, but these weights can be adjusted during the rule base simplification step if the number of rules is reduced. K denotes the number of rules in the rule base, and the aggregated output of the model, Y, is calculated by taking the weighted average of the rule consequents:

$$\hat{y} = \frac{\sum_{i=1}^{K} w_i \beta_i y_i}{\sum_{i=1}^{K} w_i \beta_i}, \quad (2.2)$$

where β_i is the degree of activation of the ith rule:

$$\beta_i = \prod_{j=1}^{n} \mu_{A_{ij}}(x_j), \quad i = 1, 2, \ldots, K, \quad (2.3)$$

and $\mu_{A_{ij}}(x_j) : \mathbb{R} \to [0, 1]$ is the membership function of the fuzzy set A_{ij} in the antecedent of R_i.

To identify the model (2.1), the regression matrix $\mathbf{X}$ and an output vector $\mathbf{y}$ are constructed from the available data:

$$\mathbf{X}^T = [x_1, \ldots, x_N], \quad \mathbf{y}^T = [y_1, \ldots, y_N]. \quad (2.4)$$

Here $N \gg n$ is the number of samples used for identification. The objective of identification is to construct the unknown non-linear function $y = F(\mathbf{X})$ from the data, where F is the TS model (2.1).

The number of rules, K, the antecedent fuzzy sets, A_{ij}, and the consequent parameters, a_i and b_i, are determined by means of fuzzy clustering in the product space of $\mathcal{X} \times \mathcal{Y}$ [1, 2, 31, 32]. Hence, the data set $\mathbf{Z}$ to be clustered is composed from $\mathbf{X}$ and $\mathbf{y}$:

$$\mathbf{Z}^T = [\mathbf{X}, \mathbf{y}]. \quad (2.5)$$

Given **Z** and an estimated number of clusters K, the Gustafson–Kessel fuzzy clustering algorithm [10] is applied to compute the fuzzy partition matrix **U** (the Gustafson–Kessel algorithm is given in Appendix A). This provides a description of the system in terms of its local characteristic behavior in regions of the data identified by the clustering algorithm, and each cluster defines a rule. Unlike the popular fuzzy c-means algorithm [5], the Gustafson–Kessel algorithm applies an adaptive distance measure. As such, it can find hyperellipsoid regions in the data that can be efficiently approximated by the hyperplanes described by the consequents in the TS model. Cluster validity measures can be applied to select K and a suitable fuzzy partition of **Z** [9].

The fuzzy sets in the antecedent of the rules are obtained from the partition matrix **U**, whose ikth element $\mu_{ik} \in [0, 1]$ is the membership degree of the data object z_k in cluster i. One-dimensional fuzzy sets A_{ij} are obtained from the multidimensional fuzzy sets defined point-wise in the ith row of the partition matrix by projections onto the space of the input variables x_j:

$$\mu_{A_{ij}}(x_{jk}) = \text{proj}_j^{\mathbb{N}n+1}(\mu ik), \qquad (2.6)$$

where proj is the point-wise projection operator [16]. The point-wise defined fuzzy sets A_{ij} are approximated by suitable parametric functions in order to compute $\mu_{A_{ij}}(x_j)$ for any value of x_j.

The consequent parameters for each rule are obtained as a weighted ordinary least-squares estimate. Let $\theta_i^T = [a_i^T; b_i]$, let $\mathbf{X}_e$ denote the matrix $[\mathbf{X}; 1]$ and let $\mathbf{W}_i$ denote a diagonal matrix in $\mathbb{R}^{N \times N}$ having the activation, $\beta_i(\mathbf{x}_k)$, as its kth diagonal element. If the columns of $\mathbf{X}_e$ are linearly independent and $\beta_i(\mathbf{x}_k) > 0$ for $1 \leq k \leq N$, then the weighted least-squares solution of $\mathbf{y} = \mathbf{X}_e \theta + \varepsilon$ becomes

$$\theta_i = [\mathbf{X}_e^T \mathbf{W}_i \mathbf{X}_e]^{-1} \mathbf{X}_e^T \mathbf{W}_i \mathbf{y}. \qquad (2.7)$$

2.4 SIMPLIFICATION AND REDUCTION OF THE INITIAL RULE BASE

Fuzzy rule-based models obtained from data are often unnecessarily complex and redundant due to the focus on quantitative aspects during their construction. Typically the modeling is driven by the minimization of some cost function, and an optimization takes place with respect to, for example, minimal quadratic error, leaving other important aspects like complexity and compactness out of consideration. Unnecessary redundancy is often present in the form of highly overlapping or compatible membership functions. In [23] we proposed applying a similarity measure to assess the pair-wise similarity of the fuzzy sets in the rule base. Similar fuzzy sets, representing compatible concepts, are aggregated in order to obtain a generalized concept represented by a common fuzzy set that can replace the similar ones in the rule base. Fuzzy sets estimated from data can also possess high similarity to a universal set, i.e. a set in which all elements of a domain have a membership close to 1. Such sets add no information to the model, and can be removed from the premise of the rules in which they occur. These operations reduce the number of fuzzy sets used in the model. Reduction of the number of rules follows as a result of this when the antecedents of some rules become equal. Such rules are aggregated into a new, general rule, which replaces the rules with the equal premise. Note that the process of aggregation is driven by the similarity among the fuzzy sets defined on the domain of the same antecedent variable, not in their product space. As such, a reduction of the model's term set is made possible without necessarily any rules being merged.

2.4.1 Similarity analysis

Many methods have been proposed to assess the similarity, or compatibility, of fuzzy concepts. A comparative analysis of different measures using human subjects was done by Zwick *et al.* [33]. The mathematical relations between the various measures was studied by Cross [8]. Setnes *et al.* [23] argued that the the fuzzy analog to the Jaccard index [12] is suitable for assessing the compatibility between fuzzy sets in a rule-based model. This similarity measure was also used by Chao *et al.* [6] in training the structure of fuzzy neural networks. The fuzzy Jaccard similarity measure is a set-theoretic measure given by

$$S(A_l, A_m) = \frac{\int (A_{lj} \cap A_{mj}) \mathrm{d}x_j}{\int (A_{lj} \cup A_{mj}) \mathrm{d}x_j}, \tag{2.8}$$

where A_{lj} and A_{mj} are fuzzy subsets of x_j. When A_{lj} and A_{mj} are defined on a discrete domain by the membership function $\mu_{A_{lj}}(x_j)$ and $\mu_{A_{mj}}(x_j)$, respectively, the fuzzy Jaccard index can be written as

$$S(A_l, A_m) = \frac{|\min(\mu_{A_{lj}}(x_j), \mu_{A_{mj}}(x_j))|}{|\max(\mu_{A_{lj}}(x_j), \mu_{A_{mj}}(x_j))|}, \tag{2.9}$$

where the min and max operators model the intersection and the union, respectively, and $|\cdot|$ denotes the cardinality of a fuzzy set, i.e. $\sum_x \mu_A(x)$. The similarity measure takes on values $s_{lm} \in [0, 1]$, where $s_{lm} = 1$ reflects strict equality, and $s_{lm} = 0$ refers to non-overlapping fuzzy sets. An example of the behavior of this similarity measure for fuzzy sets with a varying degree of overlap is shown in Figure 2.2.

2.4.2 Simplification and Reduction

The approach addressed here is based on iterative merging of pair-wise similar (compatible) fuzzy sets [23]. It requires two thresholds from the user, $\lambda, \gamma \in (0, 1)$ for merging similar

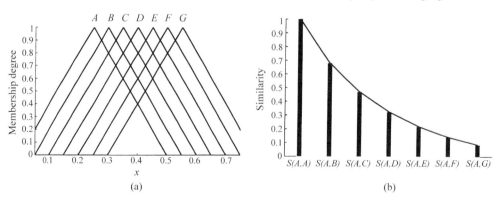

Figure 2.2 Fuzzy sets $A, B, \ldots G$ and the similarity computed for $S(A, A), S(A, B), \ldots, S(A, G)$

fuzzy sets, and removing fuzzy sets compatible with the universal set, respectively. In each iteration the similarity between all fuzzy sets in each antecedent dimension j is analyzed. The pair of fuzzy sets having the highest compatibility $S(A_{lj}, A_{mj}) > \lambda$ are merged, and a new, common fuzzy set A_{cj}, more or less representing the concepts represented by A_{lj} and A_{mj}, is created. After merging, the rule base is updated by substituting this common fuzzy set for the ones merged. The algorithm repeats, and evaluates the updated rule base, until there are no more fuzzy sets for which $s > \lambda$. Then, the resulting rule base is checked for fuzzy sets compatible with the universal set, i.e. sets for which $S(A, U) > \gamma$. These sets are removed from the rules in which they occur. The algorithm is illustrated in Figure 2.3, and is summarized in Appendix B.

The merging of fuzzy sets is accomplished by letting the support of the new, common fuzzy set A_{cj} be equal to the support of the union of the sets to be merged $\text{supp}(A_{lj} \cup A_{mj})$. This guarantees preservation of the coverage of the antecedent space. The kernel of A_{cj} is given by averaging the kernels of the sets A_{lj} and A_{mj}. If the antecedents of $p \geq 2$ rules become equal, then these p rules can be replaced by *one* common rule R_c. The consequent parameters of the reduced rule base can be re-estimated from training data (2.7), or one can calculate the parameters of R_c from the parameters of the p removed rules. The latter method does not depend on the availability of data. Let $Q \subset \{1, 2, \ldots, K\}$ be a subset of rule indices such that $A_{lj} = A_{mj}, \forall j \in \div \{1, 2, \ldots, n\}, \forall\, l, m \in Q$. $\mathbf{R}_Q$ then denotes the set of rules with equal antecedents. The rule R_c replaces the rules in $\mathbf{R}_Q$, and its antecedent part equals that of the rules $\mathbf{R}_Q$, i.e. $A_{cj} = A_{lj}$, $j = 1, 2, \ldots, n$, and $l \in Q$. The common rule R_c is made to account for all the rules $\mathbf{R}_Q$ by weighting it with the total weight of the rules $\mathbf{R}_Q$, $w_c = \sum_{i \in Q} w_i$, and by letting its consequent y_c be an average of the consequents of $\mathbf{R}_Q$. Thus, the set of rules $\mathbf{R}_Q$ is represented by a single rule R_c with weight w_c and

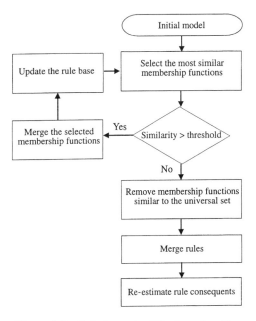

Figure 2.3 Rule base simplification algorithm

consequent parameters

$$\theta_c = \frac{1}{w_c} \sum_{i \in Q} w_i \theta_i. \tag{2.10}$$

Let $\bar{Q} = \{1, \ldots, K\} - Q$; the model output (2.2) now becomes

$$\hat{y} = \frac{\sum_{i \in \bar{Q}} w_i \beta_i y_i + w_c \beta_c y_c}{\sum_{i \in \bar{Q}} w_i \beta_i + w_c \beta_c}. \tag{2.11}$$

This substitution of $\mathbf{R}_Q$ by R_c does not alter the input–output mapping of the TS model.

2.5 MODEL PREDICTIVE CONTROL

Model predictive control (MPC) is a general methodology for solving control problems in the time domain. It is based on three main concepts:

1. Explicit use of a model to predict the process output over a prediction horizon of future discrete time instants.

2. Computation of a sequence of future control actions over a control horizon equal to or smaller than the prediction horizon, by minimizing a given objective function.

3. Receding horizon strategy; only the first control action in the sequence is applied, the horizons are moved one step ahead and optimization is repeated.

Because of the optimization approach and the explicit use of a process model, with MPC multivariable optimal control can be realized, processes with non-linearities, non-minimum phase behavior or a significant dead-time can be handled, and constraints can be implemented efficiently.

2.5.1 Basic Principles

The future process outputs are predicted over the *prediction horizon* H_p using a model of the process: $\hat{y}(k + i)$ for $i = 1, \ldots, H_p$. These values depend on the current process state, and on the future control signals $u(k + i)$ for $i = 0, \ldots, H_c - 1$, where H_c is the *control horizon*. The control signal is manipulated only within the control horizon, and remains constant afterwards, i.e. $u(k + i) = u(k + H_c - 1)$ for $i = H_c, \ldots, H_p - 1$; see Figure 2.4.

The sequence of future control signals $u(k + i)$ for $i = 0, \ldots, H_c - 1$ is computed by optimizing a given objective (cost) function, in order to bring and keep the process output as close as possible to a given reference trajectory r. Most often, the objective functions used are modifications of the following quadratic function [7]:

$$J(u) = \sum_{i=1}^{H_p} [\hat{e}(k+i)^2 + \beta \Delta u(k+i-1)^2], \tag{2.12}$$

where $\hat{e}(k + i) = r(k + i) - \hat{y}(k + i)$ accounts for minimizing the variance of the process output from the reference, while the second term represents a penalty on the control effort (related, for instance, to energy consumption). The latter term can also be expressed by using u itself, or other filtered forms of u, and the coefficient β defines the weighting of the

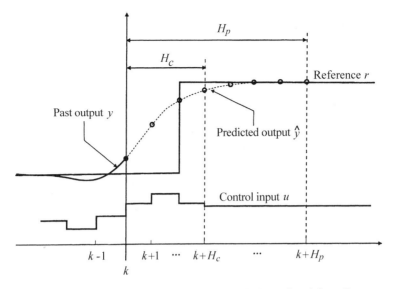

Figure 2.4 The basic principle of model predictive control. Reproduced from Sousa *et al.* [26] by permission of Elsevier Science Ltd (1997)

control effort with respect to the output error [24]. For systems with a dead time of n_d samples, only outputs from time $k + n_d$ are considered in the objective function, because outputs before this time cannot be influenced by the control signal $u(k)$. Similarly, in non-minimum-phase systems also, a delay is introduced which skips the non-minimum phase part of the response. Additional constraints like, for example, level and rate constraints on the control input or other process variables, can also be specified as part of the optimization problem. Following the receding horizon principle, only the control signal $u(k)$ is applied to the process. At the next sampling instant, the process output $y(k + 1)$ is available, and the optimization is repeated with the updated values.

The performance of MPC depends directly on the predictive accuracy of the model over the prediction horizon. As the accuracy decreases, so also does the performance of the controller. Consequently, the major part of the MPC design effort and cost is related to modeling and identification [19].

2.5.2 Optimization in MPC

For a quadratic cost function, a linear time-invariant model, and in the absence of constraints, an explicit analytic solution of the optimization problem given by (2.12), can be obtained. If convex constraints are active in the system, then the optimization problem remains convex and can be easily solved. Otherwise, in the presence of non-linearities and non-convex constraints, a non-convex optimization problem must be solved iteratively at each sampling period. This hampers the application of non-linear MPC to fast systems where many iterative optimization techniques cannot be used, owing to short sampling times and extensive computational effort. Some iterative optimization algorithms such as Nelder–Mead or sequential quadratic programming (SQP) can still be used. Such methods, however, usually converge to local minima, which result in poor solutions of the optimization problem. An alternative solution is to define the optimization problem as a search in a space with a limited number of solutions (discrete space of control actions). Then, standard techniques known from operational research and optimal control can be used, such as dynamic programming [3, 4] or the branch-and-bound method [11]. This last method can be used in a recursive way, demanding less computer memory than dynamic programming, and is presented below.

2.5.3 The Branch-and-Bound Optimization

The branch-and-bound (B&B) method is a structured search technique belonging to a general class of enumerative schemes. It is useful to solve problems for which direct solution methods either do not exist or are inefficient. The B&B method is based on the fact that, in general, only a small number of possible solutions need to be enumerated, while the remaining solutions can be discarded at an early stage. The two basic operations of the method are: *branching*—which consists of dividing possible sets of solutions into subsets; and *bounding*—the establishment of bounds on the value of the cost function over the subsets of possible solutions. These bounds eliminate those subsets that do not contain an optimal solution.

When the control actions are discretized, the B&B method can be applied to predictive control [25, 26], as shown in Figure 2.5. Let the fuzzy model predict the future outputs of the system $\hat{y}(k+1), \ldots, \hat{y}(k+H_p)$ by

$$\hat{y}(k+i) = f(x(k+i-1), u(k+i-1)), \quad i = 1, \ldots, H_p. \quad (2.13)$$

Let the input of the system be discretized such that $u(k+i-1) \in \Omega$, where

$$\Omega = \{\omega_j | j = 1, 2, \ldots, d\}. \quad (2.14)$$

At each time step (see Figure 2.5), d control alternatives are considered, yielding a search tree of maximum d branches. Let $i = 1, 2, \ldots, H_p$ denote the ith level of the tree ($i = 0$ at the

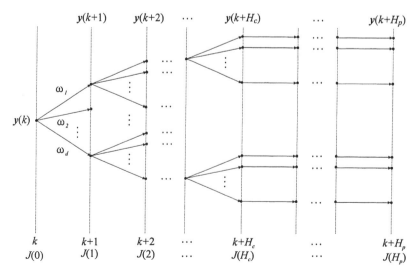

Figure 2.5 Branch-and-bound optimization in predicitive control

initial node) and let j denote the branch corresponding to the control alternative ω_j. The cumulative cost at node i, $J^{(i)}$, is given by

$$J^{(i)} = \sum_{m=1}^{i} [(r(k+m) - \hat{y}(k+m))^2 \beta (\Delta u(k+m-1))^2]. \qquad (2.15)$$

Note that no branching takes place for $i > H_c - 1$ (beyond the control horizon), i.e. the last control action $u(k + H_c - 1)$ is applied successively to the model until H_p is reached.

Application of the branching alone would result in the search of the entire tree (enumerative search), i.e. d^{H_c} possibilities. To reduce the number of alternatives, bounding is applied. A particular branch j at level i is followed only if the cumulative cost $J^{(i)}$ plus an estimated *lower bound* on the cost from the level i to the terminal level H_p, denoted $J_L^{(i)}$, is lower than an estimated upper bound of the total cost, denoted J_U:

$$J^{(i)} + J_L^{(i)} < J_U. \qquad (2.16)$$

The efficiency of the bounding mechanism depends on the quality of the bound estimates. The upper bound should be as close as possible to the optimum and the lower bound as large as possible, in order to decrease the number of new branches to be created. The availability of these estimates depends on the particular problem. If no mechanism for computing the bounds is available, then the following strategy can be applied.

The first path in the tree search follows the 'greedy' strategy of choosing the smallest $J_j^{(i)}(\omega_j)$ at each level i. When following a constant or slowly varying references, the terminal cost $J^{(H_p)}$ resulting from the 'greedy' strategy often equals the optimum or a close upper limit of it. The initial upper bound is set to this value, i.e. $J_U = J^{(H_p)}$. If at a later stage of the tree search a lower value $J^{(H_p)'} < J_U$ is found, then J_U is replaced by this $J^{(H_p)'}$. In the absence of a better estimate, the lower bound is simply set to zero, $J_L^{(i)} = 0, \forall i$. Practical experience with this algorithm shows that these 'worst-case' estimates still prevent the algorithm from exploring a large portion of the search space. One should not forget, however, that the computational complexity of the algorithm remains exponential, which makes it prohibitively expensive for large control horizons. Since a model of the plant is used to calculate the cost at each step taken in the tree, this model should be compact and computationally inexpensive. The computing time also increases with an increasing number of control alternatives, and this number should be as low as possible. However, too coarse a discretization may result in a rough control policy, inferior to those obtained with a finer discretization.

The B&B optimization technique applied to predictive control has three major advantages over other non-linear optimization methods:

1. The global discrete minimum is always found (intrinsic property of the B&B method), guaranteeing the optimality of the controller performance, in the discrete space of control alternatives.

2. The algorithm does not need any initial guess, and hence its performance cannot be negatively influenced by a poor initialization, as in the case of iterative optimization methods.

3. The B&B method implicitly deals with constraints. In fact, the presence of constraints improves the efficiency of bounding, since it restricts the search space by eliminating the control alternatives that result in violation of the constraints.

34 FUZZY MODELING FOR PREDICTIVE CONTROL

Apart from the prohibitive computational complexity for large problems, the most serious drawback of the approach described is the restriction of the possible control actions to a set of discrete control alternatives. For a continuous control space, this discretization may cause oscillations of the outputs around the reference trajectory. One way to remedy this problem is to introduce an adaptive set of control alternatives, where the discretization depends on the state of the process [27].

2.6 MODELING AND CONTROL OF AN HVAC PROCESS

In this section we apply the modeling approach and the MPC scheme previously described to the problem of modeling and controlling a heating, ventilating and air-conditioning (HVAC) system. The system consists of a fan-coil unit inside a test cell (room) under control [29]. Hot water at 65°C is supplied to the coil, which exchanges the heat between the hot water and the surrounding air. In the fan-coil unit, the air coming from the outside (primary air) is mixed with the return air from the room (recirculated or secondary air). The flows of primary and secondary air are controlled by the outside and return dampers, and by the velocity of the fan, which forces the air to pass through the coil, heating or cooling the air. The HVAC system is depicted in Figure 2.6.

The global control goal for this system is to keep the temperature of the test cell T_{in} at a certain reference value, ensuring that enough ventilation and renovated air is supplied to the room. For this purpose three different control actions can be used:

1. Velocity of the fan. The fan has three different velocities: low, medium and high.

2. Position of the dampers (outside and return). The dampers can be open in different discrete positions, controlling the amounts of air coming from outdoors and returned from the test cell.

3. Position of the heating valve. The amount of water entering the heat exchanger is controlled by the heating valve. If this valve is completely open, then the quantity of hot water supplied is maximal, and if it is closed, no hot water is supplied to the coil.

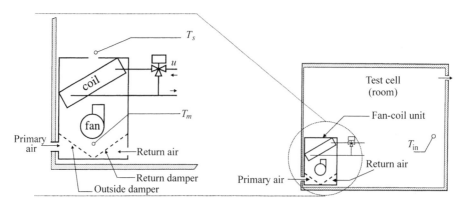

Figure 2.6 Heating, ventilation and air-conditioning system

2.6.1 Initial Modeling of the System

The global control problem is partially reduced in this example. The fan is kept at low velocity in order to increase human comfort by minimizing the noise level. Both dampers are half-open, allowing ventilation from the outside, and the return of some air from the test cell to the fan-coil. Only the heating valve is used as a control input. As shown in Figure 2.6, temperatures can be measured at different parts of the fan-coil. The supply temperature T_s, measured just after the coil, is chosen as the most relevant temperature to control.

First, an initial TS fuzzy model of the system was constructed from process measurements. The input variables were selected on the basis of correlation analysis and a physical understanding of the process. The model predicts the supply air temperature T_s based on its present and previous value, the mixed air temperature T_m, and the heating valve position u, thus:

$$x(k) = [T_s(k), T_s(k-1), u(k-1), T_m(k)]^T, \quad y(k) = T_s(k+1). \tag{2.17}$$

The model consist of ten rules, each with four antecedent fuzzy sets, of the form:

R_i : **if** $T_s(k)$ is A_{i1} **and** $T_s(k-1)$ is A_{i2} **and** $u(k-1)$ is A_{i3} **and** $T_m(k)$ is A_{i4} **Then** $T_s(k+1) = y_i$, where $y_i = \theta_i^T[x(k)^T; 1]^T$.

The antecedent membership functions and the consequent parameters were estimated from a set of input–output measurements by using fuzzy clustering and the least-squares method, as presented in Section 2.3.

The identification data set is shown in Figure 2.7. It contains $N = 800$ samples with a sampling period of 30 s. The data were collected at two different times of day (morning and afternoon), using the excitation signal u that was designed to cover the entire range of the control valve positions and to contain the important frequencies in the expected range of process dynamics. The initial model was created using the values $m = 2$ and $K = 10$ for the fuzziness parameter and the numbers of clusters, respectively. Given the fuzziness parameter m, the number of clusters was selected based on cluster validity measures. Figure 2.7(d) shows the values returned by three validity measures: average within-cluster distance [15], fuzzy hypervolume [9] and cluster flatness [1], respectively, when applied to fuzzy partitions of the identification data obtained with various numbers of clusters.

2.6.2 Validating the Initial Model

The initial model contains a total of 40 antecedent fuzzy sets, shown in Figure 2.10(a) below. A separate data set consisting of 400 observations, which was measured on another day, was used to validate the model. The supply temperatures measured and recursively predicted by the model in a free-run are compared in Figure 2.10(b) below.

The initial fuzzy model was implemented in the IMC scheme depicted in Figure 2.8 and applied to the control of the fan-coil unit. The controller's inputs are the setpoint (reference), the predicted supply temperature $\hat{T}_s$, and the filtered mixed-air temperature T_{mf}. The error signal $e(k) = T_s(k) - \hat{T}_s(k)$ is passed through a first-order low-pass digital Butterworth filter F_1. Another first-order low-pass Butterworth filter F_2 is used for the T_m to filter out the measurement noise.

36 FUZZY MODELING FOR PREDICTIVE CONTROL

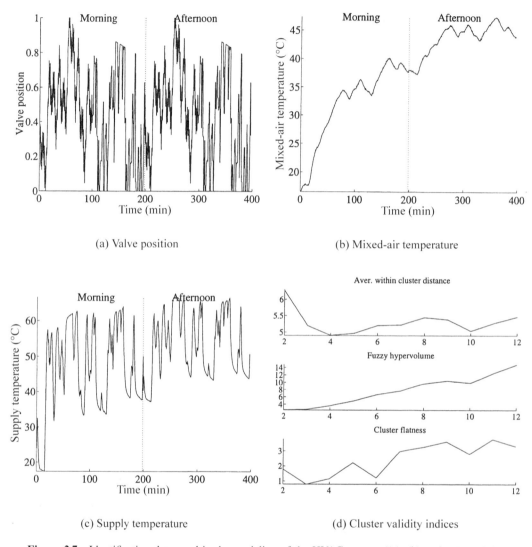

Figure 2.7 Identification data used in the modeling of the HVAC system ((a), (b) and (c)), and the three cluster validity indices (d) applied to various partitions of the identification data. All indicate a local minimum for $K = 10$

Owing to the restrictions placed on the calculation time by the 30 s sampling time, the controller could only cope with slow variations in the reference signal. As mentioned in Section 2.5.3, the B&B optimization speeds up when the process follows a slowly varying reference as the initial bound $J^{(H_p)}$ obtained with the 'greedy' search is close to the optimum. Thus, a smooth shaped reference had to be used. The control results of the real-time implementation using the initial model is presented in Figure 2.9. Here a prediction horizon $H_p = 4$ and a control horizon $H_c = 2$ were used.

MODELING AND CONTROL OF AN HVAC PROCESS

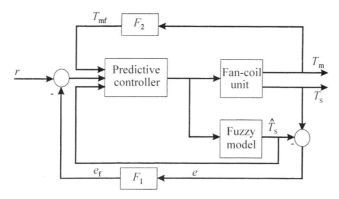

Figure 2.8 Implementation of the fuzzy predictive control scheme for the fan-coil process using an IMC structure

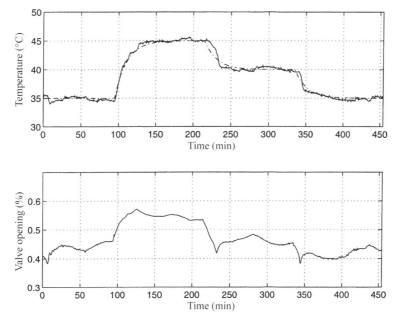

Figure 2.9 Response of the system using the predictive controller based on a complex, initial fuzzy model. The solid line is the measured output; the dash-dotted line is the reference. Reproduced from Sousa *et al.* [26] by permission of Elsevier Science Ltd (1997)

As can be seen from the results in Figure 2.9, the reference is changing very slowly. In an application of a HVAC system, one should be able to handle much faster changes in order to satisfy reference changes given by human occupants. If the delay time becomes too long, then the occupant will get annoyed and probably increase the amplitude of the reference change. As a result of this, the room will eventually get too hot or too cold, and the occupant will feel discomfort. In addition to fast response, when applied in a larger building, the HVAC system should also account for longer term predictions concerning the weather and the occupancy of the building. This will increase human comfort and productivity, and also save energy. Both longer prediction horizons and a faster changing reference will increase the

38 FUZZY MODELING FOR PREDICTIVE CONTROL

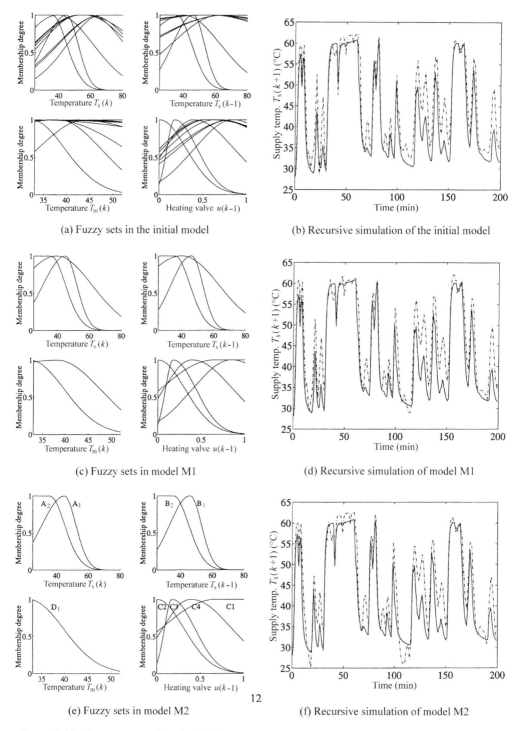

Figure 2.10 Fuzzy sets used in the initial model (a) and the simplified models M1 (c) and M2 (e), and their corresponding prediction of the validation data in a recursive simulation

computation time of the B&B optimization used here. One way to help this is to reduce the computation load of the model as it must be computed for each step in the B&B tree (see Figure 2.5).

2.6.3 Simplifying the HVAC Model

When we inspect the initial HVAC process model, we notice that there are many overlapping and similar fuzzy sets in the rule base (see Figure 2.10(a)). In order to try to reduce the complexity of the model, and thereby its computation time, we apply the rule base simplification method discussed in Section 2.4. Since the rule base simplification does not require additional knowledge or data acquisition, and no computationally intensive algorithms like fuzzy clustering are used, it is advisable to run the algorithm several times with different thresholds. In the results reported below, the threshold γ for removing fuzzy sets similar to the corresponding universal set was kept at $\gamma = 0.8$, while the threshold λ for merging fuzzy sets was varied. The consequent parameters of the rules in the obtained simplified and reduced models where re-estimated by minimizing the least squares error (2.7) using the same training data used to identify the initial model.

The various simplified models were validated on the validation data in a recursive simulation. The membership functions of the two simplified models M1 and M2, obtained with $\lambda = 0.8$ and $\lambda = 0.6$, respectively, and their validation in recursive simulation, are shown in Figure 2.10 together with the membership functions and the validation of the initial model.

From Figure 2.10 we can see that the accuracy of models M1 and M2 is just slightly lower than that of the initial model. If we consider model M2, it is significantly reduced compared with the initial model, and it consists of only four rules and nine different fuzzy sets. The rules of model M2 are given in Table 2.1.

The significant simplification of the rule base has made model M2 easy to inspect. When looking at the premiss of the rules, we notice an interesting result for the antecedent variables $T_s(k)$ and $T_s(k-1)$.

The partitionings of their domains are almost equal, as seen from the membership functions A_1, A_2 and B_1, B_2 in Figure 2.10(e). This suggests that one of the two variables could be removed from the model. If we remove the variable $T_s(k-1)$ from model M2 and re-estimate the consequent parameters, we end up with the strongly reduced model M3. The rules of this model are given in Table 2.2. Close inspection of this model reveals that not all the consequent parameters make sense when the rules are considered individually. This is due to the LS estimate (2.7) used. A further (manual) reduction of the model could be possible by, for example, removing rule R_1^{M3}.

The antecedent fuzzy sets of model M3 are the same as for model M2, shown in Figure 2.10 (e), except that the variable $T_s(k-1)$ is no longer in the model. The validation of model

Table 2.1 Premise of the simplified model M2

If	$T_s(k)$ is,	$T_s(k-1)$ is,	$u(k-1)$ is,	$T_m(k)$ is,	Then $T_s(k+1) =$
R_1^{M2}:	–	–	C_1	–	y_1
R_2^{M2}:	A_1	B_1	C_2	–	y_2
R_3^{M2}:	A_2	B_2	C_3	D_1	y_3
R_4^{M2}:	–	–	C_4	–	y_4

Table 2.2 Simplified model M3

If	$T_s(k)$ is,	$u(k-1)$ is,	$T_m(k)$ is,	Then $T_s(k+1) =$
R_1^{M3}:	–	C_1	–	$-0.4780\,T_s(k)+0.4150\,T_m(k)-0.0890u\,(k-1)+71.4320$
R_2^{M3}:	A_1	C_2	–	$0.9014\,T_s(k)+0.5324\,T_m(k)+12.2264u\,(k-1)-16.4374$
R_3^{M3}:	A_2	C_3	D_1	$0.7276\,T_s(k)+0.1832\,T_m(k)-15.5799u\,(k-1)+9.0553$
R_4^{M3}:	–	C_4	–	$1.9864\,T_s(k)-0.2764\,T_m(k)+28.1784u\,(k-1)-65.0674$

M3 on predicting the validation data in a recursive simulation is shown in Figure 2.11. The RMS error of model M3 is lower than that of the initial model. This supports the removal of the input $T_s(k-1)$ from the model, indicating that the initial structure selection was possibly not correct.

2.6.4 Control Results

The three simplified models M1, M2 and M3, and the initial model, were all implemented in the IMC scheme as depicted in Figure 2.8 and applied to the control of the fan-coil unit in a simulation experiment with a step-like reference. The prediction and control horizons were set to $H_p = 4$ and $H_c = 2$, respectively, and the incremental control signal was restricted to a discrete set of five possible changes $\Delta u \in \{-0.03, -0.015, 0, 0.015, 0.03\}$. Thus, the set of alternatives at each level of the B&B was $\Omega = \{u(i-1) - 0.03, u(i-1) - 0.015, u(i-1), u(i-1) + 0.015, u(i-1) + 0.03\}$.

All models did well with respect to the numerical performance. The result obtained with the initial model is shown in Figure 2.12. The result obtained with the simplest model, model M3, is shown in Figure 2.13. As expected from its prediction performance, the results obtained with model M3 are slightly better than those obtained with the initial model. More significantly, the FLOPS (floating-point operations) used by model M3 in the control simulation were only 15% of the FLOPS used by the initial model.

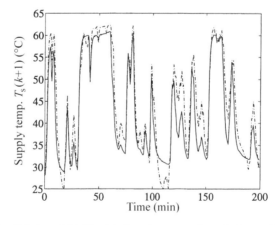

Figure 2.11 The simplified model M3 with three inputs, and a rule base with only four rules and seven fuzzy sets predicts the validation data better than the initial model with four inputs, 10 rules and 40 fuzzy sets

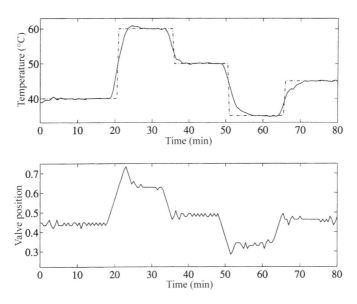

Figure 2.12 Simulation using the predictive controller based on the initial model

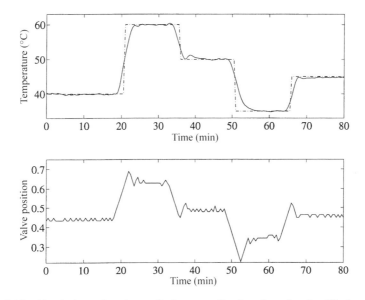

Figure 2.13 Simulation using the predictive controller based on the simplified model M3

2.6.5 Summary of Results

The results of the simplification exercise are summarized in Table 2.3. The relative performance and the computational cost of the models are also visualized in Figure 2.14. For each model, Table 2.3 lists the λ used to obtain the respective simplified model, its

42 FUZZY MODELING FOR PREDICTIVE CONTROL

Table 2.3 Results from simplification

Model	λ	Inputs	No. rules	No. mfs	FLOPS value	Error value	FLOPS control	Error control
Initial	–	4	10	40	1.0000	1.0000	1.0000	1.0000
M1	0.8	4	5	13	0.3621	1.0831	0.3807	1.0147
M2	0.6	4	4	9	0.2526	1.0367	0.2664	1.0618
M3	0.6	3	4	7	0.1660	0.8883	0.1536	0.9649

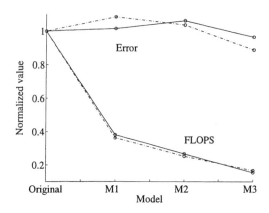

Figure 2.14 Normalized error and FLOPS used by the four models in predicting the validation data (dash-dotted line) and in the control simulation (solid line)

number of inputs, number of rules, number of membership functions, FLOPS used to predict the validation data, the validation error, FLOPS used in the control simulation, and the control error.

From the results, we notice that the error in the validation of the model and in the control simulation is more or less the same for all the models. As expected, increased error in validation means increased error in control. The same holds for the FLOPS used. This relationship is clearly illustrated in Figure 2.14. The relative FLOPS used by the model in prediction corresponds to the relative FLOPS used to solve the control problem. For a given H_p, the B&B optimization will compute the model the same number of times independently of the model used (assuming their predictions are the same). If the model is sped up twice, so also is the B&B optimization. However, owing to the exponential complexity of the B&B, the gain in FLOPS is high.

2.7 CONCLUDING REMARKS

In this chapter we have presented a way to obtain compact fuzzy rule-based models from process measurements and have shown how they can be used in model predictive control (MPC). In a set of data, fuzzy clustering techniques are used to search for relations between the process's input–output variables. From the clusters found in the data, an initial rule-based model is built, focusing mainly on the model's function-approximating properties. Thereafter, similarity-driven simplification is applied to identify and merge compatible concepts and rules in the initial rule base. In this way the number of fuzzy sets and the number of rules in the rule base are reduced. The user can introduce a tradeoff between accuracy and complexity by adjusting the similarity threshold used to merge the similar fuzzy sets. The resulting rule-based model combines good quantitative properties, inherited from the initial rule base, with compactness and increased inspectability resulting from the similarity-driven simplification.

The simplification of the fuzzy model decreases its computational demand and makes it more suitable for MPC, where not only accuracy, but also computation time is of importance. In the MPC scheme presented here, the reduced fuzzy model is used in a predictive control strategy based on discretized control actions. This enables the use of the efficient branch-and-bound algorithm to solve the non-convex optimization problem that is presented to the controller at each sample instant.

The applicability of the modeling and control scheme presented was illustrated using the modeling and control of a heating, ventilation and air-conditioning (HVAC) system. In real-time experiments, a predictive controller based on the initial fuzzy rule-based model was only able to follow slowly varying references due to the long computation time of the model. From the initial model, we derived various less complex models using the similarity-driven simplification. Not only the complexity, but also the inspectability of the model was improved. By inspection of the simplified model, a mistake in the initial structure selection was detected, leading to gains in model accuracy as well as in compactness. Various simplified models were evaluated against the initial model both in prediction and control applications. It was found that considerable reduction in computational load could be achieved without significant changes in accuracy. The simplified models can thus better cope with the real-time demands of the HVAC process, and the most reduced model, model M3, will be implemented in the controller in future studies of the system.

APPENDIX A: THE GUSTAFSON–KESSEL CLUSTERING ALGORITHM

Given the data $\mathbf{Z}$, choose the number of clusters $1 < K < N$, the fuzziness parameter $m > 1$ and the termination criterion $\varepsilon > 0$. Initialize $\mathbf{U}^{(0)}$ (e.g. random).

Repeat for $l = 1, 2, \ldots$

Step 1: Compute cluster means:

$$v_i^{(l)} = \frac{\sum_{k=1}^{N}(\mu_{ik}^{(l-1)})^m z_k}{\sum_{k=1}^{N}(\mu_{ik}^{(l-1)})^m}, \quad 1 \leq i \leq K.$$

Step 2: Compute covariance matrices:

$$F_i = \frac{\sum_{k=1}^{N}(\mu_{ik}^{(l-1)})^m (z_k - v_i^{(l)})(z_k - v_i^{(l)})^{\mathrm{T}}}{\sum_{k=1}^{N}(\mu_{ik}^{(l-1)})^m}, \quad 1 \leq i \leq K.$$

Step 3: Compute distances:

$$D_{ik}^2 = (z_k - v_i^{(l)})^{\mathrm{T}}[(\det(F_i)^{1/(n+1)} F_i^{-1}](z_k - v_i^{(l)}), \quad 1 \leq i \leq K, \quad 1 \leq k \leq N.$$

Step 4: Update partition matrix:
for $1 \leq i \leq K, 1 \leq k \leq N$,
if $D_{ik} > 0$

$$\mu_{ik}^{(l)} = \frac{1}{\sum_{j=1}^{K}(D_{ik}/D_{jk})^{2/(m-1)}},$$

otherwise

$$\mu_{ik}^{(l)} = 0 \text{ if } D_{ik} > 0, \quad \text{and } \mu_{ik}^{(l)} \in [0,1] \quad \text{with } \sum_{i=1}^{K} \mu_{ik}^{(l)} = 1.$$

until $\|U^{(l)} - U^{(l-1)}\| < \varepsilon$.

APPENDIX B: THE RULE BASE SIMPLIFICATION ALGORITHMAPP

Repeat for $j = 1, 2, \ldots, n$:

Step 1: Select most similar fuzzy sets:

$$A_{Lj} = \left\{ A_{lj} | s_{jlm} = \max_{\substack{i \neq p \\ i,p=1,\ldots,K}} (c_{jip}) \right\}.$$

Step 2: Merge selected fuzzy sets:
If $s_{jlm} > \lambda$:

$$A_{cj} = \text{Merge}(\mathbf{A}_{Lj}),$$

$$lj \in A_{Lj}, \text{ set } A_{lj} = A_{cj}.$$

Until: $s_{jlm} < \lambda$.

Step 3: Remove fuzzy sets similar to universal set:

$$s_{ij} = \frac{|A_{ij} \cap U_j|}{|A_{ij} \cup U_j|}, \quad i = 1, 2, \ldots, K,$$

where $\mu_{U_j} = 1, \forall x_j$.

If $s_{ij} > \gamma$, remove A_{ij} from the antecedent of R_i.

ACKNOWLEDGEMENT

This work was supported by the Research Council of Norway

REFERENCES

1. Babuška, R., *Fuzzy Modeling for Control*, Kluwer Academic Publishers, Boston, 1998.
2. Babuška, R. and H. B. Verbruggen, Applied fuzzy modeling, In: *Proceedings of the IFAC Symposium on Artificial Intelligence in Real Time Control*, Valencia, Spain, 1994, pp. 61–66.
3. Baldwin, J. and B. Pilsworth, Dynamic programming for fuzzy systems with fuzzy environment, *Journal of Mathematical Analysis and Applications*, **85**, 1982, 1–23.
4. Bellman, R. E. and L. A. Zadeh, Decision-making in a fuzzy environment, *Management Science*, **17**(4), 1970, 141–164.
5. Bezdek, J. C., *Pattern Recognition With Fuzzy Objective Functions*, Plenum Press, New York, 1981.
6. Chao, C. T., Y. J. Chen, and T. T. Teng, Simplification of fuzzy-neural systems using similarity analysis, *IEEE Transactions on Systems, Man and Cybernetics—Part B: Cybernetics*, **26**, 1996, 344–354.
7. Clarke, D., Advances in model-based predictive control, In: D. Clarke (ed.), *Advances in Model-Based Predictive Control*, Oxford University Press, Oxford, 1994, pp. 3–21.
8. Cross, V., An Analysis of Fuzzy Set Aggregators and Compatibility Measures, Ph.D. thesis, Wright State University, Ohio, 1993.
9. Gath, I. and A. B. Geva, Unsupervised optimal fuzzy clustering, *IEEE Transactions on Pattern Analysis and Machine Intelligence*, **11**(7), 1989, 773–781.
10. Gustafson, D. E. and W. C. Kessel, Fuzzy clustering with a fuzzy covariance matrix, In: *Proceedings of the IEEE Conference on Decision and Control*, San Diego, USA, 1979, pp. 761–766.
11. Horowitz, E. and S. Sahni, *Fundamentals of Computer Algoritms*, Computer Science Press, Inc., Maryland, 1978.
12. Jaccard, P., Nouvelles recherches sur la distribution florale', *Bulletin de la Societe de Vaud des Sciences Naturelles*, **44**, 1908, 223.
13. Kaymak, U. and R. Babuška, Compatible cluster merging for fuzzy modelling, In: *Proceedings of FUZZ-IEEE/IFES'95*, Yokohama, Japan, 1995, pp. 897–904.
14. Kosko, B., Fuzzy systems as universal approximators, *IEEE Transactions on Computers*, **43**, 1994, 1329–1333.
15. Krishnapuram, R. and C. Freg, Fitting an unknown number of lines and planes to image data through compatible cluster merging, *Pattern Recognition*, **4**(25), 1992, 385–400.
16. Kruse, R., J. Gebhardt and F. Klawonn, *Foundations of fuzzy Systems*, John Wiley and Sons, Chichester, 1994.
17. Lin, C. T., *Neureal Fuzzy Control Systems with Structure and Parameter Learning*, World Scientific, Singapore, 1994.
18. Ljung, L., *System Identification, Theory for the User*, Prentice-Hall, New Jersey, 1987.
19. Richalet, J., 'Industrial applications of model based predictive control', *Automatica*, **29**, 1993, 1251–1274.
20. Setnes, M. and U. Kaymak, Extended fuzzy c-means with volume prototypes and cluster merging, In: *Proceedings of EUFIT'98*, Aachen, Germany, 1998, pp. 1360–1364.
21. Setnes, M., R. Babuška, and H. B. Verbruggen, Complexity reduction in fuzzy modeling, *Mathematics and Computers in Simulation*, **46**(5–6), 1998, 507–516.
22. Setnes, M., R. Babuška, and H. B. Verbruggen, Transparent fuzzy modeling, *International Journal of Human—Computer Studies*, **49**(2), 1998, 159–179.
23. Setnes, M., R. Babuška, U. Kaymak, and H. R. van Nauta Lemke, Similarity measures in fuzzy rule base simplification, *IEEE Transactions on Systems, Man and Cybernetics—Part B: Cybernetics*, **28**(3), 1998, 376–386.
24. Soeterboek, R., *Predictive Control: A Unified Approach*, Prentice-Hall, New York, 1992.
25. Sousa, J., A Fuzzy Approach to Model-Based Control, Ph.D. Thesis, Delft University of Technology, Dept. of Electrical Engineering and, Control Laboratory, 1998.

26. Sousa, J., R. Babuška, and H. Verbruggen, Fuzzy predictive control applied to an air-conditioning system, *Control Engineering Practice*, **5**(10), 1997, 1395–1406.
27. Sousa, J., M. Setnes, and U. Kaymak, Adaptive decision alternatives in fuzzy predictive control, In: *Proceedings of FUZZ-IEEE'98*, Anchorage, Alaska, 1998, pp. 698–703.
28. Takagi, T. and M. Sugeno, Fuzzy identification of systems and its applications to modelling and control, *IEEE Transactions on Systems, Man and Cybernetics*, **15**, 1985, 116–132.
29. van Paassen, A. and P. Lute, Energy saving through controlled ventilation windows, In: *Proceedings of the Third European Conference on Architecture*, Florence, Italy, 1993, pp. 208–211.
30. Wang, L. X., *Adaptive Fuzzy Systems and Control: Design and Stability Analysis*, Prentice-Hall, Englewood Cliffs, 1994.
31. Yoshinari, Y. W., W. Pedrycz, and K. Hirota, Construction of fuzzy models through clustering techniques, *Fuzzy Sets and Systems*, **54**, 1992, 157–165.
32. Zhao, J., V. Wertz, and R. Gorez, A fuzzy clustering method for the indentification of fuzzy models for dynamic systems, In: *Proceedings of the Ninth IEEE Symposium on Intelligent Control*, Columbus, Ohio, 1994.
33. Zwick, R., E. Carlstein, and D. V. Budescu, Measures of similarity among fuzzy concepts: A comparative analysis, *International Journal of Approximate Reasoning*, **1**, 1987, 221–242.

3
Adaptive and Learning Schemes for Fuzzy Modeling

Gancho Vachkov
Nagoya University, Nagoya, Japan

3.1 INTRODUCTION

Fuzzy models have many engineering applications in data analysis, fault diagnosis and process control. They can be used in different control schemes as predictive models, inverse models or tuners of conventional or fuzzy controllers.

The fuzzy inference model, proposed by Takagi, Sugeno and Kang [27, 32], known as the TSK model, is a powerful tool for modeling complex non-linear relationships. Often the TSK models are referred to as universal approximators [14] due to their capability of approximating a given system with arbitrary accuracy. The general idea of the TSK fuzzy models is to decompose the input space into fuzzy regions and to approximate the system in every region by a simple (usually linear) model, also called a local model or a sub-model. The overall fuzzy model is considered to be a combination of interconnected systems with simple local models. Then the output of the overall fuzzy model is calculated as a gradual activation of the local models by using proper defuzzification schemes [23, 32, 42] based on the weighted average principle.

TSK models provide an excellent basis for developing good data-driven identification methods that require little prior knowledge of the system under investigation.

In order to identify such models two things are compulsory: a set of experimental data from the real operation of the process (system) and a respective identification algorithm. In the case of correct and dense data in the operating space, there are good analytical methods, based on the least squares technique [28, 32] to find the parameters of the singleton functions of all the rules.

Quite often the real case differs significantly from this pure situation, with data being sparsely distributed and/or highly noised. Then the fuzzy modeling problem does not have an analytical solution due to matrix singularity. Therefore robust numerical iterative algorithms for the identification of fuzzy models are of great importance to cope with this reality. They

Fuzzy Logic. Edited by S. Farinwata, D. Filev and R. Langari
© 2000 John Wiley & Sons, Ltd

are also known as learning algorithms for fuzzy modeling and could be separated into the following two categories:

1. Off-line learning for fuzzy models based on a collection of experimental data. Here the obtained fuzzy model fits in the best way the experimental data set according to a prescribed performance index and is supposed not to be changed when further used for control or simulation. Such an approach is applicable and meaningful when the data set for learning is complete enough to cover the entire operating space, so that a true fuzzy model of the system behavior can be created. The off-line learning scheme can be successfully used for mapping any kind of real experimental data onto a fuzzy model format. Some further examples in this chapter show how to replace human operator and conventional controller actions by an equivalent performance fuzzy controller.

2. Adaptive learning schemes for fuzzy models. These are usually used when the complete learning data set is not available immediately but comes in portions over time, e.g. an additional group of measurements or from a regular sampling process. In what follows two cases of adaptive learning will be considered: reinforced and real-time learning.

    A reinforced learning scheme is used when different data sets of the same process are gradually available at different times and all of them have to be utilized to gradually improve the overall fuzzy model performance. This is quite often a real case of improving knowledge about the process behavior by collecting different data sets from different parts of its operating space.

    Real-time learning algorithms for fuzzy models are typical recursive algorithms that use regular sampling data. Here each new data point is used to gradually improve the model performance.

Fuzzy models obtained by the above adaptive learning schemes can be regarded as 'fuzzy models with a memory' since they have to possess the ability to retain the partial knowledge obtained from previous data sets. This means that the new data should not destroy the current model but rather should be used as complementary information to improve the model.

In this chapter all the above fuzzy model learning schemes are described and analyzed by numerous simulated examples.

The main attractive feature of these numerical learning algorithms is their ability to work and converge quickly in the presence of any kind of noised, sparse and even contradictory data set. In particular, the so-called Local Learning for Fuzzy Models is a computationally efficient method since it decomposes the original optimization task into smaller size optimizations.

Unlike the most analytical methods known, the proposed learning algorithms are able to capture the unknown process behavior even by using a small initial data set, thus building a respective partial fuzzy model, being able to work in this narrow region. Such a feature is quite important for designing a fuzzy control system based on limited recorded data from a real control operation performed by a human operator or by another (conventional) control system. Such examples are given and discussed in this chapter. Furthermore, the ability of the adaptive learning algorithms to update the fuzzy model based on sequentially presented portions of data is very important for real applications. In these cases different portions of limited knowledge from the real working control system are presented over time and used to update the fuzzy model, thus making it gradually cover the entire operating space.

3.2 IDENTIFICATION PROBLEMS OF THE TSK FUZZY MODELS

The general structure of a TSK fuzzy model with r inputs $X_1, X_2, \ldots, X_r$ and one output Y is represented graphically in Figure 3.1.

Generally, a TS fuzzy model with L fuzzy rules $R_i (i = 1, 2, \ldots, L)$ is expressed in the following form:

R_i: **If** $(X_1$ is A_{i1} and $X2$ is $Ai2$ and ... and X_r is $A_{ir})$ **Then**

$$Y_i = P_{i0} + P_{i1} \cdot X_1 + \cdots + P_{ir} \cdot X_r, \quad \text{for } i = 1, 2, \ldots, L, \tag{3.1}$$

where $A_{i1}, A_{i2}, \ldots, A_{ir}$ are r fuzzy sets defining the antecedent (left) part of the ith fuzzy rule R_i. All fuzzy sets are characterized by their respective membership functions: $A_{ij}(X_j)$, for $i = 1, 2, \ldots, L$ and $j = 1, 2, \ldots, r$. As seen from (3.1), the consequent (right) part of each fuzzy rule has an algebraic form where $P_{i0}, P_{i1}, \ldots P_{ir}$ are $r + 1$ coefficients.

The essential point of the TSK model is its combined structure; that is, a purely linguistic left (consequent) part and an algebraic right (consequent) part which produce a single value Y_i known as a *singleton*. This kind of fuzzy model uses the linguistic information (knowledge) conveyed by the antecedents of the fuzzy rules to process the numerical information from the inputs into another numerical result at the output. Each fuzzy rule R_i represents only one local model in the overall calculation scheme.

By default we further assume that the overall fuzzy model is complete; that is, the union of the fuzzy sets used in the construction of the local models covers the entire space of the input variables. In such a case the fuzzy model can respond to any input.

All fuzzy rules $R_i, i = 1, \ldots, L$, form the fuzzy rule base in Figure 3.1. This is the structural part of the fuzzy model that expresses in a linguistic way preliminary knowledge about the process.

Another part of the fuzzy model is its data base, as shown in Figure 3.1. It contains all parameters of the fuzzy model, separated into two distinct groups: **A** and **B**. Set **A** contains all the structural parameters that define the shape and locations of all the membership functions of the fuzzy model: $A_{ij}(X_j), i = 1, 2, \ldots, L; j = 1, 2, \ldots, r$.

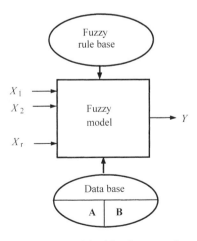

Figure 3.1 Fuzzy model with r inputs and one output

50 ADAPTIVE AND LEARNING SCHEMES FOR FUZZY MODELING

The most often used shapes of the membership functions. i.e. triangular, trapezoidal, bell-shape and Gaussian type, are shown in Figure 3.2. Each of them is characterized by two, three or four parameters.

For example, the bell-shape membership functions, (Figures 3.2(c) and (d)) are constructed by using two shapes: S and Z and contain a total of four parameters, as shown in Figure 3.2(e): A_0, A_1, B_0 and B_1, which are used to calculate the function in the following way.

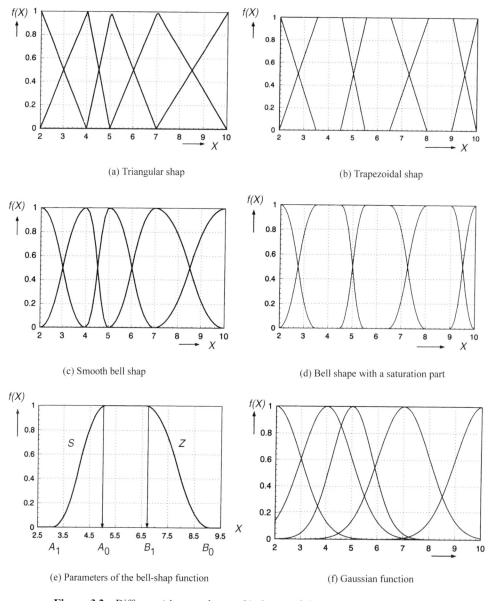

Figure 3.2 Different (sharp and smooth) shapes of the membership functions

- S-shape part of the membership function:

$$Y = 0, \text{ for } X \leq B_0;$$
$$Y = 2[(X - B_0)/D]^2, \text{ for } X > B_0 \text{ and } X \leq B;$$
$$Y = 1 - 2[(B_1 - X)/D]^2, \text{ for } X > B \text{ and } X < B_1;$$
$$Y = 1, \text{ for } X \geq B_1,$$
$$\text{where } D = B_1 - B_0; \; B = (B_0 + B_1)/2. \tag{3.2}$$

- Z-shape part of the membership function:

$$Y = 0; \quad \text{for } X \leq A_0;$$
$$Y = 1 - 2[(X - A_1)/D]^2, \quad \text{for } X > A_1 \text{ and } X \leq A;$$
$$Y = 2[(A_0 - X)/D]^2, \quad \text{for } X > A \text{ and } X < A_0;$$
$$Y = 1, \quad \text{for } X \geq A_1,$$
$$\text{where } D = A_0 - A_1; \; A = (A_1 + A_0)/2. \tag{3.3}$$

The Gaussian type of membership functions (Figure 3.2(f)) has only two parameters: location point (center) c_i and width σ_i, and are calculated as

$$A_i(x) = \exp\left(-\frac{(x - c_i)^2}{2\sigma_i^2}\right), \quad i = 1, 2, \ldots, L. \tag{3.4}$$

The smooth shapes of the membership functions (bell-shape and Gaussian type) are often preferred over the sharp shapes (triangular and trapezoidal) since they give a smooth transition in the activation process from one local fuzzy model to another and as result produce usually a better (more accurate) overall model.

The set **B** of parameters in the data base from Figure 3.1. contains all $L(r+1)$ consequent parameters for all L fuzzy rules, i.e. $P_{i0}, P_{i1}, \ldots, P_{ir}$, $i = 1, 2, \ldots, L$.

The overall (defuzzified) output Y of the fuzzy model is most often calculated as a weighted average of the outputs of all fuzzy rules, as follows:

$$Y = \frac{\sum_{i=1}^{L} f_i Y_i}{\sum_{i=1}^{L} f_i} = \frac{\sum_{i=1}^{L} f_i (P_{i0} + P_{i1} X_1 + \cdots + P_{ir} X_r)}{\sum_{i=1}^{L} f_i} \tag{3.5}$$

where f_i is the so-called activation degree (or firing strength) of the ith fuzzy rule R_i. It is calculated most often by one of the following T-norm operations:

- *Min* operation:

$$f_i = \min\{A_{i1}(X_1), A_{i2}(X_2), \ldots, A_{ir}(X_r)\}, \quad i = 1, 2, \ldots, L. \tag{3.6}$$

- *Product* operation:

$$f_i = \prod_{j=1}^{r} A_{ij}(X_j), \quad i = 1, 2, \ldots, L. \tag{3.7}$$

A normalized activation degree w_1 can be introduced as follows:

$$w_i = \frac{f_i}{\sum_{j=1}^{L} f_j}, \quad i = 1, 2, \ldots, L. \tag{3.8}$$

Then the overall output of the fuzzy model is rewritten as

$$Y = \sum_{i=1}^{L} w_i(P_0 + P_{i1}X_1 + \cdots + P_{ir}X_r). \quad (3.9)$$

In the special but often used case when all membership functions for each input overlap at the midpoint of 0.5 (as illustrated in Figure 3.2(a)–(d)), then the so-called orthogonal condition holds as follows:

$$\sum_{i=1}^{L} f_i = 1, \quad (3.10)$$

and therefore the normalization (3.8) is not needed ($w_i = f_i$).

In the identification problem of the fuzzy models it is further assumed that a data set **S** consisting of N pairs of input–output experimental data is available:

$$\mathbf{S} = \{s_1, s_2, \ldots, s_k, \ldots, s_N\}, \quad \text{with } N = |\mathbf{S}|. \quad (3.11)$$

Each experiment (data point s_k) is of the type $\{X_1(k), X_2(k), \ldots, X_r(k); d(k)\}$, for $k = 1, 2, \ldots, N$ and $d(k)$ is the real output of the process.

Identification of the fuzzy model is considered to be a calculating procedure where the parameters in the fuzzy rule base and/or in the data base (**A** and **B** from Figure 3.1) are to be determined in such a way so as to minimize a given performance index (criterion). The general scheme of the identification procedure is shown in Figure 3.3.

The identification process in which the fuzzy model is created by a numerical iterative procedure for updating the parameters of the model will be further called Learning of the Fuzzy Model.

According to the structure of the fuzzy model in Figure 3.1, there could be three different levels of identification, where different parts of the model are adjusted (separately or simultaneously), as follows:

1. *Structure identification.* The objective here is to determine the linguistic structure of the fuzzy model, namely the fuzzy rule base. It has the following components: the number of linguistic variables (membership functions) for each separate input; and the number of fuzzy rules as well as the content (completeness) of each rule, i.e. the number of inputs (linguistic variables) taking part in this rule.

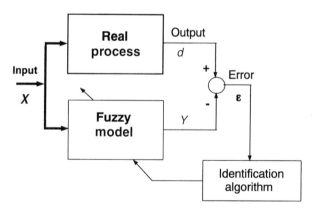

Figure 3.3 General structure for the identification of a fuzzy model

The structure identification is directly related to the linguistic knowledge of the process behavior. If a sufficient and true linguistic knowledge exists, then the structure identification can be omitted as a procedure since the linguistic structure of the model is already known. When this is not the case, the linguistic information should be extracted from the available data set according to a predefined performance criterion. Here different but usually highly combinatorial and time-consuming computation schemes are used. Their objective is to extract the most plausible fuzzy partitions (linguistic variables) and fuzzy rules that explain linguistically the process behavior by using the data set **S** and a given performance index [2, 3, 20, 27, 28, 45, 46].

2. *Antecedent parameters identification.* This is the next (lower) level of the overall identification procedure, where the parameters of the set **A** in the data base (Figure 3.1) have to be adjusted in order to minimize a given performance index. From a physical viewpoint this means that the parameters of the membership functions that define their shape and location have to be tuned so as to best fit the linguistic representation of the process. This level of identification can also be considered as a fine tuning of the structure identification from the upper level 1. From a mathematical viewpoint it is a typical non-linear optimization task and various optimization techniques have been used, such as classical optimization algorithms, neuro-fuzzy systems and genetic algorithms [11, 18, 24, 31, 42].

3. *Consequent parameters identification.* This is the bottom level of identification related only to the algebraic output part of the fuzzy model. In other words, this is a calculation procedure which assumes that the linguistic part of the model, namely the fuzzy rules and membership functions' locations and shapes, have already been defined (or identified beforehand). Then the parameters $P_{i0}, P_{i1}, \ldots, P_{ir}, i = 1, 2, \ldots, L$, in the consequent part of the fuzzy rules have to be determined in such way so as to minimize a given performance index. The total number of consequent parameters for the whole fuzzy model to be identified is $LL = (r+1) \cdot L$.

Much work has been devoted to solving this problem in the following main directions.

- Analytical solutions based on matrix operation and using least squares techniques and Kalman filter for both off-line and recursive identification [20, 23, 28, 32, 36, 37, 42]. However it should be noted that even if these methods give an accurate and non-iterative solution, they fail to find any solution in many real cases of highly noised, repeated and sparsely distributed data in the experimental space which do not cover the range of all the fuzzy rules.
- Numerical iterative learning algorithms based on gradient descent methods and adaptation [1, 9, 10, 17, 22, 35, 39, 40].
- Neural networks training approaches (backpropagation and others) in fuzzy models. This is a very popular and fast growing area of the neural-network-based fuzzy systems (neuro-fuzzy systems) [15], where the basic fuzzy operations have a neural structure representation. Then the learning process of the fuzzy models is performed on a neural network basis [8, 11, 12, 21, 25, 30, 39].

The above-mentioned three levels of identification are interrelated and therefore can be organized in one hierarchical identification scheme with the structure identification as an upper level and parameters identification (antecedent and consequent) as lower levels. There might also be different implementations of the general identification scheme, including only one or two levels with the others being fixed to preliminary given or identified values.

In many works all three levels of the fuzzy model identification are preferred to be carried out simultaneously [41]. This kind of simultaneous identification is implemented naturally in the neuro-fuzzy learning algorithms [11, 21, 38] or in the powerful evolutionary and genetic algorithms [6, 13, 26]. Here the fuzzy rules with their parameters as well as the consequent parameters of the model are identified together. Even if such an approach seems to be a universal solution to the identification problem, it has some demerits as follows:

- Putting all three kinds of parameters of the fuzzy model into one optimization group without a priority order can cause some optimization problems such as high computational cost and/or falling into a local optimum. Therefore reducing the global identification task a number of reduced-size (local) identifications could be a wiser approach.
- The high-order optimization task is quite sensitive in the final result to the initial settings of the algorithms such as: learning rates (for the neuro-fuzzy algorithms) and population size and probability of crossover (for genetic algorithms). As a result, achieving the global optimum is rather subjective and cannot be guaranteed.
- The final solution obtained by these simultaneous identification methods is a sort of departure from the initial linguistic nature of the fuzzy model—the strongest point of the fuzzy model itself. Indeed, the so-called antecedent parameters, i.e. the membership functions parameters as well as the number of fuzzy rules, are adjusted only from a purely mathematical point of view so as to fit the particular data set (often incomplete and highly sparse). Therefore the linguistic structure obtained is quite partial and covers well only this part of the data. Adding new data to the previous data causes a completely new distribution of the linguistic part of the model (membership functions and fuzzy rules).

Recently the training of the antecedent parameters of the fuzzy model is under debate [2, 12] owing to the linguistic and subjective information conveyed by them. Therefore if some knowledge of the linguistic part of the model exists, then identification of the consequent parameters only could be a sufficient solution with a real practical value.

All the learning algorithms presented in this chapter deal with this case of tuning only the consequent parameters of a fuzzy model with the predetermined antecedent part of the model.

3.3 CRITERIA AND SCHEMES FOR LEARNING AND EVALUATION OF FUZZY MODELS

The identification criterion Q obviously plays an important role in the learning process used by the fuzzy model since it directly affects the final accuracy and interpretability of the model. Two different criteria for fuzzy model learning can be principally used. They lead to two different learning schemes global and local.

3.3.1 The Global Learning Criterion, Q_G

The global criterion, Q_G, takes into account for identification only the overall (defuzzified) output Y of the fuzzy model, as shown in Figure 3.3. For a data set **S** with N points it is formulated as

$$Q_G = \sum_{k=1}^{N} [d(k) - Y(k)]^2 \rightarrow \min. \quad (3.12)$$

It is clear that in this objective function the components $Y_i(k)$, $i = 1, 2, \ldots, L$ produced by each individual fuzzy rule R_i are implicitly used to create the overall output $Y(k)$ by the weighted average defuzzification (3.2).

The general structure of global learning of a fuzzy model is shown in Figure 3.4. Here each fuzzy rule R_i, $i = 1, 2, \ldots, L$, represents one local fuzzy model shown in Figure 3.4 as $\text{LFM}_1, \ldots, \text{LFM}_L$. The global criterion (GC) (3.12) is used in the specific global learning algorithm (GLA) to update simultaneously the consequent parameters of all local models, LFM. In other words, here each data point s_k from the data set $\mathbf{S}$ is used to update simultaneously the consequent parameters of all L fuzzy rules (a total of $(r+1) \cdot L$ parameters: $P_{i0}, P_{i1}, \ldots, P_{ir}$, $i = 1, 2, \ldots, L$) or at least those relevant to this data point. Thus the global learning scheme is usually a highly multi-dimensional optimization task.

3.3.2 The Local Learning Criterion, Q_L

This objective function takes into account the output $Y_i(k)$ of each local fuzzy model (fuzzy rule R_i), and its normalized activation degree w_i, $i = 1, 2, \ldots, L$. Then the local learning criterion Q_L 'encourages' separately each fuzzy rule R_i to produce the whole of the model output in proportion to its normalized activation degree, i.e.

$$Q_L = \sum_{i=1}^{L} \sum_{k=1}^{N} w_i(k) [d(k) - Y_i(k)]^2 = \sum_{i=1}^{L} Q_i \rightarrow \min \quad (3.13)$$

Figure 3.5 gives a graphical representation of the idea for local learning of fuzzy models, with the following notations: LC—local criterion; LA—learning algorithm; DEF—defuzzufication unit. As is from this figure, each local fuzzy model LFM_i is trained separately by using its own local criterion Q_i, $i = 1, 2, \ldots, L$, and trying to fit as much as possible the real process output $d(k)$. Therefore the identification process by local learning can be viewed as a

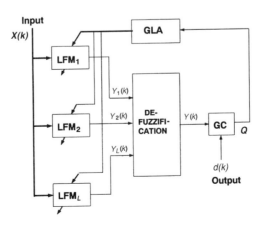

Figure 3.4 Structure of global learning by fuzzy models

ADAPTIVE AND LEARNING SCHEMES FOR FUZZY MODELING

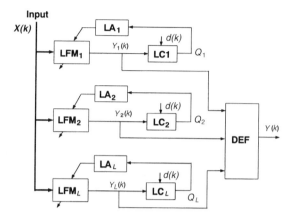

Figure 3.5 Structure of local learning by fuzzy models

decomposition of the original optimization task with size $(r+1) \cdot L$ into L independent optimization tasks each with only $(r+1)$ consequent parameters: $P_{i0}, P_{i1}, \ldots, P_{ir}$. Since the number of model inputs r is usually small, the local learning scheme leads to a significant reduction in both calculation time and complexity. As seen in Figure 3.5, the overall defuzzified model output $Y(K)$ is not directly used in the identification scheme, but the inclusion of the normalized activation degrees w_i in (3.13) guarantees a total match of the real process output d(k).

3.3.3 Evaluation Criteria

Once the fuzzy model has been identified by either global or local learning, its performance has to be analyzed by an evaluation criterion and a test data set **T** with N data (usually different from the training data set **S**). The evaluation is done by calculating the mean square error (MSE) E_2 or the mean absolute error (MAE) E_0 as follows:

$$E_2 = \frac{1}{N} \sum_{i=1}^{N} [d(k) - Y(k)]^2, \tag{3.14}$$

$$E_0 = \frac{1}{N} \sum_{i=1}^{N} |d(k) - Y(k)|. \tag{3.15}$$

When a fuzzy model is created by local learning, the so-called interpretability of the model is suggested to be evaluated by the *local mean squares error* E_L [44] as

$$E_L = \frac{1}{N} \sum_{i=1}^{L} \sum_{k=1}^{N} [d(k) - Y_i(k)]^2. \tag{3.16}$$

3.4 ALGORITHMS FOR GLOBAL LEARNING BY FUZZY MODELS

The general incremental rule for global learning by fuzzy models can be expressed in the following form [1, 10, 40]:

$$u_i^{new} = u_i^{old} + \Delta u_i, \quad i = 1, 2, \ldots, LL, \tag{3.17}$$

where u_i stands for any of the consequent parameters $P_{i0}, P_{i1}, \ldots, P_{ir}$, $i = 1, 2, \ldots, L$, of the rules. Under the assumption of constant singletons for all the fuzzy rules, i.e. $Y_i = P_{i0}$, $i = 1, 2, \ldots, L$, the identification of the fuzzy model can be significantly simplified. It leads to a faster learning process but with a certain loss of modeling accuracy.

The iterative learning scheme (3.17) is usually based on the steepest descent method and uses the gradient or its evaluation at each learning step (iteration) [1, 9, 21, 25, 38]. An approximate and inaccurate evaluation of the gradient often causes learning instability (oscillations) and/or failure to find the global optimum.

Without directly considering the gradient, the increment $\Delta u_i = \varphi(E_i, f_i, \alpha_i)$ in (3.17) can be viewed as a function of the following three parameters: the current identification error, E_i; the activation degree $0 \le f_i \ge 1$ of the rule; and the learning rate α_i. While the learning rate α_i is a matter of a subjective choice and the activation degree f_i is fixed by the predetermined antecedent parts of the rules, the identification error E_i can be calculated in various ways, thus leading to different learning algorithms [34]. This error represents the amount of the difference $Y(k) - d(k)$ between the fuzzy model output $Y(k)$ and the real process output $d(k)$ for each individual data point k. The activation degree f_i is also often included in the calculation scheme for the identification error E_i in order to reflect the individual importance of the fuzzy rule with respect to the data point. Then the learning rule (3.17) can be rewritten as follows:

$$u_i^{\text{new}} = u_i^{\text{old}} - \alpha_i \cdot E_i, \quad i = 1, 2, \ldots, L, \quad (3.18)$$

where the specially introduced weighted average rule error (WARE) E_i is calculated as

$$E_i = \sum_{k=1}^{m_i} \mu_i(k)[Y(k) - d(k)], \quad i = 1, 2, \ldots, L. \quad (3.19)$$

Here $\mu_i(k)$ represents the normalized activation degree of the ith fuzzy rule R_i, $i = 1, 2, \ldots, L$, with respect to the kth data point. It is calculated as

$$\mu_i(k) = \frac{f_i(k)}{\sum_{k=1}^{m_i} f_i(k)}, \quad i = 1, 2, \ldots, L. \quad (3.20)$$

The main idea in the learning schemes (3.18), (3.19) and (3.20) is to update the singleton u_i of the ith fuzzy rule R_i by using the information from the simulation errors of those m_i data points that activate this rule. Otherwise, if the error of all N points is used to learn this rule, then there will be a combined effect of several neighboring rules (or even of the all L rules) and this may result in a slow convergence rate and possibly oscillations.

In other words, in the proposed learning algorithm, instead of using the entire data set $\mathbf{S}$ of N points, only m_i points in the so-called *local data set* $\mathbf{S}_i$ are used to update the rule $R_i, i = 1, 2, \ldots, L$.

By definition, $\mathbf{S}_i, 1 \le i \ge L$, is called a *local data set* for the rule R_i if it contains all data points in $\mathbf{S}$, with non-zero activation degree ($f_i > 0.0$) for this rule. The number of points in $\mathbf{S}_i$ is denoted by $m_i = |\mathbf{S}_i|, i = 1, 2, \ldots, L$. Obviously the union of all consistent sets forms the complete data set $\mathbf{S}$:

$$\mathbf{S} = \bigcup_{i=1}^{N} \mathbf{S}_i. \quad (3.21)$$

Figure 3.6. gives a graphical representation of the local sets $\mathbf{S}_i$ which are usually strongly interrelated and overlap each other.

58 ADAPTIVE AND LEARNING SCHEMES FOR FUZZY MODELING

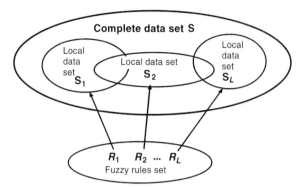

Figure 3.6 Relations between fuzzy rules and data sets

Note that when membership functions of Gaussian-type (4) are used, they cover the whole operation space of each process input so that in this particular case $m_i = N, i = 1, 2, \ldots, N$.

The normalized activation degrees (3.20) are used to consider the relative importance of each single data point from the subset $\mathbf{S}_i$ to the fuzzy rule $R_i, i = 1, 2, \ldots, L$. They actually scale the errors in (3.19) in order to calculate in a more plausible way the amount of update for this rule.

It is worth noting that even if the local data set $\mathbf{S}_i$ is considered in this learning algorithm, it still belongs to the class of global learning algorithms since in (19) the overall deffuzzified output value $Y(k)$ is used.

A simple implementation of the proposed learning algorithms (3.18), (3.19) and (3.20) is made when the learning rate α_i is equal for all the rules and is kept constant during the learning process, i.e. $\alpha_1 = \alpha_2 = \cdots = \alpha_L = \text{const}$. In what follows this version of the proposed algorithm for global learning will be called WARE1.

Obviously, the subjective way of defining the learning rate directly influences the learning speed and convergence of the learning algorithm. Therefore it is worth searching for a way to evaluate it automatically. This can be done if we look at the identification scheme from a control point of view. As shown in Figure 3.7, the iterative learning algorithms for the identification of fuzzy models can be represented as a *discrete-time control loop*. Here the 'discrete time' k is used to denote the set of all N experimental data, presented during one iteration, which is equivalent to one *learning epoch* in the neural networks training process [15]. This is an L-dimensional multiple-input – single-output (MISO) control system with L inputs (the singletons $u_1, u_2, \ldots, u_1, \ldots, u_L$) and one output (the identification error E_i). There is a strong interconnection between the L channels of such a control system, since the identification error E_i in (3.19) (the output of the 'plant') depends on several rule singletons, but not just on the singleton u_i of its own fuzzy rule R_i. The 'plant' itself has L different *gains* $G_i, i = 1, 2, \ldots, L$, for all its channels (rules) and their preliminary or *on-line* estimation is very important in order to achieve a stability and convergence of the overall identification procedure.

It is proposed here to evaluate the plant gains at each iteration step of the learning process in the following numerical incremental way:

$$G_i = \frac{\Delta E_i}{\Delta u_i}, \quad i = 1, 2, \ldots, L, \tag{3.22}$$

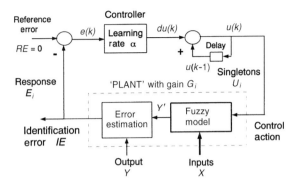

Figure 3.7 The identification process as a discrete-time feedback control loop

with Δu_i being a small predefined increment of the rule singleton and ΔE_i its respective change-of-gain, calculated by the current fuzzy model at this iteration step. Obviously it is a time-consuming process but it helps to speed up the overall learning process.

The learning rate α_i from (3.18) can be considered to be the *controller gain* of a discrete type of integral controller in Figure 3.7. Therefore it could be tuned by using knowledge about the plant gain G_i. There are indeed L different control channels, and the learning rate α_i for each individual channel (fuzzy rule) is proposed to be calculated automatically at each iteration step as $\alpha_i = 1/G_i$.

Therefore the overall learning rule in the identification procedure can be written as

$$u_i^{new} = u_i^{old} - \frac{\Delta u_i}{\Delta E_i} \sum_{k=1}^{mi} \mu_i [Y(k) - d(k)], \quad i = 1, 2, \ldots, L, \qquad (3.23)$$

with μ_i calculated by (3.20). In what follows the proposed algorithm for global learning by fuzzy models, based on (3.23) and (3.20) and the self-defining learning rate, will be called WARE2.

3.4.1 Comparison of the Learning Algorithms Using a Numerical Example

Extensive simulations on different simulated non-linear examples have shown the robust behavior and quick convergence of the proposed learning algorithms WARE1 and WARE2. The latter has a better performance because of the automatic tuning of the learning rate according to (3.23).

The following test example of a highly non-linear process with two inputs, X_1 and X_2, and one output Y has been used as in [34] to evaluate and compare the performance characteristics of the proposed learning algorithms:

$$Y = \frac{13}{33}a + 0.3 X_1 + \frac{7}{19}b + 0.2 X_2 = f(X_1, X_2), \qquad (3.24)$$

where

$$a = \sin\left(\frac{16 X_1}{3 + 4 X_1^2}\right) \quad \text{and} \quad b = \sin\left(\frac{8 X_2}{1 + 2 X_2^2}\right). \qquad (3.25)$$

A 3D plot of this example is shown in Figure 3.8.

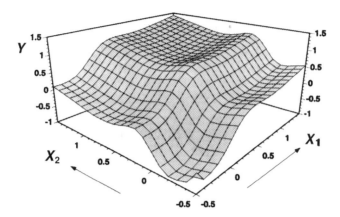

Figure 3.8 A 3D plot of the non-linear test example

To identify the fuzzy model using all the above learning algorithms a small set **S** of 49 random data in the operation space has been generated, as shown in Figure 3.9. For both inputs X_1 and X_2, five membership functions have been predefined as locations and shape (VS = Very Small; SM = Small; MD = Medium; BG = Big and VB = Very Big) and 25 fuzzy rules have been used to identify the fuzzy model using the global learning algorithms WARE1 and WARE2. The size of each local data set $S_1, S_2, \ldots, S_{25}$ is shown in Figure 3.10. As seen, the experimental data set **S** is sparse and there are no data for activating four fuzzy rules (i.e. there are four empty local data sets: S_1, S_2, S_3 and S_6). Therefore the original least squares algorithm for identification, proposed in [32] cannot produce an analytical solution. The initial values of the 25 singletons used for the learning algorithms WARE1 and WARE2 have been randomly generated, as shown in Figure 3.11.

Figure 3.12 shows the convergence rate of both learning algorithms with different learning rates α for WARE1. It is seen that the change of α in WARE1 affects only the rate of convergence, but not the final accuracy, equal to that of WARE2.

From Figure 3.13 it becomes clear that the learning algorithm WARE1 is relatively insensitive to large variations in the learning rate α, which makes the task of its subjective

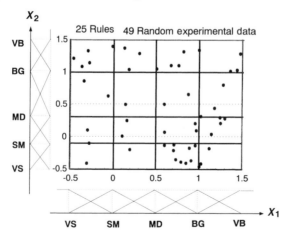

Figure 3.9 Data set of 49 random experimental data for the test example

ALGORITHMS FOR GLOBAL LEARNING OF FUZZY MODELS

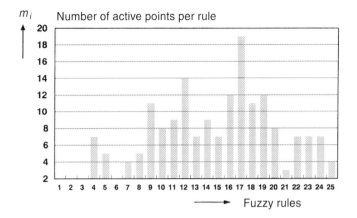

Figure 3.10 Number of active data points m_i for updating the fuzzy rules R_i, $i = 1, 2, \ldots, 25$

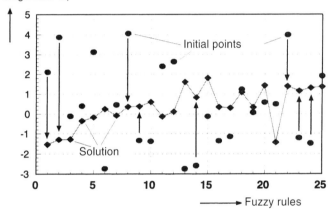

Figure 3.11 Singletons of the initial state and the solution

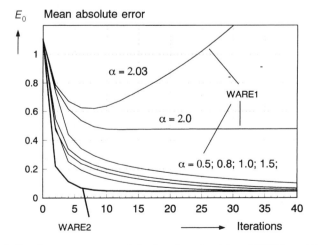

Figure 3.12 Convergence rate of the learning algorithms WARE1 and WARE2

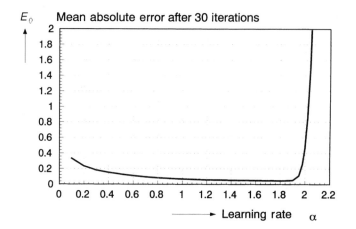

Figure 3.13 Stability of the algorithm WARE1 with different learning rates α

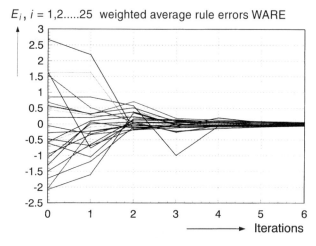

Figure 3.14 WARE2: convergence rate of the singletons u_i of all fuzzy rules with respect to the errors E_i

determination relatively simple. Of course WARE2 is more convenient and better as a learning algorithm, since it does not have an explicit tuning coefficient and gives a better convergence rate, as easily seen in Figure 3.12. Another look at the quick convergence process of WARE2 is given in Figure 3.14., which shows the individual convergence of each singleton u_i with respect to its identification error E_i, calculated by (3.19).

The advantage of the above learning algorithms over the classical least squares algorithm LSA can be found in the fact that WARE1 and WARE2 work even in the case of a sparse data set **S** which does not activate all L fuzzy rules. In such cases the above described learning algorithms produce a *partial fuzzy model,* updating only the active fuzzy rules, for which data points are available.

A weak point of the above proposed algorithms for global learning is that WARE1 and WARE2 work only with constant singletons $u_i = P_{i0}, i = 1, 2, \ldots, L$, which decreases their modeling accuracy. The extension to the case of linear-type singletons, as in (3.1), is not straightforward.

3.5 ALGORITHM FOR LOCAL LEARNING BY FUZZY MODELS

As mentioned in Section 3.3.2. and seen from (3.13) and Figure 3.5, the local learning procedure is a kind of decomposition of the overall LL-dimensional optimization task into L optimization tasks each of them with a smaller $(r+1)$ dimension. Therefore the local learning could be a proper way to cope with the high-dimensionality problem in simulation of real-life processes.

The idea of creating an overall process model based on local models and local learning is not new. It can be found at an early stage in [4] and further developed in the case of fuzzy models in [43] and [44], where combined global and local learning by fuzzy models has been proposed. In [5] and [20] the problems of non-linear system identification and adaptive predictive control have been addressed and solved by the local linear model tree algorithm LOLIMOT. These authors assume that a weighting function for each local model exists and show that the local estimation (by the least squares technique) is very fast and robust, thus decreasing the computational complexity from $O(L^3)$ in the original global case to $O(L)$ in the local modeling case. In [16] it is also shown that local estimation is more robust in the case of small noisy data sets, since it forces the linear models to represent the local surface of the unknown function. Therefore the local estimation avoids the compensation effects of the global estimation, where as the overall model has good accuracy but the interpretability at the local level is lost, as shown in [19] and [43].

The above-mentioned methods for local learning are based generally on the least squares technique and particularly on the local weighted regression (LWR), singular value decomposition (SVD) and recursive least squares (RLS) algorithms as in [7] and [44]. All of them use matrix calculations and therefore are sensitive to the data distribution and density in the operation space, as well as to the data quality in terms of contradiction, noise level and repetition of data. This fact decreases their real applicability to the relatively narrow cases of 'good quality' data sets with high density, less noise and less contradiction. Therefore the numerical iterative algorithms for local learning could be viewed as a good practical solution in the case of raw process data.

The local learning algorithm proposed in [33] is called the weighted errors balancing (WEB) method. It calculates all the resulting parameters $P_{i0}, P_{i1}, \ldots, P_{ir}$ of the ith local model, which are used to create the singletons $Y_i, i = 1, 2, \ldots, L$, of each fuzzy rule. The WEB method uses the simple idea of finding such local model parameters that *equalize* the positive and negative weighted inference errors e_i, produced by only this local model, i.e.

$$\sum_{k=1}^{m_i} e_i(k) = \sum_{k=1}^{m_i} w_i(k)[d(k) - Y_i(k)] = 0, \quad i = 1, 2, \ldots, L, \quad (3.26)$$

or

$$\sum_{i=1}^{m_i} w_i(k)[d(k) - P_{i0} - P_{i1}X_1(k) - P_{i2}X_2(k) - \cdots - P_{ir}X_r(k)] = 0, \quad i = 1, 2, \ldots, L.$$

(3.27)

Note that m_i shows the number of all data points in the ith local data set $\mathbf{S}_i$ from Figure 3.6 which activate the rule R_i. However, replacing m_i with N in (3.26) and (3.27) would not affect the result since the other points in the experimental set $\mathbf{S}$ do not activate that rule and they have zero normalized activation degrees $w_i(k)$. The calculation time would be increased.

When Gaussian-type membership functions are used, it naturaly results that $m_i = N$ for all $i = 1, 2, \ldots, L$.

In the case of constant singletons ($Y_i = P_{i0} = $ const.) it is easy to find a straightforward analytical solution of (3.27) as follows:

$$Y_i = \frac{\sum_{k=1}^{N} w_i(k) d(k)}{\sum_{k=1}^{N} w_i(k)}, \quad i = 1, 2, \ldots, L. \quad (3.28)$$

For the one-dimensional case $Y = f(X)$ this kind of solution $Y_i = $ const. is illustrated in Figure 3.15 as a bold horizontal line denoted by 'C'.

For the typical TSK fuzzy model (3.1) with linear-type singletons, equation (3.27) is not enough to determine all $r + 1$ parameters, so it is used only as an *equality constraint* in the r-dimensional optimization task. The objective function in this optimization is proposed to be the *local mean square error* (LMSE) $E2_i$ produced by the ith local model that has to be minimized:

$$E2_i = \frac{1}{N} \sum_{k=1}^{N} [D(k) - Y_i(k)]^2 \to \min. \quad (3.29)$$

An illustration of four linear local fuzzy models $Y_i = f(X) = P_{i0} + P_{i1} X_1$ for the one-dimensional case is also shown in Figure 3.15. They are shown as four lines denoted by 0, 1, 2 and 3. All of them satisfy the equality constraints (3.27) but only line 0 minimizes the criterion in (3.29) for a minimal local mean square error $E2$.

The equality constraints (3.27) are used to produce a plausible solution for the local model Y_i because they have the physical meaning of a balance (center of gravity) between the positive and negative simulation errors, produced by the respective local model. Additionally, equation (3.27) serves to decrease the dimension of the optimization task (from $r+1$ to r) thus making the learning process even less computationally expensive. For example, for a two-input process only one-dimensional optimization is needed. For a three-input process a simple two-dimensional optimization (such as the Gauss–Seidel method) is efficient and so on. In Figure 3.16. another graphical representation of this constraint optimization task for the one-dimensional case is shown.

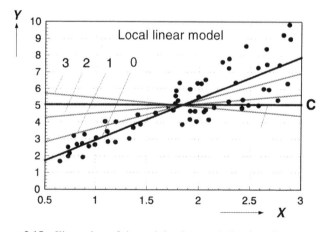

Figure 3.15 Illustration of the weighted errors balancing (WEB) method

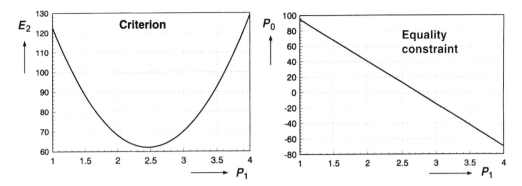

Figure 3.16 Illustration of the constraint optimization

It can be expected that the fuzzy models learned locally would be not as accurate as the models learned globally, assuming the same structure of the consequent parts of the rules (constant or linear equation). However, the numerical WEB method for local learning explained above has the advantage that it can also deal with linear-type of singletons, while the WARE1 and WARE2 algorithms for global learning are for constant singletons only. In addition, the local learning algorithms and their fuzzy models possess good interpretability of the process behavior in each local region of the process space, since they make a plausible approximation of the data in this region by an individual local model. This feature is illustrated in Figure 3.17. The test example of a non-linear one-dimensional process is taken from [23] as

$$Y = f(X) = 3X(X-1)(X-1.9)(X+0.7)(X+1.8). \qquad (3.30)$$

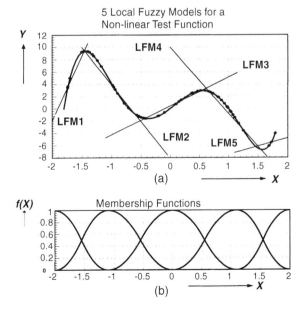

Figure 3.17 Graphical representation of the idea of *local learning* and interpretability of the local fuzzy models. (a) Five local fuzzy models for a non-linear test function; (b) membership functions

Here the randomly generated experimental data points have been used to perform local learning by the fuzzy model using the above discussed WEB algorithm with a predefined bell-shape membership function as shown in Figure 3.17(b). Since linear-type singletons are used, the solution is shown in Figure 3.17(a) in the form of five different local fuzzy models LFM1, ..., LFM5 (five straight lines). It is seen that they interpret the tendency and shape of the real process behavior (the bold line) in each local region in a quite plausible way.

3.6 REINFORCED LEARNING ALGORITHM

The next problem fuzzy model learning is the reinforcement ability of learning. As stated at the beginning of the chapter, this is a learning process in which the fuzzy model parameters are gradually updated by using a sequence of different data sets, each containing usually sparse and partial knowledge about the process. From such a viewpoint this procedure is a kind of adaptation of the fuzzy model in which its performance is gradually improved as more data are added.

In order to create a reinforcement calculation scheme, a kind of model memory should be introduced. Under the assumptions made for the local learning algorithm, it is relatively easy to create a memory for the previous model status by simply saving the balanced weight of the errors $e_i(k)$ in (3.26) with either its positive or its negative part. Then the new training set will only add some new weighted errors to the saved ones in the memory and a new run of the WEB algorithm for the updated balanced errors would create an updated fuzzy model.

Let denote the resulting data sets presented at each iteration $s = 0, 1, 2, \ldots$ by $\mathbf{D}_0, \mathbf{D}_1, \mathbf{D}_2, \ldots$. Usually each of these data sets contains different numbers of data (or even single data points). Then in the simplest case of constant singletons $P_{0i}, i = 1, 2, \ldots, L$, the pure analytical solution from (3.10) is transformed into the following recursive equation:

$$P_{0i}(s) = \frac{WD(s-1)}{W(s-1)} + \frac{WD(s)}{W(s)}, \quad s = 1, 2, \ldots, \tag{3.31}$$

where

$$WD(s) = \sum_{k=1}^{N} w_i(k)d(k); \quad W(s) = \sum_{k=1}^{N} w_i(k). \tag{3.32}$$

In order to perform this computation, two additional memories for each rule R_i are needed, namely $WD(s-1)$ and $W(s-1)$. Then the overall reinforced learning algorithm is performed in the sequence (3.31), (3.32) and the WEB algorithm from the previous section.

In what follows some simulation results obtained by the above proposed algorithm are shown on the same one-dimensional test example (3.30). This time 52 data points were randomly generated and distributed in seven data sets $\mathbf{D}_1, \mathbf{D}_2, \ldots \mathbf{D}_7$, as shown in Figure 3.18. The data sets have been presented for reinforced learning in the sequence from $\mathbf{D}_1$ to $\mathbf{D}_7$ and the fuzzy model has been adjusted gradually with its final performance shown in Figure 3.19.

It should be noted that the implementation of this reinforced learning algorithm needs some techynical details to deal with the monotonously growing weights in the model memory. One way to cope with this problem is to introduce an exponentially forgetting factor (gradually lightening the weights), which could be done mathematically in different ways.

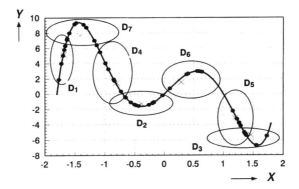

Figure 3.18 Process data distributed in seven data sets $D_1, D_2, \ldots, D_7$

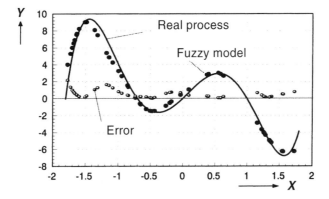

Figure 3.19 Simulation results of reinforced learning

3.7 SIMULATION RESULTS FOR CONTROL APPLICATIONS

The main advantage of the proposed algorithms for global and local learning is their ability to learn from a sparse and/or highly noised data. If this is the case, usually a partial fuzzy model is created, i.e. some of the rules are identified, but the others (the *non-active* rules) are 'frozen' at their initial settings. This is quite convenient for acquiring control knowledge from a real working control system in order to further implement a fuzzy control system that behaves in a similar way. From a theoretical viewpoint this means that the above learning algorithms could be used to map of the experimental control data into a fuzzy controller format with predetermined antecedent parts (fuzzy rules and locations). And if the reinforced learning is used further, the fuzzy controller can be gradually improved, because it would possibly contain some new control rules together with the updated existing ones.

Figures 3.20–3.25 show graphically the technology of such a mapping from data into fuzzy controller parameters. The experimental data were taken from a simulation of a classical control system with a discrete type of PID controller and a first-order process with time delay. By using eight different transient processes (Figures 3.20 and 3.22) for data collection (Figures 3.21 and 3.23), a partial fuzzy controller has been designed (Figure 3.24)

68 ADAPTIVE AND LEARNING SCHEMES FOR FUZZY MODELING

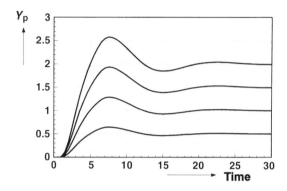

Figure 3.20 Four set-point transient processes used for collecting the experimental data

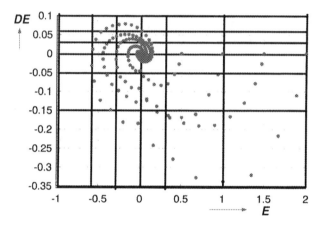

Figure 3.21 Experimental data obtained from the transient processes in Figure 3.20. Partitions correspond to the selected regions for the triangle-type fuzzy rules

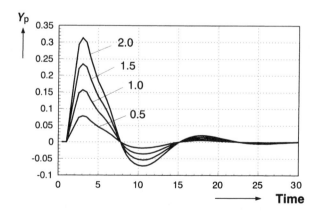

Figure 3.22 Four disturbance transient processes used for gathering experimental data

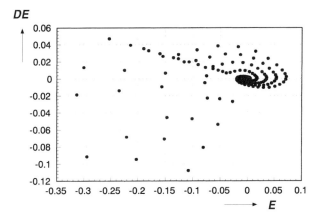

Figure 3.23 Experimental data obtained from the transient processes in Figure 3.22

PB	7	14	21	28	35	42	49
PM	6	13	20	27	34	41	48
PS	5	12	19	26	33	40	47
ZR	4	11	18	25	32	39	46
NS	3	10	17	24	31	38	45
NM	2	9	16	23	30	37	44
NB	1	8	15	22	29	36	43
		NB	NM	NS	ZR	PS	PM	PB

Figure 3.24 Partial fuzzy controller $Y = f(E, DE)$ designed by using the data generated in Figures 3.21 and 3.23

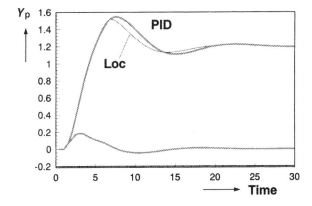

Figure 3.25 Comparison of the original PID controller with the designed linear local fuzzy controller

by using the local learning algorithm WEB. The fuzzy rules shown in bold in Figure 3.24 correspond to the active (identified) fuzzy rules after applying all data sets.

The results of the control, shown in Figure 3.25 and based on the newly designed fuzzy controller, are almost the same as those of the original control system with the digital PID controller, which proves numerically the validity of the proposed algorithm for local learning.

3.8 CONCLUSIONS

Several learning algorithms for creating TSK-type fuzzy models have been presented and analyzed in this chapter, as follows: global learning (WARE1 and WARE2), local learning (WEB) and adaptive (reinforced) learning of fuzzy models. Their main attractive feature is the ability to work and converge quickly in the presence of any kind of noised, sparse and even contradictory data set. In particular, the so-called local learning by fuzzy models (WEB) is a computationally efficient method since it decomposes the original optimization task into smaller-sized optimizations, thus being able to solve real high-dimensional problems.

Unlike the most analytical methods known, the proposed learning algorithms are able to capture the unknown process behavior even by using a small initial data set, this building a respective partial fuzzy model, still being able to work in this narrow region. Such a feature is quite important for the design of fuzzy control systems based on limited control data from a real control operation.

REFERENCES

1. Araki, S., H. Nomura, I. Hayashi and N. Wakami, A self-generating method of fuzzy inference rules, In: *Proceedings of the International Symposium IFES'91*, Yokohama, Japan, vol. 2, 1991, pp. 1047–1058.
2. Bortolan, G. and W. Pedrycz, Reconstruction problem and information granularity, *IEEE Transactions on Fuzzy Systems*, **5**(2) 1997, 234–248.
3. Chiu, S. L., Fuzzy model identification based on cluster estimation, *Journal on Intelligent and Fuzzy Systems*, **2**, 1994, 267–278.
4. Cleveland, W. S., Robust locally weighted regression and smoothing scatterplots, *Journal of the American Statistical Association*, **74**, 1979, 829–836.
5. Fisher, M., O. Nelles and R. Isermann, Adaptive predictive control based on local linear fuzzy models, In: *Proceedings of the Eleventh IFAC Symposium on System Identificaion: SYSID'97*, Japan, vol. 2, 1997, pp. 859–864.
6. Furuhashi, T., S. Matsushita and H. Tsutsui, Evolutionary fuzzy modeling using fuzzy neural networks and genetic algorithm, In: *Proceedings of the IEEE International Conference on Evolutionary Computation: ICEC'97*, 1997, pp. 623–627.
7. Golub, G. H. and C. F. Van Loan, *Matrix Computations*, 2nd ed., John Hopkins University Press, Baltimore, 1989.
8. Horikawa, S., T. Furuhashi, Y. Uchikawa and T. Tagawa, A study on fuzzy modeling using fuzzy neural networks, In: *Proceedings of the IFES'91 Symposium*, Yokohama, Japan, 1991, pp. 562–573.
9. Ichihashi, H., Iterative fuzzy modeling and a hierarchical network, In: *Proceedings of the Fourth IFSA Congress*, Brussels, 1991, pp. 49–52.

10. Isermann, R., K.-H. Lachmann and D. Matko, *Adaptive Control Systems*, Prentice-Hall, Englewood Cliffs, 1992.
11. Jang, J. R., ANFIS: Adaptive network-based fuzzy inference system, *IEEE Transactions Systems, Man and Cybernetics*, **23**, 1993, 665–685.
12. Jang, J. R., C. Sun and E. Mizutani, *Neuro-Fuzzy and Soft Computing: A Computational Approach to Learning and Machine Intelligence*, Prentice-Hall, Englewood Cliffs, 1997.
13. Kim, J., Y. Moon and B. Ziegler, Designing fuzzy net controllers using genetic algorithms, In: *IEEE Control Systems*, 1995, pp. 66–72.
14. Kosko, B., Fuzzy systems as universal approximators, In: *Proceedings of the IEEE International Conference on Fuzzy Systems*, San Diego, 1992, pp. 1153–1162.
15. Lin, C-T. and C. S. George Lee, *Neural Fuzzy Systems, A Neuro-Fuzzy Synergism to Intelligent Systems*, Prentice-Hall, Upper Saddle River, 1995.
16. Murray-Smith, R., A Local Model Network Approach to Nonlinear Modeling, Ph.D. Thesis, University of Strathclyde, 1994.
17. Murray-Smith, R. and T. A. Yohansen, Local Learning in Local Model Networks, In: *Multiple Model Approaches to Modeling and Control*, Taylor Francis Ltd, London, 1997, pp. 185–210.
18. Nakamori, Y. and M. Ruoke, Identification of fuzzy prediction models through hyper-ellipsoidal clustering, *IEEE Transactions on Systems, Man and Cybernetics*, **24**, 1994, 1153–1173.
19. Nelles, O., On the identification with neural networks as series–parallel and parallel models, In: *Proceedings of the International Conference on Artificial Neural Networks (ICANN)*, Paris, France, 1995.
20. Nelles, O., O. Hecker and R. Isermann, Automatic model selection in local linear model trees (LOLIMOT) for nonlinear system identification of a transport delay process, In: *Proceedings of the Eleventh IFAC Symposium on System Identification SYSID'97*, Kitakyushu, Japan, vol. 2, 1997, pp. 727–732.
21. Nomura, H., I. Hayashi and N. Wakami, A self-tuning method of fuzzy control by descent method, In: *Proceedings of the IEEE International Conference on Fuzzy Systems*, San Diego, 1992, pp. 203–210.
22. Otto, P., Fuzzy Modelling of Nonlinear Dynamic Systems by Inductive Learned Rules, In: *Proceedings of the Third European Congress EUFIT'95*, Aachen, Germany, vol. 2, 1995, pp. 858–864.
23. Pedrycz, W. and M. Reformat, Rule-based modeling of nonlinear relationships, *IEEE Transactions on Fuzzy Systems*, **5**(2), 1997, 256–269.
24. Rovatti, R. and R. Guerrieri, Fuzzy sets of rules for system identification, *IEEE Transactions on Fuzzy Systems*, **4**(2) 1996, 89–102.
25. Shi, Y., M. Mizumoto, N. Yubazaki and M. Otani, A learning algorithm for tuning fuzzy rules based on the gradient descent method, In: *Proceedings of the Fifth IEEE International Conference on Fuzzy Systems (FUZZ-IEEE'96)*, New Orleans, vol. 1, 1996, pp. 55–61.
26. Shimojima, K., T. Fukuda and Y. Hasegawa, Self-tuning fuzzy modeling with adaptive membership functions, rules and hierarchical structure based on genetic algorithm, *Fuzzy Sets and Systems*, **71**, 1995, 295–309.
27. Sugeno, M. and G. T. Kang, Structure identification of fuzzy model, *Fuzzy Sets and Systems*, **28**, 1988, 15–30.
28. Sugeno, M. and K. Tanaka, Successive identification of a fuzzy model and its applications to prediction of a complex system, *Fuzzy Sets and Systems*, **42**(3), 1991, 315–334.
29. Sugeno, M. and T. Yasukawa, A fuzzy-logic-based approach to qualitative modeling, *IEEE Transactions on Fuzzy Systems*, **1**, 1993, 7–31.
30. Tachibana, K. and T. Furuhashi, A hierarchical fuzzy modeling using fuzzy neural networks which enable uneven division of input space, In: *Proceedings of the Fourteenth Fuzzy Symposium SOFT'98*, Gifu, Japan, WF2-2, 1998, pp. 305–308.

31. Takagi, H. and I. Hayashi, NN-driven fuzzy reasoning, *International Journal of Approximate Reasoning*, **5**, 1991, 191–212.
32. Takagi, T. and M. Sugeno, Fuzzy identification of systems and its applications to modeling and control, *IEEE Transactions on Systems, Man and Cybernetics*, **SMC-15**(1), 1985, 116–132.
33. Vachkov, G., Algorithms for global and local learning of fuzzy models In: *Proceedings of the Fourteenth Fuzzy Symposium SOFT'98*, Gifu, Japan, WF3-3, 1998, pp. 321–324.
34. Vachkov, G. and K. Hirota, Comparison of some learning algorithms for identification of fuzzy models, In: *Proceedings of the ICONIP'97, New Zealand, Progress in Connectionist-Based Information Systems*, Springer, vol. 1, 1997, pp. 376–379.
35. Wang, L.-X., *Adaptive Fuzzy Systems and Control, Design and Stability Analysis* Prentice-, Hall, New Jersey, 1994.
36. Wang, L-X., *A Course in Fuzzy Systems and Control*, Prentice-Hall, Englewod Cliffs, 1996, p. 448.
37. Wang, L. and R. Langari, Complex systems modeling via fuzzy logic, *IEEE Transactions on Systems, Man and Cybernetics*, **2**, 1996, 100–106.
38. Wang, L.-X. and J. M. Mendel, Fuzzy basis functions, universal approximation and orthogonal least squares learning, *IEEE Transactions on Neural Networks*, **3**, 1992, 807–814.
39. Wang, L.-X. and J. M. Mendel, Back-propagation fuzzy systems as nonlinear dynamic system identifiers, In: *Proceedings of the IEEE International Conference on Fuzzy Systems*, San Diego, 1992, pp. 1409–1416.
40. Widrow, B., *Adaptive Signal Processing*, Prentice-Hall, Englewood Cliffs, 1985.
41. Yager, R. and D. Filev, United structure and parameter identification of fuzzy models, *IEEE Transactions on Systems, Man and Cybernetics*, **23**(4), 1993, 1198–1205.
42. Yager, R. and D. Filev, *Essentials of Fuzzy Modeling and Control*, John Wiley and Sons, New York, 1994.
43. Yen, J. and W. Gillespie, Integrating global and local evaluation for fuzzy model identification using genetic algorithms, In: *Proceedings of the Sixth IFSA Congress, IFSA'95*, Sao Paulo, Brazil, 1995.
44. Yen, J., L. Wang and W. Gillespie, Improving the interpretability of TSK fuzzy models by combining global learning and local learning, *IEEE Transactions on Fuzzy Systems*, **6**(4), 1998, 530–537.
45. Yoshinari, Y., W. Pedrycz and K. Hirota, Construction of fuzzy models through clustering techniques, *Fuzzy Sets and Systems*, **54**, 1993, 157–165.
46. Zadeh, L., Fuzzy sets and information granularity, In: M. Gupta, R. Ragade and R. Yager (eds.), *Advances in Fuzzy Sets Theory and Applications*, North Holland, Amsterdam, 1979, pp. 3–18.

4

Fuzzy System Identification with General Parameter Radial Basis Function Neural Network

Daouren F. Akhmetov and **Yasuhiko Dote**
Muroran Institute of Technology, Muroran, Japan

4.1 INTRODUCTION

Recently, the fuzzy approach has become one of the most intensively developing and attractive fields in modern intelligent control theory and practice. Many successful industrial applications have demonstrated the validity and the high efficiency of fuzzy systems along with remarkable economic benefits. However, there has been no systematic method developed for fuzzy system design.

The capability of fuzzy systems to solve a wide class of mapping problems has been approved by control theorists and practical engineers. One of the most important design issues of fuzzy systems is the construction of an appropriate fuzzy rules set. There are two common approaches: manual rule generation and automatic rule generation. The first is very time consuming, and its applicability may be problematic for the high-dimensional fuzzy rule set case or a poor domain knowledge database. Therefore, many researchers have been involved with the automatic fuzzy rule generation (extraction) problem [1–5].

The basic idea behind these approaches is to estimate fuzzy rules through learning from input–output sample data. For example, Kosko [2] has developed fuzzy associative memory (FAM), a class of neural networks, in order to store a set of fuzzy rules and to process them in parallel. The FAM learns fuzzy rules by clustering the sample data in the product space or the linguistic space. In this approach each cluster represents a fuzzy rule. The problem here is to overcome the possible conflicts between the generated rules [3]. Wang and Mendel [5] have proposed a method to resolve the conflict problem, while assigning a degree to each rule.

The direct usage of artificial neural networks (ANNs) to identify fuzzy rules and to tune their membership functions has been also reported [6]. But the inherent intractability of the neural network approach complicates the understanding of the identified fuzzy rules due to

Fuzzy Logic. Edited by S. Farinwata, D. Filev and R. Langari
© 2000 John Wiley & Sons, Ltd

the distributed way of information processing with ANNs. To resolve the difficulty, a fuzzy multilayer perceptron has been proposed in order to generate a number of fuzzy rules which can be obtained by the connection weights computed in the training phase [7]. Fuzzy rules make it possible to infer the output class membership values of an unknown input and to estimate a measure of certainty of that decision in the testing phase.

Recently, a class of radial basis function networks (RBFNs) have been used extensively to explicitly extract fuzzy rules from the given data. The RBFN proposed by Moody and Darken [8] is a type of neural network that utilizes local receptive fields, called *basis functions*, for function mapping. The RBFNs have demonstrated their usefulness in a variety of applications including classification, prediction, and system modeling.

The functional equivalence between fuzzy systems and the RBFNs has been shown, and an adaptive network-based fuzzy inference system, which is trained by the backpropagation algorithm, has been developed by Jang and Sun [9, 10]. Following the idea that fuzzy If–Then rules are naturally related to the basis functions, Wang and Mendel [11] have represented fuzzy systems as series expansions of fuzzy basis functions.

It is important to note that the above architecture can only realize the inference schemes of either the Sugeno type (the consequence is a linear combination of inputs) or a special case of the Sugeno type schemes (the consequence is a constant) [4]. To extend types of fuzzy reasoning other than the Sugeno type, Horikawa *et al.* [1] have proposed fuzzy neural networks that implement the process of fuzzy reasoning whose consequence part has fuzzy variables.

However, the above-mentioned fuzzy neural networks require a proper initial rule base that can be created using the expert knowledge. It is easy to see that creating the initial rule base may be difficult, as the number of inputs becomes high. Furthermore, in both of the above methods [1,10], it is necessary to determine the number of fuzzy rules to be identified before training since the networks are trained by the backpropagation algorithm.

Fuzzy systems with an adaptive capability to extract fuzzy implicative rules from sample data by a learning algorithm have already been proposed [12, 13]. Three different architectures of the RBF-based adaptive fuzzy system (AFS) are developed in order to accommodate not only the Sugeno type but also a more general fuzzy inference scheme, i.e. when the consequence is fuzzy. The proposed RBF based AFS has several advantages over other methods trained by the backpropagation algorithm. It does not require the initial rule base. Fuzzy rules are generated by recruiting the basis function units until the desired accuracy is achieved, therefore it is not necessary to specify the number of fuzzy rules to be identified. Along with these, the RBF-based AFS is compatible with existing fuzzy systems.

In this chapter, the problem of a fuzzy rule set generation through neural network representation is formulated, and some already elaborated methods are discussed. The automatic fuzzy rule generation problem is treated as the RBFN structure identification problem [26–31]. The structure modeling (SM) principle, also known as the group method of data handling (GMDH) [29, 30], providing the desired approximation accuracy and canceling the model overfitting effect, is presented. The modification of the MS, the so-called unbiasedness criterion using distorter (UCD), is described [31].

Common characteristics of the Gaussian radial basis function neural networks are presented and learning procedures for their training are studied. A novel structure RBFN and advanced learning procedure for its training are proposed which is based on the general parameter (GP) method for complex system identification [14–20]. Theoretical backgrounds of the GP approach are studied, including its convergence (stability) analysis. The main features of the proposed approach may be generalized as:

- an artificial reduction in the dimensionality of the adjusting model parameter space,
- a high (initial) convergence rate which does not depend on the dimensionality of the original (sample) model parameter vector,
- identification accuracy by the GP statistics calculation in the steady state of the learning procedure,
- the GP-model is a flexible type of model with an adjustable structure,
- a quasi-optimal model structure may be derived with some decision-making multi-step procedure based on learning period information.

Finally, a UCD-based GP RBFN AFS is proposed with the capability to extract fuzzy If–Then rules from input and output sample data. In conclusion, simulations are presented and research results are discussed.

4.2 FUZZY SYSTEMS THROUGH NEURAL NETWORKS

The fuzzy system maps an input fuzzy set X into an output fuzzy set Y in accordance with fuzzy relation R describing the fuzzy rule set (4.1):

$$Y = X \circ R, \tag{4.1}$$

where $\circ$ is a compositional rule of inference [21]. The fuzzy rule generation problem may be considered as the problem of determining the fuzzy relation R from numerical data, e.g. [1], [3], [5].

Following [3], let us consider a system with a total of n fuzzy variables (input and output variables). Suppose that each fuzzy variable can assume m fuzzy-set values. Therefore, there are m^n fuzzy cells (i.e. fuzzy rules) in the input–output product space R^n. Any set of these fuzzy cells is equivalent to some unique fuzzy system. So, the total number of possible fuzzy systems is 2^{m^n}. In practice, only a small part of this number is reproduced by an expert's efforts or estimation. The problem considered in this chapter is the estimation of the number of 'effective' fuzzy rules, i.e. those rules from the overall number m^n, which cannot be ignored in the qualitative reconstruction of real plant mapping.

Output fuzzy set B of the fuzzy system is calculated by

$$B = \sum_j w_j B'_j \quad j = 1 \ldots m^n, \tag{4.2}$$

where B'_j is the individual fuzzy set of the jth rule and w_j is the jth rule's weight. In fact, many of the weights have the value $w_j = 0$, which means that the corresponding rules can be omitted without degrading the approximation ability of the fuzzy system. The weight value $w_j = 1$ corresponds to articulation of the rule number j. According to the definition [25], for some continuous fuzzy set A defined on a continuous universe of discourse X the fuzzy membership function $\mu_A(x) : X \to [0, 1]$ assigns to each element $x \in X$ a real number $\mu_A(x)$ in the interval [0, 1]. Any fuzzy membership function, used to calculate the individual fuzzy sets B'_j, is characterized by a localized sensitive zone. In the trivial case, this zone may be presented by the whole space of the fuzzy variable variation although it is impracticable. Therefore, we encounter the following common problems with fuzzy system design:

1. clustering of the product space of the system fuzzy variables variation, and
2. determination of the set of clusters (fuzzy rules) to be used finally in the fuzzy system under construction.

The approximation abilities of neural networks [22] make it possible to reformulate the fuzzy mapping problem (4.1) and to state a neural network based mapping problem instead, which is of the form

$$\hat{Y} = F(X), \qquad (4.3)$$

where $\hat{Y}$ is an estimate of the fuzzy system (4.1) output set Y and F is the input–output mapping operator of the neural network. Therefore, the initial problem of the reconstruction of the fuzzy relation R may be considered as a problem of the neural network structure St and parameter Par identification in order to minimize a square norm:

$$\|Y - \hat{Y}\|^2 \rightarrow \min_{\langle St, Par \rangle}, \qquad (4.4)$$

where V is a measurable set in $\mathcal{R}^N$. Theoretically rigorous usage of the criterion (4.4) for the model quality is impossible due to the necessity to calculate the system input and output variable variations over the whole domain V. There are several approaches to cope with the problem. Among them, for example, the Akaike Information Criterion (AIC) [26] and its modification for neural networks the Network Information Criterion (NIC) [27], the Minimum Description Length (MDL) [28], etc.

One of the common principles used in practice to determine the model structure is the cross-validation principle of the statistics. According to this principle, all sample data are divided into two data sets, say A and B. These two are then used for model training and to judge the quality of the model structure. Several so-called external criteria were proposed to estimate the model structure quality [29, 30]. Their common feature is independence of the model internal characteristics (for example, dimensionality of the model parameter vector). Some of the external criteria are

$$\|\hat{\theta}_A - \hat{\theta}_B\|, \qquad (4.5)$$

$$\|F(\hat{\theta}_A) - F(\hat{\theta}_B)\|, \qquad (4.6)$$

$$\|Y - F(\hat{\theta}_A)\|_B, \qquad (4.7)$$

$$\|Y - F(\hat{\theta}_B)\|_A, \qquad (4.8)$$

where $A + B = W$ is the sample data set, $\hat{\theta}_A$ and $\hat{\theta}_B$ are the model parameter vectors, estimated over the data subsets A and B, respectively, Y is the teaching signal, and $\|\cdot\|_A$ and $\|\cdot\|_B$ are the norms calculated over the data subsets A and B, respectively.

For any of the above described criteria the training data set is a subset of the full data set $A \subset W, B \subset W$. Therefore, only one part of the sample data is available for model training. In some cases, with an insufficient number of training data, the quality of the model approximation may be crucially degraded. To increase the efficiency of the teaching data utilization an Unbiasedness Criterion using Distorter (UCD) has been proposed [31]. This approach cancels the model structure over-fitting problem and provides good generalization performance. The UCD is a heuristic criterion for the model structure selection. The key idea of this approach is a comparison of the performance of two models of the same but variable structure. The first model is adjusted with real, say, original sample data, while the teaching data set for the second model is constructed artificially, i.e. by some non-linear transformation of the original data. Thus, the whole sample data set is available for model training, while the test-set role is played by the non-linearly transformed original data set. The same non-linear transformation is applied to the second model output. During learning both model outputs are registered and compared with each other by means of the UCD calculation. The learning procedure is repeated for different model structures. The model structure with the least UCD value is assumed to be the solution to the structure identification problem. Simulation results of the UCD have demonstrated the superiority of the UCD model's generalization ability over, for example, the AIC and NIC approaches [31].

4.2.1 Radial Basis Function Neural Networks

Radial basis function neural networks (RBFNs) belong to the class of multilayer neural networks [2, 8, 24]. They have been widely used for data modelling, control and other tasks that deal with high-dimensional, substantially non-linear mapping. Generally, radial basis functions may be expressed as

$$\varphi_i(\tilde{x}) = f(\|\tilde{c}_i - \tilde{x}\|_2), \quad (4.9)$$

where $\tilde{c}_i$ is the K-dimensional vector denoting the center of the ith basis function, $\|\cdot\|_2$ is the Euclidian norm, and $f(\cdot)$ is a univariate function of N variables. Let denote $r = \|\tilde{c} - \tilde{x}\|_2$. Then some examples of $f(\cdot)$ are the following:

- $f(r) = r$—the radial linear function,
- $f(r) = \exp(-r^2/(2\sigma^2))$—the Gaussian function,
- $f(r) = r^2 \log(r)$—the thin plate spline function,
- $f(r) = (r^2 + \sigma^2)^{0.5}$—the multi-quadratic function
- $f(r) = \log(r^2 + \sigma^2)$—the shifted logarithm function,

where σ is a constant. The Gaussian function is the only radial basis function which can be written as a product of univariate functions [25]:

$$f_i(x) = \exp\left(-\frac{\|\tilde{c}_i - \tilde{x}\|_2^2}{2\sigma_i^2}\right) = \prod_{j=1}^{K} \exp\left(-\frac{(c_{ij} - x_j)^2}{2\sigma_i^2}\right). \quad (4.10)$$

Thus, the Gaussian RBFN may be easily interpreted from a fuzzy viewpoint with the product operator implementing a fuzzy AND, and each of the univariate Gaussian functions representing the fuzzy membership function $\mu_A(x)$.

In this chapter the RBFNs are used to cluster the product space of the fuzzy system variables. The number and the form of the Gaussian functions involved determine the structure of the fuzzy associative matrix. That is why under some conditions any RBFN learning procedure may be used for the original problem of generating fuzzy rules. Therefore, the problem is the RBFN structure and parameter identification given with the sample data. The RBFN and its learning procedure are instruments for solving the stated task. The functional equivalence between a radial basis function network and a fuzzy inference system can be obtained under the following conditions [9]:

78 FUZZY SYSTEM IDENTIFICATION

- the number of hidden nodes in the RBFN is equal to the number of fuzzy If–Then rules;
- the membership function of each rule is a Gaussian function;
- the *t*-norm operator used for computing the firing strength of each rule is multiplication;
- the output of each fuzzy If–Then rule is composed of the constant under consideration using the same method (defuzzification method) to derive output.

The output x of the RBFN is described by

$$x = \sum_{i=j}^{K} w_i R_i(\tilde{u}) + w_0, \qquad (4.11)$$

where K is the number of basis functions; w_i are the weights of the network; $R_i(\tilde{u})$ are radial basis functions of the network input $\tilde{u}$; $\tilde{u} = [u_1, u_2, \ldots, u_N]^T$ is an input vector of dimensionality N; and w_0 is the bias term. For example, the Gaussian functions

$$R_i(\tilde{u}) = \exp(-\|\tilde{u} - \tilde{c}_i\|^2 / \tilde{\sigma}_i^2), \quad i = \overline{1, K}, \qquad (4.12)$$

where $\tilde{c}_i = [c_{i1}, c_{i2}, \ldots, c_{iN}]^T$ is the center of the *i*th function in the input signal space and $\tilde{\sigma}_i^2 = [\sigma_{i1}^2, \sigma_{i2}^2, \ldots, \sigma_{iN}^2]^T$ is a receptive field width vector.

If the above-mentioned conditions of the RBFN and the fuzzy inference system are satisfied, then any RBFN learning rule can be applied to adapt fuzzy systems [12]. Some of authors using RBFN for approximation problems propose network training procedures with weights $w_i, i = \overline{0, K}$, adjusting while the centers $\tilde{c}_i$ and widths $\tilde{\sigma}_i^2$ are fixed. Alternative methods are based on simultaneous corrections of all the network parameters. The latter allows us to reduce the number of units and to simplify the network structure, resulting in the same approximating abilities. However, learning procedure complexity significantly increases.

There are well-known methods of RBFN parameter determination, both supervised and unsupervised. For example, the K-means clustering algorithm is used to search the cluster centers. Appropriate widths are calculated using nearest neighbor heuristics [8]. The gradient descent method was also utilized for this problem [23]. Although RBFNs are characterized by substantially faster training than, for example, backpropagation networks, there is a problem in decreasing the RBFN learning time for the high input space dimensionality case. The general parameter approach, considered in the next section, is one possible way to improve the RBFN's training features and to simplify their estimation accuracy analysis [14–20].

4.3 GENERAL PARAMETER RADIAL BASIS FUNCTION NETWORK (GP RBFN)

There is a trade-off between network approximation ability and learning procedure convergence speed [24]. Recurrent learning procedures for artificial neural networks (ANNs) are based on adjusting model schemes with parallel, serial or mixed (parallel–serial) types of models. ANN is a model-free estimator in the sense that its structure choice is not based on the natural (physical, chemical, etc.) law analysis of the plant's behavior. The 'black-box' approach is utilized in most of ANN-based systems. Only the input–output sample data are assumed available for the network design. However, a priori analysis of the problem is a necessary step in the optimal choice of ANN type.

It is worthwhile to note that even in the case of the 'black-box' approach the network designer has to assume the existence of the 'best' (ideal) network inside the chosen class (type of ANN). The input and output variables of such an ideal network (call it the 'plant network' or 'plant') are the real sample data to be used for the 'model network' (or just 'model') training. From this point of view, the classical identification theory, with its model-based estimation, is directly applicable to the 'model-free' neural network approach.

4.3.1 General Parameter Method for System Identification

The choice of RBFN for the stated problem-solving has been argued in the preceding section. Keeping in mind the above discussion, assume that the plant is non-linear on input variables:

$$x(t) = \tilde{a}^T \tilde{f}(t) + \eta(t), \tag{4.13}$$

where $\tilde{a}$ is an unknown parameter vector of dimensionality N; $\tilde{f}(t) = [\tilde{f}_1(t), \ldots, \tilde{f}_m(t)]^T$ is a vector function of input variables; $\tilde{f}_i(t) = [f_{i1}(\tilde{u}), \ldots, f_{ip_i}(\tilde{u})]^T$ is a subvector function of input variables; $\tilde{u}(t)$ is the input vector; and $\sum_{i=1}^{m} P_i = N, 1 \leq m \leq N$. The GP model of the plant (4.13) is of the form

$$\hat{x}(t) = (\hat{a}(0) + Q\tilde{\beta}(t-1))^T \tilde{f}(t), \tag{4.14}$$

where $\hat{x}(t)$ is the model output; $\hat{a}(0) = [\hat{a}_1(0), \cdots, \hat{a}_m(0)]^T$ and $\hat{a}_i(0) = [\hat{a}_{i1}(0), \cdots, \hat{a}_{ip_i}(0)]^T$ and the initial model parameter vector and its ith subvector; $Q = \mathrm{diag}(q_1 \tilde{I}_1, \ldots, q_m \tilde{I}_m)$ is an $(N \times m)$ matrix; $\tilde{I}_i = [11 \ldots 1]^T$ is a $(p_i \times 1)$ vector of units; q_i are weight coefficients; and $\tilde{\beta}(t) = [\beta_1(t), \ldots, \beta_m(t)]^T$ is a general parameter vector adjusted with the algorithm

$$\tilde{\beta}(t) = \tilde{\beta}(t-1) + (x(t) - \hat{x}(t))\Gamma(t) Q^T \tilde{f}(t), \tag{4.15}$$

where $\Gamma(t)$ is an $(m \times m)$ diagonal matrix of gains. The number of model (4.14) parameters to be adjusted, i.e. the number $m(1 \leq m \leq N)$ of general parameter vector $\tilde{\beta}(t) = [\beta_1(t), \ldots, \beta_m(t)]^T$ components, is reduced to increase the learning convergence rate. The following theorem gives the algorithm's (4.15) convergence conditions of the common form.

Theorem 1 *For plant (4.13) and model (4.14) the convergence conditions of the algorithm (4.15) are of the form (4.16), where $\Gamma * (t)$ is a diagonal matrix with elements equal to the corresponding diagonal elements of the matrix $Q^T E\{\tilde{f}(t)\tilde{f}^T(t)\}Q$, where $E\{\cdot\}$ is the expectation operator:*

$$0 < \Gamma(t) < \Gamma^{-1} * (t). \tag{4.16}$$

It was shown (e.g. in [18]) that, for fixed values of the signal-to-noise ratio, the quadratic form $S(t) = E\{\Delta\tilde{\beta}^T(t) Q^T Q \Delta\tilde{\beta}(t)\}$ of the general parameter error vector $\Delta\tilde{\beta}(t) = \tilde{\beta}(t) - C$, where C is the general parameter vector expectation at the steady state (asymptotic value), may be considered to be the estimated accuracy measure. Moreover, the convergence rate (the rate of decrease in the value $S(t)/S(0)$) depends on the general parameter vector dimensionality m, not on the model dimensionality N.

80 FUZZY SYSTEM IDENTIFICATION

The general parameter approach is one possible way to improve ANN's training features. The main characteristics of the GP approach are the possibility of evaluating the estimation accuracy during the training stage and an improved convergence rate due to the artificial reduction in the dimensionality of the adjusting model parameter space, especially useful for initial steps when the approximation error is large.

4.3.2 GP RBFN Training Algorithm

First, let us consider the GP approach to adjusting the RBFN weights. As soon as the RBFN is linear in weights, the GP method may be implemented in a straightforward manner. The equation describing the GP RBFN for a single output network is

$$x = \sum_{j=1}^{K}(w_j^0 + \beta)\exp\left[-\sum_{i=1}^{N}(u_i - c_j)^2/\delta_j^2\right] + (w_0^0 + \beta)], \quad (4.17)$$

where w_j^0 are the fixed initial values of network weights and β is a scalar general parameter to be adjusted with the following algorithm:

$$\beta(t) = \beta(t-1) + \gamma\beta(x^* - x) \times \left\{\sum_{j=1}^{K}\exp\left[-\sum_{i=1}^{N}(u_i - c_j)^2/\delta_j^2\right] + 1\right\} \quad (4.18)$$

where $\gamma\beta$ is the adjusting gain and x^* is the desired network output. The simplest GP RBFN structure is shown in Figure 4.1.

Simulations were performed to investigate the convergence characteristics of the algorithm (4.18). Two schemes were modeled. In both cases the parallel model identification approach was realized with the 'sample' RBFN as the plant. The structure of the 'plant' was assumed fixed and known while the plant's parameters (the 'sample' RBFN's weights) were to be estimated by the 'model', i.e. the RBFN of the same structure as the plant.

The parameters of the model (the 'model' RBFN's weights) were adjusted independently by two algorithms starting from the same initial conditions. In the first case, a conventional RBFN was used with individual adjusting of its weights. In the second case, GP RBFN was simulated with the learning algorithm (4.18). The efficiency of both methods was compared

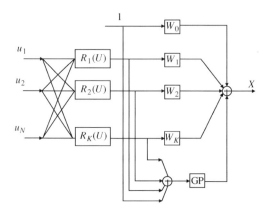

Figure 4.1 Simplest general parameter radial basis function network (GP RBFN) structure

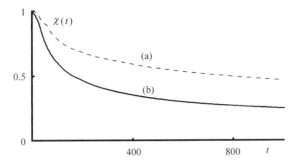

Figure 4.2 Convergence of the RBFN learning algorithm: (a) conventional RBFN, (b) GP RBFN

by using the following measure of the convergence speed:

$$\chi(t) = \frac{E\{\|\tilde{r}(t)\|^2\}}{\|\tilde{r}(0)\|^2}, \qquad (4.19)$$

where $\tilde{r}(t) = \tilde{w} - \tilde{w}(t)$ is the parameter error vector at moment t; $\tilde{w}$ is the vector of 'true' parameters (the vector of the 'plant' RBFN's weights); $\tilde{w}(t)$ is an estimate of the vector $\tilde{w}$ at moment t (the vector of the 'model' RBFN's weights); E is the expectation operator; and $\|\cdot\|$ is the Euclidean norm. Measure (4.19) is an average estimate of the learning algorithm's convergence speed in the sense of a relational reduction of the square norm of the parameter error vector. Simulation results under the same initial conditions for both methods are presented in Figure 4.2. Owing to the artificial reduction in the dimensionality of the adjusting parameter space, the learning speed of the GP RBFN has increased relatively to the conventional RBFN.

4.4 GP RBFN ADAPTIVE FUZZY SYSTEMS (AFSs)

4.4.1 Basic Algorithm

The RBFN to be used in adaptive fuzzy systems (AFSs), in common case, is assumed to be trained by means of the determining the minimum necessary number of rules (hidden unit number) and by adjusting the mean and variance vectors of individual hidden nodes as well as their weights. In this section, the simplest GP RBFN based adaptive fuzzy system for automatic fuzzy rule number determination is proposed (Figure 4.3). Only the network weights have been assumed to be adjusted by the GP algorithm, while the centers and widths of unit sensitive zones were completely determined using the network input signal range and unit number during each training epoch.

A 'sample' fuzzy system has been presented by RBFN with an 'unknown' number of hidden units (i.e. fuzzy rules). Starting from the single-unit GP RBFN, network learning has been performed by the adjustment scalar general parameter in the Learning Procedure block. GP Statistics Estimator has calculated the steady-state general parameter expectation $E\{\beta\}$ and variance $D\{\beta\}$. The approximation quality criterion (4.20) was evaluated for the current GP RBFN structure, and the decision to change the network structure (additional unit

FUZZY SYSTEM IDENTIFICATION

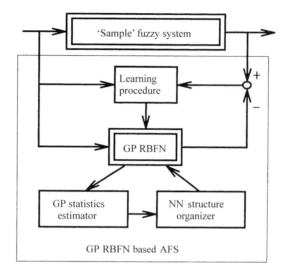

Figure 4.3 The GP RBFN adaptive fuzzy system block diagram

recruiting) was made by NN Structure Organizer:

$$Q = \frac{D\{\beta\}}{E\{\beta\}}. \tag{4.20}$$

Therefore, the GP RBFN AFS determines the 'true' fuzzy rule number by incrementally recruiting the radial basis function units and continuous estimation of the approximation quality through evaluation of the criteria (4.20) for each fixed GP RBFN structure. The network to be determined is the network with smallest Q value, and its unit number is assumed to be equal to the fuzzy rule number of the 'sample' fuzzy system.

Let consider the proposed procedure in detail for the simplest case of the GP RBFN AFS with a scalar input signal.

1. For a given maximum value u_{max} of the centered input signal u ($E\{u\} = 0$) and known number of Gaussian units q (for the first stage, $q = 1$) the sensitive zone center coordinates are calculated by

$$C_i^q = -u_{max} + \frac{u_{max}}{q}(2i - 1), \quad i = \overline{1, q}, \tag{4.21}$$

where i is the current unit number. For $q = 1$ and $i = 1$, for example, one can receive $C_1^1 = 0$.

2. The initial (basic) sensitive zone width equal for all network units is calculated as

$$\delta_0^q = \frac{u_{max}}{q\sqrt{-\ln\rho}}, \quad 0 < \rho < 1, \tag{4.22}$$

where ρ is the Gaussian function value at the boundary point of the sensitive zone. Variations in ρ provide different degrees of conjugent sensitive zone (membership function) overlapping. This results in system fuzziness variations. The meaning of the above-mentioned parameters is explained by Figure 4.4 for RBFN with three units ($q = 3$).

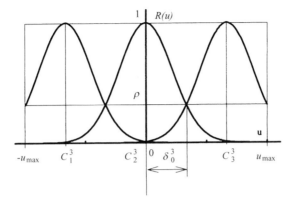

Figure 4.4 Definition of the basic parameters of the adaptive fuzzy system

3. For the network structure determined during the above two steps the scalar general parameter learning procedure is performed based on the input–output sample data according to the algorithm (4.18). Simultaneously the general parameter expectation $E\{\beta\}$ and variance $D\{\beta\}$ are estimated with some conventional method, e.g. the moving average calculation.
4. After stabilization of the GP statistics the criteria (4.16) are evaluated and memorized.
5. If $q = 1$, then GOTO 8.
6. The modified system criteria value Q^{q+1} is compared with its preceding value Q^q: if $Q^{q+1} < Q^q$, then GOTO 8 else.@ $q = q + 1$.
7. Stop.
8. The structure of GP RBFN is modified by recruiting one more Gaussian unit $q = q + 1$. Steps 1–6 are repeated.

The result of implementing the algorithm 1–8 is a number of Gaussian function units in GP RBFN which provide the best approximating accuracy. In terms of fuzzy system theory this means a solution to the fuzzy rule number determination problem.

4.4.2 Unbiasedness Criterion for the GP RBFN AFS

The problem of the reliability of the derived model is one of the most important problems that arises during the identification task solving. Preventing the model from over-fitting is a crucial point for many practical implementations. As was discussed in the preceding sections, there are several approaches to cope with this difficulty. In this section, the Unbiasedness Criterion using Distorter (UCD) approach [31] is used, which has been shown to provide improved features in comparison with conventional methods, such as the Akaike Information Criterion (AIC) [26] and its modification for neural networks Network Information Criterion (NIC) [27], and Minimum Description Length (MDL) [28].

Let us consider the UCD method as applied to the GP RBFN AFS. The system (Figure 4.5) consists of two identification subsystems of the same structure. Both of them include GP RBFN with the learning procedure described in the Section 4.4.1. The same signals are fed into the network inputs. The difference is in the way of the teaching signal is used. While the teaching signal is fed into the lower loop without any changes, the upper network is trained by 'distorted', i.e. non-linearly transformed, sample data. The output of the upper

84 FUZZY SYSTEM IDENTIFICATION

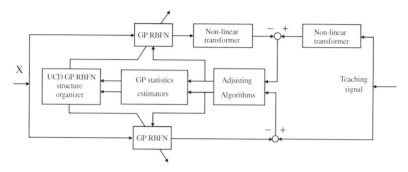

Figure 4.5 UCD GP RBFN based adaptive fuzzy system

network is also changed by the transformer of the same transfer function as for the teaching signal.

The original UCD criterion of the network structure optimality is derived [31], which is of the form

$$\text{UCD} = \sum_{j=1}^{n} [F_1(\tilde{x}^j) - F_2(\tilde{x}^j)]^2, \quad (4.23)$$

where $\tilde{x}^j$ is the jth set (vector) of the network input data, n is the overall number of network input data sets, and $F_1(\cdot)$ and $F_2(\cdot)$ are the output variables of both networks. The network structure with the smallest value of the criterion (4.23) is assumed to be the solution to the problem.

The same approach has been implemented into the GP RBFN based AFS. The key idea was the fact that the GP variance, estimated after the identification system has achieved steady state (stabilization of the GP expectation and variance values), provides information about the approximation accuracy of the model. It was expected also that the difference between the GP variances of the two RBFN (with distorter and without it) is minimal for the GP RBFN with a 'true' number of units. The UCD for the GP RBFN (UCD GP) has been derived in the form

$$\text{UCD GP} = \sum_{j=1}^{n} [D_{\beta 1}(\tilde{x}^j) - D_{\beta 2}(\tilde{x}^j)]^2, \quad (4.24)$$

where $D_{\beta 1}(\cdot)$ and $D_{\beta 2}(\cdot)$ are the variances of the general parameters of both RBFN, estimated after stabilization of both GP expectations.

4.5 SIMULATION RESULTS

The proposed fuzzy rule number algorithm determined by using the GP RBFN AFS has been studied through computer simulation. RBFN with an 'unknown' number of units was used as a 'sample' or 'plant' network for teaching data generation. Both the number of 'sample' network units and the number of inputs were chosen equal to five. All weights of the 'sample' network were of the same value 1 to indicate that the corresponding basis functions are involved in the output value calculation. In terms of the fuzzy approach this means that the membership function related to non-zero weight is actually playing its role in the input–output mapping.

Three different systems were studied. In the first case, the AFS with a single GP RBFN was analyzed (Figure 4.3). The number of the GP RBFN units q_m was changed during the learning procedure corresponding to the Q-criterion (20) variation trend. In the second case, the system with two 'model' GP RBFNs was simulated (Figure 4.5), while model quality was estimated by the UCD-criterion (4.23). Finally, the system (Figure 4.5) was investigated using the UCD-GP criterion (4.24).

In all cases the initial values $w_j^0, j = \overline{1, q_m}$, of the 'model' networks' weights were assigned equal to 0 (no a priori information available about 'valid' membership functions). However, the GP RBFNs' input number was fixed and equal to that of the 'sample' network. The centers of the Gaussian unit's sensitive fields were assumed to be uniformly distributed in the input signal space with the latter known for the AFS. The sensitive widths of all units were chosen to be equal. The cases of different degrees of sensitive zone overlapping were studied. During the simulations the value of ρ was constant: $\rho = 0.01 (\delta_0^q = 0.093)$, i.e. the basic level of overlapping was 1%. The real sensitive zone width values δ^q were calculated as follows:

$$\delta^q = \alpha \delta_0^q, \tag{4.25}$$

where α is the the overlapping coefficient. In Figure 4.6 some examples of the different degree of overlapping Gaussian functions are given (network with three units).

Two different types of input signals were used for network training. In the first simulation series it was a sine function. In the second series, the pseudo-random sequence with uniform distribution in the region of $[-1, 1]$ was used.

Some simulation results of the proposed AFS with a single GP RBFN (the first case) are shown in Figure 4.7–4.10. An example of three successive initial steps in the construction of the aggregative membership function $R(u)$ is shown in Figure 4.7. The simplest case of scalar input signal u with known boundaries $u_{\min}$ and $u_{\max}$ was considered. The centers of Gaussian functions were assumed to be uniformly distributed in the input signal space. The Gaussian

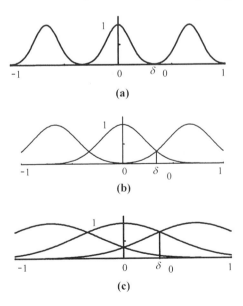

Figure 4.6. Different degree of overlapping Gaussian functions (number of units is three): (a) $\alpha = 1$; (b) $\alpha = 2$; (c) $\alpha = 4$

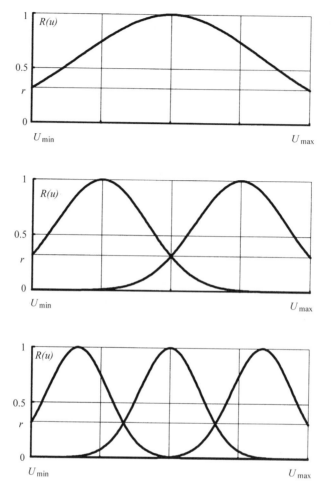

Figure 4.7 Aggregative membership function (AMF) construction procedure (three successive initial steps)

function value at the cross point of any two conjugate sensitive zones was constant and equal to r.

In Figure 4.8 the GP RBFN learning curves are shown for sinusoidal inputs and different numbers of hidden units. The steady-state (asymptotic value) error is 0 only for unit number 5 (the same as for the 'sample' network) (Figure 4.8(c)). Analogous results have been gained for random input signals (Figure 4.9) with the best approximation accuracy for the five-unit GP RBFN (Figure 4.9(c)).

The optimal unit number of judgment abilities (the optimal number of fuzzy rules) of the GP RBFN based AFS with a single model are demonstrated in Figure 4.10. The proposed system has been analyzed under different learning conditions. Two factors were determined to have a significant influence GP RBFN AFS performance, namely the overlapping coefficient α and the GP adjusting algorithm gain γ_β.

The upper Q criteria curves in Figure 4.10 correspond to the case when $\gamma_\beta = 0.6$. The sensitivity of the proposed system increases with an increase in the adjusting gain γ_β

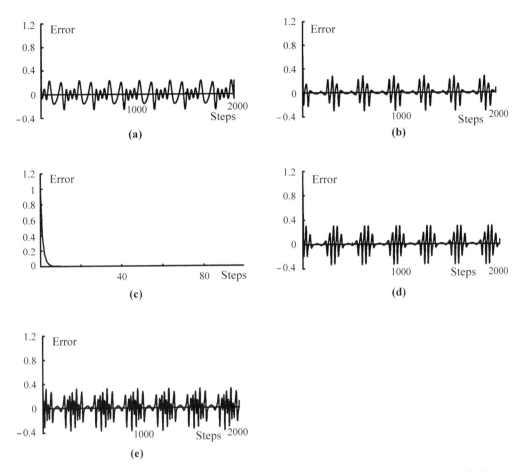

Figure 4.8 GP RBFN learning with sinusoidal input ($\alpha = 10$): (a) $q_m=1$; (b) $q_m=4$; (c) $q_m=5$; (d) $q_m=6$; (e) $q_m=9$

($\gamma_\beta = 0.1-0.6$), but the latter is bounded by the convergence conditions of the GP algorithm (Theorem 1). Therefore, this theorem gives an upper boundary estimate of the adjusting gain which provides the applicability of the GP approach to the problem stated.

The influence of the overlapping coefficient α is seen from experimental curves plotted for $\delta_0 = 0.093 (\rho = 0.01)$ and $\alpha = 1-10$ (Figure 4.10(a)–(f), respectively). For higher values of α the dependence of the system performance on γ_β value was reduced and the 'true' rule number determination was clearer.

Analogous simulations have been performed for the AFS with a two-channel structure shown in Figure 4.5. First, the conventional UCD criterion (4.23) was used to judge the RBFN unit number. Then the UCD GP criterion (4.24) was investigated.

Some typical numerical results are given in Tables 4.1–4.3. Absolute values of all of the criteria were calculated for the same conditions: overlapping coefficient $\alpha = 10$, adjusting gain $\gamma_\beta = (0.01 - 0.06)$. The cells with minimal value of the corresponding criterion for fixed γ_β (table strokes) are shown in grey. The Q criterion based structure identification of the fuzzy system, which was presented by the RBFN with five units, provided more

88 FUZZY SYSTEM IDENTIFICATION

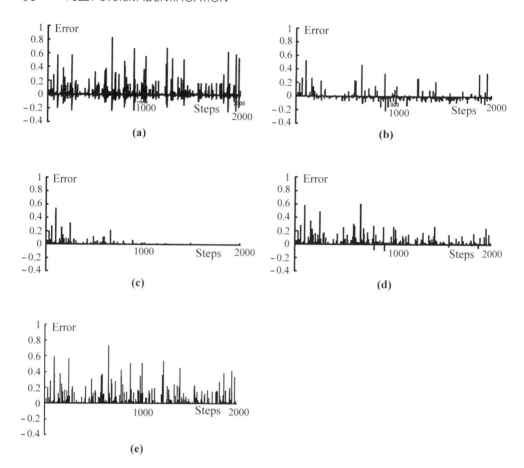

Figure 4.9 GP RBFN learning with sinusoidal input ($\alpha = 10$): (a) $q_m=1$; (b) $q_m=4$; (c) $q_m=5$; (d) $q_m=6$; (e) $q_m=9$

reliable results than the conventional UCD approach, giving biased estimates. Moreover, the conventional UCD approach demonstrated significantly high sensitivity to the learning algorithm parameter variations—the degree of Gaussian function overlapping and the adjusting algorithm gain. The best results were observed for UCD GP criterion. It was

Table 4.1 UCD criterion for overlapping degree $\alpha = 10$

		RBFN unit number									
		1	2	3	4	5	6	7	8	9	10
γ_β	0.01	8505	8336	8284	8282	8296	8315	8336	8356	8374	8390
	0.02	8613	8361	8313	8298	8296	8299	8305	8311	8317	8323
	0.03	8546	8346	8312	8299	8296	8297	8298	8301	8304	8306
	0.04	8482	8332	8308	8299	8296	8295	8296	8297	8299	8301
	0.05	8436	8322	8305	8298	8296	8295	8296	8298	8299	8300
	0.06	8404	8316	8303	8298	8296	8295	8296	8296	8297	8298

SIMULATION RESULTS

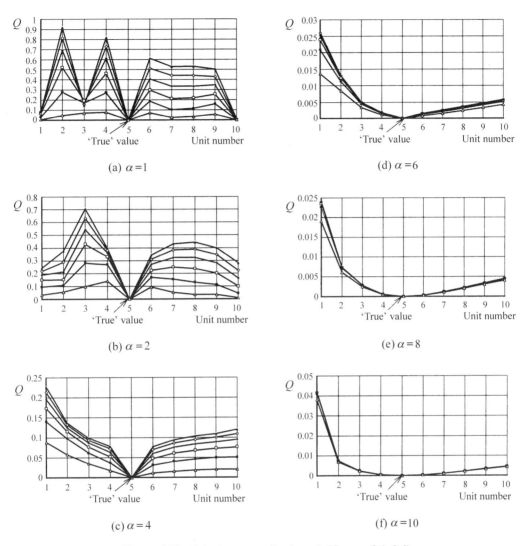

Figure 4.10 Criterion curves for ($\alpha = 1\text{-}10, \gamma_\beta = 0.1\text{-}0.6$)

Table 4.2 Q Criterion ($\times 10^{-3}$) for overlapping degree $\alpha = 10$

		\multicolumn{10}{c}{RBFN unit number}									
		1	2	3	4	5	6	7	8	9	10
	0.01	4.815	0.902	0.428	0.137	0	0.128	0.460	0.922	1.458	2.031
	0.02	13.77	2.557	1.028	0.27	0	0.218	0.755	1.471	2.227	3.116
γ_β	0.03	21.31	3.959	1.466	0.352	0	0.264	0.898	1.727	2.643	3.586
	0.04	26.64	4.943	1.759	0.404	0	0.293	0.986	1.881	2.863	3.864
	0.05	30.34	5.608	1.954	0.439	0	0.311	1.044	1.984	3.010	4.052
	0.06	32.95	6.062	2.087	0.462	0	0.325	1.084	2.057	3.114	4.184

Table 4.3 UCD GP criterion ($\times 10^{-3}$) for overlapping degree $\alpha = 10$

		\multicolumn{10}{c}{RBFN unit number}
		1	2	3	4	5	6	7	8	9	10
γ_β	0.01	1284	24.40	2.094	0.072	0	0.024	0.231	0.751	1.593	2.704
	0.02	1243	24.36	1.845	0.054	0	0.014	0.128	0.384	0.757	1.201
	0.03	797.9	15.16	1.149	0.034	0	0.009	0.082	0.247	0.488	0.774
	0.04	483.8	8.787	0.681	0.02	0	0.006	0.052	0.161	0.323	0.519
	0.05	296.7	5.179	0.412	0.013	0	0.004	0.034	0.106	0.216	0.352
	0.06	187.3	3.175	0.259	0.008	0	0.002	0.023	0.072	0.147	0.242

characterized by unbiased estimation and higher resolution ability in comparison with the Q criterion. Therefore, for the stated problem of fuzzy rule number determination through RBFN structure identification the GP RBFN AFS with a two-channel identification procedure is superior owing to the fast learning ability and higher judgment reliability.

4.6 CONCLUSION

In this chapter a fuzzy system identification problem using the artificial neural network (ANN) approach is considered. The unknown fuzzy relation reconstruction problem is treated as identification of the structure and parameters of the neural network. The functional equivalence between some classes of fuzzy systems and radial basis function networks (RBFNs), namely their localized sensitivity to the input value, is the background to the proposed approach.

The improved structure and advanced learning feature RBFN is developed based on the general parameter (GP) method of complex system identification. The developed system is based on a GP training procedure with the following features:

- an artificial reduction in the dimensionality of the adjusting model parameter space,
- high (initial) convergence rate which does not depend on the dimensionality of the original (sample) model parameter vector,
- possible identification accuracy estimation by calculating the GP statistics in the steady state of the learning procedure,
- the GP model is a flexible type of model with an adjustable structure,
- a quasi-optimal model structure may be derived with a decision-making multi-step procedure based on learning period information.

The criterion of the GP RBFN (General Parameter Radial Basis Function Network) structure optimality is derived using the GP steady-state statistics. The derived criterion is then used to develop the GP RBFN structure self-organization procedure. The general parameter radial basis function neural network based adaptive fuzzy system (GP RBFN AFS) with the ability to learn fuzzy rules from input–output sample data is proposed. The reliability and resolution ability of the proposed method have been increased by implementation of the two-network identification system. The GP RBFN based Unbiasedness Criterion using Distorter (UCD GP) has been derived. The background of the proposed procedure is the Group Method of Data Handling (GMDH) used to provide the desired approximation accuracy and to cancel the over-fitting effect.

The method developed may be easily be implemented into various types of artificial neural network structures to solve a wide class of identification, signal processing, control and related problems.

REFERENCES

1. Horikawa, S., T. Furuhashi, and Y. Uchikawa, On fuzzy modeling using fuzzy neural networks with the backpropagation algorithm, *IEEE Transactions on Neural Networks,* **3**, 1992, 801–806.
2. Kosko, B., Fuzzy associative memories, In: A. Kandel (ed.) *Fuzzy Expert Systems.*
3. Kosko, B., *Neural Networks and Fuzzy Systems*, Prentice-Hall, Englewood Cliffs, 1992.
4. Takagi, T. and M. Sugeno, Fuzzy identification of systems and its application to modeling and control, *IEEE Transactions on Systems, Man and Cybernetics,* **15**, 1985, 116–132.
5. Wang, L.-X. and J. M. Mendel, Generating fuzzy rules by learning from examples, *IEEE Transactions on Systems, Man and Cybernetics,* **22**, 1992, 1414–1427.
6. Yamaguchi, T., N. Imasaki, and K. Haruki, A reasoning and learning method for fuzzy rules with associative memory, *IEE Japan Transactions,* **110-C**, 1990, 207–215.
7. Mitra, S. and S. K. Pal, Fuzzy multi-layer perceptron, inferencing and rule generating, *IEEE Transactions on Neural Networks,* **6**, 1995, 51–63.
8. Moody, J. and C. Darken, Fast learning in networks of locally tuned processing units, *Neural Computing,* **1**, 1989, 281–294.
9. Roger Jang, J.-S. and C. T. Sun, Functional equivalence between radial basis function networks and fuzzy inference systems, *IEEE Transactions on Neural Networks,* **4**, 1993, 156–159.
10. Roger Jang, J.-S. and C. T. Sun, Predicting chaotic time series with fuzzy if-then rules, In: *Proceedings of the IEEE International Conference on Fuzzy Systems*, San Francisco, 1993, pp. 1079–1084.
11. Wang, L.-X. and J. M. Mendel, Fuzzy basis functions, universal approximation and orthogonal least squares learning, *IEEE Transactions on Neural Networks,* **3**, 1992, 807–814.
12. Kwang, B. C. and B. H. Wang, Radial basis function based adaptive fuzzy systems and their applications to system identification and prediction, *Fuzzy Sets and Systems,* **83**, 1996, 325–339.
13. Lee, S. and R. M. Kil, A Gaussian potential function network with hierarchically self-organizing learning, *Neural Networks,* **4**, 1991, 207–224.
14. Akhmetov, D. F., Elaboration of Strip Width Forming Adaptive Control System in Rough Stand Group of Hot Rolling Mill, Candidate for a Technical Science Degree Dissertation, Kazakh Polytechnical Institute, Almaty, Kazakhstan., 1989.
15. Akhmetov, D. F. and Y. Dote, Novel 'on-line' identification procedure using artificial neural network, In: *Proceedings of the IFAC Workshop on On-Line Fault Detection and Supervision in the Chemical Process Industries*, Newcastle, 1995, pp. 90–97.
16. Akhmetov, D. F. and Y. Dote, General parameter neural networks with fuzzy self-organization, In: *Proceedings of the Artificial Neural Networks in Engineering (ANNIE'96)*, St Louis, Missouri, 1996, pp. 191–196.
17. Akhmetov, D. F. and Y. Dote, Fault diagnosis with general parameter radial basis function neural network, In: *Proceedings of the IEEE/ASME International Conference on Advanced Intelligent Mechatronics (AIM'97)*, Tokyo, Japan, 1997, p. 83.
18. Akhmetov, D. F., Y. Dote, and M. S. Shaikh, System identification by the general parameter neural networks with fuzzy self-organization, *Prepared for the Eleventh IFAC Symposium on System Identification (SYSID'97)*, Kitakyushu, Japan, vol. 2, 1997, pp. 829–834.

19. Akhmetov, D. F. and Y. Dote, General parameter time series analysis method for fault diagnosis, In: *Proceedings of the Second Asian Control Conference (ASCC'97)*, Seoul, Korea, vol. 3, 1997, pp. 87–90.
20. Dote, Y., D. F. Akhmetov, and M. S. Shaikh, Control and diagnosis for power electronics systems using soft computing, In: *Proceedings of the Second Asian Control Conference (ASCC'97)*, Seoul, Korea, vol. 3, 1997, pp. 491–494.
21. Zadeh, L.A., Fuzzy sets as a basis for a theory of possibility, *Fuzzy Sets and Systems*, **1**, 1978, 3–28.
22. Hornik, K., M. Stinchcombe, and H. White, Multilayer feedforward networks are universal approximators, *Neural Networks*, **2**, 1989, 359–366.
23. Sorsa, T. and H. N. Koivo, Application of artificial neural networks in process fault diagnosis, *Automatica*, **29**(4), 1993, 843–849.
24. Poggio, T. and F. Girosi, *Extensions of a Theory of Networks for Approximation and Learning: Dimensionality Reduction and Clustering*, Artificial Intelligence Lab. Memo. 1167, MIT, Cambridge, 1990.
25. Brown, M. and C. Harris, *Neurofuzzy Adaptive Modeling and Control*, Prentice-Hall International (UK) Limited, Hemel Hempstead, 1994.
26. Akaike, H., A new look at the statistical model identification, *IEEE Transactions on Automated Control*, **AC-19**, 1974, 716–723.
27. Murata, N., S. Yoshizawa, and S. Amari, Network information criterion-determining the number of hidden units for an artificial neural network model, *IEEE Transactions on Neural Networks*, **5**(6), 1994, 865–872.
28. Rissanen, J., A universal prior for integers and estimation by minimum description length, *The Annals of Statistics*, **11**(2), 1983, 416–431.
29. Ivakhnenko, A. G., Problems of simulation of complex systems and of applied mathematical statistics, *Soviet Automated Control*, **4**, 1971, 1–6.
30. Ivakhnenko, A. G., *Inductive Method of Complex System Self-Organization*, Naukova Dumka, Kiev, 1982 (in Russian).
31. Tabuchi, J., *et al.*, A heuristic model selection criterion using distorter and its application to determination of the number of hidden units in RBFN, *The Journal of the Japanese Society for System Control and Information*, **11**(2), 1998, 61–70 (in Japanese).

ANALYSIS

5
Lyapunov Stability Analysis of Fuzzy Dynamic Systems

J. R. Layne
AFRL/SNAT, Wright Patterson AFB, USA

K. M. Passino
The Ohio State University, Columbus, USA

5.1 INTRODUCTION

In system design and analysis, the establishment of a system model is quite fundamental. In the cases where uncertainty exists, models are generally assumed to be in the form of stochastic differential or discrete-time difference equations. However, it is sometimes difficult to propagate the uncertainty through the system dynamics according to the axioms of probability theory, especially for non-linear systems. Owing to this difficulty the analysis of stability properties of uncertain non-linear systems is, in general, a difficult problem. In this chapter we show how to model uncertain systems with fuzzy dynamic system models and develop a Lyapunov stability theory for the analysis of the stability properties of these models.

The definitions of stability presented in this chapter are directly analogous to conventional notions of Lyapunov and asymptotic stability as presented in, for instance, [1]–[3]. However, conventional methods generally deal with linear vector spaces and hence systems with states that are n-dimensional vectors (instead of multi-variable membership functions as we must for fuzzy dynamic systems) and only a single equilibrium point rather than an invariant set. The concepts presented here are simply a generalization of these conventional concepts. Also, they are analogous to the stochastic stability concepts presented in [4], except there they are concerned with using the properties of a Lyapunov function stated in terms of an expected value to guarantee that system trajectories stay in a neighborhood of the origin with a probability that is dependent on how close the trajectory was to the origin when it started.

The characterizations of stability presented in this chapter are most similar to those of De Glas [5] where he discusses a Lyapunov function method for continuous-time fuzzy dynamic systems. He suggests finding a Lyapunov function and taking the derivative to

Fuzzy Logic. Edited by S. Farinwata, D. Filev and R. Langari
© 2000 John Wiley & Sons, Ltd

determine if it is negative definite. Furthermore, he defines some 'soft' stability concepts based on 'α-cuts' remaining in some invariant set. Tong [6] and Jianqin and Laijiu [7] analyze the stability of equilibrium points of a fuzzy dynamic system (e.g. the equilibrium of $X(k+1) = X(k) \circ R$ is X_e such that $X_e = X_e \circ R$ where '$\circ$' is the standard fuzzy composition that is carefully defined later in this chapter). They study the dynamics of the 'degree of stability', defined as $\alpha(X(k), X_e) = 1 - X(k) \circ X_e$, where the smaller $\alpha(X(k), X_e)$ becomes, the nearer $X(k)$ gets to X_e. Kiska et al. [8] give stability results that are based on finding the energy content of a fuzzy relation R that describes the fuzzy dynamic system. Since $X(k) = X(0) \circ R^k$ they study higher orders of R and determine if the energy is increasing (meaning unstable) or decreasing (meaning stable). Furthermore, limit cycles can be found if the energy is cycling. This method of stability analysis was applied by Chand and Hansen [9] to the analysis of a roll-controller for a flexible wing aircraft.

Relatively little work has been done on the stability of fuzzy dynamic systems. The current results are theoretical, and almost no attention has been given to their application. Moreover, the full range of stability concepts (e.g. exponential stability, uniform boundedness, and uniform ultimate boundedness) have not been studied. In this chapter we hope to fill some of the gaps by studying a fuller range of stability concepts *and* their application to a simple problem; more complex applications are treated in [10].

5.2 MATHEMATICAL PRELIMINARIES

To establish notation, first we review some basic fuzzy system concepts. We begin by defining the triangular norm and co-norm [11].

Triangular norm The triangular norm, denoted $\otimes : [0, 1] \times, \ldots, \times [0, 1] \to [0, 1]$, is an operation defined for n membership function values. Commonly used triangular norms in fuzzy system design include the following:

$$\otimes \{x_1, x_2, \ldots, x_n\} = \min\{x_1, x_2, \ldots, x_n\}, \tag{5.1}$$

$$\otimes \{x_1, x_2, \ldots, x_n\} = x_1 \cdot x_2 \cdot \ldots \cdot x_n, \tag{5.2}$$

$$\otimes \{x_1, x_2, \ldots, x_n\} = \max\{0, x_1 + x_2 + \cdots + x_n - n + 1\}. \tag{5.3}$$

Triangular co-norm The triangular co-norm, denoted $\oplus : [0, 1] \times, \ldots, \times [0, 1] \to [0, 1]$, is an operation defined for n membership function values. Commonly used triangular co-norms in fuzzy system design include the following:

$$\oplus \{x_1, x_2, \ldots, x_n\} = \max\{x_1, x_2, \ldots, x_n\}, \tag{5.4}$$

$$\oplus \{x_1, x_2, \ldots, x_n\} = x_1 + x_2 + \cdots + x_n - x_1 x_2 - x_1 x_3 - \cdots - x_{n-1} x_n$$
$$+ x_1 x_2 x_3 + \cdots + x_{n-2} x_{n-1} x_n - \cdots \pm x_1 x_2 \cdots x_n, \tag{5.5}$$

$$\oplus \{x_1, x_2, \ldots, x_n\} = \min\{1, x_1, x_2, \ldots, x_n\}. \tag{5.6}$$

Using the definitions of the triangular norm, the operations of union and intersection of ordinary sets can be extended to fuzzy sets. Consider the fuzzy sets U^1 and U^2 defined on the universe of discourse $\mathcal{U}$ with membership functions $\mu_{U^1}(u)$ and $\mu_{U^2}(u)$, respectively. The most widely accepted definitions of fuzzy set operations are as follows.

Union The union of U^1 and U^2, denoted $U^1 \cup U^2$, is a fuzzy set with a membership function defined by

$$\mu_{U^1 \cup U^2}(u) = \oplus\{\mu_{U^1}(u), \mu_{U^2}(u) : u \in \mathcal{U}\}. \tag{5.7}$$

Intersection The intersection of U^1 and U^2, denoted $U^1 \cap U^2$, is a fuzzy set with a membership function defined by

$$\mu_{U^1 \cap U^2}(u) = \otimes \{\mu_{U^1}(u), \mu_{U^2}(u) : u \in \mathcal{U}\}. \tag{5.8}$$

The symbols $\otimes$ and $\oplus$ are used to denote any triangular norm or co-norm, respectively. Now, consider fuzzy set operation over more than one universe of discourse such as the Cartesian product.

Cartesian product If $U_1, U_2, \ldots, U_r$ are fuzzy sets defined as subsets of the universes of discourse $\mathcal{U}_1, \mathcal{U}_2, \ldots, \mathcal{U}_r$, respectively, then the Cartesian product is a fuzzy set, denoted $U_1 \times U_2 \times \ldots \times U_r$, with a membership function defined by

$$\mu_{U_1 \times U_2 \times \ldots \times U_r}(u_1, u_2, \ldots, u_r) = \otimes \{\mu_{U_1}(u_1), \mu_{U_2}(u_2), \ldots, \mu_{U_r}(u_r)\}, \tag{5.9}$$

where $u_1, u_2, \ldots, u_r$ are elements of $\mathcal{U}_1, \mathcal{U}_2, \ldots, \mathcal{U}_r$ respectively.

The Cartesian product of two or more fuzzy sets is called a *fuzzy relation*. Next, we can define the *fuzzy composition* and the *fuzzy extension* principle. These operations will be used later in this chapter to determine functions of fuzzy sets.

Fuzzy composition Assume that U and R are fuzzy relations defined on $\mathcal{U}_1 \times \mathcal{U}_2 \times \ldots \times \mathcal{U}_r$ and $\mathcal{U}_1 \times \mathcal{U}_2 \times \ldots \times \mathcal{U}_r \times \mathcal{Y}$, respectively. The 'fuzzy composition' of U and R is a fuzzy relation denoted by $U \circ R$ and has a membership function defined as

$$\begin{aligned} \mu_{U \circ R}(y) = \oplus \{\otimes \{\mu_U(u_1, u_2, \ldots, u_r), \mu_R(u_1, u_2, \ldots, u_r, y) : \\ (u_1, u_2, \ldots, u_r) \in \mathcal{U}_1 \times \mathcal{U}_2 \times \ldots \times \mathcal{U}_r \text{ and } y \in \mathcal{Y}\} : \\ (u_1, u_2, \ldots, u_r) \in \mathcal{U}_1 \times \mathcal{U}_2 \times \ldots \times \mathcal{U}_r\}. \end{aligned} \tag{5.10}$$

Fuzzy extension principle Let U be a fuzzy set defined on the universe $\mathcal{U} = \mathcal{U}_1 \times \mathcal{U}_2 \times \ldots \times \mathcal{U}_r$ with membership function $\mu_U(u_1, u_2, \ldots, u_r)$. Let T be a mapping from $\mathcal{U}_1 \times \mathcal{U}_2 \times \ldots \times \mathcal{U}_r$ to $\mathcal{Y}_1 \times \mathcal{Y}_2 \times \ldots \times \mathcal{Y}_s$ such that $y = T(u_1, \ldots, u_r)$. The extension principle states that the fuzzy set $Y = T(U)$ has a membership function given by

$$\mu_Y(y) = \begin{cases} \oplus \{\mu_U(u_1, \ldots, u_r) : u_1, u_2, \ldots, u_r, y = T(u_1, \ldots, u_r)\} \\ \mu_Y(y) = 0, \quad \text{if } T^{-1}(y) = \emptyset, \end{cases} \tag{5.11}$$

where $T^{-1}(y)$ is the inverse image of y and $\emptyset$ denotes the null set.

The fact that a triangular norm is used in the definition of the fuzzy composition and extension principle is important. The triangular inequality ensures that we are no more certain about the 'outputs' (y) than we are about the inputs (u). Hence the basic mechanism behind the mapping of fuzzy sets makes some intuitive sense. Now that we have preliminary knowledge of fuzzy systems, we now consider the construction of fuzzy dynamic systems.

5.3 CONSTRUCTION OF FUZZY DYNAMIC MODELS FROM DISCRETE-TIME STOCHASTIC MODELS

Assume that the system's dynamics are expressed by a discrete-time stochastic equations of the form

$$x_{k+1} = f(x_k, u_k, w_k), \tag{5.12}$$

$$z_k = g(x_k, v_k), \tag{5.13}$$

where $x_k \in \mathbb{R}^n$ is the system state, $u_k \in \mathbb{R}^r$ is the system input, $w_k \in \mathbb{R}^p$ is a white noise input disturbance, $z_k \in \mathbb{R}^s$ is the process output, and $v_k \in \mathbb{R}^l$ is a white noise output disturbance. In general $f : \mathbb{R}^n \times \mathbb{R}^r \times \mathbb{R}^p \to \mathbb{R}^n$ and $g : \mathbb{R}^n \times \mathbb{R}^l \to \mathbb{R}^s$ are non-linear functions.

A generalized discrete time fuzzy dynamic system model is given by the following equations:

$$X_{k+1} = F(X_k, U_k, W_k), \qquad (5.14)$$
$$Z_k = G(X_k, V_k), \qquad (5.15)$$

where k is the time index, X_k defined on $\mathcal{X}_1 \times \ldots \times \mathcal{X}_n$ is a fuzzy relation for the state, U_k defined on $\mathcal{U}_1 \times \ldots \times \mathcal{U}_r$ is a fuzzy relation for the process input, W_k defined on $\mathcal{W}_1, \ldots, \mathcal{W}_p$ is a fuzzy relation for the input disturbance, Z_k defined on $\mathcal{Z}_1 \times \ldots \times \mathcal{Z}_s$ is the process output, V_k defined on $\mathcal{V}_1 \times \ldots \times \mathcal{V}_l$ is a fuzzy relation for the output disturbance, and F and G are functions that map fuzzy relations to fuzzy relations.

In this section we will look at how to indirectly generate fuzzy dynamic systems, such as that given in equations (5.14) and (5.15), by using a stochastic dynamic system model, such as presented in equations (5.12) and (5.13). In this way we exploit the wealth of accumulated modeling experience for these conventional models.

5.3.1 Construction of Fuzzy Dynamic Models Via Fuzzy Composition

To use fuzzy composition it is convenient to generate a list of fuzzy rules that characterizes the behavior of the stochastic dynamic behavior in equation (5.12). A rule base is generated so that it contains enough rules to ensure *completeness* (i.e. there exist enough rules that at least one is activated for all realizations of the vectors x_k, u_k, and w_k). Suppose the ith rule is defined so that it is activated in the region of space around the set of values $x_k = x^i \in \mathbb{R}^n$, $u_k = u^i \in \mathbb{R}^r$, and $w_k = w^i \in \mathbb{R}^p$. Here we define the following fuzzy relations:

$$X_k^i = \mathcal{F}(x^i), \qquad (5.16)$$
$$U^i = \mathcal{F}(u^i), \qquad (5.17)$$
$$W^i = \mathcal{F}(w^i), \qquad (5.18)$$
$$X_{k+1}^i = \mathcal{F}(f(x^i, u^i, w^i)), \qquad (5.19)$$
$$U(k) = \mathcal{F}(u_k), \qquad (5.20)$$

where $\mathcal{F}$ represents the process of fuzzification around a single element. The subscripts on X_k^i and X_{k+1}^i are used only to distinguish cases, not to indicate that they are changing in time.

To illustrate the fuzzification operator $\mathcal{F}$, consider the vector $x^i = [x_1^i, \ldots, x_n^i]^T$. One possible fuzzification scheme might be a normalized jointly Gaussian function

$$\mu_{X^i}(x) = \exp\{-\tfrac{1}{2}[(x - x^i)^T C^{-1}(x - x^i)]\}, \qquad (5.21)$$

where C is a symmetric positive definite matrix. Hence the fuzzification operator simply turns crisp values into fuzzy sets.

Using the fuzzy relations defined in equations (5.16)–(5.20), one can generate a fuzzy knowledge base such that the *i*th rule is given by

If $X(k)$ is X_k^i and $U(k)$ is U^i and $W(k)$ is W^i **Then** $X(k+1)$ is X_{k+1}^i,

where $X(k)$ and $X(k+1)$ are fuzzy relations characterizing the current and the next state, respectively, $W(k)$ is a fuzzy relation characterizing the process disturbance input, and $U(k)$ is a fuzzy relation characterizing the process input.

This rule can be denoted mathematically by a fuzzy relation of the form

$$R^i = X_k^i \times U^i \times W^i \times X_{k+1}^i. \tag{5.22}$$

Assuming there exist many such rules, the overall fuzzy relation describing the entire rule base is denoted by the union of these rules:

$$R = \bigcup_i R^i. \tag{5.23}$$

Thus the fuzzy dynamic system model can be expressed by the following:

$$X(k+1) = [X(k) \times U(k) \times W(k)] \circ R. \tag{5.24}$$

The choice of the fuzzy relation $W(k)$ is based on a fuzzy characterization of the process disturbance at time kT. One possible choice for the membership function of $W(k)$ is simply a normalized version of the joint probability density function of w_k. In many cases the characteristics of the process disturbance input will not change in time, so often the fuzzy relation $W(k)$ will be constant.

5.3.2 Construction of a Fuzzy Dynamic Model Via the Fuzzy Extension Principle

Another way to construct a fuzzy dynamic model is by transforming the stochastic model in equations (5.12) and (5.13) to a fuzzy model via the fuzzy extension principle. First consider equation (5.12). For time kT define $X(k)$ to be the fuzzy relation describing the state, $U(k)$ is the fuzzy relation for the process input, and $W(k)$ is the fuzzy relation for the process input disturbance. Hence the fuzzy relation for the process state at time $(k+1)T$ can be found via the fuzzy extension principle

$$\mu_{X_{k+1}}(x_{k+1}) = \begin{cases} \oplus \{\mu_X(k) \times U(k) \times W(k)(x_k, u_k, w_k) : x_k, u_k, w_k, x_{k+1} = f(x_k, u_k, w_k)\}, \\ 0, \text{ if } f^{-1}(x_{k+1}) = \emptyset. \end{cases} \tag{5.25}$$

In a similar manner, equation (5.13) can be converted to a fuzzy system of the form of equation (5.15).

5.4 STABILITY ANALYSIS OF FUZZY DYNAMIC SYSTEMS

In this section we characterize and analyze the stability properties of fuzzy dynamic systems. First, we define the convergence properties for a sequence of fuzzy sets. Then we define Lyapunov, asymptotic, and exponential stability. We also define uniform the boundedness and uniformly ultimately boundedness of fuzzy dynamic systems for bounded sets. We provide some theorems for analyzing the stability of fuzzy dynamic systems using Lyapunov's direct method. These form a practical approach to the stability analysis of fuzzy dynamic systems. Finally, we illustrate the application of the stability results on a first order fuzzy dynamic system.

5.4.1 Convergence in Fuzzy Dynamic Systems

A fundamental aspect of stability definitions is the concept of convergence of sequences. Therefore, we begin by briefly discussing some basic convergence properties of fuzzy set sequences. First, however, we felt it was useful to identify analogous convergence concepts for stochastic dynamic systems. Forms of stochastic convergence include convergence in probability, convergence in the rth moment, convergence with probability one, and convergence in distribution [4]. Here, we focus on convergence in distribution.

Definition 1 (Convergence in probability distributions) A random sequence $X(n)$ with a probability distribution function $F_n(x)$ converges in probability distribution to the random variable with probability distribution function $F(x)$ if

$$\lim_{n \to \infty} F_n(x) = F(x).$$

Note that if the convergence of $F_n(x) \to F(x)$ is uniform, then this is the same as convergence in probability density. This is easily seen by the following:

$$\frac{d}{dx} \lim_{n \to \infty} F_n(x) = \frac{dF(x)}{dx},$$

$$\lim_{n \to \infty} \frac{dF_n(x)}{dx} = \frac{dF(x)}{dx},$$

$$\lim_{n \to \infty} f_n(x) = f(x),$$

where the second line holds only if $F_n(x) \to F(x)$ uniformly. The analogous concept for fuzzy dynamic systems is convergence in membership functions.

Definition 2 (Fuzzy convergence in membership functions) A fuzzy relation $X(n)$ with a membership function $\mu_{X(n)}(x)$ converges in membership function to the fuzzy relation with membership function $\mu(x)$ if

$$\lim_{n \to \infty} \mu_{X(n)}(x) = \mu(x).$$

It is this type of convergence that will form the foundation for our stability concepts for fuzzy dynamic systems.

5.4.2 Stability of Fuzzy Dynamic Systems

The following definitions are proposed to study the stability properties of systems represented by a fuzzy dynamic system given by

$$X(k+1) = \bar{F}(X(k)), \tag{5.26}$$

where $X(k)$ and $X(k+1)$ are the state (fuzzy relations) at time k and $k+1$, respectively, and $\bar{F}$ is a function on fuzzy sets.

Let $\rho : \mathcal{X} \times \mathcal{X} \to \mathbb{R}$ denote a metric on $\mathcal{X}$, and $\{\mathcal{X}, \rho\}$ a metric space. For example, if X and $\tilde{X}$ denote two fuzzy relations defined on $\mathcal{X} = \mathcal{X}_1, \ldots, \mathcal{X}_n$ with membership functions defined by $\mu_X(x_1, \ldots, x_n)$ and $\mu_{\tilde{X}}(x_1, \ldots, x_n)$, respectively, then one possible metric is

$$\rho_p(X, \tilde{X}) = \left[\int_{x_1 \in \mathcal{X}_1} \cdots \int_{x_n \in \mathcal{X}_n} |\mu_X(x_1, \ldots, x_n) - \mu_{\tilde{X}}(x_1, \ldots, x_n)|^p dx_1 \ldots dx_n \right]^{1/p}, \quad (5.27)$$

where $1 \leq p < \infty$ and where $x_1 \in \mathcal{X}_1, \ldots, x_n \in \mathcal{X}_n$. Another possible metric is

$$\rho_\infty(X, \bar{X}) = \sup\{|\mu_X(x_1, \ldots, x_n) - \mu_{\bar{X}}(x_1, \ldots, x_n)| : x_1 \in \mathcal{X}_1, \ldots, x_n \mathcal{X}_n\}. \quad (5.28)$$

Let $\mathcal{X}_z \subset \mathcal{X}$ and $\rho(X, \mathcal{X}_z) = \inf\{\rho(X, X') : X' \in \mathcal{X}_z\}$ denote the *distance* from X to the fuzzy set $\mathcal{X}_z$. Next we give several important properties of functions used in stability analysis.

Definition 3 (Positive definite function) A continuous function $f : \mathbb{R}^n \to \mathbb{R}$ is said to be positive definite if

(i) $f(0) = 0$, and
(ii) $f(x) > 0$ for all $x \neq 0$.

Definition 4 (Positive semidefinite function) A continuous function $f : \mathbb{R}^n \to \mathbb{R}$ is said to be positive semidefinite if

(i) $f(0) = 0$, and
(ii) $f(x) \geq 0$ for all x.

Definition 5 (Negative definite function) A continuous function $f : \mathbb{R}^n \to \mathbb{R}$ is said to be negative definite if $-f$ is a positive definite function.

Definition 6 (Negative semidefinite function) A continuous function $f : \mathbb{R}^n \to \mathbb{R}$ is said to be negative semidefinite if $-f$ is a positive semidefinite function.

Definition 7 (Decrescent function) A continuous function $f : \mathbb{R}^+ \times \mathbb{R}^n \to \mathbb{R}$ is said to be negative decrescent if there exists a positive definite function $w : \mathbb{R}^n \to \mathbb{R}$ such that

$$|f(t, x)| \leq w(x), \quad \forall t \geq 0 \text{ and } \forall x \in B(r) \text{ for some } r > 0.$$

Definition 8 (Class K functions) A continuous function $f : [0.r_1] \to \mathbb{R}^+$ is said to be belong to class K, i.e. $f \in K$, if

(i) $f(0) = 0$, and
(ii) f is strictly increasing on $[0, r_1]$.

Definition 9 (Class KR functions) A continuous function $f : \mathbb{R}^+ \to \mathbb{R}^+$ is said to be belong to class KR, i.e. $f \in KR$ if

(i) $f \in K$, and
(ii) $\lim_{r \to \infty} f(r) =$ does not exist.

Following [3] we give the basic concepts and stability definitions.

Definition 10 (r-Neighborhood) The r-neighborhood of an arbitrary set $\mathcal{X}_z \subset \mathcal{X}$ is denoted by the set $S(\mathcal{X}_z, r) = \{X \in \mathcal{X} : 0 \leq \rho(X, \mathcal{X}_z) < r\}$, where $r > 0$.

We see that an r-neighborhood is the set of all multi-dimensional membership functions that are within a distance r of the set of multi-dimensional membership functions in $\mathcal{X}_z$.

Definition 11 (Invariant set) A set $\mathcal{X}_m \subset \mathcal{X}$ is called invariant with respect to equation (5.26) if from $X(0) \in \mathcal{X}_m$ it follows that $X(k) \in \mathcal{X}_m$ for all $k \in \mathbb{N}$.

Hence, $\mathcal{X}_m$ is invariant if when the state, which is a membership function, starts in $\mathcal{X}_m$ and stays in $\mathcal{X}_m$ as the membership function evolves through time.

Definition 12 (Stable in the sense of Lyapunov) An invariant set $\mathcal{X}_m \subset \mathcal{X}$ of equation (5.26) is stable in the sense of Lyapunov if for any $\varepsilon > 0$ there exists $\delta > 0$ such that when $\rho(X(0), \mathcal{X}_m) < \delta$ we have $\rho(X(k), \mathcal{X}_m) < \varepsilon$ for all $k \in \mathbb{N}$.

Definition 13 (Asymptotic stability) If equation (5.26) is stable in the sense of Lyapunov and if $\rho(X(k), \mathcal{X}_m) \to 0$ as $k \to \infty$, then the invariant set $\mathcal{X}_m$ of equation (5.26) is asymptotically stable.

Here, we see that we quantify stability in terms of convergence of a multi-dimensional membership function to a given set of membership functions. Using membership functions as analogous to a probability density function it is clear that this type of convergence is analogous to convergence in distribution.

Definition 14 (Unstable in the sense of Lyapunov) An invariant set $\mathcal{X}_m \subset \mathcal{X}$ of equation (5.26) is unstable in the sense of Lyapunov if it is not stable in the sense of Lyapunov.

Definition 15 (Region of asymptotic stability) If the invariant set $\mathcal{X}_m \subset \mathcal{X}$ of equation (5.26) is asymptotically stable then the set $\mathcal{X}_a$ of all states $X(0) \in \mathcal{X}$ with the property $\rho(X(k), \mathcal{X}_m) \to 0$ as $k \to \infty$ is called the region of asymptotic stability.

Definition 16 (Asymptotically stable in the large) The invariant set $\mathcal{X}_m \subset \mathcal{X}$ of equation (5.26) with region of stability $\mathcal{X}_a$ is called asymptotically stable in the large if $\mathcal{X}_a = \mathcal{X}$.

In this case, no matter where the initial state of the fuzzy system starts, as time evolves it will approach $\mathcal{X}_m$.

Definition 17 (Exponentially stable) The invariant set $\mathcal{X}_m \subset \mathcal{X}$ of equation (5.26) is called exponentially stable if there exists an $\alpha > 1$, and for every $\varepsilon > 0$, there exists $\delta(\varepsilon) > 0$ such that

$$\rho(X(k), \mathcal{X}_m) \leq \varepsilon \alpha^{-k}, \quad \forall k \geq 0, \tag{5.29}$$

whenever $\rho(X(0), \mathcal{X}_m) < \delta(\varepsilon)$.

This type of stability is most often used to help quantify the rate of convergence of the state.

Definition 18 (Uniformly bounded) The invariant set $\mathcal{X}_m \subset \mathcal{X}$ of equation (5.26) is called uniformly bounded if for any $\alpha > 0$, there exists a $\beta = \beta(\alpha) > 0$ such that if $\rho(X(0), \mathcal{X}_m) < \alpha$, then $\rho(X(k), \mathcal{X}_m) < \beta$ for all $k \geq 0$.

Definition 19 (Uniformly ultimately bounded (with bound B)) The invariant set $\mathcal{X}_m \subset \mathcal{X}$ of equation (5.26) is called uniformly ultimately bounded (with bound B) if there exists a $B > 0$ and if corresponding to any $\alpha > 0$, there exists $aT = T(\alpha) > 0$ such that if $\rho(X(0), \mathcal{X}_m) < \alpha$, then this implies that $\rho(X(k), \mathcal{X}_m) < B$ for all $k \geq T$.

Both of these types of bounds can be important in different settings. Following the work of [1] it is possible to extend all of these definitions to allow for random inputs.

5.4.3 The Direct Lyapunov Method for Fuzzy Dynamic Systems

We are now in a position to specify precise statements of some of the more important stability results for fuzzy dynamic systems given by equation (26). These results are readily apparent extensions to the *direct Lyapunov method* [3].

Theorem 1 (Stability in the sense of Lyapunov) *If there exists a positive definite and decrescent function $v(X(k)) : \mathbb{R} \to \mathbb{R}^\dagger$ such that $v(X(k+1)) - v(X(k))$ is negative semidefinite, then the invariant set $\mathcal{X}_m$ is stable in the sense of Lyapunov.*

Theorem 2 (Asymptotic stability in the sense of Lyapunov) *If there exists a positive definite and decrescent function $v(X(k)) : \mathbb{R} \to \mathbb{R}$ such that $v(X(k+1)) - v(X(k))$ is negative definite, then the invariant set $\mathcal{X}_m$ is asymptotically stable.*

Theorem 3 (Exponential stability) *If there exist a function $v(X(k)) : \mathbb{R} \to \mathbb{R}$ and three positive constants $c_1, c_2,$ and c_3 such that*

$$c_1(\rho^2(X(k), \mathcal{X}_m)) \leq v(X(k)) \leq c_2(\rho^2(X(k), \mathcal{X}_m)) \tag{5.30}$$

and

$$v(X(k+1)) - v(X(k)) \leq -c_3(\rho^2(X(k+1), \mathcal{X}_m)) \tag{5.31}$$

for all $k \geq 0$ and for all $X(k) \in B(r)$ for some $r > 0$, then the invariant set $\mathcal{X}_m$ is exponentially stable.

Theorem 4 (Uniformly bounded) *If there exists a function $v(X(k))$ defined on $\rho(X(k), \mathcal{X}_m) \geq R$ (where R may be large) and $0 \leq k < \infty$, and if there exist strictly increasing functions $\psi_1,$ and ψ_2 such that*

$$\psi_1(\rho(X(k), \mathcal{X}_m)) \leq v(X(k)) \leq \psi_2(\rho(X(k), \mathcal{X}_m)) \tag{5.32}$$

and

$$v(X(k+1)) - v(X(k)) \leq 0 \tag{5.33}$$

for all $\rho(X(k), \mathcal{X}_m) \geq R$ and $k \geq 0$, then the invariant set $\mathcal{X}_m$ is uniformly bounded

Theorem 5 (Uniformly ultimately bounded) *If there exists a function $v(X(k))$ defined on $\rho(X(k), \mathcal{X}_m) \geq R$ (where R may be large) and $0 \leq k < \infty$, and if there exist strictly increasing functions $\psi_1, \psi_2,$ and ψ_3 such that*

$$\psi_1(\rho(X(k), \mathcal{X}_m)) \leq v(X(k)) \leq \psi_2(\rho(X(k), \mathcal{X}_m)) \tag{5.34}$$

and

$$v(X(k+1)) - v(X(k)) \leq -\psi_3(\rho(X(k), \mathcal{X}_m)) \tag{5.35}$$

---

†The notation used here is intended to mean that v is a function of the membership function of $X(k)$, not the fuzzy set itself. This abuse of notation will be used in the rest of this chapter.

for all $\rho(X(k), \mathcal{X}_m) \geq R$ and $k \geq 0$, then the invariant set $\mathcal{X}_m$ is uniformly ultimately bounded

5.5 APPLICATION—FIRST-ORDER FUZZY DYNAMIC SYSTEM

In this section we consider an application of the Lyapunov stability result presented in Section 5.4 on a first-order fuzzy dynamic system. Consider the following linear stochastic system:

$$x_{k+1} = ax_k + w_k, \tag{5.36}$$

where k is the time index, $x_k \in \mathbb{R}$ and $x_{k+1} \in \mathbb{R}$ are the system state, $w_k \in \mathbb{R}$ is a disturbance input, and $a \in \mathbb{R}$ is a constant. We will use the extension principle to create a fuzzy dynamic system from this model.

Suppose at any time k that x_k and w_k have Gaussian membership functions, i.e.

$$\mu_{X(k)}(x_k) = \exp\left\{\frac{-(x_k - \tilde{x}_k)^2}{2p_k}\right\}, \tag{5.37}$$

$$\mu_{W(k)}(w_k) = \exp\left\{\frac{-w_k^2}{2q}\right\}. \tag{5.38}$$

We seek to find a model of the form given in equations (5.14) and (5.15). To do this we want to use the model in equation (5.36) that is in the form of equations (5.12) and (5.13) to find F and G of equations (5.14) and (5.15) by applying the extension principle to the functions f and g of equations (5.12) and (5.13) for the case in equation (5.36). For our case the extension principle is

$$\mu_{X(k+1)}(x_{k+1}) = \oplus\{\mu_{X(k) \times W(k)}(x_k, w_k) : x_k, w_k, x_{k+1} = ax_k + w_k\}. \tag{5.39}$$

From equations (5.37) and (5.38), the Cartesian product $X(k) \times W(k)$ has a membership function given by

$$\mu_{X(k) \times W(k)}(x_k, w_k) = \exp\left\{-\frac{(x_k - \tilde{x}_k)^2}{2p_k} - \frac{w_k^2}{2q}\right\}, \tag{5.40}$$

where the product was used for the triangular norm. If we apply the fuzzy extension principle and use the max operator for the triangular co-norm, from equation (5.39) we see that the problem of finding $\mu_{X(k+1)}(x_{k+1})$ reduces to the following: maximize equation (5.40) via the choice of x_k and w_k subject to the constraint in equation (5.36). To do this we form the Lagrangian

$$L = \exp\left\{-\frac{(x_k - \tilde{x}_k)^2}{2p_k} - \frac{w_k^2}{2q}\right\} - \lambda[ax_k + w_k - x_{k+1}]. \tag{5.41}$$

Take partial derivatives and set them equal to zero to obtain

$$\frac{\partial L}{\partial x_k} = \frac{-(x_k - \tilde{x}_k)}{p_k} \exp\left\{-\frac{(x_k - \tilde{x}_k)^2}{2p_k} - \frac{w_k^2}{2q}\right\} - \lambda a = 0, \tag{5.42}$$

$$\frac{\partial L}{\partial w_k} = \frac{-w_k}{q} \exp\left\{-\frac{(x_k - \tilde{x}_k)^2}{2p_k} - \frac{w_k^2}{2q}\right\} - \lambda = 0, \tag{5.43}$$

$$\frac{\partial L}{\partial \lambda} = -[ax_k + w_k - x_{k+1}] = 0. \tag{5.44}$$

From equations (5.42) and (5.43) we get

$$\frac{-\lambda a p_k}{x_k - \tilde{x}_k} = \frac{-\lambda q}{w_k}. \tag{5.45}$$

Rearranging this equation yields

$$x_k - \tilde{x}_k = \frac{ap_k}{q} w_k. \tag{5.46}$$

From equations (5.44) and (5.46)

$$x_k - \tilde{x}_k = \frac{ap_k}{q}[x_{k+1} - ax_k]. \tag{5.47}$$

After some manipulation of equation (5.47) we get

$$x_k = \frac{ap_k x_{k+1} + q\tilde{x}_k}{a^2 p_k + q}. \tag{5.48}$$

Furthermore,

$$w_k = x_{k+1} - ax_k = x_{k+1} - \frac{a^2 p_k x_{k+1} + aq\tilde{x}_k}{a^2 p_k + q}. \tag{5.49}$$

The values of x_k and w_k found in equations (5.48) and (5.49), respectively, are those that maximize $\mu_{X(k) \times W(k)}(x_k, w_k)$ subject to the given constraint in equation (5.36). If we substitute these into $\mu_{X(k) \times W(k)}(x_k, w_k)$ and use the extension principle we get

$$\mu_{X(k+1)}(x_{k+1}) = \exp\left\{-\left(\frac{ap_k x_{k+1} + q\tilde{x}_k}{a^2 p_k + q} - \tilde{x}_k\right)^2 \frac{1}{2p_k} - \left(x_{k+1} - \frac{a^2 p_k x_{k+1} + aq\tilde{x}_k}{a^2 p_k + q}\right)^2 \frac{1}{2q}\right\}. \tag{5.50}$$

After some manipulation of equation (5.50) we obtain

$$\mu_{X(k+1)}(x_{k+1}) = \exp\left\{\frac{-(x_{k+1} - a\tilde{x}_k)^2}{2(a^2 p_k + q)}\right\}. \tag{5.51}$$

Note that this is in Gaussian form where

$$\tilde{x}_{k+1} = a\tilde{x}_k \tag{5.52}$$

and

$$p_{k+1} = a^2 p_k + q. \tag{5.53}$$

So by induction, if initial uncertainties are Gaussian shaped, then they will remain Gaussian for this system for all k. Furthermore, note that the 'center' and 'spread' of the membership functions propagate in a similar manner to the mean and variance in a stochastic linear system.

The equilibrium can be found by setting $x_{k+1} = x_k$ in equation (5.52) and $p_{k+1} = p_k$ in equation (5.53) to obtain

$$x_e = 0, \tag{5.54}$$

$$p_e = \frac{q}{1-a^2}. \tag{5.55}$$

Thus the equilibrium fuzzy set (or invariant set), denoted X_e, has a membership given by the following:

$$\mu_{X_e}(x_k) = \exp\left\{\frac{-(x_k - x_e)^2}{2p_e}\right\}. \tag{5.56}$$

Let the metric be

$$\rho(X_k, X_e) = \sup_{x_k} |\mu_{X_k}(x_k) - \mu_{X_e}(x_k)|, \tag{5.57}$$

$$= \sup_{x_k}\left|\exp\left\{\frac{-(x_k - \tilde{x}_k)^2}{2p_k}\right\} - \exp\left\{\frac{-(x_k - x_e)^2}{2p_e}\right\}\right|, \tag{5.58}$$

where X_k is the state of the fuzzy dynamic system. Choose the Lyapunov candidate function to be

$$v_k = \ln(u_{X(k)}) + \frac{(x_k - \tilde{x}_k)^2}{2p_k} + (\tilde{x}_k - x_e)^2 + (p_k - p_e)^2, \tag{5.59}$$

$$= (\tilde{x}_k - x_e)^2 + (p_k - p_e)^2, \tag{5.60}$$

$$= \tilde{x}_k^2 + \left(p_k - \frac{q}{1-a^2}\right), \tag{5.61}$$

which is positive semidefinite and equal to zero only at the equilibrium. Next we show that v_k can be bounded by two strictly increasing functions ψ_1 and ψ_2 such that $\psi_1(\rho(X_k, X_e)) \leq v_k \leq \psi_2(\rho(X_k, X_e))$. To do this we use Rolle's theorem and the mean value theorem, presented next.

Theorem 1 (Rolle's theorem) *Let f be continuous on $[a,b]$ and differentiable on (a,b). If*

$$f(a) = f(b) = 0, \tag{5.62}$$

then there is at least one number c in (a,b) such that $f'(c) = 0$.

Consider a second-order function $u(x, y)$. Also, consider a function $f(z)$ that is equal to $u(x,y)$ when (x, y) are along some line in $\mathbb{R}^2$ as shown in Figure 5.1. In other words,

$$f(z) = \{u(x, y) : z = \sqrt{(x - x_0)^2 + (y - y_0)^2}\}, \tag{5.63}$$

where (x_0, y_0) is some point in $\mathbb{R}^2$, where $z = 0$. Take the derivative of $f(z)$:

$$f'(z) = \lim_{\Delta z \to 0} \frac{f(z + \Delta z) - f(z)}{\Delta z}. \tag{5.64}$$

This is the same as taking the derivative of $u(x, y)$ in the direction of z:

$$u'_z(x, y) = \lim_{\Delta x \to 0, \Delta y \to 0 \text{ in the direction of } z} \frac{u(x + \Delta x, y + \Delta y) - u(x, y)}{\sqrt{\Delta x^2 + \Delta y^2}} = f'(z). \tag{5.65}$$

APPLICATION—FIRST-ORDER FUZZY DYNAMIC SYSTEM

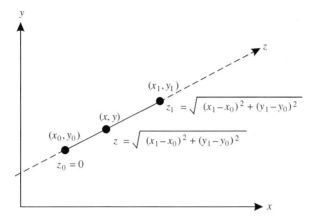

Figure 5.1 Plot illustrating $u(x, y)$ and $f(z)$

Theorem 2 (Mean value theorem) *If f is continuous on $[z_0, z_1]$ and differentiable on (z_0, z_1), then there exists a number c_z in (z_0, z_1) such that*

$$f(z_1) - f(z_0) = f'(c_z)(z_1 - z_0). \quad (5.66)$$

Equation (5.66) can be re-written in terms of $u(x, y)$ and $u'_z(x, y)$ by the following:

$$u(x_1, y_1) - u(x_0, y_0) = u'_z(c_x, c_y)\sqrt{(x_1 - x_0)^2 + (y_1 - y_0)^2}, \quad (5.67)$$

where x_1 and y_1 correspond to z_1, x_0 and y_0 correspond to z_0, and c_x and c_y correspond to c_z. Let $x_1 = \tilde{x}_k, y_1 = p_k, x_0 = x_e, y_0 = p_e, c_x = c_{\tilde{x}_k}$, and $c_y = c_{p_k}$. Furthermore, let $u(x, y) = \exp\{-(x_k - x)^2/2y\}$. Substituting these into equation (67) yields

$$\exp\left\{\frac{-(x_k - \tilde{x}_k)^2}{2p_k}\right\} - \exp\left\{\frac{-(x_k - x_e)^2}{2p_e}\right\} = u'_z(c_{\tilde{x}_k}, c_{p_k})\sqrt{(\tilde{x}_k - x_e)^2 + (p_k - p_e)^2}. \quad (5.68)$$

Take the magnitude of both sides of equation (5.68) to obtain

$$\left|\exp\left\{\frac{-(x_k - \tilde{x}_k)^2}{2p_k}\right\} - \exp\left\{\frac{-(x_k - x_e)^2}{2p_e}\right\}\right| = |u'_z(c_{\tilde{x}_k}, c_{p_k})|\sqrt{(\tilde{x}_k - x_e)^2 + (p_k - p_e)^2}. \quad (5.69)$$

Take the $\sup_{x_k}$ of both sides of equation (5.69)

$$\sup_{x_k}\left|\exp\left\{\frac{-(x_k - \tilde{x}_k)^2}{2p_k}\right\} - \exp\left\{\frac{-(x_k - x_e)^2}{2p_e}\right\}\right| = \sup_{x_k}|u'_z(c_{\tilde{x}_k}, c_{p_k})|\sqrt{(\tilde{x}_k - x_e)^2 + (p_k - p_e)^2}. \quad (5.70)$$

Square both sides of equation (5.70) to obtain

$$\sup_{x_k}\left|\exp\left\{\frac{-(x_k - \tilde{x}_k)^2}{2p_k}\right\} - \exp\left\{\frac{-(x_k - x_e)^2}{2p_e}\right\}\right|^2 = \sup_{x_k}|u'_z(c_{\tilde{x}_k}, c_{p_k})|^2\{(\tilde{x}_k - x_e)^2 + (p_k - p_e)^2\}. \quad (5.71)$$

Now we can take the $\sup_{x,y}$ of the derivative on the right-hand side of equation (5.71) to obtain

$$\sup_{x_k}\left|\exp\left\{\frac{-(x_k-\tilde{x}_k)^2}{2p_k}\right\}-\exp\left\{\frac{-(x_k-x_e)^2}{2p_e}\right\}\right|^2 \le \sup_{x,y}\{\sup_{x_k}|u'_z(x,y)|^2\}$$
$$\{(\tilde{x}_k-x_e)^2+(p_k-p_e)^2\}. \tag{5.72}$$

Since $u(x,y)$ is continuous, its derivative is bounded in all directions. Furthermore, since $u(x,y)$ is not constant for all $(x,y)\in\mathbb{R}^2$ its derivatives is not zero for some $(x,y)\in\mathbb{R}^2$. Therefore, we can divide both sides of equation (5.72) by $\sup_{x,y}\{\sup_{x_k}|u'_z(x,y)|^2\}$ to obtain

$$\left\{\frac{1}{\sup_{x,y}\{\sup_{x_k}|u'_z(x,y)|^2\}}\right\}\sup_{x_k}\left|\exp\left\{\frac{-(x_k-\tilde{x}_k)^2}{2p_k}\right\}-\exp\left\{\frac{-(x_k-x_e)^2}{2p_e}\right\}\right|^2$$
$$\le \{(\tilde{x}_k-x_e)^2+(p_k-p_e)^2\}. \tag{5.73}$$

This can be rewritten as

$$\left\{\frac{1}{\sup_{x,y}\{\sup_{x_k}|u'_z(x,y)|^2\}}\right\}\rho^2(X_k,X_e)\le v_k. \tag{5.74}$$

We can define $\psi_1(\rho(X_{\tilde{k}},X_e))$ by

$$\psi_1(\rho(X_k,X_e))\triangleq\left\{\frac{1}{\sup_{x,y}\{\sup_{x_k}|u'_z(x,y)|^2\}}\right\}\rho^2(X_k,X_e). \tag{5.75}$$

So we get

$$\psi_1(\rho(X_k,X_e))\le v_k. \tag{5.76}$$

Next we will try to find a strictly increasing function $\psi_2(\rho(X_k,X_e))$ such that $v_k\le\psi_2(\rho(X_k,X_e))$.

We will show that $v_k\le\psi_2(\rho(X_k,X_e))$ for some arbitrarily large region in $\mathbb{R}^2$ for values of $\tilde{x}_k$ and x_e and values of p_k and p_e. Recall equation (5.69)

$$\left|\exp\left\{\frac{-(x_k-\tilde{x}_k)^2}{2p_k}\right\}-\exp\left\{\frac{-(x_k-x_e)^2}{2p_e}\right\}\right|=|u'_z(c_{\tilde{x}_k},c_{p_k})|\sqrt{(\tilde{x}_k-x_e)^2+(p_k-p_e)^2}. \tag{5.77}$$

Taking the $\sup_{x_k}$ of the left-hand side of equation (5.77) yields

$$\sup_{x_k}\left|\exp\left\{\frac{-(x_k-\tilde{x}_k)^2}{2p_k}\right\}-\exp\left\{\frac{-(x_k-x_e)^2}{2p_e}\right\}\right|\ge|u'_z(c_{\tilde{x}_k},c_{p_k})|\sqrt{(\tilde{x}_k-x_e)^2+(p_k-p_e)^2}. \tag{5.78}$$

Let $M \in \mathbb{R}^+$ be some large positive real number. We can let $x_k = M$ in the derivative on the right-hand side of equation (5.78) to obtain

$$\sup_{x_k} \left| \exp\left\{ \frac{-(x_k - \tilde{x}_k)^2}{2p_k} \right\} - \exp\left\{ \frac{-(x_k - x_e)^2}{2p_e} \right\} \right| \geq |u_z'(c_{\tilde{x}_k}, c_{p_k})|_{x_k=M}|$$

$$\sqrt{(\tilde{x}_k - x_e)^2 + (p_k - p_e)^2}. \qquad (5.79)$$

Square both sides of equation (5.79) to yield

$$\sup_{x_k} \left| \exp\left\{ \frac{-(x_k - \tilde{x}_k)^2}{2p_k} \right\} - \exp\left\{ \frac{-(x_k - x_e)^2}{2p_e} \right\} \right|^2 \geq |u_z'(c_{\tilde{x}_k}, c_{p_k})|_{x_k=M}|^2$$

$$\times \{(\tilde{x}_k - x_e)^2 + (p_k - p_e)^2\}. \qquad (5.80)$$

Now take the $\inf_{x,y}$ of the derivative on the right hand side of equation (79) to yield

$$\sup_{x_k} \left| \exp\left\{ \frac{-(x_k - \tilde{x}_k)^2}{2p_k} \right\} - \exp\left\{ \frac{-(x_k - x_e)^2}{2p_e} \right\} \right|^2 \geq \inf_{x,y} |u_z'(x, y)|_{x_k=M}|^2$$

$$\{(x_k - x_e)^2 + (p_k - p_e)^2\}. \qquad (5.81)$$

Since $u(x, y)$ is continuous, its derivative is bounded in all directions. Further more, since $u(x, y)$ does not have a local maximum or minimum on the arbitrarily large region $-M + 1 \leq x \leq M - 1$ and $y > 0$, its derivative is not zero in this region. Therefore we can divide both sides of equation (5.81) by $\inf_{x,y} |u_z'(x, y)|_{x_k=M}|^2$ to obtain

$$\left\{ \frac{1}{\inf_{x,y} |u_z'(x, y)|_{x_k=M}|^2} \right\} \sup_{x_k} \left| \exp\left\{ \frac{-(x_k - \tilde{x}_k)^2}{2p_k} \right\} - \exp\left\{ \frac{-(x_k - x_e)^2}{2p_e} \right\} \right|^2$$

$$\geq (\tilde{x}_k - x_e)^2 + (p_k - p_e)^2. \qquad (5.82)$$

This can be rewritten as

$$\left\{ \frac{1}{\inf_{x,y} |u_z'(x, y)|_{x_k=M}|^2} \right\} \rho^2(X_k, X_e) \geq v_k.$$

We can define $\psi_2(\rho(X_k, X_e))$ by

$$\psi_2(\rho(X_k, X_e)) \triangleq \left\{ \frac{1}{\inf_{x,y} |u_z'(x, y)|_{x_k=M}|^2} \right\} \rho^2(X_k, X_e). \qquad (5.83)$$

Therefore we get

$$\psi_2(\rho(X_k, X_e)) \geq v_k \qquad (5.84)$$

Thus we have shown that there exist strictly increasing $\psi_1(\rho(X_k, X_e))$ and $\psi_2(\rho(X_k, X_e))$ such that $\psi_1(\rho(X_k, X_e)) \leq v_k \leq \psi_2(\rho(X_k, X_e))$.

110 LYAPUNOV STABILITY ANALYSIS OF FUZZY DYNAMIC SYSTEMS

Now we can show that the system is stable by finding the Lyapunov function difference equation by the following:

$$v_{k+1} - v_k = \tilde{x}_{k+1}^2 + \left(p_{k+1} - \frac{q}{1-a^2}\right)^2 - \tilde{x}_k^2 - \left(p_k - \frac{q}{1-a^2}\right)^2 \quad (5.85)$$

$$= a^2 \tilde{x}_k^2 + \left(a^2 p_k + q - \frac{q}{1-a^2}\right)^2 - \tilde{x}_k^2 - \left(p_k - \frac{q}{1-a^2}\right)^2 \quad (5.86)$$

$$= (a^2 - 1)\tilde{x}_k^2 + a^4 p_k^2 + q^2 + \frac{q^2}{(1-a^2)^2} + 2a^2 q p_k - \frac{2a^2 q p_k}{1-a^2} - \frac{2q^2}{1-a^2}$$

$$- p_k^2 - \frac{q^2}{(1-a^2)^2} + \frac{2 q p_k}{1-a^2} \quad (5.87)$$

$$= (a^2 - 1)\tilde{x}_k^2 + (a^4 - 1)p_k^2 + q^2 + 2a^2 q p_k + \frac{-2a^2 q p_k - 2q^2 + 2q p_k}{1-a^2} \quad (5.88)$$

$$= (a^2 - 1)\tilde{x}_k^2 + (a^4 - 1)p_k^2$$
$$+ \frac{q^2 - a^2 q^2 + 2a^2 q p_k - 2a^4 q p_k - 2a^2 q p_k - 2q^2 + 2q p_k}{1-a^2} \quad (5.89)$$

$$= (a^2 - 1)\tilde{x}_k^2 + (a^4 - 1)p_k^2 + \frac{-q^2 - a^2 q^2 - 2a^4 q p_k + 2q p_k}{1-a^2} \quad (5.90)$$

$$= (a^2 - 1)\tilde{x}_k^2 + (a^4 - 1)p_k^2 + \frac{2q p_k(a^4 - 1)}{(a^2 - 1)} + \frac{q^2(a^2 + 1)}{(a^2 - 1)} \quad (5.91)$$

$$= (a^2 - 1)\tilde{x}_k^2 + (a^4 - 1)\left[p_k^2 + \frac{2q p_k}{(a^2 - 1)} + \frac{q^2}{(a^2 - 1)^2}\right] \quad (5.92)$$

$$= (a^2 - 1)\tilde{x}_k^2 + (a^4 - 1)\left[p_k + \frac{q}{(a^2 - 1)}\right]^2 \quad (5.93)$$

$$= (a^2 - 1)\tilde{x}_k^2 + (a^4 - 1)\left[p_k + \frac{q}{(1 - a^2)}\right]^2. \quad (5.94)$$

Note that for $|a| \leq 1$ equation (5.94) is negative semidefinite and equal zero only at the equilibrium. Hence for $|a| \leq 1$ the equilibrium for the system is stable in the sense of Lyapunov. Furthermore, note that when $|a| < 1$ equation (5.94) is negative definite and equal to zero only at the equilibrium. Hence the equilibrium is asymptotically stable under this condition. Finally, note that if $c_1 = c_2 = 1$ and $c_3 = (1 - a^2)$ where $|a| \leq 1$, then it is easy to see that the equilibrium for the system is also exponentially stable.

5.6 CONCLUDING REMARKS

In this chapter we defined the following types of stability and boundedness: stability in the sense of Lyapunov, asymptotic stability, exponential stability, uniform boundedness, and uniform ultimate boundedness. We established sufficient conditions, in terms of Lyapunov functions, for each of these stability types. Also, we studied various applications of stability for systems modeled via fuzzy dynamic systems. These include a first-order fuzzy dynamic system.

REFERENCES

1. Passino, K., A. Michel and P. Antsaklis, Lyapunov stability of a class of discrete event systems, *IEEE Transactions on Automatic Control*, **39**, 1994, 269–279.
2. Khalil, H., *Nonlinear Systems*, Macmillan, Basing stoke, 1992.
3. Miller, R. and A. Michel, *Ordinary Differential Equations*, Academic Press, New York, 1982.
4. Kushner, H., *Introduction to Stochastic Control*, Holt, Rinehart, and Winston, New York, 1971.
5. de Glas, M., Invariance and stability of fuzzy systems, *Journal of Mathematical Analysis and Applications*, **99**, 1984, 299–319.
6. Tong, R. M., Some properties of fuzzy feedback systems, *IEEE Transactions on Systems, Man and Cybernetics*, **10**, 1980, 327–330.
7. Jianqin, C. and C. Laijiu, Study on stability of fuzzy closed loop control systems, *Fuzzy Sets and Systems*, **57**, 1993, 159–168.
8. Kiszka, J., M. Gupta and P. Nikiforuk, Energetistic stability of fuzzy dynamical systems. *IEEE Transactions on Systems, Man and Cybernetics*, **15**, 1985, 783–792.
9. Chand, S. and A. Hansen, Energy based stability analysis of a fuzzy roll controller design for a flexible aircraft wing, In: *Proceedings of the 1989 IEEE Conference on Decision and Control*, Tampa, Florida, 1989, pp. 705–709.
10. Layne, J. and K. Passino, Stability of fuzzy dynamic systems, *Submitted for publication*, 1997.
11. Klir, G. and T. Floger, *Fuzzy Sets, Uncertainty, and Information*, Prentice-Hall, Englewood Cliffs, 1988.

6

Passivity and Stability of Fuzzy Control Systems

R. Gorez
University of Louvain, Louvain-la-Neuve, Belgium

G. Calcev
Land Mobile Product Systems, Motorola, USA

6.1 INTRODUCTION

In the two last decades fuzzy control has generated a substantial amount of excitement in the control engineering community. This interest for fuzzy control found its origin mainly from practical applications where fuzzy controllers have proved satisfactory results. However, since fuzzy control systems are often based on some human expert knowledge, which is difficult to be translated in a mathematical analytic form, theoretical developments in this field are far from being finished. Recently, the issue of the stability of fuzzy control systems has been considered in a non-linear stability framework. Several approaches have been proposed; in particular, approaches based on the Lyapunov direct method [15, 26, 30] or on the theory of hyperstability and/or passivity concepts [6, 8, 33], approaches using numerical and graphical techniques such as phase portrait, cell-to-cell mapping, trajectory reversing methods [10, 11, 14 and 16], also approaches using traditional tools of non-linear analysis such as describing functions, etc. [1, 3, 45, 50]. Here we propose a unifying approach to the stability analysis of fuzzy control systems which is based on the theory of *passivity*. This approach is determined by the following well-known facts:

- control loops with fuzzy controllers are non-linear control systems;
- such fuzzy control systems are characterized by intrinsic uncertainties derived both from their 'model-free' specific features as well as from the fuzziness of human decisions.

Using a passivity-based approach one is able to formulate stability results for some general classes of plants and the corresponding classes of controllers, including non-linear controllers and/or non-linear plants with uncertainties. This passivity framework gives the possibility to point out some fundamental common features among various implementations of fuzzy control systems and it can be effective for both Mamdani and Takagi–Sugeno fuzzy control systems.

Fuzzy Logic. Edited by S. Farinwata, D. Filev and R. Langari
© 2000 John Wiley & Sons, Ltd

As a matter of fact, the theory of *passivity*, which nowadays is widely used in stability analysis of non-linear systems [5, 13, 21, 22, 31, 38, 49], is strongly related to the concept of *hyperstability* introduced by the Rumanian scientist V. M. Popov in the late 1950s [36]. This concept played an important role in the theory of non-linear control systems through the 1960s to the early 1980s. This concept indeed allows designers of control systems to deal with systems whose characteristics can be poorly defined, for example by means of sectorial bounds, and it offers elegant solutions for the proof of the stability of such systems. For this reason the framework of the passivity theory is very convenient for the analysis of the stability of control loops involving fuzzy control systems. One of the first attempts to apply the hyperstability concept to fuzzy systems was in [37]; however, in that paper only fuzzy controllers with very simple rules were considered. The same approach was also used in [27] for the analysis of a feedback loop consisting of a single input fuzzy controller and a non-linear controlled plant. The use of the passivity theory as a tool for stability analysis of more general fuzzy control systems was proposed for the first time in [6] and [7]. Independently, a similar approach was suggested in [33] and recently, similar tools were used in [30], based on a state–space formulation. A systematic approach to the analysis of the stability of control loops involving Mamdani-type fuzzy controllers is given in [8]. This analysis is based on the passivity theory and input–output descriptions of both the fuzzy controller and the plant to be controlled. Here it is shown that the same approach can be used for the analysis of the stability of fuzzy control systems based on Takagi–Sugeno fuzzy models. Some ideas and results presented in [8] are resumed and extended, in a unifying presentation, to Takagi–Sugeno fuzzy control systems as well as to control loops with Mamdani fuzzy controllers. Only continuous-time systems are considered here, but the extension of this approach to discrete-time control systems would be straightforward.

This chapter is organized as follows. Section 6.2 recalls some features of Mamdani and Takagi–Sugeno controllers. Section 6.3 introduces the basic concepts of stability, passivity and real positivity for non-linear systems, and also some definitions of dynamic systems and polytopic differential inclusions, and it investigates the passivity properties of fuzzy controllers. These passivity properties are used in the Section 6.4 to derive results on the stability of fuzzy control loops. Section 6.5 deals with fuzzy control of linear systems and of Euler–Lagrange systems. Finally, Section 6.6 offers some conclusions and comments.

6.2 FUZZY CONTROL SYSTEMS

6.2.1 Mamdani Fuzzy Controllers

Since Zadeh introduced the theory of fuzzy sets and fuzzy logic in his seminal paper published in 1965 [53], it has been applied to the control of complex or ill-defined processes and a number of successful applications can be found in the literature. One of the first applications of fuzzy logic to process control was proposed by Mamdani and his team [29]. Their controller, known as the Mamdani fuzzy controller, is based on some human expertise, obtained from control engineers and/or plant operators. This expertise is translated into logical rules, and fuzzy logic with appropriate fuzzification and defuzzification interfaces is used to generate from these rules a real control value which can be applied to the control input of the plant to be controlled. In this type of controllers the control rules are expressed as logical statements such as:

$$R_i : \textbf{If } z_1 \text{ is } S_{1i} \text{ and } z_2 \text{ is } S_{2i} \text{ and } \ldots \text{ and } z_n \text{ is } S_{ni} \textbf{ Then } z_0 \text{ is } S_{0i},$$

where $z_1, z_2, \ldots, z_n$ are real input variables of the controller, $S_{1i}, S_{2i}, \ldots, S_{ni}$ and S_{oi} are fuzzy sets and z_0 is an output *fuzzy variable*. Basically, these rules are that of a multivalued logical controller. Such a controller can be characterized by the following features:

- given the real input and output variables of the controller either the range of a variable is divided into a finite number of real intervals or the variable is allowed to take only a finite number of real values;
- there is a set of decision rules expressed as 'If–Then' statements that specify a value of the controller output for given values or intervals of the controller inputs;
- the rule which is fired at a given time determines the actual value of the controller output at that time.

If the range of each input variable was divided into adjacent intervals with no overlap, then such a type of logical control would be discontinuous. Allowing adjacent intervals to overlap, introducing membership functions to characterize each interval and using fuzzy logic turns the discontinuous logical controller into a fuzzy controller, with generally a continuous input–output mapping. In other words, the transitions between different values of the output of a logical controller can be smoothed thanks to the use of fuzzy logic, due to the interpolative nature of the latter. Therefore, a Mamdani controller with its fuzzification and defuzzification interfaces is equivalent to a non-linear static controller with multiple inputs and outputs. However, in many control applications, one may consider only single-input–single-output (SISO) control loops, consisting of a SISO plant controlled by a two-input–single-output (TISO) fuzzy controller, with one of the two controller input variables being proportional to the time-integral of the other or conversely the latter being proportional to the time-derivative of the former. Considering together the fuzzy controller and this relation between its input variables such controllers can generally be related to a class of SISO non-linear dynamic controllers, which here will be called QPI controllers. For many Mamdani fuzzy controllers the input–output mapping can also be associated with another class on TISO non-linear controllers, called DPS dynamic controllers, for which more powerful stability results were obtained in [8]. These two classes of non-linear controllers have passivity properties, which can be used for the stability analysis of fuzzy control loops, and will be investigated in the following section.

6.2.2 Takagi–Sugeno Fuzzy Control Systems

In 1985, with the same objective in mind of using fuzzy sets and fuzzy logic for control applications, Takagi and Sugeno proposed a new type of fuzzy control system based on a multiple model representation of the plant to be controlled [40]. The latter system is represented by a collection of models, which can be viewed as *local models*, with each one being valid in a particular operating region. Each operating region is defined by a *conditional part* similar to that of the fuzzy rules in a Mamdani controller. However, the difference with a Mamdani controller is that now the *consequent part* of each rule is an analytical expression describing the corresponding local model, for example:

$$R_i : \textbf{If } z_1 \text{ is } S_{1i} \text{ and } z_2 \text{ is } S_{2i} \text{ and} \ldots \text{and } z_n \text{ is } S_{ni}$$
$$\textbf{Then } y_i = f(p_{1i}, p_{2i}, \ldots, p_{ni}, z_1, z_2, \ldots, z_n),$$

where $z_1, z_2, \ldots, z_n$ are real input variables of the collection of models, $S_{1i}, S_{2i}, \ldots, S_{ni}$ are fuzzy sets and y_i and p_{ij} are the real output variable and parameters of the ith local model, respectively. This collection of models together with an aggregation procedure are the basic ingredients of a Takagi–Sugeno model of the system to be controlled. In general, the aggregation procedure consists of a convex combination of the outputs of the different models:

$$y(t) = \sum_{i=1}^{i=r} \lambda_i y_i(t),$$

with $\lambda_i = w_i / \sum_{j=1}^{j=r} w_j$, $i = 1, 2, \ldots, r$, being *fuzzy weights* calculated from the *firing strengths* w_i of the different rules of the model, so that, whatever the way the firing strengths are obtained, $\lambda_i \in [0, 1]$, $i = 1, 2, \ldots, r$, and $\sum_{i=1}^{i=r} \lambda_i = 1$. Since the model inputs may be either input and state variables of the plant, or input and output variables with some of their derivatives, the fuzzy weights actually are dependent on all these variables, hence a Takagi–Sugeno model is a true non-linear model of the plant. Techniques for identifying such models are proposed in the literature, for example techniques based on fuzzy clustering [54, 55]. In general, local models are linear and time-invariant, so that to each of them one can associate a linear *local controller* designed through various design methods of the classical linear control theory [24, 41, 47]. The real output of the global controller then is computed by combining that of the local controllers using the same aggregation formula as that of the Takagi–Sugeno fuzzy model of the plant, with the same values of the fuzzy weights. This leads to the control structure of Figure 6.1. Such a controller can also be referred to as a 'gain scheduling controller' [2]; in what follows, for the sake of brevity, it will be called a Takagi–Sugeno controller. In fact, the global controller output depends on the inputs of the fuzzy model so that the controller actually is non-linear.

The stability of control loops consisting of a Takagi–Sugeno fuzzy model and a state-feedback controller associated as just said, and also the design of the controller following some performance criteria and robustness issues, was dealt with in [43], [54] and [55] using

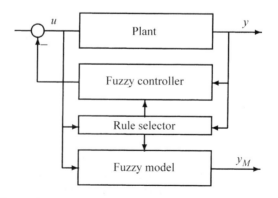

Figure 6.1 Takagi-Sugeno fuzzy feedback control system

the technique of Linear Matrix Inequalities. Here Takagi–Sugeno controllers will be considered as dynamic systems. If the local models and/or the associated local controllers are linear and time-invariant they can be represented by linear time-invariant state and output equations. Therefore, due to the properties of the fuzzy weights quoted above, a Takagi–Sugeno fuzzy model of a plant to be controlled and/or a Takagi–Sugeno fuzzy controller can be viewed as Polytopic Linear Differential Inclusions [2].

6.3 STABILITY AND PASSIVITY OF FUZZY CONTROLLERS

As seen in the previous section, both Mamdani and Takagi–Sugeno fuzzy controllers are rule-based non-linear controllers, with similar expressions for the conditional parts of the rules. Moreover, in the simplest case of Mamdani controllers the defuzzification process reduces to an aggregation formula similar to that used in Takagi–Sugeno models and controllers. However, so far there has been no common approach for the analysis of the stability of control loops that include either a Mamdani or a Takagi–Sugeno fuzzy controller. As a matter of fact, the passivity theory can be a valuable candidate for stability analysis of both Mamdani and Takagi–Sugeno fuzzy control systems. The main idea is that under some general conditions both Mamdani and Takagi–Sugeno fuzzy controllers are passive dynamic systems; Mamdani controllers because they can be viewed as QPI or DPS dynamic non-linear controllers and Takagi–Sugeno controllers because they can be imbedded in a class of Polytopic Differential Inclusions (PDI) which may have passivity properties. Therefore, as a consequence of a basic result in the passivity theory, a well-defined feedback interconnection of such a controller with a passive system results in a stable closed-loop system. Basic concepts and results of the passivity theory are recalled here, and the passivity properties of the QPI and DPS non-linear controllers associated with Mamdani controllers and also that of PDIs are investigated.

6.3.1 Basic Concepts

In the vocabulary of control scientists, a dynamic system is an object which produces an output signal for each input signal (including hidden inputs corresponding to the initial state of the system). Only n-dimensional time-invariant continuous-time dynamic systems will be considered here. The basic definition of such a dynamic system is in terms of differential equations as follows:

Definition 1 A dynamic system is a mathematical object described by the following expressions:

$$\dot{x} = f(x, u), \qquad (6.1)$$
$$y = h(x, u), \qquad (6.2)$$

where $x \in \mathbf{R}^n$ is the state vector, $u \in \mathbf{R}^m$ is the input vector, $y \in \mathbf{R}^p$ is the output vector and $f(\cdot, \cdot)$ and $h(\cdot, \cdot)$ are continuous functions.

Definition 2 An equilibrium state of a dynamical system is a state x^* such that for any input signal $u(t)$:

$$f(x^*, u) = 0. \qquad (6.3)$$

It is obvious that if for $t = t_0$ the state of a dynamic system is an equilibrium state: $x(t_0) = x^*$, then the system will not leave this equilibrium state: $x(t) = x^*$ for all $t \geq t_0$. Note that in general the above constraint determining the possible equilibrium states of a given dynamic system can be satisfied only if the input variables are kept constant: $u(t) = u^*$ for all $t \geq t_0$; in that case, an appropriate definition of the input variables resulting in $u^* = 0$ leads us to consider only equilibrium states of the *free* system $\dot{x} = f(x, 0)$.

Given an equilibrium state of a dynamic system its stability can be determined as follows. An equilibrium state is *stable* if all solutions of (6.1) starting close to this equilibrium state stay in some neighborhood of this state; otherwise, the equilibrium would be *unstable*. Furthermore, an equilibrium state is said to be *asymptotically stable* if it is stable and all solutions starting close to the equilibrium state converge onto it as time increases towards infinity [25, 28]. The stability of the equilibrium states of a given dynamic system can be analyzed via the *direct method of Lyapunov*, which involves the determination of a *Lyapunov function* V and the use of the following theorem [25], where $\dot{V}$ means the time derivative of the Lyapunov function along the lines of the solutions of (6.1) and is obtained through the function $\dot{V}(x, u) = (\partial V / \partial x)^T f(x, u)$:

Theorem 1 *Let $V : D \to \mathbf{R}$ be a continuously differentiable function defined in a domain D containing the equilibrium state x^*, with $V(x^*) = 0$ and $V(x) > 0$ in $D - \{x^*\}$. If $\dot{V} \leq 0$ for any $x \in D$, then the equilibrium state x^* is stable. Moreover, if $\dot{V}(x) < 0$ in $D - \{x^*\}$, then the equilibrium state x^* is asymptotically stable.*

In many applications the Lyapunov function is such that $\dot{V} \leq 0$ in D. Nevertheless the asymptotic stability of the equilibrium state x^* can be proved by invoking the *Lasalle invariance principle* [25] provided that the equilibrium state x^* is the unique *invariant set* of (6.1) in the subset $\{x \mid \dot{V} = 0\} \cap D$ of the state–space of the system. Another stability concept of dynamic systems, namely that of *input–output stability*, can be defined as follows: a dynamic system is stable in the input–output sense (*I–O stable*) if the norm of the output signal is smaller than the product of the norm of the input signal by some positive constant. Between the two types of stability defined above there exists a strong relationship which can be expressed in terms of observability and detectability [25].

Stability analysis plays a very important role in system theory and control engineering, since stability may be considered as a *prerequisite* of any control system. Therefore, all stability results are to be viewed as fundamentals for any further developments in these fields. The main objective of the control theory indeed may be expressed as follows: given a specific set of dynamic systems, namely the set of *plants*, design a set of dynamic systems, namely the set of *controllers*, such that through the feedback interconnection of a plant and a controller the stability of the complete system as well as some dynamic performances are achieved.

It is possible to obtain stability results for complex systems consisting of several subsystems on the basis of the passivity properties of these subsystems. First, let us introduce some definitions and results on the passivity of continuous-time dynamic systems; for details see, for example, [13], [21], [31], [36], and [49]. Assume that the input and output vectors of the system (6.1), (6.2) have the same dimension $p = m$ and the state variables have been defined in such a way that $x = 0$ is an equilibrium state of the system. The state–space and the set of admissible input signals (real-valued piecewise continuous functions of time) will be denoted by X and U, respectively.

Definition 3 The dynamic system (6.1), (6.2) is said to be dissipative if there exists a continuous non-negative real-valued storage function $V(x)$, with $V(0) = 0$, and a real-valued supply rate $W(u, y)$, such that the following dissipation inequality holds for all $t \geq t_0$ and any $u \in U$ and $x(0) \in X$:

$$V(x(t)) - V(x(t_0)) \leq \int_{t_0}^{t} W(u(\tau), y(\tau)) d\tau.$$

With the supply rate $W(u, y) = u^T y$ the system is passive; with $W(u, y) = u^T y - \varepsilon_i u^T u - \varepsilon_0 y^T y$ for some non-negative ε_i and ε_0, it is either input strictly passive if $\varepsilon_i > 0$ and $\varepsilon_0 = 0$, or output strictly passive if $\varepsilon_i = 0$ and $\varepsilon_0 > 0$, or strictly passive if both $\varepsilon_i > 0$ and $\varepsilon_0 > 0$. The above definition of input- and output-strict passivity can be extended by substituting either positive definite diagonal or symmetrical matrices for ε_i and/or ε_0, or scalar products $u^T \nu(u)$ and/or $y^T \rho(y)$ for $u^T u$ and $y^T y$ [38], on condition that such terms which in fact characterize an *excess of passivity* are positive definite functions of the input or the output vector.

The concept of passivity has a simple interpretation in terms of energy conservation: considering $V(x)$ as the energy stored in the system and $W(u, y)$ as the external power supplied to the system, it can be said that for a passive system, the difference between the final and the initial values of the energy stored over any time interval $[t_0, t]$ is less than the external energy supplied to this system during this time interval. This means that a passive system does not generate energy but rather dissipates energy. Moreover, if a system has an excess of passivity, the latter can be used to compensate for a *shortage of passivity* in another system combined with the first one. All mechanical and electromechanical systems that do not contain internal sources of energy are passive. Therefore, such systems are good candidates for control by fuzzy controllers.

It can be easily verified that any parallel combination of two passive subsystems is also passive, the storage function of the complete system being simply the sum of the individual storage functions of the two subsystems. Besides, if at least one of the two subsystems is input strictly passive, the complete system is also input strictly passive. Similar results hold for a negative feedback combination of two passive subsystems as represented in Figure 6.2. Such a closed-loop interconnection of two subsystems H_1 and H_2 is a well-defined negative feedback loop if for any given inputs d_1 and d_2 the output signals y_1 and y_2 are such that the input signal $d_1 - y_2$ generates the signal y_1 at the output of the subsystem H_1 and the input signal $d_2 + y_1$ generates the signal y_2 at the output of the subsystem H_2. The storage function of a well-defined feedback loop consisting of two passive subsystems is also defined by the sum of the individual storage functions of the two subsystems. Therefore, such a feedback loop is also passive. Moreover, it is output strictly passive either if H_1 is output strictly passive or if H_2 is input strictly passive.

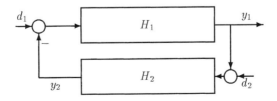

Figure 6.2 Feedback control loop

Such concepts of passivity provide a nice tool for analyzing the stability of control systems. Setting $u = 0$ in the definitions of passivity, it turns out that the zero-state equibrium, $x = 0$, of a passive system having a positive definite storage function is Lyapunov stable. Furthermore, by the Lasalle invariance principle [5], the equilibrium is asymptotically stable if the system is output strictly passive and zero-state detectable (*zero-state detectability* involves that $u(t) = 0$ and $y(t) = 0$ for all $t \geq 0 \Rightarrow \lim_{t \to \infty} x(t) = 0$). This leads to the following stability results, which are particular cases of Corollary 1 in [21]:

Theorem 2 *If two passive subsystems H_1 and H_2 are interconnected in a well-defined negative feedback connection, then the closed-loop system is stable. Asymptotic stability is guaranteed if any one of the following (non-equivalent) additional conditions is satisfied:*

1. *One of H_1 and H_2 is strictly passive.*
2. *Both H_1 and H_2 are output strictly passive.*
3. *The open-loop system $H_2 \cdot H_1$ is zero-state detectable and H_2 is output strictly passive.*
4. *The open-loop system $H_1 \cdot (-H_2)$ is zero-state detectable and either H_1 is output strictly passive or H_2 is input strictly passive.*

The previous theorem can be applied to the analysis of the stability of feedback control systems, H_1 including the controller and H_2 the plant to be controlled. In practical applications not many plants are passive, but often for stable plants, H_2 can be made passive by a parallel combination of the plant and a direct throughput with appropriate gain matrix D, possibly cascaded with a serial multiplier Z as in Figure 6.3. Then, to keep the free behavior of the closed-loop unchanged, H_1 must consist of the controller with a positive feedback diagonal matrix D and a serial multiplier Z^{-1} (Figure 6.3). If the passivity of H_1 is guaranteed or can be proved, then closed-loop stability is achieved, with asymptotic stability if one of the four additional conditions of the previous theorem is satisfied. Such a loop transformation is but the well-known *zero-shifting* technique used in stability analysis of non-linear control systems [17, 25].

Therefore, the following property will be useful for practical purposes:

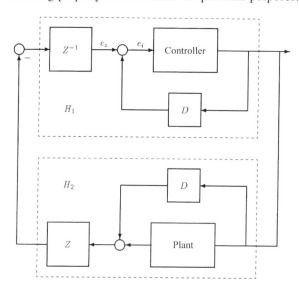

Figure 6.3 Zero-shifting loop transformation

Theorem 3 *If a system is output strictly passive, then output strict passivity or at least passivity is preserved in spite of some static positive feedback around this system.*

Proof Assume an output strictly passive system. Output strict passivity means that the supply rate to this system is in the form: $W(e_i, u) = e_i^T u - \varepsilon_o u^T u$, where the notations of Figure 6.3 are used since this property will be applied in cases where the system includes the controller. With a positive feedback with a diagonal gain matrix D, $e_i = e_z + Du$, where e_z is the input vector of the inner closed-loop. The supply rate can then be written as

$$W(e_z, u) = e_z^T u - u^T(\varepsilon_o I - D)u,$$

with I being the identity matrix. Consequently, output strict passivity holds if $\varepsilon_o I - D$ is a positive definite matrix, in other words if each diagonal entry of D is smaller than ε_o. If $\varepsilon_o I - D$ is positive semidefinite, and in particular if it reduces to the null matrix, then the closed-loop is still passive.

There is a strong relation between the *passivity* and the *real positivity* of a dynamic system [5, 36]. First, let us recall the definition of the real posivity of a continuous-time dynamic system:

Definition 4 *A continuous-time dynamic system is positive real if for all admissible u and $t \geq t_0$,*

$$\int_{t_0}^{t} u^T(\tau) y(\tau) d\tau \geq 0, \text{ whenever } x(t_0) = 0.$$

From Proposition 2.10 in [5], a passive system is positive real and conversely a positive real system is passive if it can be driven from the origin to any state by an admissible input signal, and if its available storage

$$V_a(x) = \sup_{\substack{t \geq t_0 \\ u \in \mathcal{U} \\ x(t_0) = x}} \left\{ -\int_{t_0}^{t} u^T(\tau) y(\tau) d\tau \right\}$$

is continuous. The interest of this concept for linear time-invariant (LTI) systems is that the real positivity of a stable LTI system can be expressed in terms of frequency response conditions (see, for example, [36], [44]). First of all, let us recall that a single-input–single-output (SISO) LTI system can be characterized by its transfer function $H(s)$, s denoting the Laplace variable, if and only if all the characteristic roots of the system that do not appear as poles of $H(s)$ have strictly negative real parts. For such a system the origin $x = 0$ of the state–space is obviously an equilibrium state. The latter, or more briefly the system or equivalently the transfer function $H(s)$, is stable if and only if all the poles of $H(s)$ have a non-positive real part, and asymptotically (or *strictly*) stable if and only if all the poles of $H(s)$ have a strictly negative real part. If a system is composed of one integrator and strictly stable subsystems, then there is an infinite set of stable equilibrium states aligned along a straight line in the state–space; such a system is said *neutrally stable* and its transfer function has a simple pole equal to zero, all the other poles having a strictly negative real part. The following results [25] on the real positivity of a LTI system characterized by its transfer function $H(s)$ are straightforward consequences of the Parseval theorem:

Theorem 4 *A transfer function $H(s)$ is positive real if it is stable, with the residues associated to poles with a zero real part being positive, and if $\forall \omega \geq 0$ with $j\omega$ not being a* pole of $H(s)$, $\text{Re}[H(j\omega)] \geq 0$. If $H(s)$ is strictly stable and $\forall \omega \geq 0$, $\text{Re}[H(j\omega)] > 0$ with either $\text{Re}[H(j\infty)] > 0$ or $\lim_{\omega \to \infty} \omega^2 H(j\omega) > 0$, then $H(s)$ is strictly positive real.

As stated earlier real positivity entails passivity, and strict real positivity strict passivity. Moreover, if $H(s)$ is neutrally stable and $\text{Re}[H(j\omega)] > 0$, $\forall \omega \geq 0$, such that $j\omega$ is not a pole of $H(s)$, then the system is *dissipative* (input or output strictly passive). These results can be extended to multiple-input–multiple-output systems; here only SISO systems will be considered.

6.3.2 Passivity of QPI Controllers

As noted previously, most Mamdani fuzzy controllers can be viewed as static TISO non-linear controllers with two real input variables, e_1 and e_2, and a single real output variable, u, related by an equivalent input–output bounded non-linear mapping, $u = \Phi(e_1, e_2)$. In the control applications considered here either e_1 is proportional to the time integral of e_2, or conversely e_2 is proportional to the time derivative of e_1. Including this relation between the two input variables most TISO Mamdani fuzzy controllers can be viewed as SISO non-linear dynamic controllers, here called QPI controllers. Such a controller is represented in Figure 6.4 and can be defined as follows:

Definition 5 A QPI Controller is a SISO non-linear dynamic system described by

$$\dot{e}_1 = \frac{1}{T_c} e_2, \qquad (6.4)$$

$$u = \Phi(e_1, e_2), \qquad (6.5)$$

where e_1, e_2 and u are the state, input and output variables, T_c is a time-constant and $\Phi(\cdot, \cdot)$ a real-valued continuous function which satisfies the following assumptions:

1. $\Phi(0, 0) = 0$;
2. for every pair (e_1, e_2)

$$0 \leq e_1[\Phi(e_1, e_2) - \Phi(0, e_2)] \quad \text{and} \quad 0 \leq e_2[\Phi(e_1, e_2) - \Phi(e_1, 0)]. \qquad (6.6)$$

The first property above clearly is a steady-state condition: it is obvious that a QPI controller is a neutrally stable dynamic system with $e_1 = 0, e_2 = 0$ as one of the possible equilibrium states. The second properties mean that for any given value of one of the two input variables the plot of the controller output with respect the other input lies entirely in the first and third quadrants of the Cartesian plane. These properties can be shown for a broad class of Mamdani controllers (see the Appendix); for others they can be checked by inspection of the input–output mapping of the controller. As for the controller time constant T_c, in practice its value will depend on the scale factors in the fuzzification interface of the

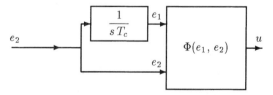

Figure 6.4 QPI controller

controller. In control applications, e_2 may be the *error signal* $e = r - y$, with r and y being the control system setpoint and the controlled process output, respectively. Then, with $\Phi(e_1, e_2) = K_c(e_1 + e_2)$ the QPI controller reduces to a linear Proportional Integral (PI) controller. The QPI controller can then be considered as a non-linear PI controller where the two actions are combined in a non-linear combination, which may be additive or not but can be characterized by the above properties. In other applications, e_2 could be proportional to the time derivative $\dot{e}$ of the error signal; in that case the QPI controller would be a non-linear Proportional-Derivative (PD) controller. The following theorem proves the passivity of QPI controllers.

Theorem 5 *With e_2 as input variable, a QPI controller is a passive dynamic system.*

Proof First, the properties of QPI controllers yield $0 \leq e_1 \Phi(e_1, 0)$ and $0 \leq e_2 \Phi(0, e_2)$, and defining $\Delta_{e_1}(e_1, e_2) = \Phi(e_1, e_2) - \Phi(0, e_2)$ and $\Delta_{e_2}(e_1, e_2) = \Phi(e_1, e_2) - \Phi(e_1, 0)$, the properties (6.6) can be written as

$$0 \leq e_1 \Delta_{e_1}(e_1, e_2) \quad \text{and} \quad 0 \leq e_2 \Delta_{e_2}(e_1, e_2).$$

Then, the passivity of a QPI controller results from the following relations:

$$\int_0^t e_2(\tau) u(\tau) d\tau = \int_0^t e_2(\tau) \Phi(e_1(\tau), e_2(\tau)) d\tau$$

$$= T_c \int_0^t \dot{e}_1(\tau) \Phi(e_1(\tau), 0) d\tau + \int_0^t e_2(\tau) \Delta_{e_2}(e_1(\tau), e_2(\tau)) d\tau$$

$$\geq T_c \int_0^t \dot{e}_1(\tau) \Phi(e_1(\tau), 0) d\tau = T_c \int_{e_1(0)}^{e_1(t)} \Phi(e_1, 0) de_1$$

$$\geq V_1(e_1(t)) - V_1(e_1(0)),$$

where

$$V_1(e_1) = T_c \int_0^{e_1} \Phi(e, 0) de \tag{6.7}$$

and hence $V(0) = 0$, is a storage function of the QPI controller.

Furthermore, if $e_2 \Delta_{e_2}(e_1, e_2) > 0$ for any $e_2 \neq 0$, then it can be said that the QPI controller is input strictly passive. This passivity property is valid with $e_2 = \dot{e}_1$ as input of the controller. The reader might note some duality of this passivity property with that of mechanical systems where passivity holds with force or torque as input and velocity as output, which is the derivative of displacement.

6.3.3 Passivity of DPS Controllers

For a broad class of Mamdani controllers [8] the (continuous) input–output mapping has the following properties:

$$\forall e_1, e_2, \quad e_1 + e_2 = 0 \Longrightarrow \Phi(e_1, e_2) = 0, \tag{6.8}$$

$$\exists K_c > 0 \text{ s.t. } \forall e_1, e_2, \quad e_1 + e_2 \neq 0 \Longrightarrow 0 \leq \frac{\Phi(e_1, e_2)}{e_1 + e_2} \leq K_c, \tag{6.9}$$

$$\forall e_1, \quad \Phi(e_1, 0) = 0 \Longrightarrow e_1 = 0. \tag{6.10}$$

124   PASSIVITY AND STABILITY OF FUZZY CONTROL SYSTEMS

From these properties the plane (e_1, e_2) can be divided into two regions by the straight line $e_1 + e_2 = 0$. The controller output is zero along this line, non-negative above, and non-positive below. This leads to the following definition:

Definition 6   A TISO non-linear static controller whose associated input–output mapping has the properties (6.8)–(6.10) is a Diagonally Positive Semidefinite (DPS) controller.

Substituting for (6.10) a stronger property,

$$\Phi(e_1, e_2) = 0 \Rightarrow e_1 + e_2 = 0, \tag{6.11}$$

leads to a subclass of DPS controllers:

Definition 7   A TISO non-linear static controller whose associated input–output mapping has the properties (6.8), (6.9) and (6.11) is a Diagonally Positive Definite (DPD) controller.

Note that DPD controllers have a strict monotonocity property in a neighborhood of the diagonal $e_1 + e_2 = 0$ in the (e_1, e_2) plane, while DPS controllers have this property only in a neighborhood of the origin.

Most often, in control applications the two inputs of the controller are proportional to the error signal e and its time derivative, respectively. In fact, the only assumptions that are needed are that in steady-state operation $e = 0 \Longleftrightarrow e_1 = 0$ and e_2 is persistently equal to zero. Therefore, more generally, as shown in [8], the two controller input signals e_1 and e_2 can be related to the error signal via filters with transfer functions $F_1(s)$ and $s\,T_c\,F_2(s)$, such that $F_1(0) = F_2(0) = 1$. For example, with e_2 given by a pseudo-derivative of e, $F_2(s) = 1/(1 + s\,T_f)$, where T_f is the time-constant of the differentiating filter. Figure 6.5(a) shows a control scheme where the error signal is fed through a linear compensator whose transfer function $C(s)$ is stable and inverse-stable. Taking into account the properties of the input–

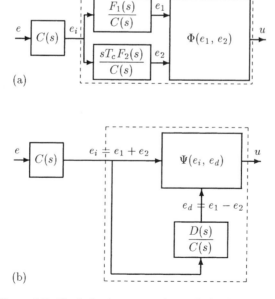

Figure 6.5   Equivalent representations of a DPS controller

output mapping of the DPS and DPD controllers, it is useful to define a SISO non-linear dynamic controller with the sum $e_i = e_1 + e_2$ as input variable, the difference $e_d = e_1 - e_2$ being an internal variable of this dynamic controller, as shown in Figure 6.5(b). This is obtained by setting

$$C(s) = F_1(s) + sT_c F_2(s), \qquad (6.12)$$

$$D(s) = F_1(s) - sT_c F_2(s), \qquad (6.13)$$

and leads to the second scheme of Figure 6.5. In the particular case where $e_1 = e$ and $e_2 = T_c \dot{e}$, hence $F_1(s) = F_2(s) = 1$, this controller is but a Proportional Derivative (PD) non-linear controller with derivative time constant T_c. More generally, it will be assumed that the two filters with respective transfer functions $F_1(s)$ and $F_2(s)$ and unit-gain ($F_1(0) = F_2(0) = 1$) are strictly stable and such that the compensator transfer function $C(s)$ is strictly stable and inverse-stable and that the internal subsystem with e_i as input and e_d as output is strictly stable and completely controllable. The properties of the input–output mapping of the TISO static controller are reflected in those of the output characteristic $u = \Psi(e_i, e_d)$ of this SISO dynamic controller: $u = \Psi(e_i, e_d)$ is a continuous function of the two variables e_i and e_d such that $\forall e_d$:

$$e_i = 0 \Longrightarrow u = 0, \qquad (6.14)$$

$$\exists K_c > 0 \text{ s.t. } \forall e_i \neq 0 \Longrightarrow 0 \leq \frac{u}{e_i} \leq K_c, \qquad (6.15)$$

$$u = 0 \text{ and } e_i = e_d \Longrightarrow e_i = 0. \qquad (6.16)$$

With a DPD static controller the property (6.16) is replaced by

$$\forall e_d, u = 0 \Longrightarrow e_i = 0, \qquad (6.17)$$

which obviously is stronger. Note, however, that in the steady state $e_d = e_i$ so that $u = 0$ then involves $e_i = 0$ with (6.16) as well as with (6.17).

For the sake of brevity the acronyms of the TISO static controllers will be used also for the associated SISO dynamic controllers, hence the following definition:

Definition 8 DPD and DPS dynamic controllers are SISO non-linear dynamic controllers whose internal dynamics is characterized by the transfer function $D(s)/C(s)$ with $C(s)$ and $D(s)$ being defined by (6.12) and (6.13), and whose output characteristics have the properties (6.14) and (6.15) and either (6.17) for DPD controllers or (6.16) for DPS controllers.

Clearly, from Figure 6.5, a DPS dynamic controller is decomposed into two parts: a static non-linear function $u = \Psi(e_i, e_d)$ and a dynamical linear subsystem whose output e_d may have some effect on the current value of the controller output u but not on the basic properties of the controller. As for the main input variable e_i, it is obtained from the control error signal through a linear compensator associated with the input filters of the controller. Then, the controller together with its input filters can be viewed as a SISO sector-bounded static non-linearity preceded by a linear compensator. Taking a storage function equal to zero and a supply rate $W(e_i, u) = e_i u - u^2/K_c$, the following passivity property is an obvious consequence of (6.15):

Theorem 6 *With e_i as input, any DPS controller is output strictly passive.*

Therefore the powerful results of the use of the hyperstability or passivity concepts in the stability analysis of nonlinear systems can be applied to control loops including DPS dynamical controllers.

6.3.4 Passivity of Polytopic Differential Inclusions

Some non-linear dynamic systems can be represented as polytopic linear differential inclusions [2]. In particular this is the case with Linear Takagi–Sugeno (LTS) models and controllers, i.e. say Takagi–Sugeno fuzzy models or controllers based on linear local models or controllers.

Definition 9 A continuous-time dynamic system is a polytopic linear differential inclusion (PDI) if the state and output equations (6.1) and (6.2) have the following form:

$$\dot{x}(t) = A x(t) + B u(t), \qquad (6.18)$$
$$y(t) = C x(t) + D u(t), \qquad (6.19)$$

where the four matrices A, B, C and D may be time- and/or state-dependent and are defined as the convex hull of a set of r given 4-tuples of constant matrices:

$$\begin{bmatrix} A & B \\ C & D \end{bmatrix} \in \mathrm{Co}\begin{bmatrix} A_i & B_i \\ C_i & D_i \end{bmatrix}, \quad i = 1, 2, \ldots, r.$$

The above definition means that the 4-tuple of the matrices A, B, C and D is given by all the convex combinations:

$$\begin{bmatrix} A & B \\ C & D \end{bmatrix} = \sum_{i=1}^{i=r} \lambda_i \begin{bmatrix} A_i & B_i \\ C_i & D_i \end{bmatrix}, \qquad (6.20)$$

with $\lambda_i \in [0, 1]$, $i = 1, 2, \ldots, r$, and $\sum_{i=1}^{i=r} \lambda_i = 1$. This is exactly the kind of representation that we can arrive at with LTS models and/or controllers. Given a Takagi–Sugeno fuzzy model with LTI local models described by state and output equations, each rule of the model is in the form:

$$R_i : \text{If } z_1 \text{ is } S_{1i} \text{ and } z_2 \text{ is } S_{2i} \text{ and} \ldots \text{and } z_n \text{ is } S_{ni}$$
$$\text{Then } \dot{x}_i(t) = A_i x(t) + B_i u(t),$$
$$y_i(t) = C_i x(t),$$

where $z_1, z_2, \ldots, z_n$ are the inputs of the Takagi–Sugeno model, $S_{1i}, S_{2i}, \ldots, S_{ni}$, $i = 1, 2, \ldots, r$, the corresponding fuzzy sets for the ith local model, x_i and y_i are the state and the output of this local model, respectively, u is the control input, and x and y are the state and the output respectively of the Takagi–Sugeno model obtained by aggregation of the local models:

$$x(t) = \sum_{i=1}^{i=r} \lambda_i x_i(t), \qquad (6.21)$$

$$y(t) = \sum_{i=1}^{i=r} \lambda_i y_i(t). \qquad (6.22)$$

Consequently, the Takagi–Sugeno model can be described by state and output equations in the form:

$$\dot{x}(t) = \sum_{i=1}^{i=r} \lambda_i [A_i x(t) + B_i u(t)], \qquad (6.23)$$

$$y(t) = \sum_{i=1}^{i=r} \lambda_i C_i x(t), \qquad (6.24)$$

with the fuzzy weights λ_i calculated from the firing strengths of the different rules, and hence depend on the inputs of the model, i.e. the state and some inputs of the actual system. Therefore, as a consequence of this complex non-linear dependence of the fuzzy weights with respect to the state and/or inputs of the system, a Takagi–Sugeno model is a non-linear model. However, since $\lambda_i \in [0, 1]$, $i = 1, 2, \ldots, r$, and $\sum_{i=1}^{i=r} \lambda_i = 1$, a LTS fuzzy model can be embedded in a class of polytopic linear differential inclusions.

Similarly, for a LTS fuzzy controller obtained from a Takagi–Sugeno model of the plant to be controlled, each rule can be written as follows:

R_i : **If** z_1 is S_{1i} and z_2 is S_{2i} and ... and z_n is S_{ni}
Then $\dot{x}_{ci}(t) = A_{ci} x_c(t) + B_{ci} e(t)$,
$u_{ci}(t) = C_{ci} x_c(t) + D_{ci} e(t)$,

where $z_1, z_2, \ldots, z_n$ are the inputs of the Takagi–Sugeno model of the plant to be controlled, $S_{1i}, S_{2i}, \ldots, S_{ni}$, $i = 1, 2, \ldots, r$, the corresponding fuzzy sets for the ith local model, x_{ci} and u_{ci} are the state and the output of the associated local controller, respectively, e is the input vector of the controller, and x_c and u_c are the state and the output respectively of the Takagi–Sugeno controller obtained by aggregation of the different local controllers:

$$x_c(t) = \sum_{i=1}^{i=r} \lambda_i x_{ci}(t), \tag{6.25}$$

$$u_c(t) = \sum_{i=1}^{i=r} \lambda_i u_{ci}(t). \tag{6.26}$$

Consequently, a controller can also be represented by state and output equations in the form:

$$\dot{x}_c(t) = \sum_{i=1}^{i=r} \lambda_i [A_{ci} x_c(t) + B_{ci} e(t)], \tag{6.27}$$

$$u_c(t) = \sum_{i=1}^{i=r} \lambda_i [C_{ci} x_c(t) + D_{ci} e(t)], \tag{6.28}$$

with the fuzzy weights λ_i calculated from the Takagi–Sugeno model of the plant to controlled. Again, the controller is non-linear but due to the properties of the fuzzy weights, it can be embedded in a class of polytopic linear differential inclusions.

Note that the output equations of the controller may contain throughput matrices D_{ci}. Throughput matrices are not allowed in the model because this would lead to *algebraic loops* in the feedback control system, and this situation must be avoided. As a matter of fact, in the actual plant there is no direct input–output connection, but throughput matrices D_i could be obtained in the model as a result of the fuzzy clustering techniques used for identification of the model. Therefore, identification of a linear Takagi–Sugeno model of the plant to be controlled must be conducted under the constraints $D_i = 0$, $i = 1, 2, \ldots, r$. Moreover, since the conditional part of the local controllers depends on the inputs of the Takagi–Sugeno of the plant to be controlled, these inputs should not include the current values of the control variables, otherwise this would raise the problem of logical consistency in the controller. This leads to additional constraints on the identification of the Takagi–Sugeno model. Another possible way to cope with problems of algebraic loops and/or logical consistency would be the insertion of additional first-order components in the measurement lines or in the control lines, but this would increase the order of the complete model.

Eventually, the passivity and stability properties obtained for a given class of polytopic linear differential inclusions hold for LTS models and/or controllers which have been embedded in this class of PDI. The passivity properties of a class of PDI can be investigated by means of the following theorem.

Theorem 7 *The dynamic system represented by the polytopic linear differential inclusion (6.18), (6.19) is passive if there exist a positive semidefinite symmetric matrix $P = P^T \geq 0$ and non-negative scalars $\eta \geq 0$ and/or $\gamma \geq 0$ such that the r following inequalities are satisfied:*

$$\begin{bmatrix} A_i^T P + P A_i + \gamma C_i^T C_i & P B_i - C_i^T/2 + \gamma C_i^T D_i \\ B_i^T P - C_i/2 + \gamma D_i^T C_i & \eta I + \gamma D_i^T D_i - (D_i + D_i^T)/2 \end{bmatrix} \leq 0, \quad i = 1, 2, \ldots, r, \quad (6.29)$$

with I being the identity matrix and the inequality meaning that the symmetric matrix on the left-hand side of (6.29) is negative semidefinite.

- *With $\eta = 0$ and $\gamma = 0$, the system is passive.*
- *With $\eta > 0$ and $\gamma = 0$, the system is input strictly passive.*
- *For systems with constant matrices $C_i = C$ and $D_i = D$:*
  —with $\eta = 0$ and $\gamma > 0$, the system is output strictly passive;
  —with $\eta > 0$ and $\gamma > 0$, the system is strictly passive.

Proof Assuming a positive definite symmetric matrix P, the positive definite quadratic form $V(x) = x^T P x$ can be used as a differentiable storage function of the system (6.18), (6.19). The latter is input or output strictly positive if there exists $\eta > 0$ or $\gamma > 0$ such that $\dot{V}(x(t)) \leq u^T(t) y(t) - \eta u^T(t) u(t) - \gamma y^T(t) y(t)$; with $\eta = 0$ and $\gamma = 0$ the system is passive. Straightforward calculations then lead to the following inequality:

$$\begin{bmatrix} A^T P + P A + \gamma C^T C & P B - C^T/2 + \gamma C^T D \\ B^T P - C/2 + \gamma D^T C & \eta I + \gamma D^T D - (D + D^T)/2 \end{bmatrix} \leq 0. \quad (6.30)$$

Since the dynamic system is a polytopic differential inclusion, the left-hand side of the above global inequality is a convex combination of the left-hand sides of the r previous inequalities (6.29). Successively setting each λ_i to 1 reduces this global inequality to the ith of the r inequalities (6.29).

In fact, what this theorem means is that if the r LTI subsystems defined by the 4-tuples of the matrices (A_i, B_i, C_i, D_i) are passive with the same storage function for all of them, then the polytopic linear differential inclusion made by all convex combinations of these subsystems has the same passivity property with the same storage function.

Note that each inequality (6.29) can be written as:

$$\eta \begin{bmatrix} O & O \\ O & I \end{bmatrix} + \gamma \begin{bmatrix} C_i^T C_i & C_i^T D_i \\ D_i^T C_i & D_i^T D_i \end{bmatrix} - \begin{bmatrix} -(A_i^T P + P A_i) & C_i^T/2 - P B_i \\ C_i/2 - B_i^T P & (D_i + D_i^T)/2 \end{bmatrix} \leq 0, \quad (6.31)$$

which can be related to the *min-max* properties of the eigenvalues of the symmetric matrices in the linear algebra [39]. It was shown in [2] that many properties of the dynamic systems represented by polytopic linear differential inclusions can be expressed in the framework of Linear Matrix Inequalities (LMIs). Let us recall that an LMI usually is written down as

$$F(\lambda) = F_0 + \sum_{i=1}^{m} \lambda_i F_i > 0, \quad (6.32)$$

where $\lambda = (\lambda_1, \lambda_2, \ldots, \lambda_m)$ is an m-tuple of real variables and $F_i = F_i^T$, $i = 0, 1, \ldots, m$, are $m + 1$ given real symmetric matrices. As usual, such an inequality means that the symmetric matrix $F(\lambda)$ is positive definite. Note that this gives a convex constraint on λ, i.e. the set $\{\lambda | F(\lambda) > 0\}$ is convex. Obviously, one can also encounter non-strict LMIs which have the form: $F(\lambda) \geq 0$, and other LMIs can be written with negative definite or non-positive definite matrices: $F(\lambda) < 0$ or $F(\lambda) \leq 0$. Many results of the use of LMIs and PDIs in system and control theory are given in [2]; some of them have been applied in [43], [54] and [55] for the synthesis of control systems based on linear Takagi–Sugeno fuzzy models. In particular, to investigate the passivity of a PDI system, one can use an appropriate optimization procedure searching for the largest *dissipation parameter* η and/or γ, subject to the constraint $P > 0$. The problem of finding the largest η or γ that satisfies the previous inequalities under the constraint $P = P^T > 0$ may be related to the standard *eigenvalue problem* (EVP) or *generalized eigenvalue problem* (GEVP) in the LMI theory [2]. These problems are expressed as

- EVP

$$\text{minimize } \eta \text{ s.t.} \begin{cases} \eta I - M(\lambda) > 0, \\ P(\lambda) > 0; \end{cases}$$

- GEVP

$$\text{minimize } \gamma \text{ s.t.} \begin{cases} \gamma N(\lambda) - M(\lambda) > 0, \\ P(\lambda) > 0, \end{cases}$$

where M, N and P are symmetric matrices that depend affinely on the optimization m-variable λ.

In the stability analysis of Takagi–Sugeno fuzzy control systems, it may be convenient to decompose each of the local controllers into a dynamic part, which will be combined with the corresponding local model of the plant to be controlled, and a static gain. This leads to the concept of polytopic linear algebraic inclusions, which are particular cases of polytopic linear differential inclusions where all the matrices A_i, B_i and C_i are null. The two following corollaries may then be useful in the stability analysis of Takagi–Sugeno fuzzy control systems.

Corollary 1 *Given a system represented by a polytopic linear differential inclusion, if the inequalities (6.29) are satisfied for some η and $\gamma = 0$, then:*

- *if $\eta > 0$, then the system is input strictly passive;*
- *if $\eta = 0$, then the system is passive;*
- *if $\eta < 0$, then the system itself is not passive, but passivity can be recovered via a parallel combination adding to this system a direct throughput of gain $(-\eta)$.*

Corollary 2 *A static system represented by a polytopic linear algebraic inclusion with positive definite diagonal matrices D_i is input and/or output strictly passive.*

Proof Each of the inequalities (6.31) can now be replaced by

$$\eta I + \gamma M D_i - D_i \leq 0,$$

where M is a positive definite diagonal matrix whose entries are upper bounds of the corresponding entries of the matrices D_i, $i = 1, 2, \ldots, r$. The above inequalities are satisfied in the following cases:

1. with $\gamma = 0$, if $\eta(>0)$ is less than the smallest diagonal entry of the positive definite matrix D_i;
2. with $\eta = 0$, if $\gamma(>0)$ is less than the inverse of the largest diagonal entry of the positive definite matrix M;
3. with positive values of η and γ less than the above values.

Therefore input strict passivity of the system is achieved with η up to a lower bound of the diagonal entries of the matrices D_i, $i = 1, 2, \ldots, r$, output strict passivity with γ up to the inverse of an upper bound of the diagonal entries of the matrices D_i, $i = 1, 2, \ldots, r$, and strict passivity for lower nonzero values of both η and γ.

The above corollary may be useful for decoupled static output feedback control of MIMO plants. Note that using the extended definition of input or output strict passivity one can replace the scalar quantities η and γ by non-negative definite diagonal matrices, which will relate the bounds on each entry of these matrices to the corresponding diagonal entries of D_i, $i = 1, 2, \ldots, r$. This corollary can also be generalized to non-diagonal matrices provided that the matrices $D_i + D_i^T$ are positive definite, input strict passivity being achieved with $\gamma = 0$ and $\eta(>0)$ less than the smallest eigenvalue of $(D_i + D_i^T)/2$, and output strict passivity with $\eta = 0$ and $\gamma(>0)$ less than the smallest eigenvalue of a generalized eigenvalue problem.

6.4 STABILITY OF FEEDBACK CONTROL WITH FUZZY CONTROLLERS

The passivity properties of fuzzy controllers established in the previous section can now be exploited in investigations of the stability of feedback control loops including fuzzy controllers. As in the previous section the three following cases will be treated separately:

- Mamdani controllers viewed as QPI controllers;
- Mamdani controllers viewed as DPS controllers;
- Takagi–Sugeno fuzzy control systems.

Considering a feedback control loop consisting of a continuous-time plant controlled by a fuzzy controller, and using as state, input and output variables of the plant and the controller the deviations in the actual variables from their values in a given steady-state operation, allows us to set to zero any exogenous signal, such as the controller setpoint, steady-state external disturbances and/or any possible control bias. Therefore, investigations of the stability of this feedback control system can be restricted to stability analysis of the zero-state equilibrium of a free control loop. The latter can then be represented by Figure 6.2 (with $d_1 = 0$ and $d_2 = 0$), or by Figure 6.3, where H_1 is a non-linear controller and H_2 is a dynamic system. Note that H_1 and H_2 may be different from the actual controller and plant to be controlled. For example, with the DPS Mamdani controller of Figure 6.5, H_1 would include the static non-linear function $\Psi(e_i, e_d)$, while the linear compensator $C(s)$ would be included in H_2 as a serial multiplier cascaded with the plant to be controlled (see Figure 6.3). More generally, if the controller can be decomposed into two cascaded parts with the first one being a stable and inverse-stable fixed dynamic subsystem, then the latter may be cancelled by a serial multiplier Z^{-1} in H_1 and recovered as a dynamic system equivalent to a serial multiplier Z cascaded with the plant in H_2. Therefore, in what follows the expression *compensated system* will be used to denote the plant to be controlled followed by a compensator whose input–output response can be represented by a serial multiplier Z. Moreover, if the part of the controller kept in H_1 is output strictly passive, then a zero-shifting loop transformation as in Figure 6.3 is also possible. Stability results can then be obtained by analysis of the passivity properties of the two subsystems H_1 and H_2 as represented in Figure 6.3.

6.4.1 Feedback Control with QPI Mamdani Controllers

The passivity of a QPI controller with e_2 as input variable was established in Theorem 5. Such a controller can then be used for H_1 in Figure 6.2, H_2 being the plant to be controlled and e_2 the control error signal. In that case the QPI controller is viewed as a fuzzy PI controller and from Theorem 2 the control loop is stable if the plant is passive. However, the QPI controller can also be viewed as a fuzzy PD controller cascaded with a multiplier $Z^{-1}(s) = 1/sT_c$ in the scheme of Figure 6.3. In this case, the multiplier $Z(s) = sT_c$ would be cascaded at the plant output in H_2 and the input e_2 is proportional to the derivative of the error signal or of the plant output. Therefore, in the latter case, again from Theorem 2, the control loop is stable if the plant is passive with respect to the derivative of its true output; this property is common to most mechanical systems and more generally to Euler–Lagrange systems [32, 34]. More generally the controller may be a QPI controller preceded by a fixed two-output compensator, which can be represented by a serial multiplier Z. The latter then is cancelled by the serial multiplier Z^{-1} in H_1 and H_2 becomes the compensated system, consisting of the plant followed by the compensator Z; e_2 is then obtained from the output of the compensated system. Therefore stability is achieved if the compensated system is passive. In all cases, from Theorem 2 asymptotic stability is achieved either with strict passivity of H_2, or with input or output strict passivity on condition of zero-state detectability of the open-loop system (the latter condition implies that $u = 0$ and $e_2 = 0$ entail $e_1 = 0$). These results are summarized in the following theorem:

Theorem 8 *A fuzzy control loop with a QPI Mamdani controller is stable in the following cases:*

1. *the plant is passive with respect to its output (the controller is used as a PI controller);*
2. *the plant is passive with respect to the derivative of its ouput (the controller is used as a PD controller);*
3. *a compensated system consisting of the plant cascaded with a stable compensator is passive.*

Asymptotic stability follows either from strict passivity of the plant (or the compensated system) or from input or output strict passivity on condition of open-loop zero-state detectability.

This result holds for SISO plants controlled by a TISO Mamdani fuzzy controller and also for MIMO plants with decoupled fuzzy control by a set of TISO Mamdani controllers.

6.4.2 Feedback Control with DPS Mamdani Controllers

An equivalent representation of DPS Mamdani controllers was given in Figure 6.5(b). Then, referring to Figure 6.3 the compensator $C(s)$ is cancelled by a multiplier $Z^{-1}(s) = C^{-1}(s)$ in

H_1 and is recovered as the multiplier $Z(s) = C(s)$ in H_2. Moreover, according to Theorem 6, the part of the controller kept in H_1 is output strictly passive. From Theorem 3 output strict passivity is preserved by a positive feedback with gain $1/K, K > K_c$; with $K = K_c$, passivity is still guaranteed. Therefore, the zero-shifting loop transformation of Figure 6.3 is allowed with $D = 1/K, K \geq K_c$, so that H_2 will consist of the parallel connection of the throughput $D = 1/K$ with the compensated system consisting of the plant cascaded with the compensator $C(s)$. Conditions for the stability of the control loop are then given in the following theorem:

Theorem 9 *A fuzzy control loop with a DPS Mamdani controller is stable if the subsystem H_2 formed by the parallel connection of the compensated system and a direct throughput with gain $1/K, K \geq K_c$, is passive. Asymptotic stability follows if any one of the following (non-equivalent) additional conditions is satisfied:*

1. *the subsystem H_2 is strictly passive;*
2. *the subsystem H_2 is output strictly passive and either $K > K_c$ or the compensated system is zero-state detectable;*
3. *either $K > K_c$ or the subsystem H_2 is input strictly passive, and the compensated system is zero-state detectable and at least neutrally stable (the condition of at least neutral stability can be removed for DPD controllers).*

Proof The stability of the control loop is a straightforward consequence of Theorem 2. The different cases of asymptotic stability given here can be related to one or the other of the four additional conditions given in Theorem 2, the first case being an obvious consequence of the first condition and the second case with $K > K_c$, and hence H_1 being output strictly passive, a consequence of the second condition. As for the second case with $K = K_c$, it is related to the third condition of Theorem 2, since $e_i = 0 \Rightarrow u = 0$ for all DPS controllers, and hence zero-state detectability of $H_2 \cdot H_1$ results from that of the compensated system. In the third case, either H_1 is output strictly passive with $K > K_c$, or H_2 is input strictly passive; therefore, in relation to the fourth additional condition of Theorem 2, the only point to be discussed is the zero-state detectability of $H_1 \cdot (-H_2)$.

If the compensated system is zero-state detectable and neutrally stable its largest invariant set for $u = 0$ reduces to time-invariant states (to the zero state in case of strict stability); in other words, $u = 0$ entails that for $t \to \infty$, the state of H_2, and consequently e_z, e_i and e_d, will go to constant values, with $e_d = e_i$ in the steady state. By (6.16) $u = 0$ and $e_d = e_i$ entail $e_i = e_d = 0$, and hence the zero-state detectability of a DPS controller and of $H_1 \cdot (-H_2)$. For DPD controllers, from (6.17) $u = 0 \Rightarrow e_i = 0$, and hence $e_z = 0$, whatever e_d, so that zero-state detectability of the compensated system results in zero-state detectability of $H_1 \cdot (-H_2)$, even for systems that have imaginary poles other than zero.

This result is an improved version of Theorem 2 in [8]. It holds for SISO plants controlled by a TISO Mamdani fuzzy controller and also for MIMO plants with decoupled fuzzy control by a set of TISO Mamdani controllers. It may be noted than the stability conditions are less strong with PDS controllers than with QPI controllers: in the case of QPI controllers, passivity of the controlled plant or of the compensated system is required; in the case of PDS controllers, what is required is the passivity of the parallel connection of a throughput $1/K, K \geq K_c$, with the compensated system. This is not surprising since the class of DPS controllers is a special class of Mamdani controllers.

6.4.3 Feedback Control with Linear Takagi–Sugeno Controllers

Assume that a LTS fuzzy model of the plant to be controlled is available and that a LTS fuzzy controller has been derived from this model. Both are represented by state and output equations, respectively, (6.23), (6.24) for the model and (6.27), (6.28) for the controller. If both the controller and the plant to be controlled are passive, then the control loop is stable according to Theorem 2, and asymptotic stability follows if one of the additional conditions of Theorem 2 is satisfied. The passivity of the controller can be checked by means of the inequalities (6.29) of Theorem 7 or the equivalent inequalities (6.31), which leads to the following lemma:

Lemma 1 *A LTS fuzzy controller represented by the state and output equations (6.27), (6.28) is passive if there exist a positive semidefinite symmetric matrix $Q = Q^T \geq 0$ and non-negative scalars $\eta_c \geq 0$ and/or $\gamma_c \geq 0$ such that the r following inequalities are satisfied:*

$$\begin{bmatrix} A_{ci}^T Q + Q A_{ci} + \gamma_c C_{ci}^T C_{ci} & Q B_{ci} - C_{ci}^T/2 + \gamma_c C_{ci}^T D_{ci} \\ B_{ci}^T Q - C_{ci}/2 + \gamma_c D_{ci}^T C_{ci} & \eta_c I + \gamma_c D_{ci}^T D_{ci} - (D_{ci} + D_{ci}^T)/2 \end{bmatrix} \leq 0, \quad i = 1, 2, \ldots, r. \tag{6.33}$$

- With $\eta_c = 0$ and $\gamma_c = 0$, the controller is passive.
- With $\eta_c > 0$ and $\gamma_c = 0$, the controller is input strictly passive.
- For controllers with constant matrices $C_{ci} = C_c$ and $D_{ci} = D_c$:
  —with $\eta_c = 0$ and $\gamma_c > 0$, the controller is output strictly passive;
  —with $\eta_c > 0$ and $\gamma_c > 0$, the controller is strictly passive.

Then the following theorem expressing the stability conditions for a feedback control loop with a LTS fuzzy controller is a straightforward consequence of Theorem 2 and of the above lemma:

Theorem 10 *A feedback control loop with a LTS fuzzy controller is stable if the controller (H_1) and the plant (H_2) to be controlled are passive. Asymptotic stability can be guaranteed by any one of the following additional conditions:*

1. *the plant is strictly passive, or the controller is strictly passive ($\eta_c > 0$ and $\gamma_c > 0$);*
2. *the plant is output strictly passive, and the controller is output strictly passive ($\gamma_c > 0$ with $\eta_c = 0$);*
3. *the controller is output strictly passive ($\gamma_c > 0$ with $\eta_c = 0$), and the open-loop system $H_2 \cdot H_1$ is zero-state detectable;*
4. *either the plant is input strictly passive or the controller is output strictly passive ($\eta_c > 0$ with $\gamma_c = 0$), and the open-loop system $H_1 \cdot (-H_2)$ is zero-state detectable.*

The zero-state detectability of the LTS controller is equivalent to that of all the local controllers. Since the latter are LTI systems, checking their zero-state detectability is easy. It may be more difficult for the plant, which is a non-linear dynamic system. However, LTS fuzzy models are *universal approximators*, in the sense that such a model is able to represent the input–output response of a given dynamic system with arbitrary accuracy over a compact set of the state–space of the system [9, 46, 51]. Therefore, the zero-state detectability of the non-linear plant could be inferred from that of its LTS model, and this can be checked easily since the LTS model is a collection of linear models.

Here also the fuzzy controller can be preceded by a fixed dynamic compensator, which may be included in H_2 as a serial multiplier cascaded with the plant to be controlled. The compensator may have been selected *a priori*, before identification of a LTS model. In that case, the LTS model to be identified will be that of the compensated system consisting of the plant and the compensator, and the LTS controller will be designed according to this model. Therefore the results of Theorem 10 can be applied to the compensated system instead of the controlled plant. In most cases, however, the compensator will be selected from a knowledge of the LTS model of the plant to be controlled. Unless identification of a new LTS model includes the compensator, the latter, assuming that it is linear and time-invariant, can be represented by state and output equations which are merged into that of all the local models. This leads to a *compensated LTS model*, consisting of a fixed compensator combined with a LTS model of the plant to be controlled, and now the LTS controller is designed according to this compensated model. If the latter and the controller are passive, then the stability of the feedback system consisting of the compensated LTS model and the LTS controller follows from Theorem 2. The passivity of the LTS controller can be checked by applying Lemma 1. The passivity of the compensated LTS model can be checked by applying the inequalities (6.29) of Theorem 7 to their state and output equations of this compensated LTS model, resulting in the following lemma:

Lemma 2 *A LTS model represented by the state and output equations (6.23), (6.24) is passive if there exist a positive semidefinite symmetric matrix $P = P^T \geq 0$ and nonnegative scalars $\eta_m \geq 0$ and/or $\gamma_m \geq 0$ such that the r following inequalities are satisfied:*

$$\begin{bmatrix} A_i^T P + P A_i + \gamma_m C_i^T C_i & P B_i - C_i^T/2 \\ B_i^T P = C_i/2 & \eta_m I \end{bmatrix} \leq 0, \quad i = 1, 2, \ldots, r. \quad (6.34)$$

- With $\eta_m = 0$ and $\gamma_m = 0$, the LTS model is passive.
- With $\eta_m > 0$ and $\gamma_m = 0$, the LTS model is input strictly passive.
- For models with constant matrices $C_i = C$:
  —with $\eta_m = 0$ and $\gamma_m > 0$, the LTS model is output strictly passive;
  —with $\eta_m > 0$ and $\gamma_m > 0$, the LTS model is strictly passive.

Conditions for the stability of a feedback system consisting of a compensated LTS model and the associated LTS controller can then be derived from Theorem 2 and expressed as follows:

Theorem 11 *A feedback system consisting of a LTS model and a LTS fuzzy controller is stable if the controller (H_1) and the model (H_2) are passive. Asymptotic stability can be guaranteed by any one of the following additional conditions assuming, in the case of (output) strict passivity, constant matrices C and D in the controller, and/or a constant matrix C in the model (the latter has no throughput: $D = 0$):*

1. *the model is strictly passive, or the controller is strictly passive:*

$$(\eta_m > 0 \text{ and } \gamma_m > 0) \quad \text{or} \quad (\eta_c > 0 \text{ and } \gamma_c > 0);$$

2. *the model is output strictly passive, and the controller is output strictly passive:*

$$(\gamma_m > 0 \text{ with } \eta_m = 0) \quad \text{and} \quad (\gamma_c > 0 \text{ with } \eta_c = 0);$$

3. *the controller is output strictly passive:*

$$(\gamma_c > 0 \text{ with } \eta_c = 0),$$

and the open-loop system $H_2 \cdot H_1$ is zero-state detectable;

4. *either the model is input strictly passive or the controller is output strictly passive:*

$$(\eta_m > 0 \text{ with } \gamma_m = 0) \quad or \quad (\gamma_c > 0 \text{ with } \eta_c = 0),$$

and the open-loop system $H_1 \cdot (-H_2)$ is zero-state detectable.

This theorem does not prove the stability of the actual control loop, which consists of the actual plant and the LTS controller. However, as previously said, LTS models are universal approximators, and it can be reasonably assumed that the input–output response of the actual plant is close to that of its LTS model. Moreover, stability conditions obtained by the Lyapunov or passivity-based approaches have intrinsic robustness features in the sense that the stability of the actual control loop is preserved in spite of small perturbations or uncertainties in the model of the plant.

In the case of static output-feedback control, a more powerful result can be obtained by combining the above theorem with Corollary 2. The LTS controller then reduces to the set of matrices D_i. Therefore, from Corollary 2 the controller is output passive with a proper selection of γ, or more generally of some diagonal matrix D_c, so that the zero-shifting transformation of Figure 6.3 is possible with $D = \gamma I$ or $D = D_c$. This leads to the following corollary:

Corollary 3 *A feedback system consisting of a LTS model and a static output-feedback LTS controller is stable if the parallel connection H_2 of the LTS model and a throughput matrix D selected as said above is passive. Asymptotic stability can be guaranteed by any one of the following additional conditions:*

1. H_2 is strictly passive, or the controller is strictly passive;
2. H_2 is output strictly passive, and the controller is output strictly passive;
3. the controller is output strictly passive, and the open-loop system $H_2 \cdot H_1$ is zero-state detectable;
4. either the model is input strictly passive or the controller is output strictly passive, and the open-loop system $H_1 \cdot (-H_2)$ is zero-state detectable.

The above result is important for all applications where the dynamic part of the controller is fixed, so that it can be merged into the compensated LTS model. The class of systems for which the corollary is applicable has been enlarged by the parallel connection with a throughput matrix.

6.5 APPLICATIONS

The stability conditions given in the previous sections involve the passivity of the plant to be controlled or of some compensated system. This passivity can be proved or checked by well-established techniques when the plant to be controlled is a LTI system or for Euler–Lagrange systems. These two cases will be considered here.

6.5.1 Control of LTI Systems by Fuzzy Controllers

It is assumed that the plant to be controlled is a stable and zero-state detectable LTI dynamic system and that the control system includes a fuzzy controller and possibly some fixed stable and inverse stable dynamic compensator cascaded between the plant output and the fuzzy controller input. This compensator then can be combined with the plant to be controlled to make what was called in the previous sections the *compensated system*. The latter can be represented by its transfer function $L(s) = C(s)P(s)$, where $P(s)$ is the transfer function of the controlled plant and $C(s)$ that of the compensator. As for the fuzzy controller it may be either a QPI or DPS Mamdani controller or a passive linear Takagi–Sugeno controller, for example a static output-feedback LTS controller. Therefore this controller is passive, and often even output strictly passive, so that stability of the control loop can be guaranteed on only the condition of passivity of the compensated system (with possibly a parallel throughput related either to some sectorial upper bound of the input–output mapping of the Mamdani controller or to some upper bound on the gains of the local controllers in a LTS controller). This stability condition can be expressed in terms of a frequency response condition derived from Theorem 4:

Theorem 12 *A control loop consisting of a passive fuzzy controller, a fixed stable and inverse stable LTI compensator $C(s)$ and a zero-detectable controlled plant $P(s)$ is asymptotically stable if:*

1. $[s\,P(s)]_{s=0} \geq 0$;
2. $\forall \omega \geq 0$, with $j\omega$ not being a pole of $P(s)$, $Re[C(j\omega)P(j\omega)] + 1/K_c > 0$, with K_c being:
   - a sectorial upper bound of the input-output mapping for a DPS Mamdani controller;
   - ∞ for a QPI Mamdani controller;
   - an upper bound on the gains of the local controllers for a static output feedback Takagi–Sugeno controller;
3. either $P(s)$ is strictly or neutrally stable or for a DPD controller, $P(s)$ is stable and all the residues of $C(s)P(s)$ associated to imaginary poles of $P(s)$ are positive.

This theorem was proven in [8] for SISO control loops with DPS Mamdani controllers. It is a straightforward consequence of the stability conditions given in the previous section, and by replacing transfer functions by transfer matrices it can be extended to decoupled control of MIMO plants by a set of TISO Mamdani controllers or by a passive MIMO static output feedback Takagi–Sugeno controller. In the case of SISO control loops this theorem has a a straightforward graphical interpretation:

The Nyquist plot of the transfer function $L(s)$ of the compensated system must stay on the right-hand side of a vertical line of abscissa $-1/K_c$.

Therefore, the compensator must be selected to distort the Nyquist plot of $L(s)$ and the abscissa of the vertical line which touches this Nyquist plot from the left-hand side determines an upper bound on the admissible values of the equivalent gain K_c of the fuzzy controller. In the frequent case of PD-like fuzzy control, where $C(s) = 1 + s\,T_c$, this stability result can be written as the well-known Popov stability condition with its nice graphical interpretation [25, 36]. However, as opposed to the classical Popov criterion, here $1 + s\,T_c$ is not the transfer function of a fictitious multiplier but that of the actual compensator: the parameters of the Popov critical line are related to some design parameters of the fuzzy controller. The previous results can be extended to the case of PI-like fuzzy control by defining $P(s) = G(s)/s$, where $G(s)$ is the transfer function of the actual plant to be controlled.

Such stability conditions allow the designer of the control system to check if stability is robust with respect to perturbations in the plant to be controlled or uncertainties in its model, in other words to check if stability holds when the transfer function $P(s)$ of the actual plant differs from the nominal model $\hat{P}(s)$ used for designing the controller. For parametric uncertainties, where some parameters of $P(s)$ can take their values within given intervals, one may draw the Nyquist plot of $C(s)P(s)$ for all perturbed systems and check that the right-hand boundary of the family of all these plots stays on the left-hand side of the vertical line with abscissa $-1/K_c$. For non-parametric uncertainties characterized by a positive scalar function $\delta(\omega)$ such that

$$\forall \omega \geq 0, \quad |P(j\omega) - \hat{P}(j\omega)| < \delta(\omega),$$

straightforward calculations lead to the following robust stability condition:

$$\forall \omega \geq 0, \quad \operatorname{Re}[C(j\omega)\hat{P}(j\omega)] + \frac{1}{K_c} > \delta(\omega)|C(j\omega)|,$$

which obviously is stronger than the condition of Theorem 12; the latter can be viewed as a particular case where $\delta(\omega)$ reduces to zero. Discussion of the use of filtered derivatives to cope with unmodelled fast dynamics is then possible.

6.5.2 Fuzzy control of Euler–Lagrange Systems

Many mechanical or electro-mechanical systems, such as robot manipulators, electrical drives, vehicles, etc. are Euler–Lagrange systems [18], which can be described by Lagrange's equations:

$$\frac{d}{dt}\left[\frac{\partial L_p(q_p, \dot{q}_p)}{\partial \dot{q}_p}\right] - \frac{\partial L_p(q_p, \dot{q}_p)}{\partial q_p} = Q_p,$$

where

- $q_p \in \mathbf{R}^n$ is the vector of plant generalized coordinates, describing the position or configuration of the system;
- $Q_p \in \mathbf{R}^n$ is the vector of external "forces" applied to this system;
- $L_p(q_p, \dot{q}_p) = T_p(q_p, \dot{q}_p) - V_p(q_p)$ is the plant Lagrangian function;
- $T_p(q_p, \dot{q}_p) = \frac{1}{2}\dot{q}_p^T M_p(q_p)\dot{q}_p$ is the kinetic energy, with $M_p(q_p) = M_p^T(q_p) > 0$ being the positive definite inertia matrix of the system;
- $V_p(q_p)$ is the potential energy, assumed to be twice differentiable and bounded from below: $V_p(q_p) + c > 0$ for some $c \in \mathbf{R}$.

In general, the external forces consist of control actions and dissipative forces:

$$Q_p = N_p u_p - \frac{\partial F_p(\dot{q}_p)}{\partial \dot{q}_p},$$

where $u_p \in \mathbf{R}^{m_p}$ is the vector of control variables, $N_p \in \mathbf{R}^{n_p \times m_p}$ is a full column matrix mapping the inputs to the generalized coordinates ($N_p = I_n$ if the system is fully actuated), and $F_p(\dot{q}_p)$ is a Rayleigh dissipation function defining a memoryless passive operator. In [32] and [34] it is shown that such an Euler–Lagrange system defines a passive operator from the inputs u_p to the actuated generalized velocities $N_p^T \dot{q}_p$, this operator being output strictly passive when the Rayleigh dissipation function defines an input strictly passive operator.

Therefore, it is possible to control Euler–Lagrange systems with fuzzy control systems; the latter may consist of a set of TISO Mamdani controllers, one for each actuated coordinate, or it may be a passive MIMO Takagi–Sugeno controller, using the generalized coordinates and velocities as controller inputs. As a consequence of the stability conditions established in the previous section, it can be shown that the zero state of Euler–Lagrange systems controlled by such fuzzy controllers is a stable equilibrium state (asymptotically stable if an additional condition of zero-state detectability is fulfilled) under any one of the two following conditions:

1. $q_p = 0$ is a strict or at least neutral equilibrium position of the plant alone;
2. the fuzzy controller is either a DPS Mamdani controller with a lower sectorial property or a Takagi–Sugeno controller, such that the sum of the potential energy of the plant and the storage function of the controller is minimum for $q_p = 0$, $\dot{q}_p = 0$.

Many applications of fuzzy control to mechanical systems are reported in the literature; see, for example, [4], [12], [19], [20], [23], [42], and [48], and also [35] which contains many useful references.

6.6 CONCLUSIONS

Many Mamdani fuzzy controllers can be reduced to a set of single-input–single-output passive dynamic non-linear controllers. This passivity property is a consequence of some features of the input–output mapping of the controller, which can be proven for some particular classes of fuzzy controllers and can be verified by inspection of the mapping for other controllers. On the other hand, control systems with linear Takagi–Sugeno controllers can be analyzed by an approach based on polytopic linear differential inclusions. The passivity of such controllers can then be investigated by some well-established techniques of linear matrix inequalities and linear algebra. Therefore the theory of passivity turns out to provide a unifying approach for stability analysis of fuzzy control loops, including either Mamdani or Takagi–Sugeno controllers. General results on the stability of feedback loops consisting of two interconnected passive subsystems can then be used to prove the stability of such fuzzy control loops. In some applications the plant to be controlled is passive; it is the case for many mechanical and electro-mechanical systems and more generally for Euler–Lagrange systems. In other applications, for example in industrial process control, the plant itself is not passive, but it is possible to insert between the plant and the fuzzy controller a fixed compensator which makes passive the compensated system consisting of the plant and the compensator. Moreover, with DPS Mamdani controllers and some Takagi–Sugeno controllers which are output strictly passive, the class of systems which can be made passive can be enlarged thanks to some zero-shifting transformation. For linear time-invariant controlled systems, this approach provides a frequency response condition, similar to the 'circle criterion' or Popov-like conditions found by other authors when the two inputs of the fuzzy controller are the error signal and its derivative; that is to say, when the compensator is a proportional-derivative filter. With this passivity-based approach, more general compensators can be considered, for example compensators providing filtered signals. Moreover, the stability condition can be strengthened to include robustness with respect to perturbations in the controlled plant or uncertainties in the model of this plant. Therefore, such a passivity-based approach may give some rigorous mathematical proof of the common statement that 'fuzzy control systems are robust and capable of coping with uncertainties and perturbations in the system to be controlled'. This statement is true in all the cases where the controlled system can be made passive by techniques such as those described in this chapter.

ACKNOWLEDGMENTS

This chapter presents research results obtained in the frame of the Belgian Program on InterUniversity Poles of Attraction initiated by the Belgian State, Prime Minister's Office for Science, Technology and Culture. The scientific responsibility rests with its authors.

APPENDIX

The input–output response of many two-input–single-output fuzzy controllers can be represented by a non-linear static mapping $\Phi(\cdot, \cdot)$ characterized by the following properties:

1 Inputs

Two scaled inputs e_1 and e_2 scaled to the same range $[-L, L]$. The latter is symmetrical with respect to zero and covered, for e_1 as well as for e_2, by $2N+1$ input fuzzy sets whose linguistic names can be arbitrarily assigned as E_i, $i = -N, \ldots, -1, 0, 1, \ldots, +N$. The associated membership functions $\mu_{E_i}(e)$ have the following properties:

for all $i, j = -N, \ldots, -1, 0, 1, \ldots, +N$,

$$0 \leq \mu_{E_i}(e) \leq 1,$$

$$\sum_{i=-N}^{i=+N} \mu_{E_i}(e) = 1,$$

$$|i - j| > 1 \Rightarrow \mu_{E_i}(e)\mu_{E_j}(e) = 0,$$

and $e > L \Rightarrow \mu_{E_N}(e) = 1$ and $e < -L \Rightarrow \mu_{E_{-N}}(e) = 1$, respectively. These properties simply mean that each input value belongs to at most two adjacent fuzzy sets, with complementary membership grades. There is no assumption on the shape of the membership functions, except a 'monotonicity' assumption: $\forall e' \neq e, \forall \lambda \in (0, 1)$,

$$\mu_{E_i}(\lambda e' + (1 - \lambda)e) \geq \min\{\mu_{E_i}(e'), \mu_{E_i}(e)\},$$

with a strict inequality for E_0 to guarantee the unicity of the zero-state equilibrium of the fuzzy control system. Therefore, except for E_0, very general membership functions, such as bell-shaped, trapezoidal or asymmetric triangular functions, are allowed.

2 Rule Base

N^2 fuzzy control rules in the form :

If $e_1(t)$ **is** E_i **AND** $e_2(t)$ **is** E_j **Then** $u(t)$ **is** $Out_{\chi(i,j)}$,

where $\chi(i,j)$ is a function whose value at i,j is an integer, relating the indexes i,j of the input fuzzy sets to the index of the output fuzzy set $Out_{\chi(i,j)}$ with center value $U_{\chi(i,j)}$. This function has the following properties:

$$\chi(i,j) = -\chi(-i,-j), \quad \forall i,j,$$
$$\chi(0,0) = 0,$$
$$j(\chi(i,j) - \chi(i,0)) > 0, \quad \forall i,j > 0,$$
$$i(\chi(i,j) - \chi(0,j)) > 0, \quad \forall i,j > 0,$$

with $U_0 = 0$, $U_i = -U_{-i}$ and $U_i > U_j$ for $i > j$.

3 Output

A single controller output whose real value is obtained (via some scale factor) by the *mean* defuzzification algorithm, using either the *min* or the *product* inference rule:

$$u = \frac{\sum_{i,j} \mu_{E_i}(e_1) * \mu_{E_j}(e_2) \cdot U_{\chi(i,j)}}{\sum_{i,j} \mu_{E_i}(e_1) * \mu_{E_j}(e_2)},$$

where $*$ denotes either the minimum or the usual product of the two input membership grades. Note that the properties of the membership functions entail that at most four rules can be fired at the same time, and hence each of the two sums in the above defuzzification formula contains at most four terms.

Remark 1 The previous assumptions include many fuzzy controllers considered in the literature (see, for example, [27], [30], [33], [52]), whose look-up table has an odd symmetry and a 'monotonicity' in linguistic terms, for example:

e_2/e_1	E_{-3}	E_{-2}	E_{-1}	E_0	E_1	E_2	E_3
E_{-3}	U_{-6}	U_{-6}	U_{-3}	U_{-2}	U_{-1}	U_{-1}	U_{-1}
E_{-2}	U_{-4}	U_{-6}	U_{-3}	U_{-2}	U_{-1}	U_0	U_{-1}
E_{-1}	U_{-4}	U_{-4}	U_{-2}	U_{-1}	U_0	U_1	U_0
E_0	U_{-1}	U_{-3}	U_{-1}	U_0	U_1	U_3	U_1
E_1	U_0	U_{-1}	U_0	U_1	U_2	U_4	U_4
E_2	U_1	U_0	U_1	U_2	U_3	U_6	U_4
E_3	U_1	U_1	U_1	U_2	U_3	U_6	U_6

It should be noted that the fuzzy controller output is bounded as a consequence of the following two facts:

1. for each input the fuzzy partitions, so the consequents $U_{f(i,j)}$ are in finite number;

2. at every time only at most four rules can be activated simultaneously, due to the simple superposition of membership functions.

Therefore one can distinguish a mesh of the mapping with a finite number of partitions. The fuzzy controller output then can be written as follows:

$$u(t) = \Phi(e_1(t), e_2(t)) = \frac{\sum_{k=-N}^{k=N} \sum_{l=-N}^{l=N} U_{\chi(k,l)} \mu_{E_k}(e_1) * \mu_{E_l}(e_2)}{\sum_{k=-N}^{k=N} \sum_{l=-N}^{l=N} \mu_{E_k}(e_1) * \mu_{E_l}(e_2)\}},$$

where $\mu_{E_i}(e_j)$ represents the value of membership function μ_{E_i} for an input value $e_j(t)$. From the above formula it is obvious that at every time the output is a convex combination of values $U_{\chi(k,l)}$ and therefore from symmetry, $u(t) \in [-M, M]$, where $M = \max_{\{k,l\}}\{U_{\chi(k,l)}\}$. Also from this convexity property, local variations in the function are bounded so that the function is locally Lipschitz. Since the number of cells is finite, the mapping is globally Lipschitz.

Continuity of the mapping can also be proved. Note that the membership functions are continuous functions and the operators min and prod maintain continuity. The denominator of the above combination is never zero, and hence the controller output can be written as an algebraic combination of continuous functions; hence it is continuous. Consequently, the non-linear input–output mapping $\Phi(\cdot, \cdot)$ is a globally Lipschitz continuous bounded function of two variables; however, in general it is not differentiable.

Now let us prove the positivity properties (6.6). First, let us assume that one of the inputs is constant and then prove the property by varying the other input. For convenience set $e_2 = \text{constant} > 0$. The symmetry properties of the look-up table allow us to assume $e_1 > 0$ without loss of generality, so that the following rules will be fired:

- **If** e_1 is E_i and e_2 is E_j **Then** u is $U_{\chi(i,j)}$;
- **If** e_1 is E_i and e_2 is E_{j+1} **Then** u is $U_{\chi(i,j+1)}$;
- **If** e_1 is E_{i+1} and e_2 is E_j **Then** u is $U_{\chi(i+1,j)}$;
- **If** e_1 is E_{i+1} and e_2 is E_{j+1} **Then** u is $U_{\chi(i+1,j+1)}$,

where $i, j \geq 0$. Thus the output of the fuzzy controller will be

$$u = \Phi(e_1, e_2) = \frac{\sum_{k=i,i+1} \sum_{l=j,j+1} U_{\chi(k,l)} \mu_{E_k}(e_1) * \mu_{E_l}(e_2)}{\sum_{k=i,i+1} \sum_{l=j,j+1} \mu_{E_k}(e_1) * \mu_{E_l}(e_2)}.$$

Then calculate the expression:

$$e_1(\Phi(e_1, e_2) - \Phi(0, e_2))$$

$$= e_1 \left(\frac{\sum_{k=i,i+1} \sum_{l=j,j+1} U_{\chi(k,l)} \mu_{E_k}(e_1) * \mu_{E_l}(e_2)}{\sum_{k=i,i+1} \sum_{l=j,j+1} \mu_{E_k}(e_1) * \mu_{E_l}(e_2)\}} - \frac{\sum_{l=j,j+1} U_{\chi(0,l)} \mu_{E_l}(e_2)}{\sum_{l=j,j+1} \mu_{E_l}(e_2)} \right)$$

$$= e_1 \left(\frac{\sum_{k=i,i+1} \sum_{l=j,j+1} (U_{\chi(k,l)} - U_{\chi(0,l)}) \mu_{E_l}(e_2) \mu_{E_k}(e_1) *, \mu_{E_l}(e_2)}{(\sum_{k=i,i+1} \sum_{l=j,j+1} \mu_{E_k}(e_1), \mu_{E_l}(e_2))(\sum_{l=j,j+1} \mu_{E_l}(e_2))} \right)$$

$$\geq e_1 \frac{\min_{k \neq 0} |U_{\chi(k,l)} - U_{\chi(0,l)}| \sum_{l=j,j+1} \mu_{E_l}(e_2) \sum_{k=i,i+1} \mu_{E_k}(e_1) * \mu_{E_l}(e_2)}{\sum_{l=j,j+1} \mu_{E_l}(e_2) \quad (\sum_{k=i,i+1} \sum_{l=j,j+1} \mu_{E_k}(e_1), \mu_{E_l}(e_2))}$$

Consequently:

$$e_1(U_{\chi(k,l)} - U_{\chi(0,l)}) \geq 0, \quad \forall e_1.$$

Moreover, if $e_1 > 0 (e_1 < 0)$, then for all $k > 0 (k < 0) U_{\chi(k,l)} - U_{\chi(0,l)} > 0 \, (U_{\chi(k,l)} - U_{\chi(0,l)} < 0)$ and hence

$$e_1(\Phi(e_1, e_2) - \Phi(0, e_2)) > 0, \quad \forall e_1 \neq 0, \forall e_2.$$

Furthermore, from the above property and from the Lipschitz property:

$$e_1(\Phi(e_1, e_2) - \Phi(0, e_2)) \leq \lambda' e_1^2, \quad \forall e_1, e_2.$$

REFERENCES

1. Atherton, D. P., A describing function approach for the evaluation of fuzzy logic control, In: *Proceedings of the American Control Conference*, 1993, pp. 765–766.
2. Boyd, S., El Ghaoui, L., Feron, E., and Balakrishnan, V., *Linear Matrix Inequalities in System and Control Theory*, SIAM, Philadelphia, 1994.
3. Brae, M. and Rutherford, D. A., Theoretical and linguistic aspects of the fuzzy logic controller, *Automatica*, **15**, 1979, 533–577.
4. Brown, S. C. and Passino, K. M., Intelligent control for an acrobot, *Journal of Intelligent and Robotic Systems: Theory and Applications*, **18**, 1997, 209–248.
5. Byrnes, I., Isidory, A. and Willems, J. C., Passivity, feedback equivalence and the global stabilization of minimum phase nonlinear systems. *IEEE Transactions on Automatic Control*, **AC-36**(11), 1991, 1228–1240.
6. Calcev, G., Stability analysis of a fuzzy control system: a frequency approach, In: *Proceedings of the Ninth CSCS*, Bucharest, Romania, 1993, pp. 252–257.
7. Calcev, G., A passivity result for fuzzy control systems, In: *Proceedings of the Thirty-fifth IEEE Conference on Decision and Control*, 1996, pp. 2727–2728.
8. Calcev, G., Gorez, R. and De Neyer, M., Passivity approach to fuzzy control systems, *Automatica*, **34**(3), 1998, 339–344.
9. Castro, J. L., Fuzzy logic controllers are universal approximators, *IEEE transactions on Systems, Man and Cybernetics*, **25**(4), 1995, 629–635.
10. Chen, Y. Y. and Tsao, R. C., A description of the dynamical behaviour of fuzzy systems, *IEEE Transactions on Systems, Man and Cybernetics*, **SMC-19**(4), 1989, 745–755.
11. Chiang, H., Hirsch, M. and Wu, F., Stability regions of nonlinear autonomous dynamical systems, *IEEE Transactions on Automatic Control*, **AC-33**(1), 1988, 16–27.
12. Chiu, S. and Chand, S., Fuzzy controller design and stability analysis for an aircraft model, In: *Proceedings of the American Control Conference*, 1991, pp. 821–826.
13. Desoer, C. A. and Vidyasagar, M., *Feedback Systems: Input–Output Properties*, Academic Press, New York, 1975.
14. De Neyer, M. and Gorez, R., Comments on 'Practical design of nonlinear fuzzy controllers with stability analysis for regulating processes with unknown mathematical models', *Automatica*, **32**(11), 1996, 1613–1614.
15. Farinwata, S. S. and Vachtsevanos, G., Robust stability of fuzzy logic control systems, In: *Proceedings of the American Control Conference*, Seattle, 1995, pp. 2267–2271.
16. Genesio, R., Tartaglia, M. and Vicino, A., On the estimation of asymptotic stability regions: State of the art and new proposals. *IEEE Transactions on Automatic Control*, **AC-30**(8), 1985, 747–755.
17. Gibson, J. E., *Nonlinear Automatic Control*, McGraw-Hill, New York, 1963.
18. Goldstein, H., *Classical Mechanics*, Addison-Wesley, Menlo Park, 1974.
19. Gorez, R. and De Neyer, M., Fuzzy control of robotic manipulators and mechanical systems, In: Tzafestas, S. G. and Venetsanopoulos, A. N. (eds.), *Fuzzy Reasoning in Information, Decision and Control Systems*, Kluwer Academic Publishers, Dordrecht, 1994, pp. 451–492.
20. Gorez, R., De Neyer, M., Godin, O., Calcev, G., Jen, C. W. and Johnson, D. A., Fuzzy control versus conventional control: Naive and experimental comparisons, *Journal* **A** *(Benelux Quarterly Journal on Automatic Control)*, **36**(3), 1995, 15–20.
21. Hill, D. J. and Moylan, P. J., Stability results for nonlinear feedback systems, *Automatica*, **13**, 1977, 377–382.

22. Hill, D. J., Dissipative nonlinear systems: Basic properties and stability analysis, In: *Proceedings of the Thirty-first IEEE Conference on Decision and Control*, 1992, pp. 3259–3264.
23. Hillsley, K. L. and Yurkovich, S., Vibration control of a two-link flexible robot arm, *Dynamics and Control*, **3**, 1993, 261–280.
24. Johansen, T. A., Fuzzy model-based control: Stability, robustness and performance issues, *IEEE Transactions on Fuzzy Systems*, **2**(3), 1994, 221–234.
25. Khalil, H. K., *Nonlinear Systems*, Prentice-Hall, Upper Saddle River, 1996.
26. Langari, R. and Tomizuka, M., Stability of fuzzy liguistic control systems, In: *Proceedings of the Twenty-ninth IEEE Conference on Decision and Control*, 1990, pp. 2185–2190.
27. Lim, J. T., Absolute stability of a class of nonlinear plants with fuzzy logic controllers, *Electronic Letters*, **28**(21), 1992, 1968–1970.
28. Lefschetz, S., *Stability of Nonlinear Control Systems*, Academic Press, New York, 1965.
29. Mamdani, E. H., Application of fuzzy algorithms for control of simple dynamic plant, *Proceedings of the IEE*, **121**(12), 1974, 1585–1588.
30. Melin, C., Stability analysis of fuzzy control systems: Some frequency criteria, In: *Proceedings of the Third ECC*, Roma, Italy, 1995, pp. 815–819.
31. Narendra, K. S. and Taylor, J. H., *Frequency Domain Criteria for Absolute Stability*, Academic Press, New York, 1973.
32. Nijmeijer, H. and van der Schaft, A., *Nonlinear Dynamical Control Systems*, Springer-Verlag, New York, 1990.
33. Opitz, H. P., Fuzzy-control and stability criteria, In: *Proceedings of EUFIT'93*, Aachen, Germany, 1993, pp. 130–135.
34. Ortega, R., Loria, A., Kelly, R. and Praly, L., On passivity-based output feedback global stabilization of Euler-Lagrange systems, *International Journal of Robust and Nonlinear Control*, (Special Issue on Control of Nonlinear Mechanical Systems). **5**(4), 1995, 313–325.
35. Passino, K. M. and Yurkovich, S., *Fuzzy Control*, Addison-Wesley, Menlo Parl, 1998.
36. Popov, V. M., *Hiperstabilitatea Sistemelor Automate*, Edit. Academiei, Bucharest, 1966.
37. Ray, K. and Majumder, D. D., Application of circle criteria for stability analysis of linear SISO and MIMO systems associated with fuzzy logic controller, *IEEE Transactions on Systems, Man and Cybernetics*, **SMC-14**(2), 1984, 345–349.
38. Sépulchre, R., Jankovic, M. and Kokotovic, P., *Constructive Nonlinear Control*, Springer-Verlag, Berlin, 1997.
39. Strang, G., *Linear Algebra and its Applications*, Harcourt Brace Jovanovich, Orlando, 1988.
40. Takagi, T. and Sugeno, M., Fuzzy identification systems and its application to modeling and control. *IEEE Transactions on Systems, Man and Cybernetics*, **SMC-15**(1), 1985, 116–132.
41. Tanaka, K. and Sugeno, M., Stability analysis and design of fuzzy control systems, *Fuzzy Sets and Systems*, **45**: 1992, 135–156.
42. Tanaka, K., Design of model-based fuzzy controller using Lyapunov's stability approach and its application to trajectory stabilization of a model car, In: Nguyen, H. Y., Sugeno, M., Tong, R. and Yager, R. R. (eds.), *Theoretical Aspects of Fuzzy Control*, Wiley, New York, 1995, pp. 31–50.
43. Tanaka, K., Ikeda, T. and Wang, H. O., Robust stabilization of a class of uncertain nonlinear systems via fuzzy control: Quadratic stabilizability, H^∞ control theory and linear matrix inequalities, *IEEE Transactions on Fuzzy Systems*, **4**(1), 1996, 1–13.
44. Vidyasagar, M., *Nonlinear Systems Analysis*, Prentice-Hall, Englewood Cliffs, 1993.
45. Wang, L., Zhang, H. and Xu, W., PAD-analysis of fuzzy control stability, *Fuzzy Sets and Systems*, **38**, 1990, 27–42.
46. Wang, L. X., Fuzzy systems are universal approximators, In: *Proceedings of the Thirty-first IEEE Conference on Decision and Control*, 1992, pp. 1163–1170.
47. Wang L. X., *Adaptive Fuzzy Systems and Control: Design and Stability Analysis*, Prentice-Hall, Englewood Cliffs, 1994.

48. Widjaja, M. and Yurkovich, S., Intelligent control for swing up and balancing of an inverted pendulum system, In: *Proceedings of the IEEE Conference on Control Applications*, 1995, pp. 534–542.
49. Willems, J. C., Mechanisms for the stability and instability in feedback systems, *Proceedings of the IEE*, **64**(1), 1976, 24–35.
50. Yamashita, T., Katoh, R., Singh, S. and Hori, T., Stability analysis of fuzzy control systems applying conventional methods, In: *Proceedings of IROS'91: International Workshop on Intelligent Robots and Systems*, 1991, pp. 1579–1584.
51. Ying, H., General fuzzy systems are functions approximators, In: *Proceedings of the Thirty-second Conference on Decision and Control*, 1993, pp. 1739–1742.
52. Ying, H., Practical design of nonlinear fuzzy controllers with stability analysis for regulating processes with unknown mathematical models, *Automatica*, **30**, 1994, 1185–1195.
53. Zadeh, L. A., Fuzzy sets, *Information and Control*, **8**, 1965, 338–353.
54. Zhao, J., System Modeling, Identification and Control using Fuzzy Logic, Ph.D. dissertation, University Louvain, Belgium, 1995.
55. Zhao, J., Gorez, R. and Wertz, V., Synthesis of fuzzy control systems based on linear Takagi–Sugeno fuzzy models, In: Murray-Smith, R. and Johansen, T. A. (eds.), *Multiple Model Approaches to Modelling and Control*, Taylor & Francis, London, 1997, pp. 307–336.

7
Frequency Domain Analysis of MIMO Fuzzy Control Systems

Anibal Ollero, Francisco Gordillo and **Javier Aracil**
Escuela Superior de Ingenieros
Camino de los Descubrimientos s/n, Sevilla-41092, Spain

7.1 INTRODUCTION

The design of fuzzy logic controllers (FLCs) is an open problem in which heuristics play an important role. In this way, many of the available design methods do not give fully satisfactory results because they lack the analytical background that would allow a deep analysis that could ensure good performance in all situations (or at least, in a wide variety of them).

Stability is the most important feature of feedback systems. The classical problem of feedback system design is to compensate the destabilizing effects of the combined interaction of closed loop causality and delays. To deal with the stability problem of fuzzy control systems some of the traditional stability analysis methods for non-linear control systems have been applied [6, 14, 15]. One of the first proposals was the use of the describing function method [11].

In order to analyze the global stability of a system it is desirable to know the existence of other equilibria different from the desired operation point as well as the occurrence of limit cycles (both stable and unstable). The relevance of this analysis must be pointed out as the existence of, for example, an unstable limit cycle around the nominal operating point will lead to this point being locally but not globally stable. Thus, many of the existing stability analysis methods would not detect the loss of the global stability.

To study the existence of equilibria other than the operating point, a method was proposed in [1] and [3]. On the other hand, the simple, well-known technique of the describing function can be used to predict the existence of limit cycles. This technique has been applied previously to the case of SISO fuzzy control systems (FCSs) [2, 7]. In such a case it has been shown that the intrinsic saturations which define the universe of discourse of the input variables play an important role in the emergence of limit cycles and new equilibrium points.

Fuzzy Logic. Edited by S. Farinwata, D. Filev and R. Langari
© 2000 John Wiley & Sons, Ltd

Furthermore, the approach proposed is based on frequency response techniques and then benefits from the ability of these techniques to consider pure delays without additional complexity.

The object of this chapter is to show how to extend these techniques to analyze MIMO FCSs. This chapter considers three classical methods to analyze the global stability of MIMO FCSs. In this way, not only the stability analysis is performed but also a robustness measurement can be achieved. The combination of the describing function method with singular values has been previously applied in [10] to the design of an H_∞ optimal controller of a non-linear system. In this chapter the conjunction of singular values and the describing function method is used to predict limit cycles.

In this chapter it is assumed that the control system can be separated into linear and non-linear, memoryless parts. This decomposition can be achieved in a wide variety of cases. For example, in [5] it is shown how Takagi–Sugeno control systems can be decomposed in the desired way.

The chapter is organized as follows. In Section 7.2 the general problem of multiple equilibria in fuzzy control systems is analyzed. In Sections 7.3 and 7.4 the existence of limit cycles is studied. Section 7.5 presents some conclusions.

7.2 MULTIPLE EQUILIBRIA IN MIMO FUZZY CONTROL SYSTEMS

The existence of other equilibria apart from the desired one plays an important role in the stability of a dynamic system. Figure 7.1 shows the phase portrait of a fuzzy system which presents a locally stable equilibrium point (the origin) and two saddle points. In this way, the origin has a limited attraction basin. It can be seen that there are trajectories which tend to infinity.

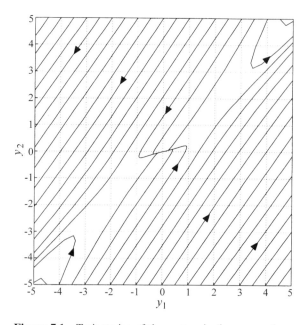

Figure 7.1 Trajectories of the system in the y_1–y_2 plane

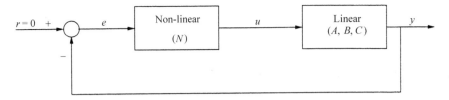

Figure 7.2 General decomposition of the system

Consider the MIMO system of Figure 7.2, where the input and the output of the linear block (u and y, respectively) have the same dimension m. The equations of the system are:

$$\dot{x} = Ax + Bu, \quad (7.1)$$
$$y = Cx, \quad (7.2)$$
$$e = -y, \quad (7.3)$$
$$u = N(e), \quad (7.4)$$

where $N(e)$ stands for the memoryless, non-linear part of the system.

The equilibria of the above system are given by the condition $\dot{x} = 0$; that is, $x = -A^{-1}Bu$, assuming A^{-1} exists. Then, at the equilibria $y = -CA^{-1}Bu$, but as $CA^{-1}B = G(0)$, the following equation is obtained:

$$e = -G(0)N(e) \quad (7.5)$$

For simplicity assume $m = 2$. In such a case equation (7.5) can be rewritten as

$$\begin{bmatrix} -e_1 \\ -e_2 \end{bmatrix} = \begin{bmatrix} g_{11}(0) & g_{12}(0) \\ g_{21}(0) & g_{22}(0) \end{bmatrix} \begin{bmatrix} n_1(e_1, e_2) \\ n_2(e_1, e_2) \end{bmatrix}. \quad (7.6)$$

As $n_1(0,0) = n_2(0,0) = 0$, then the operating point $e_1 = e_2 = 0$ is an equilibrium point. The problem at stake is whether there are more equilibria; that is, equilibria other than the operating point. That means solutions for e_1 and e_2 to equation (7.6) other than the origin. To search for these other solutions equation (7.6) will be rewritten as

$$\begin{bmatrix} n_1(e_1, e_2) \\ n_2(e_1, e_2) \end{bmatrix} = -\begin{bmatrix} \hat{g}_{11}(0) & \hat{g}_{12}(0) \\ \hat{g}_{21}(0) & \hat{g}_{22}(0) \end{bmatrix} \begin{bmatrix} e_1 \\ e_2 \end{bmatrix} \doteq -\hat{G}(0)e, \quad (7.7)$$

where $\hat{G}(0)$ stands for $G^{-1}(0)$, the inverse of $G(0)$. Equation (7.7) leads to

$$\begin{aligned} n_1(e_1, e_2) &= -\hat{g}_{11}(0)e_1 - \hat{g}_{12}(0)e_2, \\ n_2(e_1, e_2) &= -\hat{g}_{21}(0)e_1 - \hat{g}_{22}(0)e_2. \end{aligned} \quad (7.8)$$

This system of equations has a simple geometrical interpretation. The first equation can be thought of as the curve that results at the intersection of the plane $-\hat{g}_{11}(0)e_1 - \hat{g}_{12}(0)e_2 = z$ with the surface $n_1(e_1, e_2) = z$, projected onto the space (e_1, e_2). The same is true for the second equation, giving rise to another curve. If these two curves intersect each other at points other than the origin, then there are equilibria other than the operating point.

The stability of the equilibria can be analyzed by at least two methods. The simplest one is to determine the Jacobian matrix

$$J = \begin{bmatrix} \dfrac{\partial n_1}{\partial e_1} & \dfrac{\partial n_1}{\partial e_2} \\ \dfrac{\partial n_2}{\partial e_1} & \dfrac{\partial n_2}{\partial e_2} \end{bmatrix}$$

at the equilibria, and then, since at each equilibrium $J = -\hat{G}(0)$, the eigenvalues of $(A - BJC)$ will give its stability character.

7.3 FREQUENCY ANALYSIS OF LIMIT CYCLES

Another cause of the loss of global stability is the emergence of limit cycles. In what follows the analysis of limit cycles is considered, where it is assumed that the input–output relationship of the non-linear part (FLC in Fig 7.3) can be made additively separable.

Let $G(s)$ be the transfer matrix of the linear part and $N(a, \omega)$ the describing function matrix of the non-linear part where a is the vector of the input amplitudes, and ω the common frequency of all the inputs. As the non-linear part is assumed to be memoryless, $N(a, \omega)$ will not depend on ω and so will be written $N(a)$.

In order to analyze the existence of limit cycles in MIMO fuzzy systems, the harmonic balance is performed [12]. The corresponding equation is

$$(G(j\omega)N(a) + I)a = 0, \tag{7.9}$$

where $N(a)$ is a matrix of the form $N = [n_{ij}(a_j)]$. To solve this equation the method suggested by Mees [12] can be used. This method is based on the fact that equation (7.9) can only have a solution if $G(j\omega)N(a) + I$ has at least one zero eigenvalue. Assuming that $G(j\omega)$ is square and non-singular, and applying the Gershgorin theorem, the following stability condition is obtained for each ω:

$$|\hat{g}_{kk}(\omega) + n_{kk}(a_k)| > \sum_{i \neq k} |\hat{g}_{ik}(\omega) + n_{ik}(a_k)|, \quad \forall k, \tag{7.10}$$

where $[\hat{g}_{ij}(\omega)] = G^{-1}(j\omega)$. This condition is satisfied if

$$|\hat{g}_{kk}(\omega) + n_{kk}(a_k)| > \sum_{i \neq k} |\hat{g}_{ik}(\omega)| + |n_{ik}(a_k)|, \quad \forall k. \tag{7.11}$$

These stability conditions can be applied to MIMO FCSs in three different ways [13]:

- The first one is based on equation (7.11). The graphical interpretation of this condition (Figure 7.4) is that the Gershgorin bands of the linear part must not intersect a sort of Gershgorin band for the nonlinear part (in the present case the FLC).

- The second method, which is less conservative but more arduous than the previous one, is based on equation (7.10). In this case, the Gershgorin bands of $\hat{g}_{ii}(\omega) + n_{ii}(a_i)$ must not encompass the origin. Since these bands depend on a_i, a family of such bands is obtained.

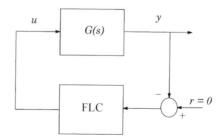

Figure 7.3 Structure of a MIMO Fuzzy Control System

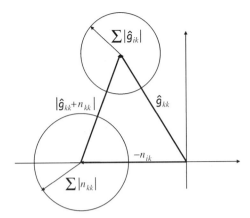

Figure 7.4 Interpretation of equation (7.11)

- The last method, which is based on the direct analysis of equation (7.9), consists in the analysis of the possible crossing of the characteristic loci of $G(\omega)N(a)$ through the point $(-1, 0)$. In fact, there is a family of characteristic loci parameterized in a. This is the hardest method, but it is more precise (with the approximations of the describing function methods) since it provides necessary and sufficient conditions for stability. Furthermore, this method does not require the transfer matrix G to be square; that is, the plant can have different numbers of inputs and outputs. With modern computers and available numerical tools (such as the Multivariable Frequency Domain Matlab Toolbox [4]) this method is affordable in many situations so it will be more suitable than the previous ones. Its drawback is that, in spite of its graphical character, this method does not supply useful robustness measurements because the eigenvalues of the open loop do not give reliable information about the robust stability of the closed loop. These measurements can be achieved by means of a different method based on the singular values, as is shown in the following section.

7.4 ROBUST ANALYSIS OF LIMIT CYCLES USING SINGULAR VALUES

For a linear transfer function $G(s)$ the singular values of $G(j\omega)$ are a function of ω. If the system contains non-linearities, then the describing function matrix can be used to obtain the singular values of the corresponding quasilinear transfer function [10].

Consider the system of Figure 7.5 in which FLC represents a fuzzy logic controller designed for the nominal model of the plant $G(s)$ and Δ represents the model error (this is usually called the multiplicative error model). Let $N(a)$ be the describing function matrix of the FLC.

Consider the feedback loop of Figure 7.5. In [8] it is proved that the describing function method does not predict limit cycles if the following conditions are satisfied:

1. there are no solutions to the harmonic balance equation for the nominal closed loop ($\Delta = 0$);

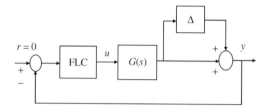

Figure 7.5 Fuzzy control loop with multiplicative error model

2. the model error satisfies

$$\bar{\sigma}(\Delta(j\omega)) < \frac{1}{\bar{\sigma}(G(j\omega)N(a)(I+G(j\omega)N(a))^{-1})}, \quad \forall \omega, \forall a. \tag{7.12}$$

Furthermore, there exists a rational transfer function matrix Δ that satisfies

$$\bar{\sigma}(\Delta(j\omega)) \leq \frac{1}{\bar{\sigma}(G(j\omega)N(a)(I+G(j\omega)N(a))^{-1})}, \quad \forall \omega, \forall a,$$

such that the harmonic balance predicts a limit cycle.

The proof is similar to the one for internal stability for linear systems (see, for example, [9]), and can be found in the above reference.

In this way a measure of robustness against limit cycles is obtained. Condition 1 can be verified with the help of the methods of Section 7.2 or by checking if $\underline{\sigma}(I + G(j\omega)N(a)) \neq 0$.

In [8] it is shown how the methods expounded in Section 7.2 can deal with the analysis of the existence of limit cycles. In that reference an example is included. Figure 7.6 shows some of the plots obtained. The value of the minimum can be considered as a measure of robustness.

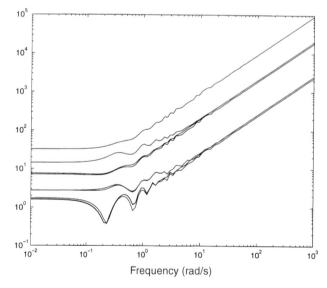

Figure 7.6 Singular value plots of $1/\bar{\sigma}(G(j\omega)N(a)(I+G(j\omega)N(a))^{-1})$

7.5 CONCLUSIONS

A frequency domain methodology to globally analyze a MIMO Fuzzy Control System has been proposed.

The stability analysis of multivariable non-linear systems with pure delays is a very difficult task. In this chapter it is assumed that the FCS can be decomposed into a linear plant with pure delays and a non-linear controller. A method to determine how far the FCS is from the emergence of a limit cycles has been presented. In this way a robustness measurement can be obtained. The analysis should be completed with the analysis of the existence of other equilibrium points.

With the information supplied by the methodology, significant consequences can be inferred regarding the global stability of the whole system. The existence of attractors other than the operating point implies that the origin's basin of attraction is bounded, so the system is at best locally, but not globally, stable. That restricts the disturbances that the system can suffer without being out of control.

ACKNOWLEDGMENTS

This work has been done in the framework of the project FAMIMO funded by the European Commission in the ESPRIT program (LTR Project 21911). The authors would also like to thank the Spanish Agency CICYT for partially supporting this work under grants TAP96-1184-C04-01 and TAP97-0553.

REFERENCES

1. Aracil, J., A. Ollero and A. García-Cerezo, Stability indices for the global analysis of expert control systems, *IEEE Transactions on Systems, Man and Cybernetics*, **23**(2), 1989, 603–606.
2. Aracil, J., F. Gordillo and T. Álamo, Global stability analysis of second-order fuzzy control systems, In: R. Palm, D. Driankov and H. Hellendorn (eds.), *Advances in Fuzzy Control*, Physica-Verlag, Heidelberg, pp. 11–31.
3. Aracil, J., F. Gordillo and A. Ollero, Multiple equilibria in MIMO fuzzy control systems, *Proceedings of the Seventh Conference on Information Processing and Management of Uncertainty in Knowledge-Based Systems (IPMU'98)*, Paris, 1998, pp. 1870–1871.
4. Boyle, J. M., M. P. Ford and J. M. Maciejowski, A multivariable toolbox for use with Matlab, *IEEE Control System Magazine*, **9**(1), 1989, 59–65.
5. Cuesta, F., F. Gordillo, J. Aracil and A. Ollero, Global stability analysis of a class of multivariable Takagi–Sugeno fuzzy control systems, *IEEE Transactions on Fuzzy Systems*, **7**(5), 1999, 505–520.
6. Driankov, D., H. Hellendorn, and M. Reinfrank, *An Introduction to Fuzzy Control*, Springer-Verlag, Berlin, 1993.
7. Gordillo, F., J. Aracil and T. Álamo, Determining limit cycles in fuzzy control systems, *Proceedings of the 1997 FUZZ-IEEE*, Barcelona, 1997, pp. 193–198.
8. Gordillo, F., J. Aracil and A. Ollero, Frequency domain analysis of multivariable fuzzy control systems, In: R. Whalley and M. Ebrahimi, (eds.), *Application of Multi-Variable System Techniques*, Professional Engineering Publishing Ltd, London and Bury St Edmunds, 1998, pp. 243–252.
9. Green, M. and D. J. N. Limebeer, *Linear Robust Control*, Prentice-Hall, Englewood Cliffs, 1995.
10. Katebi, M. R. and Y. Zhang, H_∞ control analysis and design for nonlinear systems, *International Journal of Control*, **61**(2), 1995, 459–474.

11. Kickert, W. J. M. and E. H. Mandani, Analysis of a fuzzy logic controller, *Fuzzy Sets and Systems*, **1**, 1978, 29–44.
12. Mees, A. I., Describing function, circle criteria and multiloop feedback systems, *Proceedings of the IEE*, **120**(1), 1973, 126–130.
13. Ollero, A., J. Aracil and F. Gordillo, Stability analysis of MIMO fuzzy control systems in the frequency domain, *Proceedings of the 1998 FUZZ-IEEE*, Anchorage, 1998, pp. 49–54.
14. Wang, L. X., *A Course in Fuzzy Systems and Control*, Prentice-Hall, Englewood Cliffs, 1997.
15. Yager, R. R. and D. P. Filev, *Essentials of Fuzzy Modeling and Control*, Wiley, New York, 1994.

8
Analytical Study of Structure of a Mamdani Fuzzy Controller with Three Input Variables

Hao Ying
University of Texas Medical Branch, Galveston, USA

8.1 INTRODUCTION

Fuzzy controllers have been used worldwide in many fields [6, 9, 20]. They are usually practically constructed, instead of theoretically designed, using the trial-and-error method and computer simulation. This is due to relatively weak fuzzy control theory in existence and also due to the fact that fuzzy controllers are generally non-linear controllers that are difficult to analyze and design. In recent years, efforts have been made to understand fuzzy controller structures (e.g. [3], [5], [8], [12–15], [19], [23], [26]), to analyze fuzzy controllers (e.g. [10], [11], [16–18], [22], [25], [28], [29]) and to develop methods for determining the stability of fuzzy control systems (e.g. [4], [7], [11], [17], [25]).

One of the most important aspects of fuzzy control research is the study of fuzzy controller structure. For general SISO (single-input–single-output) and MIMO (multiple-input–multiple-output) fuzzy controllers, we have found their analytical structures and limiting structures [24, 27]. The fuzzy controllers can always be decomposed into the sum of a global non-linear controller and a local non-linear controller. The structure of the global and local controllers depends on the configuration of the fuzzy controllers and hence is different from one controller to another. We have investigated some specific configurations in which the fuzzy controllers use error and rate change of error (rate, for short) as input variables. We have analytically proved that the structure of the non-linear fuzzy controllers that use triangular input fuzzy sets, singleton output fuzzy sets, linear fuzzy control rules, the Zadeh fuzzy logic AND operator, the Lukasiewicz fuzzy logic OR operator and the centroid defuzzifier is the sum of a global two-dimensional multilevel relay and a local nonlinear PI controller [23]. We have extended this result to other SISO fuzzy controllers and also to some MIMO fuzzy controllers [26], all of which use two input variables.

Fuzzy Logic. Edited by S. Farinwata, D. Filev and R. Langari
© 2000 John Wiley & Sons, Ltd

154 MAMDANI FUZZY CONTROLLER WITH THREE INPUT VARIABLES

The most popular controller in industry is the linear PID (proportional-integral-derivative) controller, which uses error, rate and derivative of rate change of error (d_rate, for short) of process output as input variables. In contrast, the majority of fuzzy control research and applications focuses only on fuzzy controllers that use error and rate as input variables; d_rate is hardly utilized. Consequently, there exist few results on the analytical structure of non-linear fuzzy controllers that use the same input variables as the PID controller does.

In this chapter we investigate the structure of a non-linear fuzzy controller using these three input variables. The controller employs triangular input fuzzy sets, trapezoidal output fuzzy sets, linear fuzzy control rules, the product fuzzy logic AND operator, the Lukasiewicz fuzzy logic OR operator, the Mamdani minimum inference method and the centriod defuzzifier. The chapter is organized as follows. In Section 8.2, the configuration and components of the fuzzy controller are defined. In Section 8.3, the analytical structure of the fuzzy controller is derived, and the characteristics of the structure are analyzed. For comparison, the structure of the fuzzy controller using error and rate as input variables is also revealed. The limiting structure of the fuzzy controller when the number of the input fuzzy sets approaches infinity is studied.

8.2 CONFIGURATION OF THE FUZZY CONTROLLER

The fuzzy controller in this study uses the process output, denoted $y(nT)$, where T is the sampling period and nT is the sampling time, to calculate its three scaled input variables. They are scaled error, scaled rate and scaled d_rate, as expressed mathematically by the following:

$$e^* = GE \cdot e(nT) = GE(SP(nT) - y(nT)),$$
$$r^* = GR \cdot r(nT) = GR(e(nT) - e(nT - T)),$$
$$d^* = GD \cdot d(nT) = GD(r(nT) - r(nT - T)),$$

where $SP(nT)$ is the target of process output, and $nT - T$ is the previous sampling time. GE, GR and GD are the scaling factors for error, rate and d_rate, respectively.

Each of the scaled input variables is fuzzified by $N = 2J + 1 (J \geq 1)$ input fuzzy sets. Among them, J fuzzy sets are for positive scaled input variables, another J fuzzy set for negative scaled input variables and one fuzzy set for near zero scaled input variables. We designate E_i, R_j and D_k ($-J \leq i, j, k \leq J$) as a fuzzy set for e^*, r^* and d^*, respectively. Their membership functions are identical and triangular-shaped. Note that the identity of the membership functions is with respect to e^*, r^* and d^* (the scaled input variables), not with respect to $e(nT)$, $r(nT)$ and $d(nT)$ (the physical input variables). Thus, by choosing different values for GE, GR and GD, the membership functions are different in terms of $e(nT)$, $r(nT)$ and $d(nT)$. In other words, the identity requirements on the membership functions of the input fuzzy sets are actually not restrictive.

The definitions of these membership functions are shown graphically in Figure 8.1. The central values for E_{-J}, R_{-J} and D_{-J} are defined as $-L$ and those for E_J, R_J and D_J as L. The central values of the remaining $2J - 1$ fuzzy sets are required to be equally spaced with the space between two adjacent sets being

$$S = \frac{L}{J}.$$

CONFIGURATION OF THE FUZZY CONTROLLER   155

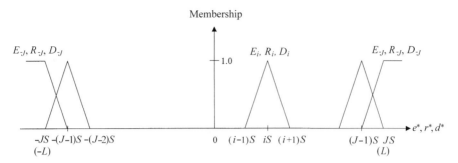

Figure 8.1 Graphical definitions of the triangular input fuzzy sets. There are a total $2J + 1$ of them

The central values of E_i, R_j and D_k are, therefore, $i \cdot S, j \cdot S$ and $k \cdot S$, respectively. That is, the membership value of the triangular fuzzy sets is one at the central values. The triangular membership functions in $[-L, L]$ are as follows: the membership function of E_i (or R_i or D_i) is zero at $(i-1)S$ and increases linearly to one at $i \cdot S$ and then decreases linearly to zero at $(i+1)S$; the membership function is zero elsewhere. In $(-\infty, -L]$ the membership functions of E_{-J}, R_{-J} and D_{-J} are always equal to one, and in $[L, +\infty)$ the membership functions of E_J, R_J and D_J are always equal to one. In these two intervals, the membership functions of all the other fuzzy sets are zero.

We designate the incremental output of the fuzzy controller at nT as $\Delta u(nT)$. There are $6J+1$ (i.e. $3N-2$) output fuzzy sets for it, and their definitions are illustrated in Figure 8.2. Among them, $3J$ output fuzzy sets are for positive $\Delta u(nT)$, another $3J$ output fuzzy sets for negative $\Delta u(nT)$ and one output fuzzy set for near zero $\Delta u(nT)$. We designate ΔU_m, where $-3J \leq m \leq 3J$, as an output fuzzy set. The central values for ΔU_{-J} and ΔU_J are defined as $-H$ and H, respectively. The central values for the rest of the output fuzzy sets are required to be equally spaced and the space between two neighboring output fuzzy sets is

$$V = \frac{H}{3J}.$$

Obviously, the central value for ΔU_m is $m \cdot V$. The membership functions of the ΔU_m's are identical and trapezoidal-shaped with the upper-side being $2A$ and the lower-side being $2V$. To define the shape of these trapezoids, a parameter

$$\theta = \frac{A}{V}$$

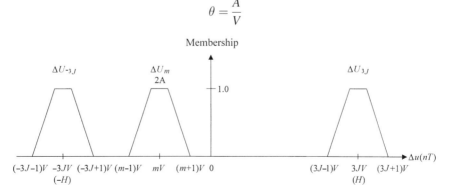

Figure 8.2 Illustrative definitions of the trapezoidal output fuzzy sets. There are a total of $6J + 1$ of them. Note that $2A$ and $2V$ are, respectively, the upper-side and lower-side of the trapezoids

156 MAMDANI FUZZY CONTROLLER WITH THREE INPUT VARIABLES

is employed. We apply the following constraint to this parameter:

$$\theta \leq 0.5$$

to avoid overlap between the upper-sides of two adjacent output fuzzy sets. In $[-H, H]$ the membership function of ΔU_m increases linearly from zero at $(m-1)V$ to one at $m \cdot V - A$ and stays at one until $m \cdot V + A$ where it begins to decrease linearly to zero at $(m+1)V$. The membership function is zero elsewhere. In $(-\infty, -H]$ the membership function of ΔU_{-3J} is always equal to one, and in $[H, +\infty)$ the membership function of ΔU_{3J} is always equal to one. The membership functions of all the other output fuzzy sets are zero in these two intervals.

N^3 control rules are necessary in order to cover $N \times N \times N$ possible combinations of the input fuzzy sets. In this study we use the following linear control rules [23]:

IF e^* is E_i **AND** r^* is R_j **AND** d^* is D_k **Then** $\Delta u(nT)$ is ΔU_{i+j+k}. (8.1)

The control rules are said to be linear because the linear function is used to relate the indices of E, R and D to the index of ΔU.

The widely-used Mamdani minimum inference method, whose definition is illustrated in Figure 8.3, is employed to infer output fuzzy sets from input ones in the control rules. We denote by $\mu_{i,j,k}(\Delta u)$ as membership for ΔU_{i+j+k}, which is calculated by evaluating the memberships in the rule antecedents in (8.1) using the product fuzzy logic AND operator. More specifically, $\mu_{i,j,k}(\Delta u) = \mu_i(e^*)\mu_j(r^*)\mu_k(d^*)$. The shaded area in Figure 8.3 is

$$S(\mu_{i,j,k}(\Delta u)) = (2 - \mu_{i,j,k}(\Delta u) + \theta \cdot \mu_{i,j,k}(\Delta u))\mu_{i,j,k}(\Delta u) \cdot V. \quad (8.2)$$

Some fuzzy rules generate memberships for the same output fuzzy sets. In these cases, the Lukasiewicz fuzzy logic OR operator (i.e. x OR y OR $z = \min(x+y+z, 1)$) is used to obtain a combined membership because the conditions being ORed are maximally negatively correlated [21].

The prevalent centriod defuzzifier is employed to defuzzify the output fuzzy sets. Since the shapes of the output fuzzy sets are identical, the global centroid can be calculated using the local centroid, which are the products of the central values and the partial trapezoids of the

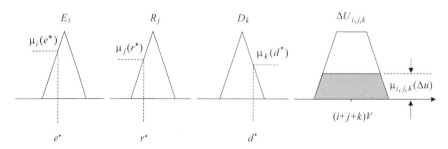

Figure 8.3 Illustrative definition of the Mamdani minimum inference method. We denote $\mu_{i,j,k}(\Delta u)$ as the membership function for the output fuzzy set ΔU_{i+j+k}. $\mu_{i,j,k}(\Delta u)$ is calculated by using the product fuzzy logic AND operator, from the memberships of E_i, R_j and D_k and the linear control rule (8.1). And $\mu_{i,j,k}(\Delta u) = \mu_i(e^*)\mu_j(r^*)\mu_k(d^*)$. The shaded area can be calculated using (8.2). Note that $2A$ and $2V$ are, respectively, the upper-side and lower-side of the trapezoidal $\Delta U_{i+j+k}(-J \leq i,j,k \leq J)$

output fuzzy sets involved. As a result, the scaled incremental output of the fuzzy controller is

$$GU \cdot \Delta u(nT) = GU \frac{\sum_{\mu_{i,j,k}(\Delta u) \neq 0} S(\mu_{i,j,k}(\Delta u)) \cdot (i+j+k)V}{\sum_{\mu_{i,j,k}(\Delta u) \neq 0} S(\mu_{i,j,k}(\Delta u))}, \quad (8.3)$$

where GU is a scaling factor for $\Delta u(nT)$.

8.3 ANALYTICAL STUDY OF THE FUZZY CONTROLLER STRUCTURE

Theorem 1: *The structure of the fuzzy controller using error, rate and d_rate as input variables is the sum of a global three-dimensional multilevel relay and two local non-linear controllers, one of which is a non-linear PID controller with variable gains.*

Proof Without loss of generality, we assume

$$i \cdot S \leq e^* \leq (i+1)S,$$
$$j \cdot S \leq r^* \leq (j+1)S,$$
$$k \cdot S \leq d^* \leq (k+1)S,$$

where

$$-L \leq e^*, r^*, d^* \leq L$$

By fuzzifying e^*, r^* and d^*, the memberships for $E_i, E_{i+1}, R_j, R_{j+1}, D_k$ and D_{k+1} are

$$\mu_i(e^*) = -\frac{E^* - 0.5S}{S}, \quad \mu_{i+1}(e^*) = \frac{E^* + 0.5S}{S}, \quad (8.4)$$

$$\mu_j(r^*) = -\frac{R^* - 0.5S}{S}, \quad \mu_{j+1}(r^*) = \frac{R^* + 0.5S}{S}, \quad (8.5)$$

$$\mu_k(d^*) = -\frac{D^* - 0.5S}{S}, \quad \mu_{k+1}(d^*) = \frac{D^* + 0.5S}{S} \quad (8.6)$$

where

$$E^* = e^* - (i+0.5)S, \quad R^* = r^* - (j+0.5)S \quad \text{and} \quad D^* = d^* - (k+0.5)S. \quad (8.7)$$

Note that

$$\mu_i(e^*) + \mu_{i+1}(e^*) = 1 \quad \mu_j(r^*) + \mu_{j+1}(r^*) = 1, \quad \mu_k(d^*) + \mu_{k+1}(d^*) = 1.$$

The membership value for all the other input fuzzy sets is zero. Consequently, only the following eight (i.e. 2^3) fuzzy rules are executed:

If e^* is E_{i+1} AND r^* is R_{j+1} AND d^* is D_{k+1} Then $\Delta u(nT)$ is $\Delta U_{i+j+k+3}$, (r_1)
If e^* is E_{i+1} AND r^* is R_{j+1} AND d^* is D_k Then $\Delta u(nT)$ is $\Delta U_{i+j+k+2}$, (r_2)
If e^* is E_{i+1} AND r^* is R_j AND d^* is D_{k+1} Then $\Delta u(nT)$ is $\Delta U_{i+j+k+2}$, (r_3)
If e^* is E_{i+1} AND r^* is R_j AND d^* is D_k Then $\Delta u(nT)$ is $\Delta U_{i+j+k+1}$ (r_4)
If e^* is E_i AND r^* is R_{j+1} AND d^* is D_{k+1} Then $\Delta u(nT)$ is $\Delta U_{i+j+k+2}$ (r_5)
If e^* is E_i AND r^* is R_{j+1} AND d^* is D_k Then $\Delta u(nT)$ is $\Delta U_{i+j+k+1}$ (r_6)
If e^* is E_i AND r^* is R_j AND d^* is D_{k+1} Then $\Delta u(nT)$ is $\Delta U_{i+j+k+1}$ (r_7)
If e^* is E_i AND r^* is R_j AND d^* is D_k Then $\Delta u(nT)$ is ΔU_{i+j+k}. (r_8)

158 MAMDANI FUZZY CONTROLLER WITH THREE INPUT VARIABLES

Using the product fuzzy logic AND operator and the membership functions (8.4–8.6), the memberships for the rule consequent in (r_1 - r_8) can be obtained as analytical expressions. Note that the rules r_2, r_3 and r_5 produce memberships for the same output fuzzy set, $\Delta U_{i+j+k+2}$, and that r_4, r_6 and r_7 generate memberships for the same output fuzzy set, $\Delta U_{i+j+k+1}$. The Lukasiewicz fuzzy logic OR operator is used to calculate the combined memberships. The combined membership for $\Delta U_{i+j+k+2}$ is

$$\min(\mu_{i+1}(e^*)\mu_{j+1}(r^*)\mu_k(d^*) + \mu_{i+1}(e^*)\mu_j(r^*)\mu_{k+1}(d^*) + \mu_i(e^*)\mu_{j+1}(r^*)\mu_{k+1}(d^*), 1). \tag{8.8}$$

Because

$$\mu_{j+1}(r^*)\mu_k(d^*) + \mu_j(r^*)\mu_{k+1}(d^*) = \frac{1}{2} - \frac{2R^*D^*}{S^2} \leq 1$$

and

$$\mu_i(e^*)\mu_{j+1}(r^*)\mu_{k+1}(d^*) \leq \mu_i(e^*),$$

we get

$$\mu_{i+1}(e^*)\mu_{j+1}(r^*)\mu_k(d^*) + \mu_{i+1}(e^*)\mu_j(r^*)\mu_{k+1}(d^*) + \mu_i(e^*)\mu_{j+1}(r^*)\mu_{k+1}(d^*)$$
$$= \mu_{i+1}(e^*)(\mu_{j+1}(r^*)\mu_k(d^*) + \mu_j(r^*)\mu_{k+1}(d^*)) + \mu_i(e^*)\mu_{j+1}(r^*)\mu_{k+1}(d^*)$$
$$\leq \mu_{i+1}(e^*) + \mu_i(e^*)\mu_{j+1}(r^*)\mu_{k+1}(d^*) \leq \mu_{i+1}(e^*) + \mu_i(e^*) = 1.$$

Hence, the result of (8.8) is always the memberships being ORed. Similarly, it can be proved that this is also the case for the combined membership for $\Delta U_{i+j+k+1}$.

Substituting all the membership expressions for the output fuzzy sets into the defuzzifier (8.3) and then mathematically manipulating the resulting expressions, the analytical structure of the fuzzy controller is obtained as follows:

$$GU \cdot \Delta u(nT) = (i + j + k + 1.5) GU \cdot V$$
$$+ \left[K_i \left(e(nT) - \frac{(i+0.5)S}{GE} \right) + K_p \left(r(nT) - \frac{(j+0.5)S}{GR} \right) \right.$$
$$\left. + K_d \left(d(nT) - \frac{(k+0.5)S}{GD} \right) \right] + K \cdot E^* \cdot R^* \cdot D^*,$$

where

$$K_p = \frac{\beta_1 \cdot GR \cdot GU \cdot H}{3L}, \quad K_i = \frac{\beta_1 \cdot GE \cdot GU \cdot H}{3L}, \quad K_d = \frac{\beta_1 \cdot GD \cdot GU \cdot H}{3L},$$

$$K = \frac{\beta_2 \cdot GU \cdot H}{3LS^2},$$

$$\beta_1 = \frac{2(7+\theta)S^6 - 8(1-\theta)(E^*R^* + E^*D^* + R^*D^*)S^4}{(15+\theta)S^6 - 4(1-\theta)\{16(E^*R^*D^*)^2 + 4S^2[(E^*R^*)^2 + (E^*D^*)^2 + (R^*D^*)^2] + S^4(E^{*2}+R^{*2}+D^{*2})\}}.$$

and

$$\beta_2 = \frac{8(1-\theta)[3S^2 - 4(E^*R^* + E^*D^* + R^*D^*)]S^4}{(15+\theta)S^6 - 4(1-\theta)\{16(E^*R^*D^*)^2 + 4S^2[(E^*R^*)^2 + (E^*D^*)^2 + (R^*D^*)^2] + S^4(E^{*2}+R^{*2}+D^{*2})\}}.$$

Designating
$$\Delta u_G(nT) = (i+j+k+1.5)GU \cdot V,$$

$$\Delta u_{L1}(nT) = \left[K_i\left(e(nT) - \frac{(i+0.5)S}{GE}\right) + K_p\left(r(nT) - \frac{(j+0.5)S}{GR}\right) + K_d\left(d(nT) - \frac{(k+0.5)S}{GD}\right)\right],$$

and
$$\Delta u_{L2}(nT) = K \cdot E^* \cdot R^* \cdot D^*,$$

then
$$GU \cdot \Delta u(nT) = \Delta u_G(nT) + \Delta u_{L1}(nT) + \Delta u_{L2}(nT).$$

Here $\Delta u_G(nT)$ is a three-dimensional multilevel relay with respect to its inputs i, j and k. Note that

$$\Delta u_G(nT) = (i+j+k+1.5)GU \cdot V = [(i+0.5)S + (j+0.5)S + (k+0.5)S]\frac{GU \cdot H}{3L}.$$

This means that $\Delta u_G(nT)$ is calculated according to $(i+0.5)S$, $(j+0.5)S$ and $(k+0.5)S$, which are the coordinates of the center of the cube configured by $[iS, (i+1)S], [jS, (j+1)S]$ and $[kS, (k+1)S]$ in which the current scaled input variables lie, with respect to the origin of the scaled input state space, (0, 0, 0). Therefore, the relay is called a global relay [23].

$\Delta u_{L1}(nT)$ is a non-linear PID controller with variable gains and its steady-state is $((i+0.5)S/GE, (j+0.5)S/GR, (k+0.5)S/GD)$, which changes with input variables. The gains, K_p, K_i and K_d vary with time because β_1 changes with, E^*, R^* and D^*. This non-linear PID controller is said to be a local controller because its control action is calculated based on E^*, R^* and D^*, which are the differences between the current scaled input variables and the center of the cube in which they lie (see (8.7)).

$\Delta u_{L2}(nT)$ is a local non-linear controller in the form of the cross product of E^*, R^* and D^*. The non-linear gain, K, also varies locally with E^*, R^* and D^*. □

The maximum control action from each of these three controllers can be determined quantitatively, as stated by the following theorem.

Theorem 2

1. $|\Delta u_G(nT)| \leq |\Delta u_G(nT)|_{\max} = \frac{N-2}{N-1} GU \cdot H;$

2. $|\Delta u_{L1}(nT) + \Delta u_{L2}(nT)| \leq |\Delta u_{L1}(nT) + \Delta u_{L2}(nT)|_{\max} = \frac{GU \cdot H}{N-1}.$ (8.9)

Proof

1. When $i = j = k = J - 1$ or when $i = j = k = -J, |\Delta u_G(nT)|$ reaches its maximum, which is

$$|\Delta u_G(nT)|_{\max} = (3J - 1.5)GU \cdot V = \frac{N-2}{N-1} GU \cdot H.$$

Therefore,
$$|\Delta u_G(nT)| \leq |\Delta u_G(nT)|_{\max}.$$

160 MAMDANI FUZZY CONTROLLER WITH THREE INPUT VARIABLES

2. When $E^* = R^* = D^* = 0.5S$ or $E^* = R^* = D^* = -0.5S, |\Delta u_{L1}(nT) + \Delta u_{L2}(nT)|$ reaches its maximum, which is

$$|\Delta u_{L1}(nT) + \Delta u_{L2}(nT)|_{\max} = \frac{GU \cdot H}{N-1}.$$

Hence,

$$|\Delta u_{L1}(nT) + \Delta u_{L2}(nT)| \leq \frac{GU \cdot H}{N-1}. \qquad \square$$

Theorem 2 reveals that when $N > 3$, the three-dimensional multilevel relay plays a dominant role in the fuzzy control action in comparison with the two local non-linear controllers. The larger N, the more control action from the global relay and the less from the local controllers. The work mechanism of the fuzzy controller is that the control action of the global relay is adjusted by the local non-linear controllers. The maximum amount of adjustment, however, is subject to the limitation described in (8.9). The farther the values of the scaled input variables are away from $((i+0.5), (j+0.5)S, (k+0.5)S)$, the larger the magnitude of the adjustment. When $E^* = R^* = D^* = 0.5S$ or $E^* = R^* = D^* = -0.5S$, the adjustment is maximized (which is, $GU \cdot H/(N-1)$). On the other hand, when $E^* = R^* = D^* = 0$, the adjustment is minimum (which is zero). When $N = 3$, the weight of the control action from the global and local controllers in the fuzzy control action is the same. When $N = 2$, the global relay will no longer exist and the fuzzy controller becomes a global non-linear PI controller with variable gains [25]. If N becomes larger and larger, the resolution of the relay increases, while the adjustment effect of the local controllers decreases. If N grows without bound, the fuzzy controller becomes a linear PID controller as the following theorem states (see [24] and [27] for generalized results).

Theorem 3 (Limit theorem) *When N approaches ∞,*

1. $\lim_{N \to \infty} (\Delta u_{L1}(nT) + \Delta u_{L2}(nT)) = 0,$

   and

2. $\lim_{N \to \infty} \Delta u_G(nT) = \dfrac{GU \cdot H}{3L} (GE \cdot e(nT) + GR \cdot r(nT) + GD \cdot d(nT)).$

$\lim_{N\to\infty} \Delta u_G(nT)$ *(i.e. $\lim_{N\to\infty} GU \cdot \Delta u(nT)$ in this case) is a linear PID controller with the proportional gain, integral gain and derivative gain being $GR \cdot GU \cdot H/3L$, $GE \cdot GU \cdot H/3L$ and $GD \cdot GU \cdot H/3L$, respectively.*

Proof

1. $\lim_{N \to \infty} |\Delta u_{L1}(nT) + \Delta u_{L2}(nT)| \leq \lim_{N \to \infty} |\Delta u_{L1}(nT) + \Delta u_{L2}(nT)|_{\max} = \lim_{N \to \infty} \dfrac{GU \cdot H}{N-1} = 0.$

   Therefore,

   $$\lim_{N \to \infty} (\Delta u_{L1}(nT) + \Delta u_{L2}(nT)) = 0.$$

2. $\lim_{N \to \infty} \Delta u_G(nT) = \lim_{N \to \infty} (i + j + k + 1.5) GU \cdot V = \lim_{N \to \infty} \dfrac{i+j+k}{J} \cdot \dfrac{GU \cdot H}{3}.$

Note that

$$\frac{i}{j} \leq \frac{e^*}{L} \leq \frac{i+1}{J}, \quad \frac{j}{J} \leq \frac{r^*}{L} \leq \frac{j+1}{J} \quad \text{and} \quad \frac{k}{J} \leq \frac{d^*}{L} \leq \frac{k+1}{J}.$$

Hence,

$$\lim_{N \to \infty} \frac{i}{J} = \frac{e^*}{L}, \quad \lim_{N \to \infty} \frac{j}{J} = \frac{r^*}{L} \quad \text{and} \quad \lim_{N \to \infty} \frac{k}{J} = \frac{d^*}{L}.$$

As a result,

$$\lim_{N \to \infty} \Delta u_G(nT) = \frac{GU \cdot H}{3L} (GE \cdot e(nT) + GR \cdot r(nT) + GD \cdot d(nT)). \quad \square$$

A practical implication of Theorem 3 is that when linear fuzzy control rules are adopted, the fuzzy controller designer should select a relatively small number of input fuzzy sets in order to take advantage of the non-linearity of the fuzzy controller. The more input fuzzy sets are used, the less non-linear the fuzzy controller will be.

In our previous work we have studied some fuzzy controllers that use two input variables, error and rate, and linear fuzzy control rules. Our analytically derived results have shown that these fuzzy controllers are the sum of a global two-dimensional multilevel relay and a local non-linear PI controller (e.g. [23], [26]). Note that the configurations of these fuzzy controllers are different from those in this chapter. Nevertheless, the conclusion on the controller structure is applicable to the fuzzy controller in the present chapter when it uses only two input variables, as stated by the following theorem.

Theorem 4 *When the fuzzy controller uses error and rate as input variables instead of the three input variables, its structure becomes the sum of a global two-dimensional multilevel relay and a local non-linear PI controller.*

Proof The proof of this theorem is similar to that of Theorem 1. When d_rate is excluded from the input variables, the total number of output fuzzy sets reduces to $4J + 1$ (i.e. $2N - 1$) and the space between the central values of two adjacent output fuzzy sets becomes

$$V = \frac{H}{2J}.$$

Also, only N^2 linear control rules will be needed:

If e^* is E_i AND r^* is R_j Then $\Delta u(nT)$ is ΔU_{i+j}.

The structure of the fuzzy controller is derived as follows:

$$GU \cdot \Delta u(nT) = (i+j+1)GU \cdot V$$
$$+ \left[K_i \left(e(nT) - \frac{i+0.5)S}{GE} \right) + K_p \left(r(nT) - \frac{j+0.5)S}{GR} \right) \right],$$

where

$$K_p = \frac{\beta_1 \cdot GR \cdot GU \cdot H}{2L}, \quad K_i = \frac{\beta_1 \cdot GE \cdot GU \cdot H}{2L},$$

and

$$\beta_1 = \frac{2(3+\theta)S^4 - 8(1-\theta)E^*R^*S^2}{(7+\theta)S^4 - 4(1-\theta)\{(E^*S)^2 + 4(E^*R^*)^2 + (R^*S)^2\}}.$$

Obviously, $GU \cdot \Delta u(nT)$ consists of a global two-dimensional multilevel relay and a local non-linear PI controller. □

It can easily be proved that the absolute value of the maximal outputs of the global two-dimensional relay and the local non-linear PI controller are $(N-2)GU \cdot H/(N-1)$ and $GU \cdot H/(N-1)$, respectively. As the theorem below states, the global two-dimensional multilevel relay approaches a global linear PI controller as $N \to \infty$, whereas the local non-linear PI controller disappears (see also [1], [2], [23], [24] for related results].

Theorem 5 *If the fuzzy controller uses error and rate as input variables, then*

$$\lim_{N \to \infty} GU \cdot \Delta u(nT) = \frac{GU \cdot H}{2L} (GE \cdot e(nT) + GR \cdot r(nT)).$$

Proof See the proof of Theorem 3.

8.4 CONCLUSION

The explicit structure of the fuzzy controller using error, rate and d_rate of process output as input variables has been derived. The result is the sum of a global three-dimensional multilevel relay and two local non-linear controllers. One of the local non-linear controllers is a non-linear PID controller with variable proportional gain, integral gain and derivative gain. We have shown that, when $N > 3$, the major portion of the control action of the fuzzy controller is contributed by the global relay while the local non-linear controllers merely fine tune the output of the global controller. The role of the global and local controllers in the fuzzy control action have been quantitatively determined. Additionally, it has been proved that the limit structure of the fuzzy controller is determined by the global multilevel relay which becomes a global linear PID controller as the number of input fuzzy sets grows without bound. In the meantime, the two local non-linear controllers disappear.

As we have shown in our previous paper [23], the structure of the fuzzy controller using linear control rules and error and rate as input variables is the sum of a global two-dimensional multilevel relay and a local non-linear PI controller. Logically, one would expect that when the additional input variable d_rate is involved, the fuzzy controller structure should be the sum of a global three-dimensional multilevel relay and a local non-linear PID controller. We show that this is not the case for the fuzzy controller studied in this chapter. Whether this conjecture holds for other fuzzy controllers that use the three input variables remains a future research topic.

ACKNOWLEDGMENT

This research was supported in part by an Advanced Technology Program Grant from Texas Higher Education Coordinating Board.

REFERENCES

1. Bouslama, F. and A. Ichikawa, Fuzzy control rules and their natural control laws, *Fuzzy Sets and Systems*, **48**, 1992, 65–86.

2. Buckley, J. J. and H. Ying, Fuzzy controller theory: Limit theorems for linear fuzzy control rules, *Automatica*, **25**, 1989, 469–472.
3. Buckley, J. J. and H. Ying, Linear fuzzy controller: it is a linear nonfuzzy controller, *Information Sciences*, **51**, 1990, 183–192.
4. Chen, G.-R. and H. Ying, BIBO stability of nonlinear fuzzy PI control systems, *Journal of Intelligent and Fuzzy Systems*, **5**, 1997, 245–256.
5. Hajjaji, A. E. and A. Rachid, Explicit formulas for fuzzy controller, *Fuzzy Sets and Systems*, **62**, 1994, 135–141.
6. Kosko, B., *Neural Networks and Fuzzy systems: A Dynamical Systems Approach to Machine Intelligence*, Prentice-Hall, Englewood Cliffs, 1991.
7. Langari, R. and M. Tomizuka, Stability of fuzzy linguistic control systems, *Proceedings of the IEEE Conference on Decision and Control*, Hawaii, 1990.
8. Langari, G., A nonlinear formulation of a class of fuzzy linguistic control systems, *Proceedings of the American Control Conference*, Chicago, Illinois, 1992.
9. Lee, C. C., Fuzzy logic in control system: fuzzy logic controller, part I and II, *IEEE Transactions on Systems, Man and Cybernetics*, **20**, 1990, 408–435.
10. Lewis, F. L. and K. Liu, Towards a paradigm for fuzzy logic control, *Automatica*, **32**, 1996, 167–181.
11. Malki, H. A., H. D. Li and G. Chen, New design and stability analysis of fuzzy PF controllers, *IEEE Transactions on Fuzzy Systems*, **2**, 1994, 245–254.
12. Matia, F., A. Jimenez, R. Galan and R. Sanz, Fuzzy controllers: Lifting the linear–nonlinear frontier, *Fuzzy Sets and Systems*, **52**, 1992, 113–129.
13. Mizumoto, M., Realization of PID controls by fuzzy control methods, *Proceedings of the First IEEE International Conference on Fuzzy Systems*, San Diego, California, 1992, pp. 709–715.
14. Pok, Y. M. and J. X. Xu, An analysis of fuzzy control systems using vector space, *Proceedings of the Second IEEE International Conference on Fuzzy Systems*, San Francisco, California, 1993, pp. 363–368.
15. Siler, W. and H. Ying, Fuzzy control theory: The linear case, *Fuzzy Sets and Systems*, **33**, 1989, 275–290.
16. Wang, H. O., K. Tanaka and M. F. Griffin, An approach to fuzzy control of nonlinear systems: Stability and design issue, *IEEE Transactions on Fuzzy Systems*, **4**, 1996, 14–23.
17. Wang, L.-X., *Adaptive Fuzzy Systems and Control—Design and Stability Analysis*, Prentice Hall, Englewood Cliffs, 1994.
18. Wang, P. Z., H. M. Zhang and W. Xu, Pad-analysis of fuzzy control stability, *Fuzzy Sets and Systems*, **38**, 1990, 27–42.
19. Wong, C. C., C. H. Chou and D. L. Mon, Studies on the output of fuzzy controller with multiple inputs, *Fuzzy Sets and Systems*, **57**, 1993, 149–158.
20. Yen, J., R. Langari, and L. A. Zadeh (eds.), *Industrial Applications of Fuzzy Logic and Intelligent Systems*, IEEE Press, New York, 1995.
21. Ying, H., W. Siler and D. Tucker, A new type of fuzzy controller based upon a fuzzy expert system shell FLOPS, *Proceedings of the IEEE International Workshop on Artificial Intelligence for Industrial Applications*, Hitachi City, Japan, 1988.
22. Ying, H., W. Siler and J. J. Buckley, Fuzzy control theory: A nonlinear case, *Automatica*, **26**, 1990, 513–520.
23. Ying, H., A nonlinear fuzzy controllers with linear control rules is the sum of a global two-dimensional multilevel relay and a local nonlinear proportional-integral controller, *Automatica*, **29**, 1993, 499–505.
24. Ying, H., General analytical structure of typical fuzzy controllers and their limiting structure theorems, *Automatica*, **29**, 1993, 1139–1143.
25. Ying, H., The simplest fuzzy controllers using different inference methods are different nonlinear proportional-integral controllers with variable gains, *Automatica*, **29**, 1993, 1579–1589.

26. Ying, H., Analytical structure of a two-input two-output fuzzy controller and its relation to PI and multilevel relay controllers, *Fuzzy Sets and Systems*, **63**, 1994, 21–33.
27. Ying, H., Structure decomposition of the general MIMO fuzzy systems, *Journal of Intelligent Control and Systems*, **1**, 1996, 327–337.
28. Ying, H., The Takagi–Sugeno fuzzy controllers using the simplified linear control rules are nonlinear variable gain controllers, *Automatica*, **34**, 1998, 157–167.
29. Ying, H., Constructing nonlinear variable gain controllers via the Takagi–Sugeno fuzzy control, *IEEE Transactions on Fuzzy Systems*, **6**, 1998, 226–234.

9
An Approach to the Analysis of Robust Stability of Fuzzy Control Systems

Shehu S. Farinwata
Ford Research Laboratory, Dearborn, Michigan, USA

9.1 INTRODUCTION

In this chapter, the practical robust stability of a closed-loop, model-based fuzzy control system with respect to parameter variation is considered. The class of systems considered consists of linear and non-linear systems controlled by the cell-space fuzzy control method [51, 58]. For non-linear systems, the model dynamics and output functions are assumed to be smooth so that a cell-wise linearization can be performed. Thus, the global nature of the robust stability result obtained is necessarily the union of similar local results of the cells. A simple quadratic performance index is formulated for the system's implicit states. This index is chosen such that it has the property of an energy-like function, which then allows a Lyapunov-like constraint to be imposed on its time evolution. Furthermore, an expression for the system's sensitivities is formulated in terms of fuzzy quantities and incorporated into the time derivative of the energy function. By employing a stability requirement on this function, one is able to obtain inequality bounds on the system's sensitivities, error and parameter deviations in terms of fuzzy quantities. The theory is developed in two parts. First, it is assumed that the states are decoupled in the sense that a parameter perturbation in a particular state has no effect on other states. The main robust stability result is developed based on this assumption. In the second part, an interaction is allowed to exist between states as a result of parameter variation. Estimates of the measure of the interactions are derived on stability grounds. A more general result for robust stability is then given. The point of stability convergence is the system's specified target point. Thus, stability convergence to this point, in the presence of parameter perturbations and external disturbance, inherently addresses the robust performance of the closed-loop system. A measure of robustness is then formulated using singular values. Finally, a robustness measure and an optimum robust set are defined. To illustrate the concept, a previously designed fuzzy logic controller for an automotive engine idle speed was used as a case study.

Fuzzy Logic. Edited by S. Farinwata, D. Filev and R. Langari
© 2000 John Wiley & Sons, Ltd

9.2 PERSPECTIVE

The role of stability in systems design in control engineering theory and practice cannot be overemphasized. Much attention is now being given to the issue of performance assessment of an aspect of the so-called emerging technologies, i.e. fuzzy logic control, neural networks, genetic algorithms and other forms of data mining that are used in conjunction with these. The aspect addressed here is fuzzy control as a major component of an intelligent control paradigm. For useful, practical systems, the rigorous analytical design of stable controllers and the analysis of the resulting closed-loop system are often complimented by rigorous testing. In many situations this is both a necessity and a requirement. Consider at the following two scenarios that a system designer faces. On the one hand, a system that is apparently stable over a long period of time, though does not seem to exhibit the perfect performance according to some specifications, say steady state-error. On the other hand, another system that seems to meet the ideal performance specs at the outset, but blows up only seconds after it has been activated. It would not be surprising if a system designer sat back and jubilated more over the more stable, longer duration system than over the well-performed but short-lived system, either in real-life testing or during a computer simulation. The incorporation of both stability and desired performance requirements into system design is not a trivial task. Often, some compromise must be made. In general, if one has a choice to make, one might opt for a highly stable system with only a reasonably satisfactory performance behavior over a not-so-stable one with a high degree of performance attainment. Thus, another way to put it is that sustained stability should not be compromised. Nevertheless, in certain practical applications such as precision engineering, e.g. in biomedical engineering and others where performance may be measured in almost infinitesimal units, performance and stability may even become indistinguishable. Now, we have mentioned performance very much in quite loose, household terms. For typical inertial systems, a category that most physical systems fit, a system's trajectory may be selected by specifying the damping, rise-time, peak overshoot while the steady-state behavior may be specified via settling time and steady-state error. These are nominal design performance specifiers in control engineering. The issue of robustness is the issue of robust stability and robust performance, i.e. the maintaining a stable operation and the meeting of a specified performance objective by a wide range of closely related family of plants, in the presence of bounded variations in the internal and external inputs. Practitioners in control engineering know that there is one more important element in the design process, namely sensitivity. As systems age, wear and tear creep in, stretching set-points for components and subsystems away from their nominal values, and even stretching physical design tolerances for these systems. Temperature, humidity, mechanical vibration, etc. also play a major role in aiding the degradation of a system's performance. A system may continue to operate in a stable manner while suffering from minor to substantial drift in the nominal parameter values until a point is reached where the calibration is totally messed up (or out of whack!). The system may concurrently exhibit a degrading performance and may ultimately fail. Thus, it is also an understatement to say that the issue of a system's sensitivity to parameter variation is of paramount importance in control engineering and deserves adequate attention. In all realistic formulations of robustness measures, sensitivity information is incorporated as an integral part. It is not difficult to see how the parameters of a system can change the behavior of the system by modifying the intended operational envelope, but not necessarily the structure of the system. This is a much better temporary comfort than if the system's structure were to change as a result of parameter variations. This is an important point to remember. However, new members of the original system's family may be generated in the process as the parameters vary in their specified set. The family's structure is not changed. However, unattended failures due to excessive parameter variations away from the nominal values could become catastrophic and may lead to changes in the system's structure. For the originally intended purpose, the system has now become virtually useless and unsafe. The recovery effort may now require a restructuring of any salvageable part of the system, and reconfiguring the controls then, perhaps, hopefully, will contend with a less than optimum, if not precarious, performance.

9.3 THE NOMINAL FUZZY CONTROL PROBLEM

The objective is to study the asymptotic stability convergence of the closed-loop system with the fuzzy control inputs being continually generated by the fuzzy controller in the system. The convergence point is a specified set-point. Figure 9.1 shows the general non-linear system, f, under fuzzy control. The desired set-point is $x_d \in \mathbb{R}^n$ and the error is $e = x - x_d$. The control u is generated by the fuzzy controller.

Consider the system to be controlled expressed approximately as

$$\dot{x}(t) = f(x(t), u(t), \alpha), \qquad y = h(x), \tag{9.1}$$

where $x(t) \in \mathbb{R}^n$ is the vector of the states, $u(t) \in \mathbb{R}^m$ the input vector, $\alpha \in \mathbb{R}^l$ the parameter vector, and $r \in \mathbb{R}$ is a scalar disturbance input. Let f be a non-linear smooth mapping in general, and is completely determined as: $f: \mathbb{R}^n \times \mathbb{R}^m \times \mathbb{R}^l \to \mathbb{R}^n$, and $h: \mathbb{R}^n \to \mathbb{R}^p$, for $t \in [0, +\infty)$.

For the purpose of preserving existence of the above solution we assume continuity of $x(t)$ with respect to α. So f is smooth in x, and at least piecewise smooth in u and α, for $t \geq 0$.

Let $x(t) \in X, u(t) \in U, y(t) \in Y$, and $\alpha \in D$. These sets are also the feasible or admissible sets which are all subsets of their corresponding spaces, as already specified. Global discussion will therefore pertain to these subsets. In the case of the fuzzifiable quantities, the universes of discourse are also considered to be the same subsets alluded to.

Given the set of desired trajectories $\{x_d\} \subset F(X)$, a fuzzy controller is sought that will generate bounded $u(t)$ or $\{u_n\}, n \in \mathbb{N}$ such that a given $x(t_0)$ at time t_0, is moved to $x(t)$, for $t > t_0$, by which time $x(t)$ is kept arbitrarily close to x_d, with both t_0 and $t \in [0, +\infty)$. This is the fuzzy control problem in a very general sense. Indeed, corresponding to $\lim x(t) = x_d$ as $t \to \infty$, is a control $u(t) \equiv u^*$, or that the fuzzily generated sequence $\{u_n\}$ converges to some u^*. In a host of practical control problems, u^* is often pre-calculated or is specified a priori. In a such a case, the set $\{(x_d, u^*)\}$ is necessarily a fixed point for (9.1), for some nominal α [54].

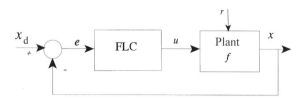

Figure 9.1 Fuzzy control system structure

168 ROBUST STABILITY OF FUZZY CONTROL SYSTEMS

In what follows we re-state a more general fuzzy control algorithm in the form of **If–Then** rules below.

R_i: **If** e_1^i is *small* and e_2^i is *big*,..., and e_n^i is *positive medium* **Then** u_1^i is *negative medium* and u_2^i is *positive large*, ..., and u_m is *negative medium*.

The totality of the fuzzy control rules, $\bigcup R_i$, constitutes the fuzzy control rule base, or the fuzzy controller. The closed-loop fuzzy control system may be represented as

$$\Sigma : \left(\bigcup_{i=1}^{s} R_i; f(\{x_k\}, \{u_k\}) \right),$$

where $f(\ldots)$ is the model of the process to be controlled with its actual crisp states sequence $\{x_k\}$ and crisp input sequence, $\{u_k\}$; s is the number of rules.

9.4 EQUILIBRIUM POINTS FOR FUZZY CONTROLLED PROCESSES

In classical Lyapunov stability of a system of the type in (9.1), the unforced system $f(x_e, 0)$ is studied with $f(x_e, 0) = 0$, where x_e is the equilibrium state. $x_e = 0$ is often employed with, actually, some loss of generality. In a variety of physical processes the equilibrium point $x_e = 0$ may not be convenient or even feasible. In such a case, it is appropriate to define the equilibrium point as the error $e_1 = x_1 - x_{1d}$, and $e_2 = x_2 - x_{2d}$. For instance, x_1 and x_2 may be an automobile's engine idling speed and the engine's manifold pressure, in which case $e_1 = e_2 = 0$ is not only feasible but has the proper physical significance as well. In general, one can always define a new equilibrium point via a shift of origin. This freedom should always be exercised when dealing with a real, physical process. These newly defined states will be referred to as the system's *implicit states*. The introduction of membership functions for the system's physical variables of interest necessarily induces a partition of the feasible space of the variables. The size of the partitioned space may correspond to the product of the cardinalities of the membership functions of any two variables. Such a space is then generally made up of cells of arbitrary shapes in general. However, this need not be the case. For computational reasons, rectangular cells are often used where the boundaries can be easily specified, even though the cell itself is a *covering* of one or more linguistic labels. Thus, the whole state space may look like a huge rectangular grid system. The application considered here employs techniques similar to this. A detailed discussion on this can be found in [58].

9.5 FUZZY ROBUSTNESS ANALYSIS

Recently, there has been much effort focused on stability and robustness analysis of fuzzy control systems. Several of these have been based on the Takagi–Sugeno (T-S) fuzzy system because of the availability of analysis tool from linear control theory [18, 54, 59–62, 67]. The linear matrix inequality (LMI) approach also included in these references has gained a lot of applicability for a variety of model-based fuzzy control systems. While all these works have been pouring in, little effort has been made to incorporate sensitivity factors into the analysis. Much of the effort has capitalized only on stability margins without providing inequality bounds on interactions, error deviations and parameter variations. Fewer still have attempted to express the inequality bounds in terms of fuzzy quantities such as the membership function. As reported in [8], the energy of a fuzzy set depends on its support and shape of its membership function.

Thus, it is important that any discussion of stability and therefore of robustness of a fuzzy system incorporates these elements, whether the analysis is carried out entirely in a fuzzy domain or in a hybrid domain. By hybrid, we mean both fuzzy and non-fuzzy (deterministic), to distinguish this from another usage which considers hybrid to be fuzzy and stochastic. The essence of fuzziness of the system needs to be captured and incorporated into the formulation, possibly via a direct utilization of the membership functions of the pertinent systems' variables. However, if the membership functions are too complicated to be handled directly in the analysis, the resulting crisp output of the closed-loop system should suffice for analysis. This is true as long as a fuzzy controller is used to close the loop. This controller should also be the *de facto* stabilizer of the system under the various operating conditions. Such an approach is not only relevant and practical, but also establishes a baseline for comparison with other types of controllers that could be used in the same framework to stabilize the system.

9.5.1 Robustness Problem Statement

The block diagram presented in Figure 9.2 shows the paradigm that was employed in this discussion. In the diagram, f is the plant, whose approximate mathematical model is available in the form given in (9.1), and belongs to a family P of closely related plants: $P = \{f(x, \alpha, r)\}$. $\Delta f \triangleq \{\Delta \alpha, \Delta x, r\}$ represents the perturbation set. It is a lumped expression of the uncertainty in the system which, generally speaking, could be in the form of unmodelled dynamics, $d\Delta x/dt$, parameter variation, $\Delta \alpha$, or external disturbance and noise. Furthermore, let u, α, x and $d \in$ some normed linear space, and such that $\|u\| < M_u, \|\alpha\| < M_\alpha, \|x\| < M_x$ and $\|d\| < M_d$, with M_u, M_α, M_x and M_d all $< \infty$. As before, $x \in F(x) \subset X$ is the set of feasible states, and $u \in F(U) \subset U$ is the set of feasible inputs. The set of admissible parameters will be denoted by $F(\alpha)$. Similarly, the set of physically allowable parameter perturbation is denoted by $F(\Delta \alpha)$. Since one is concerned with the behavior of the system under the mappings f and h, for all u, and a specified α, it is only appropriate to assume that f and h can vary in a neighborhood of some α_0, the nominal parameter value. The controller, K, is the fuzzy logic controller (FLC), that generates u_F to drive the uncertain system. We further consider that we have a set of performance

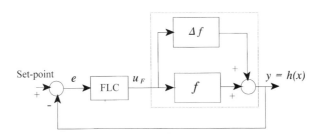

Figure 9.2 Uncertain fuzzy feedback system

characteristics or properties, Θ, that we want the system to assume under fuzzy control. Putting all these together, we have what will be referred to as the robustness set, $\Re = \{P, K, \Theta\}$. Any of these elements can be generic in nature. For instance, in this case, K is an FLC. It is a set of fuzzy **If–Then** rules that has been designed for a given plant $f \in P$ using the predominantly essential procedures: fuzzification, inferencing using an appropriate knowledge base and defuzzification [1, 3]. With this, we now state the robustness problem or objective.

The Problem Given a FLC in the robustness set $\Re$, the following two things are required.

a) The FLC to stabilize the nominal plant, $f \in P$, where all parameters are fixed at their nominal values.
b) The FLC to stabilize the perturbed plant $f + \Delta f$, where $|\Delta f|$, be it unmodelled dynamics, parameter variation, or noise, is bounded and is due to all reasonably and all practically possible perturbations in the system, including the onset of bounded external disturbances of known origins.

As a result, we want to obtain an approximate, quantitative measure of the system's robustness, including bounds on the perturbations and the error sensitivities.

9.5.2 Concepts of Sensitivity and Robustness

It is not uncommon to find sensitivity and robustness being used (sometimes misused) synonymously. The two are different. On the one hand, the sensitivity of a system is usually discussed with respect to a certain parameter of the system. In other words, it is a statement of how the system as a whole, or in part, responds to a parameter variation. It is often expressed as the ratio of the fractional change of an observed system's variable to a fractional change in the parameter that has caused the change. Therefore, in discussing sensitivity, nothing is usually said about the stability of the system. On the other hand, robustness is always associated with a system's stable operation in the presence of internal and external perturbations. The internal perturbation invariably contains the case of parameter variations mentioned above. Robustness is therefore often used as a measure of a system's performance. This sub-section introduces some definitions and terminology that will be utilized in the development of robustness measures of a closed-loop system under fuzzy control. Again, the dynamic system may be represented as a 6-tuple:

$$\sum : (f, h; x, \alpha, u, \mathrm{d}), \tag{9.2}$$

where f is a non-linear operator, $x \in \mathbb{R}^n$ is the observable state vector, $\alpha \in \mathbb{R}^r$ is the real parameter vector, $u \in \mathbb{R}^m$ is the input vector, $r \in \mathbb{R}$ is a scalar disturbance and h is the non-linear output map. The system, as an input–output map, is expressed as

$$\sum : f = \dot{x} - F(x, u, \alpha, \mathrm{d}) = 0, \quad y = h(x, u, \alpha, \mathrm{d}). \tag{9.3}$$

Definition 1: Insensitivity The system is considered insensitive with respect to $\alpha = \alpha_0 + \Delta\alpha$ perturbation if $\forall u_0 \in F(U), y$ depends continuously on α. That is, if $\|y - y_0\|$ is small, $\forall u_0$ fixed, and $\alpha \neq \alpha_0$, then the system's output is insensitive with respect to $\alpha + \alpha_0$.

Figure 9.3 illustrates the setting of this definition.

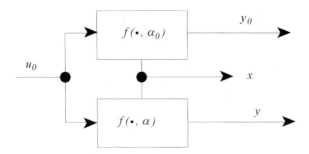

Figure 9.3 Parameter sensitivity

Definition 2: Robustness The system is robust to parameter variation if $\forall \alpha \in$ the ball $B(\alpha_0, \Delta\alpha)$, the system is stable from an input–output point of view. That is, the system is stable at $\alpha, \forall \alpha \in B(\alpha_0, \Delta\alpha)$, if $\|y - y_0\|$ is small for $\|\alpha - \alpha_0\|$ small and such that (9.3) holds for fixed $u \in F(U)$.

Usually, stability of the system at α means that $\|y_1 - y_2\|$ is small whenever $\|u_1 - u_2\|$ is small, such that (9.3) is satisfied for u_1 and u_2, and α. Clearly, this pertains to some continuity of the input–output maps, f and h. These concepts will be utilized when, in addition to $\Delta\alpha$, an external disturbance $d \in F(D)$ is applied. In this case the system will be considered robust if it is stable for $\alpha \in B(\alpha_0, \Delta\alpha)$ and $d \in F(D)$, and such that some design specification, say the error $e = x - x_d$, is minimized.

We conclude that if the above closed-loop system is stable for all bounded ΔF and all physically possible bounded inputs such that the desired design goal is met, then the FLC is robust. This may furthermore be interpreted as referring to robust stability or robust performance.

9.5.3 Formulation of Fuzzy System Robustness

The task of the fuzzy control system may be broken down into stability and meeting a desired objective. This says that the fuzzy controller should guarantee a stable operation as a specified objective is met. The class of systems considered is given *implicitly* by

$$\dot{x}' = f(x', u, \alpha), \qquad x = e = x' - x_d \in \Re^n, \tag{9.4}$$

where $u \in \mathbb{R}^m, \alpha \in \mathbb{R}^l$ and $f \in \mathbb{R}^r$. x is the implicit state vector, and x' is the actual state vector.

Note that the terminology and definitions developed in the previous sub section are general. In what follows different symbols will sometimes be used; these are clarified without loss of generality:

$x = $ a system's states with their universes of discourse, X;
$\alpha_i = [\alpha_{\min}, \alpha_{\max}]$, with nominal value, α_{i0};
$\Delta\alpha_i = $ a small perturbation in α_i about the nominal value;
$u = $ the input vector.

Define the error as

$$e = x' - x_d \in \mathbb{R}, \tag{9.5}$$

where x_d is the desired state vector.

In what follows we will consider that $r \leq n$, i.e. there are at most as many parameters as there are states. This will allow us to use square matrices as will be done shortly. This is a minor restriction which can be removed.

For simplicity and clarity, we will consider the case of only two parameters and two states i.e., $r = n = 2$. The *implicit* states the error coordinates are then defined as $x_1 = e_1 = x'_1 - x_{1d}$ and $x_2 = e_2 = x'_2 - x_{2d}$. The generalization to $\mathbb{R}^n$ is fairly straightforward.

Assumption 1 *Suppose that the parameter α_i is predominantly a characteristic of the state i dynamics, $\dot{x}'_i$, and such that 'cross-coupling' or interaction between channels is negligible. In other words, a perturbation in the parameter α_i has a negligible effect on the growth of error $e_j, i \neq j$. That is, e_i is not directly sensitized by the perturbation, $\Delta \alpha_j$. Therefore, one parameter can be fixed as the other is varied. This hypothesis is not difficult to satisfy on most engineered systems where some ad hoc procedures are usually employed in practical sensitivity analysis.*

The result is extended later on to the case where a very weak interaction is assumed to exist, and which is therefore considered to be vanishingly small. As a consequence of above assumption, the following definitions are stated.

Definition 3 *Define a unit parameter perturbation of a state (or channel) as*

$$\frac{\Delta \alpha_i}{\alpha_i} = k_{ii} \in \mathbb{R}, \quad i = 1, 2, \ldots, r. \tag{9.6}$$

Definition 4 *The fuzzy sensitivity of the real output function $e(x, \alpha_i)$ with respect to the real parameters $\alpha_i, i = 1, \ldots, r$, is expressed by [40]*

$$S^e_\alpha = \frac{1 - \mu_{\Delta e}}{1 - w_1 \mu_{\Delta \alpha_1} - w_2 \mu_{\Delta \alpha_2} - \cdots - w_r \mu_{\Delta \alpha_r}}, \quad \sum_{i=1}^{i=r} w_i = 1, \quad w_i \in [0, 1], \tag{9.7}$$

where the μ's are membership functions of the deviations and the w's are weights that are heuristically attached to signify the importance of a parameter. For two parameters jointly considered and of equal importance, the above becomes

$$S^e_\alpha = \frac{1 - \mu_{\Delta e}}{1 - w_1 \mu_{\Delta \alpha_1} - w_2 \mu_{\Delta \alpha_2}}, \quad w_1 = w_2 = 0.5. \tag{9.8}$$

If the parameter α_i is attributed to state i only, then this can be viewed as the case of a single parameter per state, which can be varied independently. That is

$$S^{e_i}_{\alpha_i} = \frac{1 - \mu_{\Delta e_i}}{1 - \mu_{\Delta \alpha_i}}, \quad i = 1, \ldots, n, \; t \geq 0. \tag{9.9}$$

From this and Definition 3, the set below is formed for $n = 2$:

$$\{S^{e_1}_{\alpha_1}, S^{e_2}_{\alpha_2}; k_{11}, k_{22}\}, \quad t \geq 0. \tag{9.10}$$

This set gives all the relevant sensitivities and perturbations for the $r = n = 2$ case.

However, whenever cross-coupling exists, the single parameter expression will still be used but with the parameter α_i also giving rise to a change in e_j, for $i \neq j$. In the general case the set in (9.10) assumes the more general form below:

$$\{S^{e_i}_{\alpha_i}, S^{e_j}_{\alpha_i}; k_{ii}\}, \quad t \geq 0, \tag{9.11}$$

where $S^{e_1}_{\alpha_1}$, for example, measures the sensitivity of e_1 due to a change in α_1.

9.5.4 The Main Result

Theorem 1 *Let the fuzzy controller K stabilize $f(\cdot, \alpha_{i0}) \in F, i = 1, 2$ (nominal stability), and suppose that Assumption 1 holds.*

Robust stability. *The fuzzy controller is robust with respect to F if the matrix formed as*

$$P = [p_{ij}] = [S^{e_j}_{\Delta\alpha_i} k_{ii}], \quad i = 1, \ldots, l = n, \quad (9.12)$$

is negative semi-definite, for $t \geq 0$.

The matrix, P, will be referred to as the per unit sensitivity matrix. For $r = n = 2$, this is

$$P = \begin{bmatrix} S^{e_1}_{\Delta\alpha_1} k_{11} & 0 \\ 0 & S^{e_2}_{\Delta\alpha_2} k_{22} \end{bmatrix} \leq 0, \quad \forall t > 0. \quad (9.13)$$

9.5.5 Derivation of the Main Result

Let us define the performance measure as a lower-bounded function given by

$$\mathfrak{J} \triangleq V(e) = \tfrac{1}{2}(e_1^2 + e_2^2) \quad \text{or} \quad V(x_1, x_2) = \tfrac{1}{2}(x_1^2 + x_2^2). \quad (9.14)$$

This needs to be minimized. A global minimizer for this is $x_1 = x_2 = 0$.

Suppose that $x_1 = x_2 = 0$ is the only equilibrium point of the nominal plant. From a stability standpoint, the FLC is required to ensure that $\dot{V} = dV(x_1, x_2)/dt \leq 0$, for all $t > 0$, or $dV(x_1, x_2) \leq 0$. From (9.12) we have

$$dV(x_1, x_2) = \tfrac{1}{2}(2x_1 dx_1 + 2x_2 dx_2) = x_1 dx_1 + x_2 dx_2. \quad (9.15)$$

But

$$dx_1 = \frac{\partial x_1}{\partial \alpha_1} \Delta\alpha_1 + \frac{\partial x_1}{\partial \alpha_2} \Delta\alpha_2, \quad (9.16)$$

and

$$dx_2 = \frac{\partial x_2}{\partial \alpha_1} \Delta\alpha_1 + \frac{\partial x_2}{\partial \alpha_2} \Delta\alpha_2,$$

with $\Delta\alpha \approx d\alpha$ (small changes).

Since we are concerned with a local evolution of the system in the cell state space, let us consider the most general notion of sensitivity from control theory. The sensitivity of an observed variable e which is a function of the parameter α is defined as the percentage change in the quantity e divided by the percentage change in the parameter α that caused the change in e [95]. The most commonly used expression is

$$S^e_\alpha = \frac{\partial e/e}{\partial \alpha/\alpha} = \frac{\partial e}{\partial \alpha} \frac{\alpha}{e}. \quad (9.17)$$

This should give a reasonably good estimate of the sensitivity for small changes. In all cases, fuzzy or non-fuzzy, the objective is to keep the sensitivity as small as possible. Ideally, a system that is completely insensitive to all parameter variation is said to have a sensitivity of zero. We will use (9.17) to find an approximate sensitivity expression in terms of fuzzy quantities. Thus, comparing (9.9) and (9.17) gives

$$S_\alpha^e = \frac{\partial e}{\partial \alpha}\frac{\alpha}{e} \approx \frac{1-\mu_{\Delta e}}{1-\mu_{\Delta \alpha}}. \tag{9.18}$$

From this, it is approximated that

$$\frac{\partial e}{\partial \alpha} \approx \left(\frac{1-\mu_{\Delta e}}{1-\mu_{\Delta \alpha}}\right)\frac{e}{\alpha}. \tag{9.19}$$

This is a very important expression for it allows the partial derivative to be approximated in terms of fuzzy quantities. For notational simplicity, the deviation from the parameter α will be denoted by, $\tilde{\alpha}$, instead of $\Delta \alpha$, with a membership function $\mu_{\tilde{\alpha}}$. Using (9.16)–(9.19), and using x_1 and x_2 in place of e_1 and e_2, in the expression for dV in (9.15), the expression becomes:

$$dV(x_1, x_2) = x_1 \left[\frac{1-\mu_{\tilde{x}_1}}{1-\mu_{\tilde{\alpha}_1}}\frac{\Delta \alpha_1}{\alpha_1}x_1 + \frac{1-\mu_{\tilde{x}_1}}{1-\mu_{\tilde{\alpha}_2}}\frac{\Delta \alpha_2}{\alpha_2}x_1\right]$$
$$+ x_2\left[\frac{1-\mu_{\tilde{x}_2}}{1-\mu_{\tilde{\alpha}_1}}\frac{\Delta \alpha_1}{\alpha_1}x_2 + \frac{1-\mu_{\tilde{x}_2}}{1-\mu_{\tilde{\alpha}_2}}\frac{\Delta \alpha_2}{\alpha_2}x_2\right]. \tag{9.20}$$

Using the set in (9.11), the above expression becomes

$$dV(x_1, x_2) = x_1[S_{\tilde{\alpha}_1}^{\tilde{x}_1}k_{11}x_1 + S_{\tilde{\alpha}_2}^{\tilde{x}_1}k_{22}x_1] + x_2[S_{\tilde{\alpha}_1}^{\tilde{x}_2}k_{11}x_2 + S_{\tilde{\alpha}_2}^{\tilde{x}_2}k_{22}x_2]. \tag{9.21}$$

Letting the cross terms $S_{\tilde{\alpha}_1}^{e_2} = 0 = S_{\tilde{\alpha}_2}^{e_1}$ and rearranging, this becomes:

$$dV(x_1, x_2) = [x_1 \ x_2]\begin{bmatrix} S_{\tilde{\alpha}_1}^{x_1} & 0 \\ 0 & S_{\tilde{\alpha}_2}^{x_2} \end{bmatrix}\begin{bmatrix} k_{11} & 0 \\ 0 & k_{22} \end{bmatrix}\begin{bmatrix} x_1 \\ x_2 \end{bmatrix} \tag{9.22}$$

or

$$dV(x_1, x_2) = [x_1 \ x_2]\begin{bmatrix} S_{\tilde{\alpha}_1}^{x_1}K_{11} & 0 \\ 0 & S_{\tilde{\alpha}_2}^{x_2}k_{22} \end{bmatrix}\begin{bmatrix} x_1 \\ x_2 \end{bmatrix} \triangleq x^T P x. \tag{9.23}$$

Then $dV(x_1, x_2) \leq 0 \Rightarrow P \leq 0$. □

9.6 GENERALIZATION OF THE ROBUST STABILITY RESULT

In reality, a system may not be easily decoupled. An interaction is bound to exist between certain modes of the system, especially at high frequencies. Even if this is not the case, sometimes certain modes of the system are just naturally coupled and any attempt to decouple them is usually done only to simplify design or analysis. Such couplings may need to be taken into account, depending on the particular system, its complexity and the required degree of operational precision. In what follows it is assumed that state interactions exist as a result of parameter variations. If indeed a given parameter, α_j, is considered to be a dominant characteristic of only the state x_j, then whenever α_j assumes a value other than its nominal value, one would expect, in general, a more than proportionate change in the $\dot{x}_j$ dynamics than in the $\dot{x}_i$ dynamics and vice versa, depending of course on how x_i and x_j are actually coupled. If this is not the case, then the interactions may not be assumed to be weak. As is often advisable in practical robust parameter designs, we will not attempt to

GENERALIZATION OF THE ROBUST STABILITY RESULT

measure the root cause of the interaction, but rather its effect. Thus, rather than embark upon measuring such quantities as k_{ij} for $i \neq j$, we will proceed to estimate the strength of any resulting interactions indirectly as cross sensitivities. The cross sensitivities must therefore be taken into account and reflected in the P-matrix.

9.6.1 Virtual Interactions Based on Stability

In what follows an expression for an approximate measure of the interaction will be derived in terms of the decoupled or state-characteristic unit perturbations k_{ii} and the sensitivities $S_{\alpha_i}^{e_i}$. Let us denote by ε_{ij} (or $S_{\alpha_i}^{e_j}$) the virtual interactions. These are interactions that resulted because of the effect of varying α_i on x_j. In other words, it is the sensitivity function of e_j due to α_i. For n states, there will be $n^2 - n$ such interactions. This is reflected in the augmented matrix given below:

$$\begin{bmatrix} S_{\alpha_1}^{x_1} & 0 & \cdots & 0 \\ 0 & S_{\alpha_2}^{x_2} & & \vdots \\ \vdots & & \ddots & 0 \\ 0 & 0 & \cdots & S_{\alpha_n}^{x_n} \end{bmatrix} + \begin{bmatrix} 0 & \varepsilon_{12} & \cdots & \varepsilon_{1n} \\ \varepsilon_{21} & \ddots & & \varepsilon_{2n} \\ \vdots & & & \vdots \\ \varepsilon_{n1} & \varepsilon_{n2} & \cdots & 0 \end{bmatrix} = \begin{bmatrix} S_{\alpha_1}^{x_1} & \varepsilon_{12} & \cdots & \varepsilon_{1n} \\ \varepsilon_{21} & S_{\alpha_2}^{x_2} & & \varepsilon_{2n} \\ \vdots & & \ddots & \vdots \\ \varepsilon_{n1} & \varepsilon_{n2} & \cdots & S_{\alpha_n}^{x_n} \end{bmatrix}. \quad (9.24)$$

The implication of this augmentation is that superposition can be used to determine the overall effect of the individual interactions. Specifically, ε_{ij} (due to α_i) can be determined by keeping $k_j = \Delta \alpha_j / \alpha_j = 0$. In order to realize the full effect of the perturbations, the actual perturbations $\Delta \alpha_i$ will be used instead of the per unit quantities, k_{ii}. Identical results are obtainable either way. The matrix in (9.22) is now augmented to obtain the *total sensitivity matrix* expressed as

$$dV = [x_1 \ldots x_n] \begin{bmatrix} S_{\alpha_1}^{x_1} & \varepsilon_{12} & \cdots & \varepsilon_{1n} \\ \varepsilon_{21} & S_{\alpha_2}^{x_2} & & \varepsilon_{2n} \\ \vdots & & \ddots & \vdots \\ \varepsilon_{n1} & \varepsilon_{n2} & \cdots & S_{\alpha_n}^{x_n} \end{bmatrix} \begin{bmatrix} \Delta\alpha_1 & 0 & \cdots & 0 \\ 0 & \Delta\alpha_2 & & \vdots \\ \vdots & & \ddots & 0 \\ 0 & 0 & \cdots & \Delta\alpha_n \end{bmatrix} \begin{bmatrix} x_1 \\ \vdots \\ x_n \end{bmatrix} \overset{\Delta}{=} x^T \tilde{P} x, \quad \text{where } \tilde{P} = [P_{ij}]. \quad (9.25)$$

We now proceed to expand (9.25) in quadratic form as

$$dV(x_1, \ldots, x_n) = \sum_{i=1}^{n} \sum_{j=1}^{n} P_{ij} x_i x_j,$$

where

$$P_{ij} = \begin{cases} S_{jj} \Delta \alpha_j, & i = j, \\ \varepsilon_{ij} \Delta \alpha_j, & i \neq j. \end{cases} \quad (9.26)$$

For stability convergence in the presence of interactions between the states, $dV \leq 0$, gives the set of inequalities below:

$$dV \leq 0 \Rightarrow x_j \Delta \alpha_j \varepsilon_{ij} x_i + x_j S_{\tilde{\alpha}_j}^{x_j} \Delta \alpha_j x_j \leq 0, \quad \text{with } k_{ii} = 0 \tag{9.27}$$

From these, the measures of the virtual interactions are bounded as

$$\varepsilon_{ij} < -S_{\tilde{\alpha}_j}^{x_j} \frac{x_j}{x_i}, \quad \forall t > 0. \tag{9.28}$$

The above set of inequalities will be revisited shortly. Meanwhile, let ε_{ij} be replaced by the respective infima of the right-hand side of the inequalities as

$$\varepsilon_{ij}^* = \inf_{\alpha_j} \left(-S_{\tilde{\alpha}_j}^{x_j} \frac{x_j}{x_i} \right) \tag{9.29}$$

From Section 9.5, the fuzzy sensitivity of e_j with respect to the parameter α_i may be expressed as

$$S_{\tilde{\alpha}_i}^{x_j} \stackrel{\Delta}{=} \frac{1 - \mu_{e_j}}{1 - \mu_{\tilde{\alpha}_i}} = \frac{1 - \mu_{\tilde{\alpha}_j}}{1 - \mu_{e_i}} S_{\tilde{\alpha}_j}^{x_j} S_{\tilde{\alpha}_i}^{x_i}. \tag{9.30}$$

Using this and (9.29), the interactions are bounded as

$$S_{\tilde{\alpha}_i}^{x_j} \leq \varepsilon_{ij}^*. \tag{9.31}$$

With these, the expression for dV from (9.25) becomes

$$dV = [x_1 \ldots x_n] \begin{bmatrix} S_{\tilde{\alpha}_1}^{x_1} & \varepsilon_{12}^* & \cdots & \varepsilon_{1n}^* \\ \varepsilon_{21}^* & S_{\tilde{\alpha}_2}^{x_2} & & \varepsilon_{2n}^* \\ \vdots & & \ddots & \vdots \\ \varepsilon_{n1}^* & \varepsilon_{n2}^* & \cdots & S_{\tilde{\alpha}_n}^{x_n} \end{bmatrix} \begin{bmatrix} \Delta \alpha_1 & 0 & \cdots & 0 \\ 0 & \Delta \alpha_2 & & \vdots \\ \vdots & & \ddots & 0 \\ 0 & 0 & \cdots & \Delta \alpha_n \end{bmatrix} \begin{bmatrix} x_1 \\ \vdots \\ x_n \end{bmatrix} \stackrel{\Delta}{=} x^T \tilde{P} x, \quad \text{where } \tilde{P} = [p_{ij}]. \tag{9.32}$$

Equation (9.32) is now the performance function that needs to be minimized in order to obtain the 'best' values for $S_{\tilde{\alpha}_i}^{x_i}$ and $\Delta \alpha_i$ such that, in addition, $dV \leq 0$ for asymptotic stability convergence.

In what follows we obtain robust stability bounds for the interactions. The inequality bounds for the couplings have already been determined in (9.28)–(9.31). Thus the coupling terms are to be chosen such that

$$\varepsilon_{ij} \leq \varepsilon_{ij}^*, \quad \forall t > 0, \forall i \neq j, \tag{9.33}$$

where the quantities with asterisks are the infima given in (9.29). An inherent restriction is that the sensitivities are not ideal, thus making both interactions non-zero as desired. For a specified perturbation, $\Delta \alpha_i$, the corresponding interaction, ε_{ij} or ε_{ji}, can be determined upon knowing the profiles of the sensitivities. The infima of the interactions are given by

$$\varepsilon_{ij}^* \leq -S_{\tilde{\alpha}_j}^{x_j} |_{\Delta \alpha_j} \inf \left(\frac{x_j}{x_j} \right), \quad \forall i, j, i \neq j, \forall t > 0. \tag{9.34}$$

GENERALIZATION OF THE ROBUST STABILITY RESULT   **177**

The bounds on the parameter perturbations are then obtained from the determinant conditions in (9.32). These bounds are not hard and fast even though they are written in precise inequalities. It may be difficult to satisfy them automatically, as the expression for dV is multi-parameter. Some heuristic procedure may have to be employed iteratively to resolve any conflicting requirements.

9.6.2 General Result for Robust Stabilization

A general version of the main result stated in Theorem 1 in Section 9.5.3 now follows. Note that in general the $\tilde{P}$ is not symmetric. Thus, in order to derive the general result, it needs to be transformed into symmetric form. This is given below as

$$dV(x_1, \ldots, x_n) = \sum_{i=1}^{n} \sum_{j=1}^{n} p_{ij} x_i x_j, \quad \text{where } \tilde{P}|_{\text{sym}} = [p_{ij}]_{\text{sym}},$$

with

$$P_{ij} = \begin{cases} s_{ij} \Delta \alpha_j, & i = j, \\ \dfrac{\varepsilon_{ij} \Delta \alpha_j + \varepsilon_{ji} \Delta \alpha_i}{2} & i \neq j. \end{cases} \quad (9.35)$$

Theorem 2 *The fuzzy logic controller will stabilize the whole of the plant family, P, in Section 9.5.1 without Assumption 1 if the augmented matrix, $\tilde{P}|_{\text{sym}}$, in (9.32) is negative semi-definite that is, if its leading principal minors are;*

$$\Delta_1 = p_{11} = S_{\tilde{\alpha}_1}^{x_1} \Delta \alpha_1 \leq 0, \quad \Delta_{12} = \det \begin{pmatrix} P_{11} & P_{12} \\ P_{21} & P_{22} \end{pmatrix} \leq 0,$$

$$\Delta_{123} = \det \begin{pmatrix} P_{11} & P_{12} & P_{13} \\ P_{21} & P_{22} & P_{23} \\ P_{31} & P_{32} & P_{33} \end{pmatrix} \leq 0, \ldots, \quad \Delta_{12\ldots n} = \Delta(\tilde{P}_{\text{sym}}) \leq 0. \quad (9.36)$$

9.6.3 Minimizing dV

Being the performance index, dV needs to be minimized. Now that P is symmetric, a host of tools from optimization can be applied. Suppose the errors are normalized as $\|x\|^2 = 1$. The minimization problem can be stated as $\min x^T \tilde{P} x$ subject to $x^T x = 1$ or $(x^T x - 1) = 0$. This constraint can be adjoined to dV via a Lagrange multiplier, λ.

The problem is then restated as

$$\min(dV = x^T \tilde{P}_{\text{sym}} x) - \lambda(x^T x - 1). \quad (9.37)$$

This requires that the derivative with respect to each component x_i vanish. The result is a set of eigenvalue eigenvector equations $Ax - \lambda x = 0$. If x satisfies this condition, then

$$dV = x^T \tilde{P}_{\text{sym}} x = \lambda x^T x = \lambda \quad \text{as } x^T x = 1).$$

$$\therefore dV_{\min} = \lambda_{\min} (\text{and } dV_{\max} = \lambda_{\max}).$$

Remarks The conditions stated for robust stabilization by the FLC are all time-dependent and are therefore very strong; this is a consequence of the fuzzy sensitivities which are time-dependent. Thus, in general, dV will have to be monitored with respect to time to observe its dynamic behavior. Also, from the matrix conditions, one obtains specific bounds in the form of inequalities on the desirable values of $\mu_{\tilde{x}}$ and $\mu_{\tilde{\alpha}}$ in their respective universes of discourse.

Even though one has been able to minimize the expression for dV, this does not tell us how to select the optimum sensitivities, perturbations and interactions. Without this minimization though, dV will be very conservative. So far, no bounds have been imposed on the sensitivities directly, other than the matrix conditions. Ideally, it would be desirable that each sensitivity be very small. In other words, it is undesirable that small parameter perturbations should lead to large trajectory errors. This then raises the question about the size of $\|\tilde{P}_{\text{sym}}\|$. Precisely how large can the perturbations be before the FLC ceases to stabilize the system? Therefore, bounds on the sensitivity and the parameter variation needs to be derived in terms of inequalities. Before this is considered, some concepts pertaining to the ideal and worst case bounds on these quantities will be introduced, appropriately, in fuzzy terms.

9.7 FUZZY EXTREMES OF PERTURBATIONS

The analysis being developed addresses the case of small parameter perturbation, $\Delta\alpha$, where $\Delta\alpha \in [-\Delta\alpha_m, \Delta\alpha_m]$, α_0 = nominal. In terms of an arbitrary membership function shape, the fuzzy set for 'small' is illustrated in Figure 9.4. Note that the origin in the universe of discourse corresponds to no parameter deviation, i.e. $\alpha = \alpha_0$, the nominal value. Thus, any $|\Delta\alpha| > \Delta\alpha_m \neq$ 'small' and is not considered.

Definition 5 *Worst case sensitivity* is such that a 'small' parameter perturbation leads to the 'largest' change in output. That is, if

$$(1 - \mu_{\Delta\alpha}) \to 0 \Rightarrow \mu_{\Delta\alpha} \to 1 \ni (1 - \mu_{\Delta e}) \to 1, \text{ or } \mu_{\Delta e} \to 0.$$

This is explained in Figure 9.5. That is, if $\forall \Delta\alpha \in I_\alpha(\varepsilon_0), \mu_{\Delta\alpha} \to 1, \Delta e \in F(\Delta e) \setminus \text{Supp}(\mu_{\Delta e})$ such that $1 - \mu_{\Delta\alpha} \to 0$, then the result is therefore $S_\alpha^e k_{ij} \to \infty$.

Definition 6 *Ideal Sensitivity* is obtained when S_α^e is smallest, i.e.

$$(1 - \mu_{\Delta e})/(1 - \mu_{\Delta\alpha}) \sim 0.$$

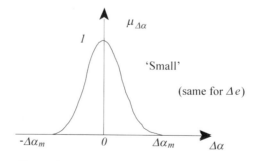

Figure 9.4 Fuzzy set for small perturbations

FUZZY EXTREMES OF PERTURBATIONS 179

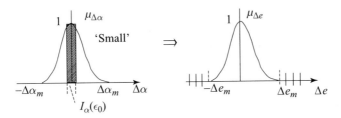

Figure 9.5 Worst case sensitivity

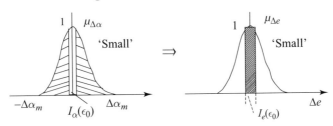

Figure 9.6 Ideal case sensitivity

In other words, the output level is unaffected by parameter perturbations. This situation is illustrated in terms of fuzzy quantities in Figure 9.6. The Condition for ideal sensitivity is such that if $\forall \Delta\alpha \in \mathrm{Supp}(\mu_{\Delta\alpha})\ I_\alpha(\varepsilon_0) \Rightarrow \Delta e \in aI_e(\varepsilon_0)$ on Δe, where *Supp* denotes *support*, then the result is $S_\alpha^e k_{ij} \to 0$ (the smallest or ideal case).

Robustness Concerns:

- How large can k_{ij} be?
- How large can S_α^e be?

Or, on combining these, how large can $S_\alpha^e k_{ij}$ be before the system is destabilized, i.e. before the *P*-matrix fails to be ≤ 0?

9.7.1 A Measure of Fuzzy Robustness

The matrix conditions in (9.36) were obtained by applying Sylvester's rule for negative definiteness. The second condition is a determinant condition. However, it is known that the determinant of a matrix is not always a good indication of how close the matrix comes to being singular. The eigenvalues are hardly a better measure. A better measure of the near-singularity of a matrix is the set of singular values of the matrix. For the *real* matrix M, the singular values are given by

$$\sigma_i = [\lambda_i(M^\mathrm{T} M)]^{1/2}. \tag{9.39}$$

These are non-negative, real numbers ordered as $\sigma_1 \leq \sigma_2 \ldots \leq \sigma_p$. The largest singular value is denoted by $\bar{\sigma}$ and the smallest, by $\underline{\sigma}$. The smallest singular value is a measure of the singularity of the matrix. The smaller it is, the closer the matrix gets to being singular. A small $\underline{\sigma}$ is therefore an indication that the matrix is nearly singular [103, 104].

So, for the fuzzy total sensitivity matrix, the supremum of the maximum singular value gives a measure of the size of this augmented sensitivity matrix. This is given by

$$\|\tilde{P}\|_\infty = \sup_t \bar{\sigma}(\tilde{P}, t). \tag{9.40}$$

As a measure of robustness, the condition number of the matrix will be used when this is reasonably small, i.e. when the smallest singular value is different from zero. The condition number is defined as $\bar{\sigma}/\underline{\sigma}$, i.e. the ratio of the largest to the smallest singular value. The larger this ratio the faster the matrix approaches singularity. When an eigenvalue becomes very small, the condition number becomes excessively big and will not be used as the robustness measure; instead, the next smallest singular value will be used. For instance, in the two-dimensional case, this is just the maximum singular value. Because these singular values are time-varying, the robustness measure is obtained from the above as the smallest of the ratio and is called the *fuzzy robustness measure (FRM)* given by

$$FRM = \frac{\sup_t \bar{\sigma}_i(\tilde{P},t)}{\inf_t \underline{\sigma}_i(\tilde{P},t)} = \frac{\|\tilde{P}\|_\infty}{\inf_t \sigma_i(\tilde{P},t)}, \quad (9.41)$$

where $FRM \geq 1$. Consider the feasible universe of discourse for the fuzzy robustness measure to be $[FRM_{min}, FRM_{max}]$. The computed value of FRM in (9.41) can have membership in this interval which can be described linguistically in the fuzzy set in Figure 9.7.

As already mentioned, whenever $\underline{\sigma} \sim 0$, the FRM will be considered to be $\bar{\sigma}$. This will be required to be as small as possible for good robustness. The same linguistic description given in Figure 9.7 will still be used.

9.7.2 Comments

Various ways to derive inequality bounds on sensitivities and perturbations for robust stability have been shown in this chapter. Many of these do not directly impose any size restriction on the augmented sensitivity matrix. However, the use of singular values gives a measure of the size of this matrix. This, therefore, provides an additional measure that the designer may consider, in attempting to be as less conservative as possible, in the selection of the bounded design quantities. The FRM using (9.40) or (9.41) can also be used to get a 'feel' for when the design is getting out of control, and therefore consider a possible redesign. Redesign here refers to possibly using smaller perturbation in a particular parameter, or coping with a less than optimum sensitivity. Whatever the case, the designer should exercise engineering prudence in utilizing the hosts of inequality bounds in selecting the appropriate design parameters and the region of the universes of discourse of the relevant variables which provide a satisfactorily robust performance. The theory has been developed to include the case of both coupled and decoupled state interactions. By setting

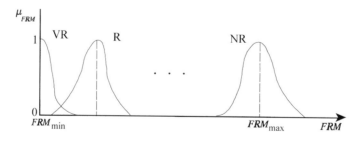

Figure 9.7 Fuzzy robustness mMeasure

$\varepsilon_{ij} = 0, \forall i \neq j$, the corresponding results under Assumption 1 are obtained. The matrix conditions are made possible because one has succeeded in putting dV in quadratic form. This is only an unusual temporary comfort, so to say. In general, this may not be possible if the number of parameters is not the same as the number of states. The result is easily extended to include an external disturbance by considering this as one of the parameters. However, this decreases the possible number of parameters by one. If desired, the scalar expression for dV, instead of the matrix form, may still be used with a little loss of generality, and only some difficulty in achieving the explicit matrix conditions.

9.7.2.1 A Compromise between dV_{min} and FRM_{min}

When attempting to select the optimum parameters for robust performance, a compromise will have to be made between a high precision stability convergence to the set-point and what is merely an indication of a good performance behavior. Since dV is required to be zero at the origin, the closer to zero this is at the steady state the better the convergence to the set-point. The minimum point of this does not necessarily correspond to the minimum value of the fuzzy robustness measure. The designer should therefore decide which of the options is more desirable for the situation.

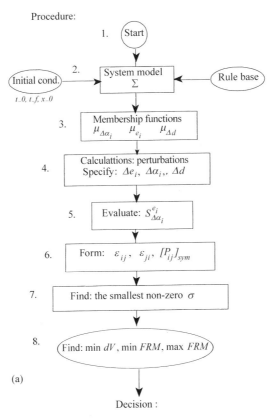

(a)

Figure 9.8(a) Procedure for fuzzy robustness calculations

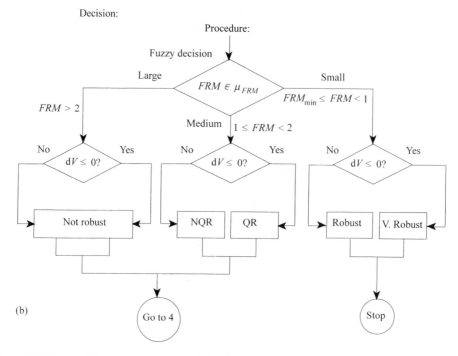

Figure 9.8(b) Decision process for determining FRM (QR = quite robust, NQR = not quite robust, V = very robust)

The procedure for calculating the relevant fuzzy robustness quantities are outlined in Figure 9.8(a). Figure 9.8(b) explains the decision process for determining the linguistic robustness measures for the fuzzy control system.

9.8 APPLICATION EXAMPLE

Consider the test-bed fuzzy idle speed controller for the two-state engine model reported in [58]. The simplified model is described by

$$\dot{x}_1 = k_p[a_0 x_2 + a_1 x_1 - a_2 x_1 x_2 - a_3 x_2 x_1^2 + g(x_1)a_4] + [k_p g(x_1) a_5] u_1$$
$$+ k_p[g(x_1) a_6 u_1^2] \triangleq F_1(x, u, \alpha_1, d),$$

$$\dot{x}_2 = k_N \left[-a_7 + a_8 \frac{\dot{m}_{a0}}{x_2} + a_{12} x_2 - (a_{13} + a_{14}) x_2^2 \right] + [a_{10} x_2 + a_{11}] k_N u_2$$
$$+ [-a_9 k_N] u_2^2 + a_{15} T_d \triangleq F_2(x, u, \alpha_2, d). \tag{9.42}$$

In actual variables, the dynamics are expressed by

$$\dot{p} = k_p(\dot{m}_{ai} - \dot{m}_{a0}),$$
$$\dot{N} = k_N(T_i - T_L),$$

where

$$\dot{m}_{ai} = (1 + 0.907u_1 + 0.0998u_1^2)g(x_1),$$

$$\dot{m}_{a0} = -0.0005968x_2 - 0.1336x_1 + 0.0005341x_2x_1 + 0.000001757x_2x_1^2,$$

$$T_i = -39.22 + \frac{325024}{120x_2}\dot{m}_{a0} - 0.0112u_2^2 + 0.000675u_2x_2(2\pi/60)$$

$$+ 0.635u_2 + 0.0216x_2(2\pi/60) - 0.000102x_2^2(2\pi/60)^2,$$

$$T_L = (x_2/263.17)^2 + T_d,$$

$$g(x_1) = \begin{cases} 1, & x_1 < 50.66, \\ 0.0197(101.325x_1 - x_1^2)^{1/2}, & x_1 \geq 50.66, \end{cases}$$

P = manifold pressure, N = engine speed, $u_1 = \delta$ = spark advance, $u_2 = \theta$ = throttle angle, m_{ai} = mass flow into the manifold, m_{a0} = mass flow out of the manifold and into the cylinder, T_i = developed torque, T_L = load torque and $g(p)$ is the pressure influence function.

We define $x = (x_1 \; x_2)^T = (P \; N)^T, u = (u_1 \; u_2)^T = (\delta \; \theta)^T, \alpha_1 = kp$, and $\alpha_2 = kN$. The errors in P and N are given by $e_1 = x_1 - x_{1d}$ and $e_2 = x_2 - x_{2d}$; $d = T_d$ is a known scalar disturbance such as the air condition.

It is important to reiterate that the objective in both the design of the reported fuzzy controller and the robustness analysis on this simplified test-bed model is only to demonstrate applications of the developed methodologies and the mechanics involved. There is no claim whatsoever in both cases that fuzzy control has tackled the world's most difficult problems.

The coefficients, a_i, and their values are given below:

$a_0 = 0.0005968$ $a_1 = 0.1336$ $a_2 = 0.0005341$
$a_3 = 0.000001757$ $a_4 = 1$ $a_5 = 0.907$
$a_6 = 0.0998$ $a_7 = 39.22$ $a_8 = 2708.533$
$a_9 = 0.0112$ $a_{10} = 0.000070686$ $a_{11} = 0.635$
$a_{12} = 0.002306$ $a_{13} = 0.0000011856$ $a_{14} = (1/263.17)^2$
$a_{15} = -k_N = -54.26$
$k_p = 42.4, \; d = T_d \in [0, 61]$Nm.

Control Objective To design a fuzzy controller to maintain the engine's idle speed at 750 rev/min in the presence of a multitude of disturbances, the largest being the onset of an air condition. Again, details of this controller have been reported in [58] and [110] and will not be repeated here.

The general form of the linguistic rules for this controller as given in [58] is

If (N is F_1) and (P is F_2) **Then** ($\Delta\delta$ is G_1) and ($\Delta\theta$ is G_2),

where F_i and G_i are the consequent and premise linguistic terms, respectively. The rulebase consisted of 56 rules as reported in [58]. It is only instructive here to show the same rulebase after clustering. This has reduced the number of rules to only 7 as shown in Figure 9.9. The result of this is a great reduction of computational burden in the analyses.

184 ROBUST STABILITY OF FUZZY CONTROL SYSTEMS

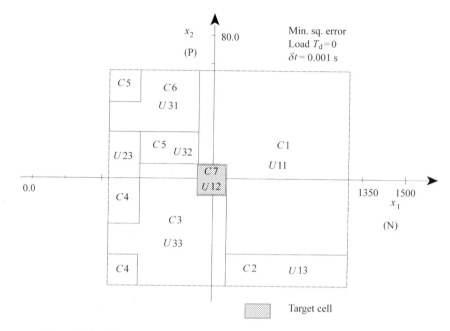

Figure 9.9 Clustered rulebase for the idle speed controller (load $r = T_d = 0$)

As a matter of interest, the cell clustering alluded to is given below:

C1: $814.151 \leq x_1 \leq 1500$ and $9.00 \leq x_2 \leq 80.0$,
$689.226 \leq x_1 \leq 815.151$ and $38.995 \leq x_2 \leq 80.0$;

C2: $814.5 < x_1 \leq 1500$ and $-0.996 < x_2 \leq 9.00$;

C3: $439.377 < x_1 < 814.151$ and $-0.996 < x_2 \leq 28.998$,
$439.377 < x_1 \leq 689.226$ and $28.998 < x_2 \leq 38.995$,
$314.452 < x_1 \leq 439.377$ and $9.00 < x_2 \leq 18.990$;

C4: $314.457 < x_1 \leq 439.377$ and $18.990 < x_2 \leq 48.994$,
$314.458 < x_1 \leq 439.377$ and $-0.996 < x_2 \leq 9.00$;

C5: $439.377 < x_1 \leq 689.226$ and $38.995 < x_2 \leq 48.995$,
$314.458 < x_1 \leq 439.377$ and $58.992 < x_2 \leq 80.0$;

C6: $439.377 < x_1 \leq 689.226$ and $48.994 < x_2 \leq 80.0$,
$314.452 < x_1 \leq 439.377$ and $48.994 < x_1 \leq 58.992$.

Target cell, C7: $689.226 < x_1 < 814.151$ and $28.998 < x_2 < 38.995$.

The cells, C_i, are *covered* by finite linguistic labels for the appropriate variable's membership function.

The application example will be illustrated for initial conditions in the regions defined by C1, C2 and C7, thus allowing the firing of at least two fuzzy rules. For good robust performance under the parameter variations, the output deviation from the set-point should be in a small neighborhood of the center of cell C7. This acceptable neighborhood should be supplied by the designer. In this example, a speed deviation of less than 1% will be specified.

This is not considered too small at all. Even in the presence of a major disturbance such as the onset of an air conditioning unit (21 Nm), the speed varied only by about 3.5% of the set-point of 750 rev/min (750 to 776 rev/min), while a new pressure was maintained that was about 38% above the set-point of 34 kPa. However, to operate at this new higher set-point, an additional control authority was required. In the case of the parameter variations, it is required that all perturbations be such that an additional control activity will not be needed. In other words, the fuzzy controller can hold its own up to a certain level of parameter variations, with all trajectories remaining in the specified small neighborhood of the set-point. This will be the general view-point taken in the theory developed, whether one is analyzing the robustness of an engine fuzzy controller or one for regulating the dynamic behavior of demand and supply about the equilibrium quantity and price, in a competitive market situation. In the case of the test-bed fuzzy controller, $x \in \mathbb{R}^2, \alpha \in \mathbb{R}^2$, so $n = r = 2$. The parameters are multiplicative, each entering the system through a particular dynamics.

Nominal values:

- $\alpha_{10} = kp = 42.4$.
- $\alpha_{20} = kN = 54.26$.
- $x_{1d} = 34, x_{2d} = 750, r = T_d = 0$.

'*Small*' *perturbations*:

- $\Delta\alpha_1, \Delta\alpha_2 \in [-11, 11]$.
- Set-point deviation: $e_1 \in [-10, 10], (30\%); e_2 = [-12, 12], (1.6\%)$.

9.8.1 Problem Statement

The aim is to study the robustness properties of the fuzzy controller designed for the above nominal system with respect to parameter variations about their nominal values, and in a small neighborhood of the set-point. The goal of the analysis is to be able to establish the best setting parameter for the most robust operation, i.e. with the least set-point deviation. At this point also, the corresponding estimates of the measures of interactions and sensitivities are obtained. The analysis will proceed in the following manner.

1. Investigate assumption 1 in Section 9.5 by varying α_1 with $\Delta\alpha_2 = 0$, and vice versa, and observe the interactions.
2. Determine the best parameter variation for the most robust operations.
3. Determine the fuzzy robustness measure, by first determining FRM_{min} and FRM_{max}. For this, one needs only determine the FRM beyond which the set-point deviation is not acceptable, even though the system remains stable. At such a point, an additional control compensation may be needed as already discussed.
4. Select a set of values of the parameters, sensitivities and interactions during the 'most robust' operation as found in 2.
5. Verify the system's operation by selecting parameters from the optimum set.

For $r = n = 2$, the indices i and j range from 1 to 2. The following are obtained:

$$dV = [x_1 x_2] \begin{bmatrix} S_{\alpha_1}^{x_1} & \varepsilon_{12} \\ \varepsilon_{21} & S_{\alpha_2}^{x_2} \end{bmatrix} \begin{bmatrix} \Delta\alpha_1 & 0 \\ 0 & \Delta\alpha_2 \end{bmatrix} \begin{bmatrix} x_1 \\ x_2 \end{bmatrix} \triangleq x^T \tilde{P} x, \quad \text{where } \tilde{P} = [P_{ij}]. \quad (9.43)$$

The symmetrical $\tilde{P}$-matrix is immediately formed from above as

$$\tilde{P}|_{sym} = \begin{bmatrix} S_{\tilde{\alpha}_1}^{x_1} \Delta\alpha_1 & \dfrac{\varepsilon_{12}\Delta\alpha_2 + \varepsilon_{21}\Delta\alpha_1}{2} \\ \dfrac{\varepsilon_{12}\Delta\alpha_2 + \varepsilon_{21}\Delta\alpha_1}{2} & S_{\tilde{\alpha}_2}^{x_2} \Delta\alpha_2 \end{bmatrix}, \quad \text{with } \sigma_i = \sqrt{\lambda_i(\tilde{P}^T \tilde{P})}, \quad (9.44)$$

where the sensitivities are in given in the set below:

$$\left\{ S_{\tilde{\alpha}_1}^{e_1} = \dfrac{1 - \mu_{e_1}}{1 - \mu_{\Delta\alpha_1}}, S_{\tilde{\alpha}_2}^{e_2} = \dfrac{1 - \mu_{e_2}}{1 - \mu_{\Delta\alpha_2}}, t; \Delta\alpha_1, \Delta\alpha_2 \right\}. \quad (9.45)$$

The interactions are given by

$$\varepsilon_{12} < -S_{\tilde{\alpha}_2}^{x_2} \dfrac{x_2}{x_1} \quad \text{and} \quad \varepsilon_{21} < -S_{\alpha_1}^{x_1} \dfrac{x_1}{x_2}, \quad \forall t > 0. \quad (9.46)$$

$$\text{Note}: e_i \stackrel{\Delta}{=} x_i = x'_i - x_{id}.$$

9.8.1.1 Membership Functions

The intervals already specified for the perturbations will be considered as their universes of discourse. The linguistic labeling is done heuristically. A more systematic way to construct the membership functions for the errors is to use a probablistic means to convert the time-varying signals $e_i(t)$ to membership functions using the method of windowing. In this example the errors are assumed to be uniform in the 'small' neighborhood of the set-point. This assumption may be valid if the set-point is the stable equilibrium for the system, for as $t \to \infty$ the error transients become progressively of bounded and diminishing variations in this neighborhood. Thus, $\exists t_0$ such that corresponding to the interval $[t_0, t_\infty]$, is a sequence of positive, negative numbers, including the number zero, which converge uniformly to zero at the set-point. It is from this sequence that the membership functions for the 'small' errors are derived. Among all possible shapes, the triangular membership functions are employed for ease of computation. These are shown in Figure 9.10.

9.8.2 Simulation Studies and Results

Several simulations have been conducted with various combinations of the two parameters. Variations that are both below and above the nominal values of these parameters have been considered. However, only a handful of representative cases are discussed here. The essence is to show particularly that the relevant fuzzy quantities used in the formulation of the theory can be computed. In order to simplify the notation, the sensitivities S_{ii} are used in place of $S_{\tilde{\alpha}_i}^{e_i}$, for all $i(i = 1, 2)$. In the first set of simulations (Figures 9.11–9.14), the parameters were varied such that α_1 was fixed at its nominal value, $\alpha_{10} = 42.40$, i.e. $\Delta\alpha_1 = 0$, while α_2 was varied below its nominal value, $\alpha_{20} = 54.26$. The fuzzy controlled system was simulated for 2 s and the relevant quantities $S_{11}(t), S_{22}(t), \varepsilon_{12}(t), \varepsilon_{21}(t), dV/dt$ and $FRM(t)$ were computed and plotted as functions of time. The second set of simulations consisting of Figures 9.15–9.18 was essentially the same as the first set except that this time

APPLICATION EXAMPLE 187

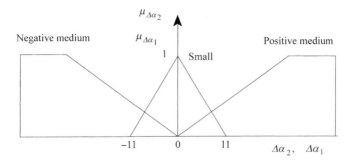

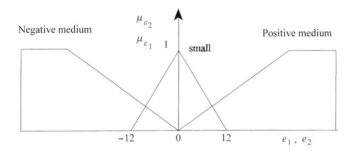

Figure 9.10 Membership functions for (a) $\Delta\alpha_i$ and (b) e_i

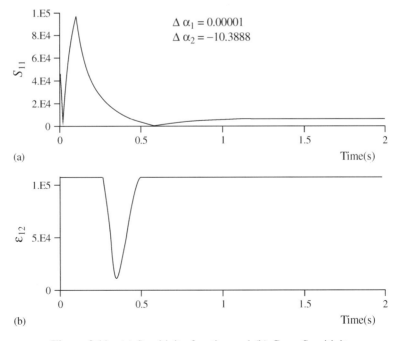

Figure 9.11 (a) Sensitivity function and (b) Cross Sensitivity

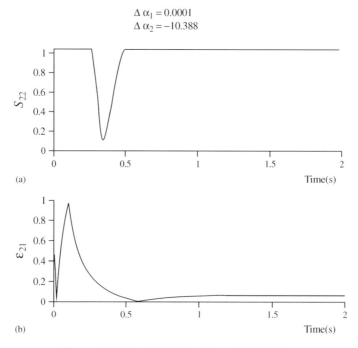

Figure 9.12 Sensitivity functions: (a) S_{22} (b) ε_{21}

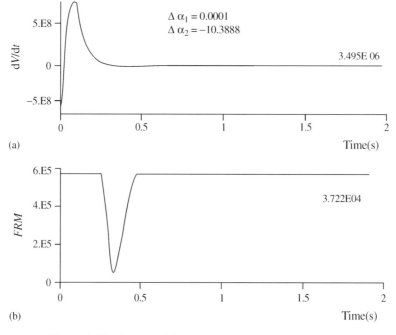

Figure 9.13 dV (a) and fuzzy robustness measure (*FRM*) (b)

APPLICATION EXAMPLE

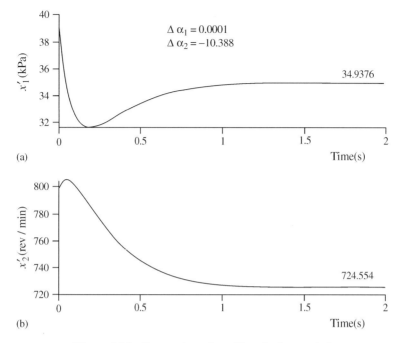

Figure 9.14 State trajectories with only $\Delta\alpha_2$ varied

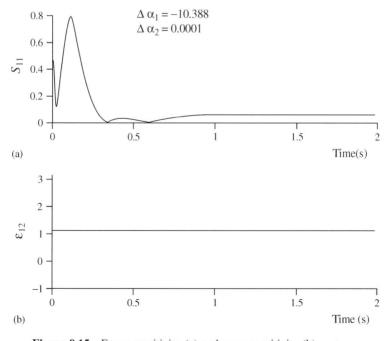

Figure 9.15 Fuzzy sensitivity (a) and cross-sensitivity (b) w.r.t. α_1

190 ROBUST STABILITY OF FUZZY CONTROL SYSTEMS

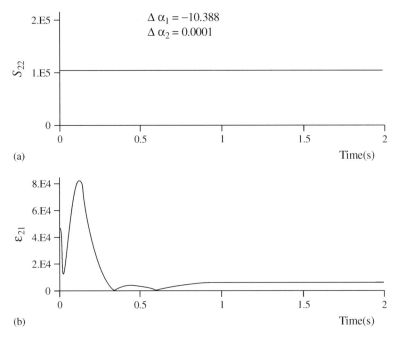

Figure 9.16 Fuzzy sensitivity (a) and cross-sensitivity (b) w.r.t. $\Delta\alpha_1$

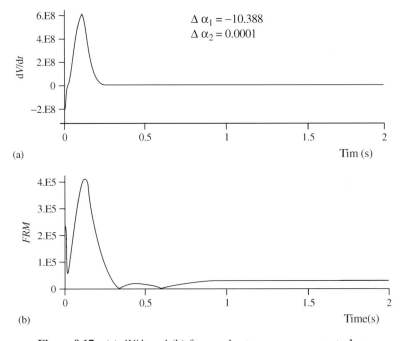

Figure 9.17 (a) dV/dt and (b) fuzzy robustness measures w.r.t. $\Delta\alpha_1$

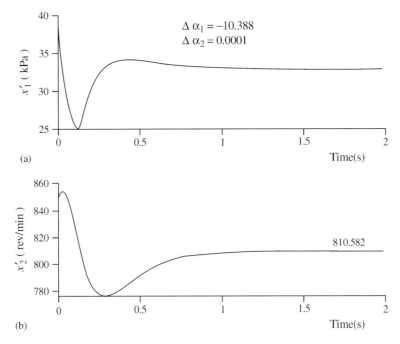

Figure 9.18 The actual state trajectories: pressure (a) and speed (b)

the parameter α_1 was varied by the same amount below its nominal value as in the first set while $\Delta\alpha_2 = 0$. The reason for this procedure was to observe if the interactions between the states might be neglected and the system considered decoupled under the effect of the parameter variations, which would allow a much simpler analysis. In the third set of simulations, Figures 9.19–9.23, an optimum fuzzy robustness set was sought which consisted of the individual sensitivities, the cross-sensitivities, dV/dt and FRM such that the set-point deviations are minimum. The last set of simulations (Figures 9.24 and 9.25) was conducted so as to address the issue of compromise between dV_{min} (or minimum steady-state errors) and FRM_{min}. By further varying α_2 lower than in the third set, a slightly different robustness set was obtained. The *minimum* set of the optimum robust elements may described as

$$\Re_\alpha = \{S_{11}, S_{22}, \varepsilon_{12}, \varepsilon_{21}, \Delta\alpha_1, \Delta\alpha_2; dV, FRM\}.$$

For the two cases (a) and (b) considered, the sets are:

$$\begin{aligned}\Re_\alpha : (a) &= \{0.10845, 1.1724E-2, 2.3883E-2, 5.3240E-2,\\ &\quad -5.10, -10.3888; -0.2452, 0.6751\},\\ \Re_\alpha : (b) &= \{0.10849, 1.1466E-2, 2.3356E-2, 5.3261E-2,\\ &\quad -5.10, -10.3889; -0.2434, 0.6726\}.\end{aligned} \quad (9.47)$$

The justification as to whether to use parameters from the set (a) or (b) depends on the designer and the particular application. In particular, the question that needs to be answered is whether the slight authority in further varying α_2 from the value in (a) to the value in (b) is worth the better convergence of x_2 (Figure 9.21 and Figure 9.24) at the expense of a lesser

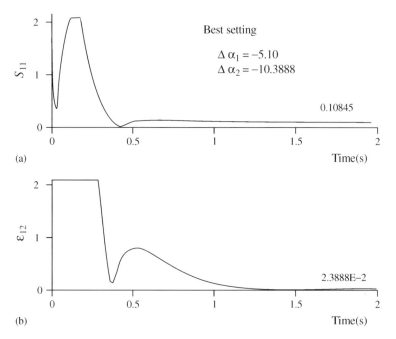

Figure 9.19 Sensitivities under the 'best' setting (both parameters varied)

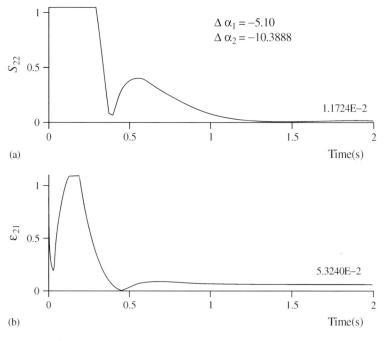

Figure 9.20 Sensitivities under the 'best' setting (both parameters varied)

APPLICATION EXAMPLE

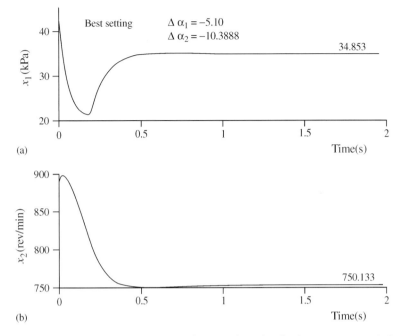

Figure 9.21 Stability convergence of state trajectories (both parameters varied)

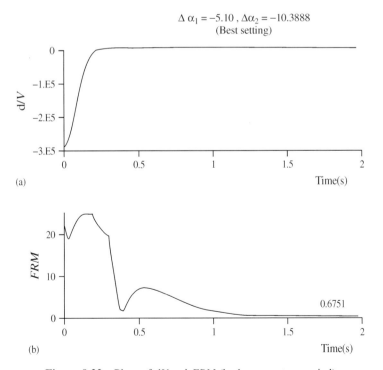

Figure 9.22 Plots of dV and FRM (both parameters varied)

194 ROBUST STABILITY OF FUZZY CONTROL SYSTEMS

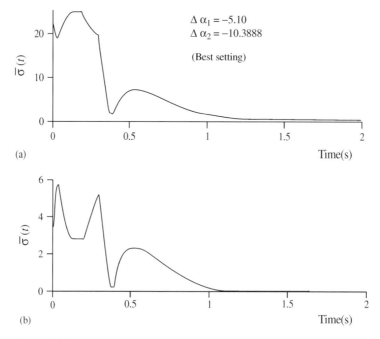

Figure 9.23 Singular values of the 'best' setting (both parameters varied)

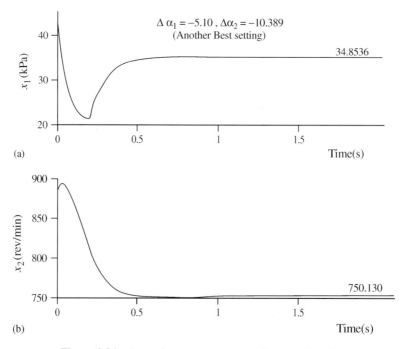

Figure 9.24 An optimum convergence of state trajectories

APPLICATION EXAMPLE 195

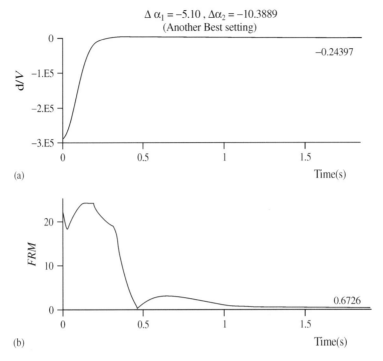

Figure 9.25 Plots of dV and FRM for a different optimum setting

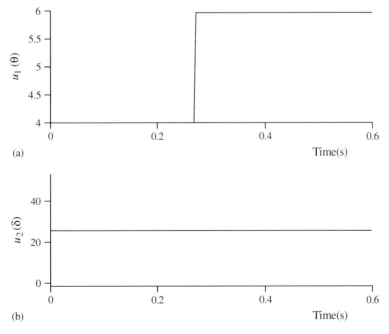

Figure 9.26 The fuzzily derived controls in C1, C3 and C7

robustness (higher *FRM*) that resulted. In both cases, the fuzzy control system is 'Very Robust' (VR). Note that during all of these parameter variations, $\Delta\alpha_1 = -5.10$ was the 'best' found. Finally, the fuzzy controls for the regions tried in the simulations are shown in Figure 9.26.

9.8.3 Discussion

The results of the simulations suggested that for this fuzzy control system the parameters should be varied below their nominal values in order to attain a 'very robust' performance. The variation of α_1 was in accordance with the stability requirement as stated in the theory, that $\Delta\alpha_1 < 0$. It was found that with α_2 fixed and α_1 varied, the effect on e_2 is $\varepsilon_{12} = 1.0965$, was small but non-zero. Compared with the individual sensitivity $S_{11} = 6.9192 \times 10^{-2}, \varepsilon_{12}/S_{11} = O(10^2)$ and not $\neq 0$. Thus ε_{12} may not be neglected. On the other hand, by keeping α_1 fixed, that is $\Delta\alpha_1 = 0$, and varying α_2, the effect on e_1 as measured by ε_{21} is 6.0677×10^{-2}, while $S_{22} = 1.0589$. In this case ε_{21}/S_{22} is quite small. However, the decision whether or not to neglect at this time depends on the particular design requirement. Overall, the theory developed provides a systematic framework for analyzing the robustness of the fuzzy idle speed controller.

9.9 CONCLUSIONS

The theory introduced and developed in this chapter addresses the issue of robustness of an input–output map with respect to parameter perturbations, and external disturbances in general. The general notions of sensitivity and robustness were introduced followed by a statement of the robustness problem. The systematic formulation of the robustness measure developed here allows a class of both linear and linearizable non-linear systems operating under fuzzy logic rules to be analyzed for robust performance. The result of the theorem gives the required inequalities on the sensitivities S_α^e and therefore on the membership function μ_e, and also on the allowable perturbations $\Delta\alpha$. An estimate of the coupling of parameters is first derived and a similar but more general result for robust performance is derived. Based on these, the sensitivity matrix, $\tilde{P}$, can be made as 'tight' as possible using singular values such that the conclusion on dV continues to hold. A fuzzy control is considered the only stabilizing element in the closed-loop system. In the development, it is assumed that the system's states are available for measurement, and that the system's outputs are some of these states. In order to incorporate the key design specification, the analysis has been developed in terms of the trajectory errors, which needed to be minimum. The type of stability in question is therefore an asymptotic stability convergence of the bounded outputs, to the prescribed desired output, to within a small, finite, non-zero tolerance. This should hold for all bounded, continually fuzzily generated control inputs and bounded perturbations. Finally, the fuzzy robustness measure is formulated quantitatively, and is also expressible in linguistic terms such as robust, very robust, not robust, and so on. As an application example, a test-bed fuzzy idle speed controller for an automotive engine was used. The theory was seen to be a valuable tool for systematic robustness analysis for the class of fuzzy control systems considered. Although in the example robustness around the equilibrium point was analyzed, this could equally well be done in other regions of the state space where the system's robust operation is of prime concern. In such a case, membership functions are to be constructed for universes of discourse in the particular locality. The final result in the analysis is the determination of the optimum robust parameter set or sets. The choice of the design elements from this set should therefore lead to a 'very robust' design. In much of modern control robust controller design a similar procedure to the one discussed herein is pursued in an iterative manner until an optimum controller is found which minimizes some normed performance measure. The approach here, though seemingly tedious, is actually along the lines of handling real, engineered systems, and is more than the usual exercise in analytical formalism.

BIBLIOGRAPHY

1. Zadeh, L. A., Fuzzy sets concepts, *Information and Control*, **8**, 1965, 338–353.
2. Zadeh, L. A., Outline of a new approach to the analysis of complex systems and decision processes, *IEEE Transactions on Systems, Man and Cybernetics*, **SMC-3**(1), 1973, 28–44.
3. Zadeh, L. A., K. S. Fu, K. Tanaka and M. Shimura, *Fuzzy Sets and their Applications to Cognitive and Decision Processes*, Academic Press, New York, 1975.
4. Lee, C. C., Fuzzy logic in control systems: Fuzzy logic controller—Part I (and Part II), *IEEE Transactions on Systems, Man and Cybernetics*, **20**(2), 1990, 404–435.
5. Chen, Y. Y. and T. C. Tsao, A description of the dynamical behavior of fuzzy systems, *IEEE Transactions on Systems, Man and Cybernetics*, **19**(4), 1989, 745–755.
6. Sugeno, M., An introductory survey of fuzzy control, *Information Sciences*, **36**, 1985, 59–83.
7. Pedrycz, W., An approach to the analysis of fuzzy systems, *International Journal of Control*, **34**(3), 1981, 404–421.
8. Kiszka, J. B., M. M. Gupta and P. N. Nikiforuk, Energetistic stability of fuzzy dynamic systems, *IEEE Transactions on Systems, Man and Cybernetics*, **SMC-15**(6), 1985, 783–791.
9. Ray, K. S. and D. D. Majumder, Application of circle criteria for the stability analysis of linear SISO and MIMO systems associated with fuzzy logic controller, *IEEE Transactions on Systems, Man and Cybernetics* **SMC-14**(2), 1984, 345–349.
10. Chand, S. and S. Hansen, Energy based stability analysis of a fuzzy roll controller design for a flexible aircraft wing, *Proceedings of the Twenty-eighth IEEE Conference or Decision and Control*, Tampa, Florida, 1989, 705–709.
11. Tong, R. M., Some properties of fuzzy feedback systems, *IEEE Transactions on Systems, Man and Cybernetics*, **SMC-10**(6), 1980, 327–330.
12. Mamdani, E. H., Applications of fuzzy algorithms for control of simple dynamic plant, *Proceedings of the IEE*, **121**(12), 1974, 1585–1588.
13. Kickert, W. J. M. and E. H. Mamdani, Analysis of a fuzzy logic controller, *Fuzzy Sets and Systems*, **1**, 1978, 29–44.
14. Mamdani, E. H., J. J. Ostergaard and E. Lembessis, Use of fuzzy logic for implementing rule-base control of industrial processes, Times Studies in Management Science, **20**, 1984, 429–445.
15. King, P. J. and E. H. Mamdani, The application of fuzzy control systems to industrial processes, *Automatica*, **13**, 1977, 235–242.
16. Sugeno, M. and T. Terano, Analytical representation of fuzzy systems, In: *Fuzzy Automata and Decision Processes*, Elsevier North-Holland, Amsterdam, 1977.
17. Larsen, P. M., Industrial application of fuzzy logic control, *International Journal of Man–Machine Studies*, **12**, 1980, 3–10.
18. Aracil, J. *et al.*, Stability indices for the global analysis of expert control systems, *IEEE Transactions on Systems, Man and Cybernetics*, **19**(5), 1989, 998–1007.

19. Langari, R. and M. Tomizuka, Stability of fuzzy linguistic control systems, *Procedings of the Twenty-ninth IEEE Conference on Decision and Control*, Hawaii, 1990, pp. 2185–2190.
20. Chen, Y. Y., Stability analysis of fuzzy control—A Lyapunov approach, *Proceedings of the IEEE International Conference on Systems, Man and Cybernetics*, vol 3, 1987, 1027–1031.
21. Nguyen, H. T. et al., How to control if even experts are not sure: robust fuzzy control, In: *Proceedings of the Second IFIS*, College Station, Texas, 1992, pp. 153–162.
22. Nguyen, H. T., An empirical study of robustness of fuzzy systems, *Proceedings of the second FUZZ-IEEE*, San Francisco, 1993, pp. 1340–1345.
23. Dolezal, V., Robust stability and sensitivity of input-output systems over extended spaces, Part 1. Robust stability, *Circuits, Systems, Signal Process*, **10**(3), 1991, 361–388.
24. Tanaka, K. and M. Sugeno, Stability analysis and design of fuzzy control systems, *Fuzzy Sets and Systems*, **45**, 1992, 135–156.
25. Li, Y. and Y. Yonezawa, Stability analysis of a fuzzy control system by the hyperstability theorem, *Japanese Journal of Fuzzy Theory and Systems*, **3**(2), 1991, 209–214.
26. Ray, K. S. et al., L_2-stability and the related design concept for SISO linear system associated with fuzzy logic controller, *IEEE Transactions Systems, Man and Cybernetics*, **SMC-14**(6), 1984, 932–939.
27. Kania, A. A., On stability of formal fuzziness systems, *Information Sciences*, **22**, 1980, 51–68.
28. Singh, S., Stability analysis of discrete fuzzy control system, *Proceedings of the First FUZZ-IEEE*, 1992, pp. 527–534.
29. Kawamoto, S. et al., An approach to stability analysis of second order fuzzy systems, *First FUZZ-IEEE*, 1992, pp. 1427–1434.
30. Hunt, L.R. et al., Global transformations of nonlinear systems, *IEEE Transactions on Automatic Control*, **AC-28**(1), 1983, 24–31.
31. Hsu, C. S., A discrete method of optimal control based upon the cell state space concept, *Journal of Optimization Theory and Applications*, **46**(4), 1985, 547–568.
32. Holmblad, L. P. and J. J. Ostergaard, Control of a cement kiln by fuzzy logic, In: *Fuzzy Information and Decision Processes*, North-Holland, Amsterdam, 1982.
33. Tong, R. M. et al., Fuzzy Control of the activated sludge wastewater treatment process, *Automatica*, **16**, 1980, 659–701.
34. Tong, R. M., A control engineering review of fuzzy systems, *Automatica*, **8**, 1977, 559–568.
35. Braae, M. and D. A. Rutherford, Theoretical and linguistic aspect of the fuzzy logic controller, *Automatica*, **15**, 1979, 553–577.
36. Tong, R. M., Analysis and control of fuzzy systems using finite discrete relations, *International Journal of Control*, **27**(3), 1978, 431–440.
37. Wang, P. P. et al., Fuzzy set theory: Past, Present and Future, In: *Advances in Fuzzy Sets, Possibility Theory and Applications*, Plenum Press, New York, 1983, pp. 1–11.
38. Schneider, D. E., P. Wang and M. Togai, Design of a fuzzy logic controller for a target tracking system, *Proceedings of the First IEEE International Conference on Fuzzy Systems*, 1992, pp. 1131–38.
39. Bernard, J. A., Use of a rule-based system for process control, *IEEE Control Systems*, **8**(5), 1988, 3–12.
40. Wang, P. and M. Togai, Sensitivity analysis of dynamic systems via fuzzy set theory, *Proceedings of the International Congress of Applied Systems Research and Cybernetics*, vol VI, 1980, pp. 3062–3069.
41. Kandel, A., On the modelling of uncertain systems, *Proceedings of the International Congress of Applied Systems Research and Cybernetics*, vol VI, 1980, pp. 2939–2944.
42. De Glas, M., Invariance and stability of fuzzy systems, *Journal of Mathematical Analysis and Applications*, **99**, 1984, 299–319.
43. Takagi, T. and M. Sugeno, Fuzzy identification of systems and its applications to modelling and control, *IEEE Transactions on Systems, Man and Cybernetics*, **SMC-15**(1), 1985, 116–132.

44. Dote, Y. and T. Saitoh, Stability Analysis of Variable-Structured PI Controller by Fuzzy Logic Control for Servo system, Department of Computer Science and Systems Engineering, Muroran Institute of Technology, Japan, 1991.
45. Powell, B. K. and J. Cook, Non-linear low frequency phenomenological engine modeling and analysis, *Proceedings of the 1987 American Control Conference*, 1987, pp. 332–340.
46. Cho, D. and K. Hedrick, A nonlinear controller design method for fuel injected Engines. *Journal of Engineering for Gas Turbines and Power*, **110**, 1988, 313–320.
47. Murakami, S. and M. Maeda, Automobile speed control system using a fuzzy logic controller, In: M. Sugeno (ed.), *Industrial Applications of Fuzzy Control*, Elsevier Science Publishers, Amsterdam, 1985, pp. 105–123.
48. Olbrot, A. W. and B. K. Powell, Robust design and analysis of third and fourth order time delay systems with application to automotive idle speed control, *Proceedings of the 1989 IEEE American Control Conference*, vol. 2, 1989, pp. 1029–1039.
49. Kang, H. and G.J. Vachtsevanos, Nonlinear fuzzy control based on the vector field of the phase portrait assignment algorithm, *Proceedings of the 1990 American Control Conference*, San Diego, 1990, pp. 1479–1484.
50. Kang, H. and G. J. Vachtsevanos, Fuzzy hypercubes: Linguistic learning/reasoning systems for intelligent control and identification, *Proceedings of the Conference on Decision and Control*, Brighton, 1991.
51. Vachtsevanos, G. J., S. S. Farinwata and H. Kang, A systematic design methodology for fuzzy logic control with application to automotive idle speed control, *Proceedings of the Thirty-first IEEE Conference on Decision and control*, Tucson, Arizona, 1992, pp. 2547–2548.
52. Guang-Quan, Z., Fuzzy continuous function and its properties, *Fuzzy Sets and Systems*, **43**, 1991, 159–171.
53. Chiu, S. and S. Chand, Fuzzy controller design and stability analysis for an aircraft model, *Proceedings of the 1991 IEEE American Control Conference*, 1991, pp. 821–826.
54. Farinwata, S. S. and G. Vachtsevanos, Stability analysis of the fuzzy controller designed by the phase portrait assignment algorithm, *Proceedings of the Second IEEE International Conference on Fuzzy Systems*, San Francisco, 1993, pp. 1377–1382.
55. Gupta, M. M., J. B. Kiszka and G. M. Trojan, Multivariable structure of fuzzy control systems, *IEEE Transactions on Systems, Man and Cybernetics*, **SMC-16**(5), 1986, 638–655.
56. Farinwata, S. S. and G. Vachtsevanos, A survey on the controllability of fuzzy logic systems, *Proceedings of the Thirty-second IEEE Conference on Decision and Control*, San Antonio, Texas, 1993.
57. Mamdani, E. H., 20 years of fuzzy control: Experiences gained and lessons learnt, *Proceedings of the Second IEEE International Conference on Fuzzy Systems*, San Francisco, 1993, pp. 339–344.
58. Vachtsevanos, G., S. S. Farinwata and D. K. Pirovolou, Fuzzy logic control of an automotive engine: A systematic design methodology, *IEEE Control Systems*, Special Issue on Intelligent Control, **13**(3), 1993, 62–68.
59. Tanaka, K. and H. O. Wang, Energy based fuzzy regulators and fuzzy observers: A linear matrix inequality approach, *Proceedings of the Thirty-sixth IEEE Conference on Decision and Control*, 1992, pp. 1315–1320.
60. Farinwata, S. S. and G. Vachtsevanos, Robust stability analysis of the fuzzy logic control systems, Invited paper, *Proceedings of the 1995 American Control Conference*, vol. 3, Seattle, 1995, pp. 2267–2271.
61. Langari, R. and L. Wang, Stability of fuzzy control systems via nonlinear singular perturbation theory, *Proceedings of the 1995 American Control Conference*, vol. 6, Seattle, 1995.
62. Wang, H. O., K. Tanaka and M. Griffin, Analytical framework of fuzzy modeling and control of nonlinear systems: Stability and design issues, *Proceedings of the 1995 American Control Conference*, vol. 3, Seattle, 1995, pp. 2272–2276.

63. Hsu, C. S., *Cell-to-Cell Mapping*, Springer-Verlag, New York, 1987.
64. Wang, L., *Adaptive Fuzzy Systems and Control*, Prentice-Hall, Englewood Cliffs, 1994.
65. Filev, D., Algebraic design of fuzzy logic controllers, *IEEE ISIC*, Dearborn, 1996, pp. 253–258.
66. Passino, K. and S. Yurkovich, *Fuzzy Control*, Addison-Wesley, Reading, 1998.
67. Gorez, R. and G. Calcev, Passivity and stability of fuzzy control systems, In: S. Farinwata, D. Filev and R. Larguer, (eds.), *Fuzzy Control: Synthesis and Analysis*, John Wiley, Chichester, 1999, Chapter 6 in This Volume.
68. Rosenbrock, H. H., *State-Space and Multivariable Theory*, John Wiley, New York, 1976.
69. Takagi, T. and M. Sugeno, Fuzzy identification of systems and its applications to modelling and control, *IEEE Transactions on Systems, Man and Cybernetics*, **SMC-15**(1), 1985, 116–132.
70. Jamshidi, M., *Large-Scale Systems: Modeling and Control*, North-Holland, Amsterdam, 1982.
71. Siljak, D. D., *Large-Scale Dynamic Systems*, North-Holland, Amsterdam, Englewood Cliffs, 1978.
72. Lunze, J., *Feedback Control of Large-Scale Systems*, Prentice-Hall, 1991.
73. Kwakernaak, H. and R. Sivan, *Linear Optimal Control systems*, John Wiley, New York, 1972.
74. Bhattacharya, S. P. et al., *Robust Control: The Parametric Approach*, Prentice-Hall, Englewood Cliffs, 1995.
75. Ackermann, J. et al., *Robust Control: Systems with Uncertain Physical Parameters*, Springer-Verlag, New York, 1993.
76. El-Hawary, M. E., (ed.), *Electric Power Applications of Fuzzy System*, IEEE Press, New York, 1998.
77. Zhao, J., V. Wertz and R. Gorez, Fuzzy gain scheduling controllers based on fuzzy models, *Proceedings of FUZZ-IEEE*, New Orleans, 1996, pp. 1670–1676.
78. Jadbabaie, A., A. Titli and M. Jamshidi, Fuzzy observer-based control of nonlinear systems, *Proceedings of the Thirty-sixth IEEE Conference on Decision and Control*, San Diego, 1997, pp. 3347–3349.
79. Tanaka, K., T. Ikeda and H. O. Wang, Controlling chaos via model-based fuzzy control system design, *Proceedings of the Thirty-sixth IEEE Conference on Decision and Control*, San Diego, 1997, pp. 1488–1493.
80. Hong, S. and R. Languor, Synthesis of an LMI-based fuzzy control system with guaranteed optimal H_∞ Performance, *Proceedings of the 1998 FUZZ-IEEE*, Alaska, 1998, pp. 422–427.
81. Debnath, L. and P. Mikusinski, *Introduction to Hilbert Spaces with Applications*, Academic Press, New York, 1990.
82. Kelley, C. T., *Iterative Methods for Linear and Nonlinear Equations*, vol. 16, SIAM, Philadelphia, 1995.
83. Eslami, M., *Theory of Sensitivity in Dynamic Systems—An Introduction*, Springer-Verlag, New York, 1994.
84. Zabczyk, J., *Mathematical Control Theory: An Introduction*, Birkhäser, 1992.
85. Sánchez, D. A., *Ordinary Differential Equations and Stability Theory: An Introduction*, Dover, New York, 1968.
86. Kosko, B., *Neural Networks and Fuzzy Systems: A Dynamical Systems Approach to machine Intelligence*, Prentice-Hall, Englewood Cliffs, 1991.
87. Vidyasagar, M., *Nonlinear Systems Analysis*, Prentice-Hall, Englewood Cliffs, 1978.
88. Slotine, J-J. E. and W. Li, *Applied Nonlinear Control*, Prentice-Hall, Englewood Cliffs, 1991.
89. Owens, D. H., *Feedback and Multivariable Systems*, Peter Peregrinus, 1978.
90. Doyle, J. C., B. A. Francis and A. R. Tannenbaum, *Feedback Control Theory*, Macmillan, Basingstoke, 1992.
91. Rosenbrock, H. H., *State-Space and Multivariable Theory*, John Wiley, New York, 1970.
92. Bartle, R. G., *The Elements of Real Analysis*, 2nd edn., John Wiley, New York, 1976.
93. Bell, D. J., *Mathematics of Linear and Linear and Nonlinear systems for Engineers and applied Sciences*, Oxford University Press, Oxford, 1991.

94. Terano, T., K. Asai and M. Sugeno, *Fuzzy Systems Theory and Its Applications*, Academic Press, New York, 1991.
95. Ogata, K., *Modern Control Engineering*, Prentice-Hall, Englewood Cliffs, 1970.
96. Naumov, B., *Nonlinear Control Systems*, MIR/CRC, 1990.
97. Sugeno, M., *Industrial Applications of Fuzzy Control*, North-Holland, Amsterdam, 1985.
98. Khalil, H. K., *Nonlinear Systems*, Macmillan, Basingstoke, 1992.
99. Isidori, A., *Nonlinear Control Systems*, Springer-Verlag, New York, 1989.
100. Grieffel, D. H., *Applied Functional Analysis*, Ellis Horwood, Chichester, 1981.
101. Kreyszig, E., *Introductory Functional Analysis with Applications*, John Wiley, New York, 1978.
102. Desoer, C. A. and M. Vidyasagar, *Feedback Systems: Input–Output Properties*, Academic Press, New York, 1975.
103. Barnett, S., *Matrices: Methods and Applications*, Oxford University Press, Oxford, 1990.
104. Friedland, B., *Control System Design: An Introduction to State Space Methods*, McGraw-Hill, New York, 1986.
105. Zimmermann, H.-J., *Fuzzy Set Theory and Its Applications*, Kluwer, Dordrecht, 1991.
106. Pedrycz, W., *Fuzzy Control and Systems*, John Wiley, New York, 1989.
107. D'Azzo, J. J. and C. H. Houpis, *Linear Control Systems Analysis and Design*, McGraw-Hill, New York, 1981.
108. Brogan, W. L., *Modern Control Theory*, Quantum, 1974.
109. Farinwata, S. S., "Sensitivity stabilization for a class of fuzzy systems", *18th Int. Conf. of NAFIPS*, pp. 163–168, New York, 1999.
110. Vachtsevanos, G. and S. S. Farinwata, Fuzzy logic control: A systematic design and performance assessment methodology, In: M. J. Patyra and D. M. Mlynek (eds.), *Fuzzy Logic Implementation and Applications*, John Wiley, Chichester, 1996, Chapter 2.

10
Fuzzy Control Systems Stability Analysis with Application to Aircraft Systems

Shehu S. Farinwata
Ford Research Laboratory, Dearborn, USA

Stephen L. Chiu
Rockwell Science Center, Thousand Oaks, USA

10.1 INTRODUCTION

Fuzzy control systems are generally considered applicable to processes that are mathematically ill-understood and where human experience is available for control rule synthesis. An equally important characteristic of fuzzy control is the ability to encompass qualitative and highly non-linear control objectives. This ability makes fuzzy control applicable to processes for which the control objective cannot be adequately expressed in a form amenable to standard control synthesis methods. The applications are described in Sections 10.2 and 10.3. They employ Lyapunov stability theory which applies to most physical systems that are inherently dissipative in nature. Lyapunov theory is developed, leading to a passivity theory for dissipative linear and non-linear systems. However, the applications considered here are of linear systems, continuous-time and discrete-time, that are controlled via fuzzy control. Thus, the development of the non-linear passivity theory has been made very brief so as to fit the scope of the applications. An application of the non-linear dissipative mapping for a non-linear system under fuzzy control was treated in [4].

The first application develops an input–output stability analysis on the basis of fuzzy rules, and the control input values generated by the fuzzy controller, for each 'cell partition' of the phase plane. This is then applied to a missile autopilot. The scenario considered here is where a missile autopilot is released from an aircraft. It is required that the missile maintain proper heading to the target, in the presence of various aerodynamic disturbances, especially under the wing. Here it is assumed that the rulebase has been designed for a linear plant whose

204 STABILITY ANALYSIS OF AIRCRAFT SYSTEMS

approximate model is available. Furthermore, the rules may have been derived using the operator manual-type If–Then rules, or based directly on the approximate model. The model used here can be found in a good book on aircraft dynamics and control such as [3] and [36]. The method in this example is based on formulating an input–output dissipative map, and then applying a Lyapunov-like analysis for stability convergence. For a linear system, an input–output mapping may always be formulated as an energy-like function. By proper choice of outputs, and the inputs generated by the fuzzy controller, the energy-like function is forced to be dissipative fairly readily by applying the Kalman–Yakubovich lemma.

The second application makes explicit use of Lyapunov's direct method to analyze the stability of a more complex system, Rockwell's Advanced Technology Wing (ATW) aircraft model. Here, a fuzzy roll controller design is described. The controller modulates six control surfaces on the ATW to achieve the desired roll rate while maintaining wing loads within safe bounds.

The use of fuzzy control rules enabled highly flexible control behavior that operated the system close to the constraint limits, thereby achieving excellent roll performance. The stability approach presented is also based on partitioning the state space and applying a Lyapunov stability criterion to the individual partitions. A subset of the control rules is analyzed using this method and shown to provide asymptotically stable control.

10.1.1 Fuzzy Control

At the basis of fuzzy logic is the representation of linguistic descriptions as *membership functions* [24]. The membership function indicates the degree to which a value belongs to the class labeled by the linguistic description. For example, the linguistic description BIG may be represented by the membership function $\mu_{BIG}(x)$ shown in Figure 10.1, where the abscissa is an input value and the ordinate is the degree to which the input value can be classified as BIG. In this example, the degree to which the number 80 is considered BIG is 0.5, i.e. $\mu_{BIG}(80) = 0.5$.

Fuzzy control rules have the following form:

$$\text{If } X_1 \text{ is } \mu_{A_{i,1}} \text{ and } X_2 \text{ is } \mu_{A_{i,2}} \text{ Then } U \text{ is } \mu_{B_i},$$

where X_1 and X_2 are the inputs to the controller, U is the output, μ_A and μ_B are membership functions, and the subscript i denotes the rule number. Given the input values of x_1 and x_2,

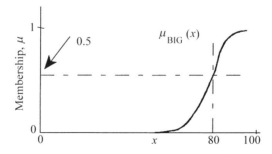

Figure 10.1 Membership function defines a linguistic description

INTRODUCTION

one way, the degree to which the antecedent of rule i is satisfied is given by

$$\alpha_i = \mu_{A_{i,1}}(x_1) \wedge \mu_{A_{i,2}}(x_2),$$

where $\wedge$ denotes the minimum operator. The value α is often called the *degree of fulfillment* of the rule. We compute the output value by

$$u = \frac{\sum_{i=1}^{s} \alpha_i \mu_{B_i}^d}{\sum_{i=1}^{s} \alpha_i},$$

where $\mu_{B_i}^d$ is the *defuzzified* value of the membership function μ_{B_i}, and s is the number of rules.

The defuzzified value of a membership function is the single value that best represents the linguistic description; typically, we take the abscissa of a membership function's centroid as its defuzzified value. In essence, each rule contributes a conclusion weighted by the degree to which the antecedent of the rule is fulfilled. The final control decision is obtained as the weighted average of all the contributed conclusions. This method of computing the output is a modified form of the standard *center of area* method [25]. This modified method is gaining popularity in control applications due to its computational and analytical simplicity.

Complementing the basic inference procedure, we can provide an optional rule weight parameter to indicate the relative importance of a rule. Suppose each rule is assigned the additional weight w_i between 0 and 1.0, then the output is computed by

$$u = \frac{\sum_{i=1}^{s} w_i \alpha_i \mu_{B_i}^d}{\sum_{i=1}^{s} w_i \alpha_i}.$$

10.1.2 Lyapunov Stability of Non-linear Fuzzy Control Systems

The objective is to study the asymptotic stability convergence of the closed-loop system with the fuzzy control inputs being continually generated by the fuzzy controller in the system. The convergence point is a specified set-point. Figure 10.2 shows the general nonlinear system, f, under fuzzy control. The desired set-point is $x_d \in \mathbb{R}^n$ and the error is $e = x - x_d$. The control u is generated by the fuzzy controller, FLC.

10.1.3 The Fuzzy Control Problem

Consider the system to be controlled is expressed approximately as

$$\dot{x}(t) = f(x(t), u(t), r), \quad y = h(x), \qquad (10.1)$$

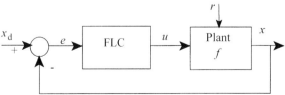

Figure 10.2 Fuzzy control system structure

where $x(t) \in \mathbb{R}^n$ is the vector of the states, $u(t) \in \mathbb{R}^m$ is the input vector, and $r \in \mathbb{R}$ is a scalar disturbance input. Let f be a non-linear smooth mapping in general, and completely be determined as $f : \mathbb{R}^n \times \mathbb{R}^m \times \mathbb{R} \to \mathbb{R}^n$ and $h : \mathbb{R}^n \to \mathbb{R}^p$, for $t \in [0, +\infty)$.

For the purpose of preserving existence of the above solution, we will assume continuity of $x(t)$ with respect to r. So f is smooth in x, and at least piecewise smooth in u and r, for $t \geq 0$.

Let $x(t) \in X, u(t) \in U, y(t) \in Y$, and $r \in D$. The feasible or admissible sets which are all subsets of their corresponding spaces as already specified, will be denoted by $F(X)$, $F(U)$ and $F(D)$. Global discussion will therefore pertain to these subsets. In the case of the fuzzifiable quantities, the universes of discourse are also considered to be the same subsets alluded to.

Given the set of desired trajectories $\{x_d\} \subset F(X)$, a fuzzy controller is sought which will generate $u(t)$ or $\{u_n\}, n \in \mathbb{N}$, such that a given $x(t_0)$ at time t_0, is moved to $x(t)$, for $t > t_0$, by which time $x(t)$ is kept arbitrarily close to x_d, with both t_0 and $t \in [0, +\infty)$. The boundedness of $u(t)$ or u_n is an issue that is of particular interest, especially in stability analysis. This is the fuzzy control problem in a very general sense. Indeed, corresponding to $\lim x(t) = x_d$ as $t \to \infty$, is a control $u(t) \equiv u^*$, or that the fuzzily generated sequence $\{u_n\}$ converges to some u^*. In a host of practical control problems, u^* is often pre-calculated or is specified a priori. In a such a case, the set $\{(x_d, u^*)\}$ is necessarily a fixed point for (10.1), for some nominal r [17].

We re-state a more general fuzzy control algorithm in the form of **If–Then** rules below.

R_i: **If** e_1^i is *small* and e_2^i is *big*,..., and e_n^i is *positive medium* **Then** u_1^i is *negative medium* and u_2^i is *positive large*, ..., and u_m^i is *negative medium*.

The totality of the fuzzy control rules, $\bigcup R_i$, constitutes the fuzzy control rule base, or the fuzzy controller. The closed-loop fuzzy control system may be represented as

$$\Sigma : \left(\bigcup_{i=1}^{s} R_i ; f(\{x_k\}, \{u_k\}) \right),$$

where $f(\cdot, \cdot)$ is the model of the process to be controlled with its actual crisp states sequence $\{x_k\}$ and crisp input sequence $\{u_k\}$; s is the number of rules.

Fundamental Assumption *The system is assumed to be dissipative so that one can argue heuristically that the total energy of the system decreases as time progresses until a state of equilibrium is reached.*

The stability analyses conducted here all allude to this fundamental tenet for physical systems. Where this physical property is not readily inherent in the system that is being controlled, the fuzzy controller is made to induce the property by proper choice of input–output mapping [4].

Lemma 1 *If the differentiable function $V(t)$ has a finite limit as $t \to \infty$, and if $\dot{V}(t)$ is uniformly continuous, then $\dot{V}(t) \to 0$ as $t \to \infty$.*

This is a statement of Barbalat's lemma. However, the class of systems considered here is autonomous, so the uniform continuity restriction may be alleviated. The consequence of this is the following Lyapunov-like lemma.

Lyapunov-Like Lemma 2 *Let the scalar function V(x,u) satisfy the following conditions:*

1. $V(x,u)$ is lower bounded.
2. $\dot{V}(x,u)$ is negative semi-definite.
3. $\dot{V}(x,u)$ is uniformly continuous in u.

Then $\dot{V}(x,u) \to 0$ *as* $t \to \infty$.

Note that the u's in the above lemma are the fuzzy control inputs that are being generated as the system evolves. These control inputs are used to induce the input–output dissipative mapping in expression 2 in Lemma 2 above, for all 'cells' in the rulebase. Note, however, that the continuity of x under u needs to be assured so that V and its time derivative will be continuous in x under u.

10.1.4 Equilibrium Points for Fuzzy Controlled Processes

In classical Lyapunov stability of a system of the type in (10.1), the unforced system $f(x_e, 0)$ is studied with $f(x_e, 0) = 0$, where x_e is the equilibrium state. $x_e = 0$ is often employed with, actually, some loss of generality.

In what follows we give some technicalities.

1. In a variety of physical processes the equilibrium point $x_e = 0$ may not be convenient or even feasible. For example, an aircraft cruising at a certain speed at a given cruising altitude can be considered as a steady-state condition. Clearly, the throttle, and the vertical stabilizer among other control surfaces, cannot be set to zero in order to study the stability of the cruise controller. In such a case it is appropriate to define the equilibrium point as the error $e_1 = x_1 - x_{1d}$ and $e_2 = x_2 - x_{2d}$, where x_1 and x_2 may be the speed and the vertical stabilizer angle, and x, d, x^2d are the predetermined or attained set-points, in which case $e_1 = e_2 = 0$ is not only feasible but has the proper physical significance as well. In general, one can always define a new equilibrium point via a shift of origin. This freedom should always be exercised when dealing with a real, physical process with non-zero equilibrium or set-points.

2. It is not always practicable to remove the 'forcing' function from a physical process that is in operation and expect it to even function at all, let alone satisfactorily. Thus, the study of $f(x_e, u)$ as $f(x_e, 0)$, with $u = 0$, is not always a physical reality as can be imagined from the aircraft scenario given above. Also, in a variety of physical devices, certain non-zero *bias* inputs are always needed to set the device in a proper operational regime. Take the case of a bipolar junction transistor, for instance. It would be inconceivable to attempt to use it as an amplifier, without the required bias stabilizing currents or voltages. There is a host of engineered processes that one can think of that command this particular kind of attention for their steady-state input–output stability analyses. This is why this particular method of stability analysis was proposed. In light of this, the stability of the fuzzy control process will be studied with $u \neq 0$ in the various regions of the phase plane, unless the fuzzy controller explicitly suggests such a zero value in a particular region. For linear fuzzy controlled systems, this zero value of the control will be shown to cause some difficulty in the analysis when $x_e = x_d \neq 0$.

10.1.5 The Partitioned State Space

The introduction of membership functions for the system's physical variables of interest necessarily induces a partition of the feasible space of the variables. The size of the partitioned space may correspond to the product of the cardinalities of the membership functions of any two variables. Such a space is then generally made up of cells of arbitrary shapes. However, this need not be the case. For computational reasons, rectangular cells are often used where the boundaries can be easily specified, even though the cell itself is a *covering* of one or more linguistic labels. Thus, the whole state space may look like a huge rectangular grid system.

The applications reported here employ techniques similar to this. More discussion on this can be found in [4]. A cell space is shown in Figure 10.3 where a typical cell is represented as L_{mn}.

10.1.6 Dissipative Mapping and Input-Output Stability

Consider the system described in Figure 10.4 where U is the input space, Y is the output space and $H: U \rightarrow Y$. The properties of U and Y are needed in the discussion of stability of the system.

Definition 1 For any fixed $p \in [1, \infty)$, $f : \mathbb{R} \rightarrow \mathbb{R}$ is said to belong to L_p iff f is locally integrable and

$$\|f\|_p = \left(\int_0^\infty |f(t)|^p \, dt \right)^{1/p} < \infty.$$

When $p = \infty$ and $\|f\|_\infty$ is finite, then $f \in L_\infty$ (uniform boundedness), where $\|f\|_\infty = \sup |f(t)|, t \geq 0$.

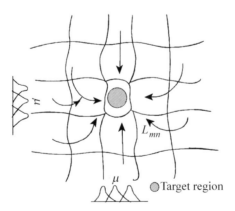

Figure 10.3 Partitioned state space

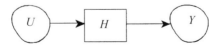

Figure 10.4 Input-output operator

INTRODUCTION

The definition allows one to talk about *L*-stability of the system by observing similar properties on the input. Specifically, the operator, H, in the block diagram is considered L_p stable if $u \in L_p$ is mapped to $y \in L_p$. It suffices to say that bounded-input–bounded-output stability results when $p = \infty$. For our purpose, we will consider the inputs and outputs to be in the same L_p-space. It may, in general, be necessary sometimes to work in an extended space, to be on the safe side so to say, for output signals that may have finite escape time even when the input signals are well behaved. Also, we consider the systems in our applications to be causal. With this in mind, the mapping $H : L_p \to L_p$ is *L*-stable if $\exists$ non-negative numbers $\gamma, \beta < \infty, \ni \|Hu\| \leq \gamma\|u\| + \beta, \forall u \in L_p$. The smallest γ for which this holds is the gain of the system, see [39].

Consider the feedback interconnection in Figure 10.5, which is made up of two input–output stable systems, H_1 and H_2. From the statement of the small gain theorem, the feedback interconnection of the two input–output systems will be input-output stable provided that the product of the system gains is less than one. Thus, under the assumption that $\gamma_1 \gamma_2 < 1$, then for all $u_1, u_2 \in L_p$, the signals e_1, y_2 and $e_2, y_1 \in L_p$, and e_1 and e_2 are bounded as

$$\|e_1\| \leq \frac{1}{1 - \gamma_1\gamma_2}(\|u_1\| + \gamma_2\|u_2\| + \beta_2 + \gamma_2\beta_1),$$

$$\|e_2\| \leq \frac{1}{1 - \gamma_1\gamma_2}(\|u_2\| + \gamma_1\|u_1\| + \beta_1 + \gamma_1\beta_2).$$

A connection with Lyapunov stability is easily shown by imposing some conditions on the mappings f and h in (10.1). This is briefly highlighted in what follows. Before we do that, let us re-state the statement of exponential stability.

Definition 2 An equilibrium point 0 is said to be exponentially stable if $\exists \alpha, \lambda \in \mathbb{R}_+ \ni$

$$\|x(t)\| \leq \alpha\|x(0)\|e^{-\lambda t}, \quad \forall t > 0$$

where λ is the decay or convergence rate. It is clear from this statement that exponentially stable systems are also asymptotically stable ones.

The next thing we want to do is to find a suitable scalar, energy-like storage function $V(x)$ from which the dissipative mapping may be formed. Since it is assumed that the fuzzy controller is stable, a converse stability theorem ensures the existence of a Lyapunov function which could be the $V(x)$ we are seeking.

Consider the system in (10.1) which is stated below in a slightly modified manner:

$$\dot{X}(t) = f(X, U, d), \quad y = h(X), \quad X \doteq x - x_\mathrm{d}, \quad U \doteq u - u^*,$$

where the set $\Omega = \{(u^*, x_\mathrm{d})\}$ is the specified equilibrium point, for some nominal d.

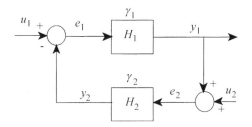

Figure 10.5 Feedback interconnection

Lemma 3 *Suppose*

1. $X=0$, $U=0$, is an equilibrium point for above, for some d;
2. $X=0$ is a globally exponentially stable equilibrium point for $f(X, 0)$;
3. $\partial f/\partial x|_U, \partial f/\partial U|_U$ are globally bounded; and
4. $\|h(X)\| \leq k\|X(t)\|$.

Then it can be shown that $\exists k_1, k_2$, *and* k_3, *non-negative such that*

$$\|X(t)\| \leq k_1 \exp(-\lambda t)X(0) + k_2\|U(t)\| + k_3.$$

Using the condition on h above we have that

$$\|y(t)\| \leq kk_1\exp(-\lambda t)\|X(0)\| + kk_2\|U(t)\| + kk_3.$$

Note that the left-hand side, $\|y(t)\|$, combined with the first term in the right-hand side in the above is essentially what we get from asymptotic stability in the sense of Lyapunov, while combined with the second and third terms, we get what is akin to input–output stability, with kk_2 as the gain of the system.

10.1.7 Dissipative Mapping for the Fuzzy Control System

Let $V(x)$ be such a lower bounded scalar function. Note that this is what makes this only a Lyapunov-like analysis already alluded to as $V(x)$ may only be positive semi-definite. It is given by

$$\dot{V} = \frac{\partial V}{\partial e}\dot{e} = \frac{\partial V}{\partial e}f(x, u), \quad \text{with } e \triangleq x - x_d, \tag{10.2}$$

where x_d is a constant vector. Consider the output mapping $y = h(x)$. We will suppose that h is at least $C^1(X)$ so that we can perform a cell-wise linearization of equation (10.1) if we so desire. So h is a differentible map. Furthermore, since f in (10.1) has an asymptotically stable equilibrium point, a converse argument allows us to see that f is also locally injective. For our purpose in general, we do not need h to be a diffeomorphism. Thus h has an inverse, $h^{-1}(y) = x$, for some fuzzy control. The smoothness of x lies in the argument of the existence of a solution of (10.1) under fuzzy control. This is not too difficult to see as the state space, which is compact and contractive, is made up of a finite collection of compact cells which are compact in the usual cartesian coordinates. However, as already alluded to, in 'fuzzy coordinates' each cell is a finite covering of linguistic labels. The cells are therefore contractive under the appropriate fuzzy control. That is why we consider h to be at least cell-wise smooth. The output–input map becomes

$$\dot{V}(h^{-1}(y), \tilde{u}) = \frac{\partial V}{\partial e}\dot{e} = \frac{\partial V}{\partial e}|_{(x_d, u^*)} f(h^{-1}(y), \tilde{u}), \tag{10.3}$$

where $\tilde{u}$ are the fuzzy control inputs. To be dissipative is to satisfy condition 2 in Lemma 2 in all cells with their respective fuzzy controls. Incidentally, this is closely related to the notion of non-linear passivity, the discussion of which will take us too far afield and will be beyond the intended scope of the applications in this chapter. However, some discourse may be found in [34] and [38]. For a very complex expression, it may be convenient to simulate the expression (10.3) for various cells in the phase plane in order to determine the definiteness. Finally, condition 3 of Lemma 2 needs to be checked. This can be done in a less cumbersome way by showing that d^2V/dt^2 is bounded. The conclusion of the lemma then follows.

Definition 3 We will say that the input–output mapping (10.3) for the non-linear system under fuzzy control is dissipative if

$$\dot{V}(h^{-1}(y), \tilde{u}) \leq -(\alpha_1(\|y\|) + \alpha_2(\|\tilde{u}\|)), \tag{10.4}$$

where $\alpha_1(\cdot)$ and $\alpha_2(\cdot)$ are *class-K* functions. This then ensures that the equilibrium point of (10.1) at the origin is approached asymptotically for some combinations of fuzzy control and outputs.

If a region or cell of the phase plane can be found, with its associated fuzzy control, for which dV/dt fails to be negative semi-definite, $\forall t$, then the fuzzy control system is unstable. For this particular region it would be possible to track the fuzzy control down to the membership function level and redesign the particular rule in question, rather than discard the entire rulebase. As observed, the practicality of this analysis is apparently limited when:

1. The number of regions or cells in the phase plane is large and there are different controls for each cell.
2. The expression for the non-linear process is very complex, and renders any analytical determination of the definiteness of dV/dt near impossible.

One way to circumvent this, as suggested above, is to actually simulate $dV/dt|_{\tilde{u}}$ for various points in the four quadrants of the phase plane. Another way is to prudently cluster the rulebase into a fewer number of rules [4]. In general, to be complete, this simulation will have to be exhaustive, especially for a suspicious, highly non-linear and potentially unstable process.

10.1.8 Stability of Linear Fuzzy Control Systems

Suppose the fuzzy controller is designed to control a linear plant with a transfer function $H(s)$ or state space elements $\sum(A, B, C, D)$. The system is normally expressed as

$$\sum : \dot{x} = Ax + Bu, \quad y = Cx + Du, \tag{10.5}$$

with $x \in \mathbb{R}^n$, $A \in \mathbb{R}^{n \times n}$, $B \in \mathbb{R}^{n \times m}$, $u \in \mathbb{R}^m$, $C \in \mathbb{R}^{p \times n}$, and $D \in \mathbb{R}^{p \times m}$.

The block diagram of the closed-loop fuzzy control system is shown in Figure 10.6. In this case an input–output map can be formed relatively easily. Systems with no direct feed are considered so that $D = 0$ in (10.5). Suppose the system is open-loop stable, i.e. Re $\lambda(A) < 0$.

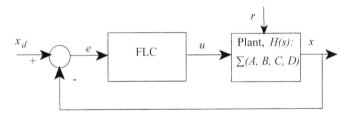

Figure 10.6 Linear fuzzy control system

If $u \in L_\infty$, then $y \in L_\infty, \mathrm{d}y/\mathrm{d}t \in L_\infty$ and y is uniformly continuous. If in addition $\lim u(t) = u_0$ as $t \to \infty, u_0 \in \mathbb{R}$, then $y(t) \to y_0 = H(0)u_0 \in \mathbb{R}$. Such (u_0, y_0) will be considered for the fuzzy control system, for various partitions of the state space. In general, if $u \in L_p$, then $y \in L_p$, for $p \in [1, +\infty)$.

10.1.9 Positive Realness and Dissipativeness

In what follows we will summarize some important results of the positive real lemma. This establishes the dissipative nature of the system and the existence of Lyapunov functions for studying the stability of the system. If the system $\sum(A, B, C, D)$ is dissipative, then this implies that the transfer function $H(s)$ is strictly positive real and that means it is physically realizable.

Definition 4 A rational function $H(s)$ of the complex variable s is positive real (PR) if:

1. $H(s)$ is real for real s
2. $\mathrm{Re}(H(s)) \geq 0$, for all $\mathrm{Re}(s) > 0$.

Also, $H(s)$ is strictly positive real (SPR) if $H(s\text{-}e)$ is PR, for some $e > 0$.

The Kalman–Yakubovich lemma (KYL) is an important link in relating strictly positive real functions to the existence of a Lyapunov function and hence the stability of the corresponding dynamic system.

Lemma 4 (Lefschetz–Kalman–Yakubovich) *Given a scalar $\gamma \geq 0$ and a vector C, and the system $\sum(A, B, C, D)$ with $\mathrm{Re}\lambda(A) < 0$. Let (A, B) be controllable and let there exist $L > 0$. Then there exist $e > 0$, and q and $P > 0$ such that*

$$A^\mathrm{T} P + PA = -qq^\mathrm{T} - \varepsilon L,$$
$$PB - C^\mathrm{T} = \sqrt{\gamma}q, \tag{10.6}$$

iff

$$H(s) = \tfrac{1}{2}\gamma + C^\mathrm{T}(sI - A)^{-1}B \tag{10.7}$$

is strictly positive real (SPR).

It is clear from above lemma that the positive realness condition imposed on $H(s)$ is necessary and sufficient for the existence of P and q that satisfy (10.6).

A version of the Kalman–Yakubovich lemma results when $\gamma = 0$. This is the version that has been employed by the first application to be considered in this chapter. Setting $\gamma = 0$ gives

$$A^\mathrm{T} P + PA = -qq^\mathrm{T} - \varepsilon L = -Q, \quad Q = Q^\mathrm{T} > 0,$$
$$PB = C^\mathrm{T}. \tag{10.8}$$

Furthermore, if the positive realness condition holds for the linear time invariant (LTI) system $\sum(A, B, C, D)$, i.e. H, then we say that the system is passive; that is,

$$\int_0^T u(t)^\mathrm{T} y(t) \mathrm{d}t \geq 0$$

for all solutions with $x(t) = 0$.

INTRODUCTION 213

Definition 5 A system with input $y(t)$ and an input $u(t)$ is said to be dissipative with dissipation $\eta \geq 0$, or has a non-negative dissipation if

$$\int_0^T (u(t)^T y(t) - \eta u(t)^T u(t)) dt > 0$$

for all trajectories with $x(0) = 0$.

P in (10.8) may be found by solving the matrix inequality:

$$A^T P + PA + (PB - C^T)(D + D^T)^{-1}(PB - C^T)^T \leq 0, \quad D + D^T > 0. \tag{10.9}$$

Using the Schur compliment, we write this as a linear matrix inequality (LMI) as

$$P > 0, \quad \begin{bmatrix} A^T P + PA & PB - C^T \\ B^T P - C & -D^T - D \end{bmatrix} \leq 0, \tag{10.10}$$

which is feasible if $\sum(A, B, C, D)$ is non-expansive, i.e. if

$$\int_0^T y^T y \, dt \leq \int_0^T u^T u \, dt, \quad \forall x(0) = 0,$$

or

$$\|H\|_\infty \leq 1, \quad \text{where } \|H\|_\infty \doteq \sup \|H\| \mid \text{Re}(s) > 0.$$

The system being open-loop stable, a converse stability theorem ensures of the existence of a Lyapunov function that fulfills condition 1 of Lemma 2. Let this scalar function be $V(x) = \frac{1}{2} x^T P x$, where P is a positive definite symmetric matrix. This ensures that $V(x)$ is also lower bounded. The time derivative is given as

$$\begin{aligned} \frac{dV}{dt} &= \tfrac{1}{2}[\dot{x}^T P x + x^T P \dot{x}] = \tfrac{1}{2}[(x^T A^T + u^T B^T) P x + x^T P A x + x^T P B u] \\ &= \tfrac{1}{2} x^T (A^T P + PA) x + \tfrac{1}{2}(x^T P B U + x^T P B u) \\ &= \tfrac{1}{2} x^T (A^T P + PA) x + x^T P B u \end{aligned} \tag{10.11}$$

For global stability, $dV/dt < 0$. The first term can be made negative definite by making A stable independent of B and C, by setting $A^T P + PA = -Q$, for some positive definite symmetric matrix, Q. Doing this results in

$$\frac{dV}{dt} = -\tfrac{1}{2} x^T Q x + x^T P B u, \quad -Q = A^T P + PA, \tag{10.12}$$

$$\frac{dV}{dt} < 0 \Rightarrow \frac{x^T Q x}{x^T x} > \frac{2 x^T P B u}{x^T x}. \tag{10.13}$$

Note that the left-hand side of the implied inequality in (10.12) is actually a Rayleigh quotient, which can be derived from the eigenvalue equation for a matrix Q that is real and symmetric. If the eigenvalues of Q are numbered such that $\lambda_1 \geq \lambda_2 \geq \cdots \geq \lambda_n$, then $\lambda_1 \geq r, \lambda_n \leq r$, where $\lambda_n \equiv \lambda_{\min}$, and $\lambda_1 \equiv \lambda_{\max}$, where r is known as the Rayleigh constant.

As a consequence of this,

$$\lambda_n \leq \frac{x^T Q x}{x^T x} \leq \lambda_1, \quad r = \frac{x^T Q x}{x^T x}, \tag{10.14}$$

$$\lambda_{\min}(Q) > 2 \frac{x^T P B u}{\|x\|^2}, \quad Q = A^T P + PA, \quad x \in L_{mn} \text{ and } u = u_c. \tag{10.15}$$

Global Stability If (10.14) holds in every region, L_{mn}, of the state space, then the system is globally, asymptotically stable.

Suppose that the desired equilibrium point $x^* = x_d$ is other than the origin, $x = 0$. Then the trajectory error $e = x - x_d$ can be formed such that the state equation (10.4) becomes the error system below:

$$\dot{e} = Ae + Ax_d + Bu. \tag{10.16}$$

Given x_d, the term Ax_d may be considered as an offset that the fuzzy controller must overcome for all L_{mn}. It will be required therefore that $Ax_d \in \rho(B)$, the range of B, $\forall u$ and $x_d \neq 0$. That is, $\exists B_{mn} \neq 0 \ni Ax_d = B_{mn} u_c$. This will be satisfied for $u_c \neq 0$. The trivial case when $x_d = 0, u = 0$ is not particularly interesting in the development of an input–output stability, for it refers to the unforced system $\dot{x} = Ax$ for which $x = 0$ is the only equilibrium point. With such B_{mn}, the error system becomes

$$\dot{e} = Ae + (B_{mn} + B)u = Ae + B'u, \tag{10.17}$$

The same analysis can now be carried out with a broader generality.

10.1.10 Verifying Dissipativeness

Consider equation (10.11) which is reproduced below for convenience:

$$\frac{dV}{dt} = x^T P B u - \tfrac{1}{2} x^T Q x.$$

By employing the Kalman–Yakubovich lemma we see that since $y = Cx$, (10.11) defines a dissipative mapping between u and y provided that B and C are related by $C^T = PB$. So, $y^T = x^T C^T = x^T PB$ and equation (10.11) becomes

$$\frac{dV}{dt} = y^T u - \tfrac{1}{2} x^T Q x. \tag{10.18}$$

Thus, by compatible choices of u and y, a family of dissipative input–output maps can be constructed. That is, given the system's fuzzy control inputs and the input distribution matrix, B, one can choose an infinity of outputs from which the linear system will appear dissipative. If expression (10.17) above fulfills condition 2 in Lemma 2 for compatible choices of u and y, then the dissipativeness of the system is therefore verified.

10.2 LINEAR CONTINUOUS-TIME MODEL APPLICATION

10.2.1 A Missile Autopilot

Consider the regulation problem of a missile autopilot's yaw axis. The linearized dynamic equation of the yaw axis under consideration is given by

$$\begin{bmatrix} \dot{r} \\ \dot{\beta} \end{bmatrix} = \begin{bmatrix} -0.791 & 0 \\ -1 & -0.013 \end{bmatrix} \begin{bmatrix} r \\ \beta \end{bmatrix} + \begin{bmatrix} -1.721 \\ 0.0213 \end{bmatrix} \delta_r \triangleq \dot{x} = Ax + Bu, \quad x \triangleq \begin{bmatrix} r \\ \beta \end{bmatrix}, u \triangleq \delta_r, \tag{10.19}$$

where β is the side slip and r is the yaw rate. These are known as the aerodynamic angles. The linearized model was determined on the basis of such quantities as dynamic pressure, Mach number, velocity, and angle of attack. δ_r is the rudder position.

It is required that the missile's reference point, $\beta = 0, r = 0$, at the moment of release from the aircraft, be maintained (Figure 10.7).

Regulation of this yaw axis has been achieved via fuzzy logic control using the systematic design methodology reported in [13]. The membership functions for r, β and the only incremented control, δ_r, are given in Figures 10.8, 10.9 and 10.10, respectively, for illustration purposes. A sample from the 40 fuzzy control rules used to control this system is also shown in Figure 10.11.

10.2.2 Analysis

First, one needs to check that the system is controllable. The controllability matrix, $U = [B \ \ AB]$ is found to be full rank, and hence it is controllable. A and B are given in (10.18). Let the observed outputs of interest be $y_1 = r = x_1$ and $y_2 = \beta = x_2$. An input–output mapping is then formed as

$$\frac{dV}{dt} = y^T PBu - \frac{1}{2} x^T Qx, \qquad (10.20)$$

Figure 10.7 A missile sutopilot's launch

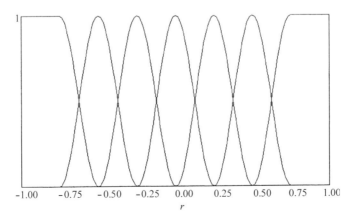

Figure 10.8 Membership function for yaw rate (δ_r)

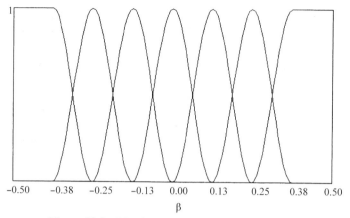

Figure 10.9 Membership function for side-slip (β)

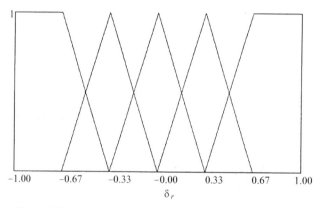

Figure 10.10 Membership function for rudder position (δ_r)

where P and Q are to be determined. The eigenvalues of A are $\lambda_1(A) = -0.013$ and $\lambda_2(A) = -0.7910$, and so the unforced system is asymptotically stable. By a converse stability theorem, $\exists P, Q > 0 \ni A^T P + PA = -Q$. On solving this so-called Lyapunov equation, it is found that

$$P = \begin{bmatrix} 122.21 & -95.67 \\ -95.67 & 76.92 \end{bmatrix} \quad \text{and} \quad Q = \begin{bmatrix} 2 & 0 \\ 0 & 2 \end{bmatrix}. \tag{10.21}$$

Expression (10.19) then becomes

$$\frac{dV}{dt} = (-212.3612 y_1 + 166.2865 y_2) u_{\text{FLC}} - x_1^2 - x_2^2. \tag{10.22}$$

This is the input–output mapping for the fuzzy controlled system. The FLC in the closed-loop configuration is to ensure that this input–output mapping is negative definite, for all regions of the state space, in order to reach a global stability conclusion, if this is what is sought; otherwise, only local stability conclusions may be reached. From (10.22) it is clear that the stability issue reduces to the regulation of the first term:

$$(-212.3612 y_1 + 166.2865 y_2) u_{\text{FLC}} \leq 0 \tag{10.23}$$

LINEAR CONTINUOUS-TIME MODEL APPLICATION

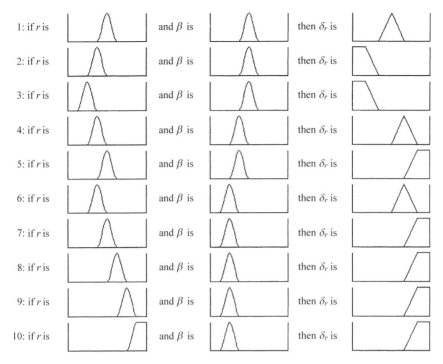

Figure 10.11 Partial fuzzy rulebase for the Yaw Axis Controller

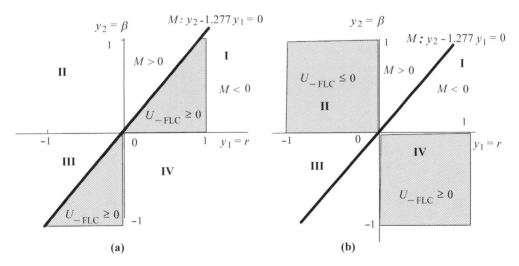

Figure 10.12 (a) The FLC's stability regions: y_1, y_2 both >0 or both <0. In the shaded regions, the FLC should produce controls that are ≥ 0. The supplementary region is the one above $M=0$. Here, the expected controls should be ≤ 0. (b) Definite sign for M: occurs when y_1 and y_2 have opposite signs. With (y_1, y_2) in quadrant II, the controls should be ≤ 0. In quadrant IV, the controls should be ≥ 0

218 STABILITY ANALYSIS OF AIRCRAFT SYSTEMS

by the FLC. If this regulation holds for all y_1 and y_2, then the FLC is globally asymptotically stable. For any initial conditions in the state space the FLC was expected to produce control inputs that will lead to an asymptotic stability of the closed-loop system. The regions of stability, based on the FLC's regulation of expression (10.22), were analyzed and shown in Figure 10.10. For all initial conditions on the line $M : y_2 - 1.277 y_1 = 0$, the fuzzy logic control system is therefore expected to be asymptotically stable, regardless of the fuzzy controller's prescribed inputs. This line is, therefore, the *primary invariant manifold* [19]. Analytically, this is seen to be obvious, since on this line, $dV/dt = -\frac{1}{2}(x^T Q x)$, which goes

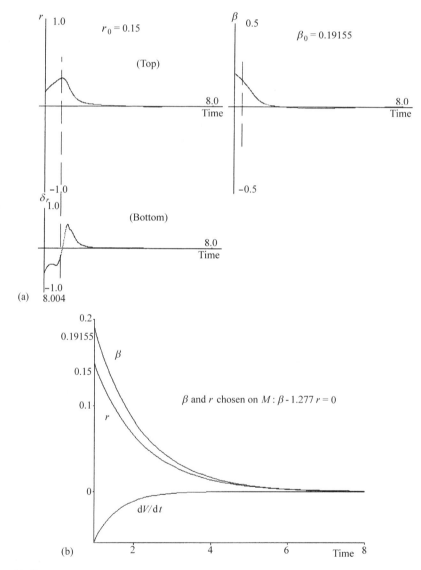

Figure 10.13 (a) Top: Trajectories for r and β with initial conditions on M. Bottom: Control profile. It has no effect on asymptotic stability for initial conditions on M. Vertical lines (dotted) are only illustrating extents of transients. (b) Plots of β, r and dV/dt for initial conditions on M

to zero, where Q already solved Lyapunov's equation for the stable, unforced system. Thus, any control profile prescribed by the FLC while the states are on this line should not contradict the known analytical result. To see this, points were taken on this line in the first quadrant (I) and the third quadrant (III) and the system was simulated with the FLC in the closed loop. The results are shown in Figure 10.13(a) and Figure 10.13(b), and Figure 10.14. Finally, a couple of sample points were taken: (1) $r = 0.15, \beta = 0.1$, to illustrate the situation in Figure 10.12(a), quadrant I, and (2) $r = 0.3, \beta = -0.15$, to illustrate the case of Figure 10.12(b), quadrant IV.

The results are shown in Figure 10.15 and Figure 10.16.

10.2.3 Simulation Studies and Results

The simulation results are shown in Figures 10.13–10.16. These are only representative results in that they have been produced solely to illustrate the theory. The actual stability analysis is detailed in Figure 10.12. Figure 10.13(a) shows the result of having points on the primary invariant manifold, M, as determined from the stability analysis. Even though a control profile was generated by the FLC for points on this line, the trajectories for r and β are seen to be asymptotically stable. This fact was further supported as depicted by the first rule in Figure 10.11; this is therefore an *invariant rule*. The same situation is depicted in Figure 10.13(b), where, in addition, dV/dt is seen to be negative definite. Note that theoretically, $dV/dt = -0.5x^T Q x$, for points on M. However, in this figure, the effect of the transients have been removed to show that dV/dt is predominantly negative definite. Figures 10.14–10.16 were produced to verify the stability requirement in Figure 10.12, on the control profile generated by the FLC. For the particular points chosen in the simulations, quadrants I, III and IV in Figure 10.12 suggest that the generated fuzzy control, $u_{FLC} \geq 0$. This is seen to be the case here in the simulation results as the generated controls become steady.

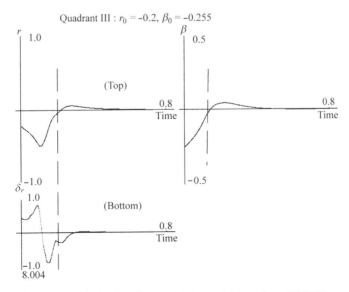

Figure 10.14 Trajectories (top) and controls (bottom) in quadrant III in Figure 10.11(a)

220 STABILITY ANALYSIS OF AIRCRAFT SYSTEMS

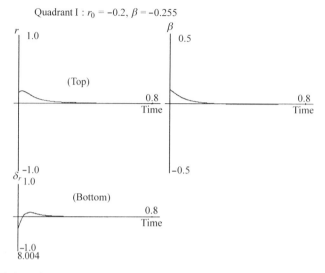

Figure 10.15 Trajectories (top) and control profile (bottom) in Quadrant III (of Figure 10.11(a) for r and β both positive)

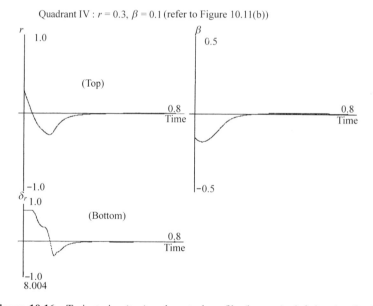

Figure 10.16 Trajectories (top) and control profile (bottom): definite sign for M

10.2.4 Conclusions

The design of the fuzzy controller for the missile autopilot was again summarized. Moreover, the main issue of this section is the proposed stability analysis of this controller. For this example, a missile autopilot's yaw axis linearized dynamics was considered. Instead of invoking the Kalman–Yakubovich lemma directly by solving the simultaneous equations

$A^T P + PA = -Q$ and $PB = C$, an input–output mapping dV/dt was first formulated which utilizes P and Q derived upon solving only Lyapunov's equation. This expression was then analyzed to determine the actions that needed to be taken by the FLC, for various points in the state space. The essence of the theorem is in the formulation of an input–output mapping. This was readily determined for the missile autopilot system which was stable in the open loop. The theory was used to determine the stability regions of the system. The results were seen to be in accord with the analytical results.

10.3 LINEAR DISCRETE-TIME MODEL APPLICATION[1]

10.3.1 Advanced Technology Wing Aircraft Model

In this section we describe a fuzzy roll controller design for Rockwell International's Advanced Technology Wing (ATW) aircraft model. The controller modulates six control surfaces on the ATW to achieve the desired roll rate while maintaining wing loads within safe bounds. The use of fuzzy control rules enabled highly flexible control behavior that operated the system close to the constraint limits, thereby achieving excellent roll performance. We also present a method of stability analysis for fuzzy control systems based on partitioning the state space and applying a Lyapunov stability criterion to the individual partitions. A subset of the control rules is analyzed using this method and shown to provide asymptotically stable control.

10.3.2 Introduction

We have designed a fuzzy roll controller for Rockwell International's Advanced Technology Wing (ATW) aircraft model. The ATW is an experimental wind-tunnel model that employs active leading and trailing edge control surfaces on a flexible wing structure to enhance aerodynamic performance. Because of the light weight, flexible wing structure, a crucial control problem is to ensure that wing loads do not exceed safe bounds during flight maneuvers. Since curtailing wing loads invariably curtails maneuver performance, the fuzzy controller does not attempt to alleviate loads if the loads are safely within the prescribed bounds. The controller achieved highly flexible, non-linear control behavior that operated the system close to the load limits, sacrificing maneuver performance only when necessary. In this section we describe the derivation of the control rules from a qualitative analysis of the ATW plant equations. We also present a rigorous method of stability analysis for fuzzy control systems based on partitioning the state space and applying the Lyapunov criterion to each partition. This approach utilizes only the minimum and maximum bounds on the control output within the partitioned regions of the state space and does not impose any

---

[1] The material in the remainder of this chapter is reprinted with permission from (a) Chui, S., Fuzzy logic for control of roll and moment for a flexible wing aircraft, *IEEE Control Systems Magazine*, June 1991, 42–48, © 1991 IEEE; (b) *Proceedings of the American Control Conference*, Boston, MA. June 1991, pp. 821–826, © 1991 IEEE.

10.3.3 The ATW Problem

Our control problem is focused on the roll maneuver of the Advanced Technology Wing (ATW), a wind-tunnel aircraft model used in experimental testing of novel concepts involving the integration of aeroservoelastic structural dynamics and control. Aerodynamic control is provided by four control surfaces on each wing: the leading edge outboard (LO), leading edge inboard (LI), trailing edge outboard (TO), and trailing edge inboard (TI) surfaces. Loading on the wing exists in three forms: torsion moment, bending moment, and hinge moment; here we consider only control of the torsion moments. The torsion moments are measured by strain gauges at the wing midspan and root, which correspond to the outboard torsion moment and the inboard torsion moment, respectively. A diagram of the ATW identifying the control surface and torsion moment locations is shown in Figure 10.17.

The task of the roll controller is to determine the appropriate control surface deflections that will achieve a commanded roll rate as well as satisfy prescribed torsion moment constraints. Although there are eight control surfaces available on the ATW, the contribution to the roll maneuver from the left and right leading edge inboard surfaces is negligible. Therefore, for our purpose they are excluded as usable actuators.

In what follows we will use the symbol δ to denote surface deflection and τ to denote torsion moment. Subscripts l and t denote leading and trailing edge; subscripts i and o denote inboard and outboard; and subscripts r and l denote right and left. For example,

δ_{lol} = leading edge outboard left control surface deflection;
δ_{tir} = trailing edge inboard right control surface deflection;
τ_{il} = inboard left torsion moment;
τ_{or} = outboard right torsion moment.

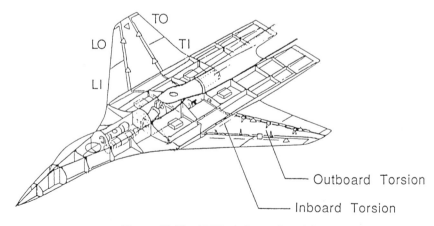

Figure 10.17 ATW wind-tunnel model

Defining the vectors

$$\boldsymbol{\delta} = [\delta_{\text{lol}} \quad \delta_{\text{tol}} \quad \delta_{\text{til}} \quad \delta_{\text{lor}} \quad \delta_{\text{tor}} \quad \delta_{\text{tir}}]^{\text{T}},$$

$$\boldsymbol{\tau} = [\tau_{\text{il}} \quad \tau_{\text{ol}} \quad \tau_{\text{ir}} \quad \tau_{\text{or}}]^{\text{T}},$$

a model of the ATW dynamics for the flight condition of Mach 0.9 and dynamic pressure (q) of $150.0/\text{ft}^2$ is

$$\dot{p}(t) = Ap(t) + B\boldsymbol{\delta}, \tag{10.24}$$

$$\boldsymbol{\tau}(t) = Cp(t) + D\boldsymbol{\delta} + E, \tag{10.25}$$

where p is the roll rate, and

$$A = [-1.49], \quad B = [-11.2 \quad 23.9 \quad 55.7 \quad 11.2 \quad -23.9 \quad -55.7],$$

$$C = [2.17 \quad -0.734 \quad -2.17 \quad 0.734]^{\text{T}}, \tag{10.26}$$

$$D = \begin{bmatrix} -17.1 & -711 & -1033 & 8.3 & 27.3 & -15.7 \\ -86.2 & -103 & -106 & -1.5 & -1.3 & 3.0 \\ 8.3 & 27.3 & -15.7 & -17.1 & -711 & -1033 \\ -1.5 & -1.3 & 3.0 & -86.2 & -103 & -106 \end{bmatrix}, \tag{10.27}$$

$$E = [-76.0 \quad -7.7 \quad -120 \quad -8.5]^{\text{T}}$$

In the above equations, p is in radian/s, $\boldsymbol{\delta}$ is in radian, and τ is in ft-lb. The physical limit of the control surface deflection is ± 0.23 radian for all surfaces.

10.3.4 Control Architecture

The fuzzy roll controller is comprised of two modules, each containing a set of fuzzy control rules. The first module receives the roll rate error and roll acceleration as the input and determines a generalized deflection command as its output. The generalized deflection command is a scalar representing the required deflection for a surface if all surfaces deflect by an equal amount in the appropriate directions. The generalized deflection command can be transformed into individual surface deflection commands by multiplying the scalar by a directional vector that relates the surface deflection directions to the roll direction.

Let δ_g be the generalized deflection command, then the deflection command vector is given by

$$\boldsymbol{\delta} = \delta_g [1 \quad -1 \quad -1 \quad -1 \quad 1 \quad 1]^{\text{T}}. \tag{10.28}$$

The second module receives the generalized deflection command and torsion moment measurements as input and adjusts the command vector of equation (10.28) to ensure the torsion moments are within given bounds. A block diagram of the roll controller is shown in Figure 10.18. In words, the first module generates a nominal deflection command vector to achieve the desired roll rate; the second module adjusts this nominal command vector to achieve the desired torsion moment constraints.

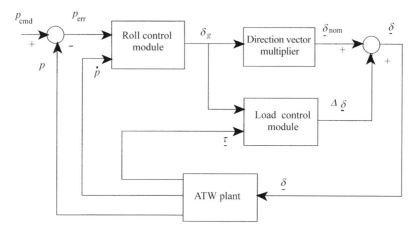

Figure 10.18 Roll controller block diagram

10.3.5 Control Rule Synthesis

The control module for determining the generalized deflection command contains a set of rules that implements a form of proportional-derivative control. The rules are listed in Figure 10.19 and the membership functions used in the rules are shown in Figure 10.20. The prefixes 'p' and 'n' in the linguistic descriptors denote 'positive' and 'negative', respectively; thus, 'p. big' means 'positive big'. Only the rules for handling the positive rate error and positive acceleration are shown; their negative counterparts are symmetric and can be obtained by simply reversing the sign of the conclusion. Hence, a total of 14 rules are used in this module, 7 for proportional control and 7 for derivative control. Note that each derivative control rule is preceded by a number which represents the weight of the rule. The derivative control rules are given a much smaller influence than the proportional control rules to avoid overdamping, in the same way that the derivative gain is typically smaller than the proportional gain in conventional linear control.

Additionally, the derivative control actions are predicated upon the condition that the rate error is near zero. The resultant behavior is that the controller will not impose damping until the vehicle approaches the commanded roll rate. This control strategy aggressively pushes

---

if p_{err} is p.big then δ_g is n.big;
if p_{err} is p.med then δ_g is n.med;
if p_{err} is p.sml then δ_g is n.sml;
if p_{err} is zero then δ_g is zero;

0.2 if p_{err} is near.zero and accel is p.big then δ_g is p.big;
0.2 if p_{err} is near.zero and accel is p.med then δ_g is p.med;
0.2 if p_{err} is near.zero and accel is p.sml then δ_g is p.sml;
0.2 if p_{err} is near.zero and accel is zero then δ_g is zero;

Figure 10.19 Roll control rules

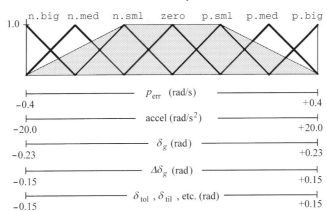

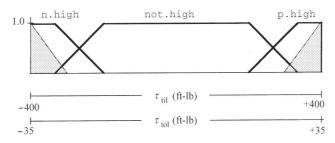

Figure 10.20 Membership functions used in control rules

the vehicle toward the commanded roll rate at maximum acceleration, and applies damping to stabilize the vehicle only during close tracking.

Figure 10.21 shows a simulation of the system response (solid lines) when the deflection command vector determined by this control module is used directly (without adjustment for torsion moment constraints). The response shown is for a step roll rate command of 2.5 rad/s and controller sampling time of 0.005 s. The figure also shows (dashed lines) the response of the system when the error-near-zero clause is removed from the derivative control rules. We see that the modulation of damping improved the response time significantly. However, because of the larger surface deflections utilized, the torsion moments experienced during the maneuver are also significantly higher.

The second control module contains a set of rules for adjusting the nominal deflection command vector to alleviate sensed as well as anticipated excessive torsion moments. No adjustments are made if the torsion moments are well within prescribed bounds.

The rules for deflection adjustment are obtained by studying the qualitative attributes of the matrices B and D as given by equations (10.26) and (10.27). From equation (10.24) we see that matrix B is an actuator gain matrix that provides a measure of the effectiveness of each control surface in generating roll acceleration. From equation (10.25) we see that matrix D is a similar gain matrix that provides a measure of the 'effectiveness' of each control

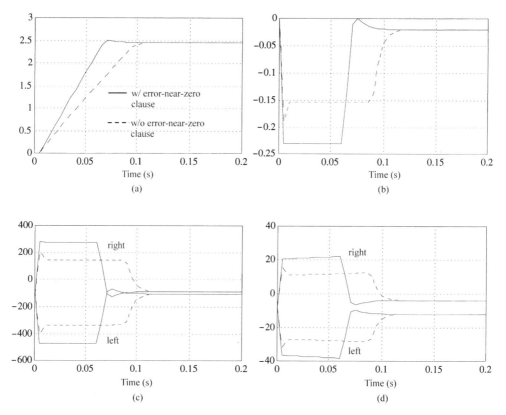

Figure 10.21 System response without torsion moment constraints. (a) Roll rate (rade/s); (b) generalized deflection (rad); (c) inboard torsion moments (ft-lb); (d) outboard torsion moments (ft-lb)

surface in generating the torsion moments. Hence, matrices B and D together describe the relative contributions of each surface to the roll power (i.e. the ability to effect roll motion) and the torsion moments, and thereby form the basis for establishing rules for trade-offs between roll power maintenance and torsion moment reduction.

We note from the B matrix that the relative contributions to the roll power from the leading edge outboard (LO), trailing edge outboard (TO), and trailing edge inboard (TI) surfaces are approximately given by the ratios $1:2:5$. Hence, if any accommodation in the surface deflections is required to alleviate torsion moment, we can sacrifice the LO surfaces with minimal loss of roll power, followed by the TO surfaces, and lastly the TI surfaces.

Consider the task of reducing the left inboard moment. We note from the D matrix that the relative contributions to the left inboard moment from the elements of the surface deflection vector are approximately given by the ratios $1:40:60:1:1:1$. These ratios show that adjustment to the left TO and left TI surfaces have the greatest impact on the left inboard moment; adjustment to all other surfaces has negligible effect. Since the relative contributions of the left TO surface versus the left TI surface is $2:5$ for roll power and $2:3$ for the torsion moment, better roll power versus load trade-off is clearly obtained by using the left TO surface as the primary means of load alleviation. The left TI surface must be adjusted with careful restraint to avoid severe reduction in roll power.

Now consider the left outboard moment. We note again from the D matrix that the relative contributions to the left outboard moment from the elements of the surface deflection vector are approximately given by the ratios $40:50:50:1:1:1$. These ratios indicate that all left side surfaces have approximately equal influence on the left outboard moment, while the right side surfaces have negligible influence. Since the LO surfaces have minimal effect on roll power, we can freely adjust the left LO surface to alleviate the torsion moment. The left TO surface can be used for supplemental adjustment at a higher cost to roll power.

The load alleviation rules derived from the above qualitative analysis are listed in Figure 10.22, with some ancillary rules for further improving performance. Only the rules for alleviating positive left inboard and positive left outboard torsion moments are shown; their negative and right side counterparts are merely symmetric rules. The membership functions used in the rules are shown in Figure 10.20, where the bounds for the inboard torsion moments are set as $[-400 +400]$ and the bounds for the outboard torsion moments are set as $[-35 +35]$. The response of the system with the load alleviation module in effect is shown in Figure 10.23. We see that the control system enforced the torsion moment bounds without any significant degradation in roll rate performance. The outboard torsion moment bounds are particularly enforced with ease because of the latitude afforded by the LO surfaces.

10.3.6 Stability Analysis

The non-linear nature of fuzzy control provides enhanced performance but also frustrates mathematical analysis. Several approaches using approximation techniques have been developed to analyze the stability of fuzzy control systems. Kickert and Mamdani [26] applied classical describing function techniques to a multi-relay model of fuzzy control. Cumani [27] uses a possibilistic approach where the system behavior is represented by time-

---

if τ_{il} is not.high then δ_{tol} is zero and δ_{til} is zero;
if τ_{il} is p.high then δ_{tol} is p.med;
if $\tau_{;il}$ is p.huge then δ_{tol} is p.big and δ_{til} is p.sml;

if τ_{ol} is not.high then δ_{lol} is zero and δ_{tol} is zero;
if τ_{ol} is p.high then δ_{lol} is p.big and δ_{tol} is p.sml;
if τ_{ol} is p.huge then δ_{lol} is p.big and δ_{tol} is p.med;
/*
Rules for suppressing large changes in the command vector that may cause large positive moments at the next sampling time.
*/
if $\Delta\delta_g$ is p.big and τ_{il} is not.n.high then δ_{tol} is p.big and δ_{til} is p.sml
if $\Delta\delta_g$ is p.big and τ_{ol} is not.n.high then δ_{lol} is p.big and δ_{tol} is p.med;
/*
Rule for modifying roll power loss by countering left side surface adjustments with beneficial right side surface adjustment
*/
if τ_{il} is p.huge and τ_{ir} is not.n.high then δ_{tir} is p.sml;

Figure 10.22 Load alleviation rules

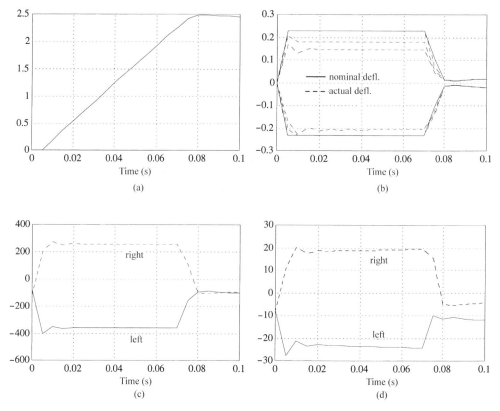

Figure 10.23 System response with torsion moment constraints. (a) Roll rate (rad/s); (b) deflections (rad); (c) inboard torsion moments (ft-lb); (d) outboard torsion moments (ft-lb)

evolving conditional possibilistic distributions. Tong [28] represents system operators as fuzzy relational matrices and analyzes the stability of a fuzzy relational difference equation. Using the fuzzy relational matrix, Kiszka *et al.* [29] formulated a Lyapunov-like energy function for a fuzzy set and examined whether the energy function was monotonically decreasing. This energy-like method has been applied to a previous fuzzy controller design for the ATW [30].

Rigorous stability analysis techniques have been developed recently to provide more reassuring proofs than those afforded by the approximate techniques. Chen [31] applies the method of *cell-to-cell mapping* [32] in which the state space is discretized into cells and possible transitions between cells are studied through exhaustive search. Langari and Tomizuka [23, 33] derive closed-form analytic criteria for single-input–single-output controllers, but impose stringent constraints on the rule structure and partitioning of fuzzy membership functions. We propose an analytic/numeric method that imposes no constraint on the fuzzy controller except for completeness of the rules (i.e. a control action can be inferred for every point in the state space). This method also partitions the state space into cells, but only to compute the minimum and maximum bounds on the controller output for the individual cells. Based on these bounds, we then attempt to verify that a Lyapunov function is monotonically decreasing in every cell.

10.3.6.1 Stability Criterion

Consider a linear discrete-time model of the controlled plant described by

$$x(k+1) = Ax(k) + Bu(k).$$

We define a Lyapunov function $V(x(k))$ of the form

$$V(x(k)) = x(k)^\mathrm{T} P x(k),$$

where P is a positive-definite, symmetric matrix. Then

$$\Delta V = V(k+1) - V(k) = x^\mathrm{T}(k+1) P x(k+1) - x^\mathrm{T}(k) P x(k)$$
$$= x^\mathrm{T}(k)[A^\mathrm{T} P A - P] x(k) + 2 u^\mathrm{T}(k) B^\mathrm{T} P A x(k) + u^\mathrm{T}(k) B^\mathrm{T} P B u(k).$$

For global asymptotic stability, the following condition must be satisfied:

$$\frac{x^\mathrm{T}(-A^\mathrm{T} P A + P)x}{x^\mathrm{T} x} > \frac{2 u^\mathrm{T} B^\mathrm{T} P A x}{x^\mathrm{T} x} + \frac{u^\mathrm{T} B^\mathrm{T} P B u}{x^\mathrm{T} x}, \qquad (10.29)$$

where the index (k) is dropped for brevity. This implies

$$\lambda_{\min}(-A^\mathrm{T} P A + P) > \frac{2 u^\mathrm{T} B^\mathrm{T} P A x}{\|x\|^2} + \frac{u^\mathrm{T} B^\mathrm{T} P B u}{\|x\|^2}. \qquad (10.30)$$

For single-input plants, the above condition simplifies to

$$\lambda_{\min}(-A^\mathrm{T} P A + P) > \frac{\|u\|^2}{\|x\|^2}\left(2 B^\mathrm{T} P A \frac{x}{u} + B^\mathrm{T} P B\right). \qquad (10.31)$$

If the condition given by equation (10.30) or equation (10.31) holds in every region of the state space, then the system is asymptotically stable. We will apply this stability criterion to the ATW control system realized by the roll control rules, excluding the action of the load alleviation rules. With only the roll control rules, the dynamics of the ATW given by equation (10.24) simplifies to

$$\frac{\mathrm{d}p(t)}{\mathrm{d}t} = [-1.49]p(t) + [-182]\delta_\mathrm{g}, \qquad (10.32)$$

where δ_g is the generalized deflection command. For the control sampling time of 0.005 s, the continuous-time model given by equation (10.32) is equivalent to the following discrete-time model:

$$p(k+1) = 0.9926 p(k) - 0.905 \delta_\mathrm{g}(k).$$

Using a state vector that contains the roll rate and roll acceleration as its elements, the state equation is

$$x(k+1) = \begin{bmatrix} 0.9926 & 0 \\ -0.0074 & 0 \end{bmatrix} x(k) + \begin{bmatrix} -0.905 \\ -0.905 \end{bmatrix} u(k), \qquad (10.33)$$

where

$$x(k) = \begin{bmatrix} x_1 \\ x_2 \end{bmatrix} \doteq \begin{bmatrix} p(k) \\ p(k) - p(k-1) \end{bmatrix} \quad \text{and} \quad u(k) \doteq \delta_\mathrm{g}(k).$$

It should be noted that $u(k)$ is a function of $x(k)$ and the acceleration axis now given in units of rad/s per sample period.

Using the Lyapunov weighing matrix

$$P = \begin{bmatrix} 68 & 0 \\ 0 & 1 \end{bmatrix},$$

substitution of the matrices of equation (10.33) into equation (10.31) yields

$$1 > \frac{\|u\|^2}{\|x\|^2}\left(-122.1\frac{x_1}{u(x)} + 56.5\right) \tag{10.34}$$

as the stability criterion for the ATW control system.

10.3.6.2 Verifying Stability

The state space is first partitioned into cells in which the minimum and maximum bounds on the control output can be readily determined. Figure 10.24 shows the minimum and maximum bounds on the control output computed for partitions of the roll rate error and acceleration space. These bounds can be readily determined from the control rules with minimal numerical calculation. Consider the upper left corner cell of Figure 10.24. The maximum control output is obtained when the rate error is exactly negative big (calling for positive big deflection) and the error-near-zero clause inhibits any consequence from the acceleration rules, thus producing an exactly positive big deflection. The minimum control output is obtained when the rate error is exactly negative medium (calling for positive medium deflection) and the acceleration error is exactly negative big (calling for negative big deflection), resulting in a weighted average deflection between positive medium and negative big that can be easily computed from the modified center of area given below:

$$u = \frac{\sum_{i=1}^{s} w_i \alpha_i \mu_{B_i}^d}{\sum_{i=1}^{s} w_i \alpha_i},$$

where $\mu_{B_i}^d$ is the *defuzzified* value of the membership function μ_{B_i}, s is the number of rules, and α is the *degree of fulfillment* of the rule. (see Section 10.2).

Control bounds for regions outside that shown in Figure 10.24 can also be determined readily from the rules handling the extrema of the input range.

For each cell, we verify equation (10.34) with conservative bounds on $u(x)$ and x. The validity of equation (10.34) can be tested via several different tacks. For example, equation (10.34) is satisfied if

$$1 > \sup_{u,x}\left(\frac{\|u\|^2}{\|x\|^2}\right) \sup_{u,x}\left(-122.1\frac{x_1}{u(x)} + 56.5\right), \tag{10.35}$$

where sup denotes the supremum value.

Consider again the upper left corner cell of Figure 10.24. Since both x and $u(x)$ are positive in this cell, a conservative value for the second supremum in equation (10.35) is found at the smallest value of x and the largest value of u, i.e. at $x = 0.2667$ and $u = 0.23$. Substituting these values into equation (10.35), we see that the right-hand side of equation (10.35) must be negative; thus, Lyapunov stability for this region of the state space is verified. Similarly, we can show Lyapunov stability for the remaining cells of the state space, except for those containing the zero roll rate state; this unresolved region is shown in Figure 10.24 by a highlighted bounding box. Conservative bounds failed to establish stability for those regions near the equilibrium. Since the bounds become less conservative as a cell becomes smaller, we can further partition the unresolved region into smaller cells, determine the new control bounds, and apply the same analysis. However, we note from the control rules that within the unresolved region the error-near-zero clause is perfectly fulfilled and the controller provides simple proportional-derivative control. It can be shown that within this region the fuzzy controller implements the following linear control law:

$$u = (K_p x_1 + 0.2 K_d x_2)/1.2, \quad \text{where } K_p = 0.575 \text{ and } K_d = 2.3,$$

or simply

$$u(t) = Kx(t), \quad \text{where } K = [0.479 \quad 0.383]. \tag{10.36}$$

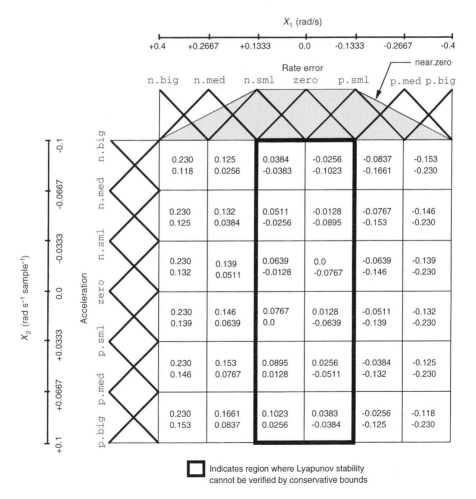

Figure 10.24 State space partitions and the associated control bounds

Substitution of the linear control law of equation (10.36) into equation (10.30) yields the stability criterion

$$\lambda_{\min}(-A^T PA + P) > \frac{x^T[2K^T B^T PA + K^T B^T PBK]x}{\|x\|^2}$$

or

$$1 > \lambda_{\max}(2K^T B^T PA + K^T B^T PBK) \qquad (10.37)$$

The matrix on the right-hand side of equation (10.37) has eigenvalues of -37.3 and 0. Therefore we conclude that the stability criterion is satisfied within the formerly unresolved region.

A much simpler stability criterion can be established in the continuous-time domain due to the simplicity of the plant model. We first assume a control law of the form

$$u(t) = K_1(p, \dot{p})p + K_2(p, \dot{p})\dot{p},$$

where K_1 and K_2 can be thought of as *effective gains*. Substitution of this control law into equation (10.32) yields

$$\ddot{p} = (-1.49 - 182K_1)p - 182K_2\dot{p}$$

or

$$\dot{p} = \frac{(-1.49 - 182K_1)}{(1 + 182K_2)} p.$$

Above is a scalar system of the form $dp(t)/dt = ap(t), a \in \mathbb{R}$, the solution of which is $p(t) = p(0)e^{at}, t \geq 0$. It is immediate then that the system is asymptotically stable if $a < 0$. That is if

$$a \doteq \frac{(-1.49 - 182K_1)}{(1 + 182K_2)} < 0.$$

This condition is satisfied if both $K_1(\cdot)$ and $K_2(\cdot)$ are positive definite. It suffices to consider them as *class-K* functions with argument $\|p\|$, where $\boldsymbol{p} = [p \quad dp/d]^T$. We can choose, for instance, $K_i = \alpha_i(\|\boldsymbol{p}\|), i = 1, 2$. However, if these gains are mere constants, then the condition degenerates to having $K_1, K_2 \geq 0$. This condition can be readily shown to be true from the control rules and the modified center of area inference procedure that we have employed.

10.3.7 Conclusions

We have designed a fuzzy controller for roll maneuver and load control of an aircraft model. Excellent response was achieved using relatively simple rules derived from a qualitative analysis of the plant equations. The use of fuzzy control enabled the realization of non-linear and effective control behavior. In particular, the adaptation of damping as a function of distance from the goal state helped to improve system response time; applying load alleviation measures only when the loads are close to the bounds ensured that roll performance is not unnecessarily compromised.

10.4 SUMMARY

The methods of stability analysis developed in this chapter pertained to the successful formulation of an input–output mapping from an energy-like expression for the driven system. A theorem was stated for guaranteeing that this energy-like function is negative definite. For non-linear systems, there is no unique way to determine this mapping. However, by identifying the outputs of the system from the available states, these are explicitly incorporated into the energy-like expression. Since the system is driven by the fuzzy logic controller in the closed loop, this expression must contain the fuzzy control variable. The task of ensuring that this energy is diminished only at the specified origin is assumed by the fuzzy controller. It was stated that for relatively simple systems the analytical proof of one of the requirements of the theory, namely the negative definiteness of dV/dt, may be easy to show. However, for a complex dynamic system, the undue approximations that would usually be employed in the proof process may be either infeasible or physically unrealistic. Thus, the expression may have to be simulated for various points in the state space. Once this is done, the remaining requirements may be easily satisfied by analytical means. Nevertheless, simulation results should reconfirm this fact. For a stable linear system, the positive real lemma (Kalman–Yakubovich) may be used to prove dissipativeness. The only difficulty is that simultaneous equations need to be solved involving Lyapunov's equation and the second condition on P, B and C. If the system is stable, then one can always solve this, but it is very difficult to have the specified output matrix, C, automatically producing a P that solves Lyapunov's equation. If this is the case, then one may have to accept the resulting input matrix, which may be different from the specified one. This can always be rescaled to recover the signals required as the output. Moreover, the problem can be cast as an LMI problem solvable in Matlab as a feasibility problem. Equivalently, it may be easier to formulate an input–output mapping for the controlled system, and the control inputs generated by the FLC should ensure that this is at least negative semi-definite.

In Section 10.3 we also presented a method of stability analysis for fuzzy control systems based on partitioning the state space and applying the Lyapunov criterion to each partition. This method is effective for regions in the state space distant from the origin, and requires increasingly finer partitioning for regions near the origin. From the standpoint of analysis, it is advantageous to design fuzzy controllers that behave linearly near the origin.

BIBLIOGRAPHY

1. Zames, G., On the input–output stability of time-varying nonlinear feedback systems: Part I: conditions derived using concepts of loop gain, conicity, and positivity, *IEEE Transactions on Automatic Control*, **AC-11**(2), 1966.
2. Zames, G., On the input–output stability of time-varying nonlinear feedback systems: Part II: conditions involving circles in the frequency plane and sector nonlinearities, *IEEE Transactions on Automatic Control*, **AC-11**(3), 1966.
3. Vukobratovic, M. and R. Stojic, *Modern Aircraft Flight Control*, Springer-Verlag, New York, 1988.
4. Farinwata, S. S., Performance Assessment of Fuzzy Logic Control Systems via Stability and Robustness Measures, Ph.D. Dissertation, Georgia Institute of Technology (Georgia Tech), School of Electrical and Computer Engineering, 1993.

5. Farinwata, S. S., D. Pirovolou and G. J. Vachtsevanos, An input–output stability analysis of a fuzzy controller for a missile autopilot's yaw axis, Proceedings of the Third *IEEE International Conference on Fuzzy Systems*, Orlando, 1994, pp. 930–935.
6. Slotine, J. J and W. Li, *Applied Nonlinear Control*, Prentice-Hall, Englewood Cliffs, 1991.
7. Narendra, K. S. and A. M. Annaswamy, *Stable Adaptive Systems*, Prentice-Hall, Englewood Cliffs, 1989.
8. Anderson, B. D. O. et al., *Stability of Adaptive Systems*, MIT Press, Cambridge, 1986.
9. Kailath, T., *Linear Systems*, Prentice-Hall, Englewood Cliffs, 1980.
10. Chef, C. T., *Linear Systems Theory and Design*, Holt, Rinehart and Winston, Cambridge, 1984.
11. Boyd, S. et al., *Linear Matrix Inequality in Systems and Control Theory*, SIAM, Philadelphia, 1994.
12. Dawson, D. M. et al., *Nonlinear Control of Electric Machinery*, Mercel Dekker, New York, 1998.
13. Vachtsevanos, G., S.S. Farinwata and D. Pirovolou, Fuzzy logic control of an automotive engine: A systematic design methodology, *IEEE Control Systems*, Special Issue on Intelligent Control, **13**(3), 1993, 62–68.
14. Patyra, M. J. and D. M. Mlyne, *Fuzzy Logic: Implementation and Applications*, John Wiley, Chichester, 1996.
15. Dudley, R. M., *Real Analysis and Probability*, Wadsworth and Brooks/Cole, 1989.
16. Bartle, G., *The Elements of Real Analysis*, John Wiley, Chichester, 1976.
17. Farinwata, S. S. and G. Vachtsevanos, Stability Analysis of the fuzzy controller designed by the phase portrait assignment algorithm, *Proceedings of the second FUZZ-IEEE*, San Francisco, 1993, pp. 1377–1382.
18. Khalil, H. K., *Nonlinear Systems*, Macmillan, Basingstoke, 1992.
19. Kang, H. and G. Vachtsevanos, Nonlinear fuzzy control based on vector field of the phase portrait assignment algorithm, *Proceedings of the 1990*, American Control Conference, San Diego, 1990, pp. 1479–1484.
20. Chen, Y. Y., Stability analysis of fuzzy control—A lyapunov approach, *Proceedings of the IEEE International Conference on Systems, Man and Cybernetics*, vol. 3, 1985, pp. 1027–1031.
21. Maiers, J. and Y. Sherif, Application of fuzzy set theory, *IEEE Transactions on Systems, Man and Cybernetics*, **SMC-15**(1), 1985, 175–189.
22. Bernard, J., Use of a rule-based system for process control, *IEEE Control Systems Magazine*, October, 1988, 3–13.
23. Langari, R. and M. Tomizuka, Analysis and synthesis of fuzzy linguistic control systems, *Intelligent Control 1990*, ASME, 1990.
24. Zadeh, L., Outline of a new approach to the analysis of complex systems and decision processes, *IEEE Transactions on Systems, Man and Cybernetics*, **SMC-3**(1), 1973, 28–44.
25. Sugeno, M., An introductory survey of fuzzy control, *Information Sciences*, **36**, 1985, 59–83.
26. Kickert, W. and E. Mamdani, Analysis of a fuzzy logic controller, *Fuzzy Sets and Systems*, **1**(1), 1978, 29–44.
27. Cumani, A., On a possibilistic approach to the analysis of fuzzy feedback systems, *IEEE Transactions on Systems, Man and Cybernetics*, **SMC-12**(3), 1982, 417–422.
28. Tong, R., Some properties of fuzzy feedback systems, *IEEE Transactions on Systems, Man and Cybernetics*, **SMC-10**(6), 1980, 327–330.
29. Kiszka, J., M. Gupta and P. Nikiforuk, Energetistic stability of fuzzy dynamic systems, *IEEE Transactions on Systems, Man and Cybernetics*, **SMC-15**(5), 1985, 783–792.
30. Chand, S. and S. Hansen, Energy based stability analysis of a fuzzy roll controller design for a flexible aircraft wing, *Proceedings of the Twenty-eighth IEEE Conference on Decision and Control*, Tampa, 1989, pp. 705–709.
31. Chen, Y. Y. The Global Analysis of Fuzzy Dynamic Systems, Ph.D. thesis, Dept. of Electrical Engin, U. C. Berkeley, 1989.
32. Hsu, C., *Cell-to-Cell Mapping: A Method of Global Analysis of Nonlinear Dynamic Systems*, Springer-Verlag, New York, 1988.

33. Langari, R. and M. Tomizuka, Stability of fuzzy linguistic control systems, *Proceedings of the Twenty-ninth IEEE Conference on Decision and Control*, Hawaii, 1990, pp. 2185–2190.
34. Levine, W. B. (ed.), *The Control Handbook*, CRC/IEEE Press, 1996.
35. Sontag, E., *Mathematical Control Theory*, Springer-Verlag, New York, 1990.
36. Stevens, B. L. and F. L. Lewis, *Aircraft Control and Simulation*, John Wiley, New York, 1992.
37. Tao, G., A simple alternative to the Barbalat lemma, *IEEE Transactions on Automatic Control*, **42**(5), 1997, 698.
38. Lin, W. and C. I. Byrnes, Passivity and absolute stabilization of a class of discrete-time nonlinear systems, *Automatica*, **31**(2), 1995, 263–267.
39. De Figueiredo, R. J. P. and Chen, G., *Nonlinear Feedback Control Systems. An Operator Theory Approach*, Academic Press, 1993.

SYNTHESIS

11

Observer-Based Controller Synthesis for Model-Based Fuzzy Systems via Linear Matrix Inequalities

Ali Jadbabaie
California Institute of Technology, Pasadena, USA

Chaouki T. Abdallah and **Mohammad Jamshidi**
University of New Mexico, Albuquerque, USA

Andre Titli
LAAS du CNRS and INSA, Tolouse, France

11.1 INTRODUCTION

There has recently been a rapidly growing interest in using Takagi–Sugeno (T–S) fuzzy models to approximate non-linear systems. This interest relies on the fact that dynamic T–S models are easily obtained by linearization of the nonlinear plant around different operating points. Once the T–S fuzzy models are obtained, linear control methodology can be used to design local state feedback controllers for each linear model. Aggregation of the fuzzy rules results in a generally nonlinear model, but in a very special form known as a Polytopic Linear Differential Inclusion (PLDI) [1]. Fortunately, the stability conditions for these types of systems can be formulated in a Linear Matrix Inequality (LMI) framework, which can then be solved using convex optimization methods. However, the resulting stability conditions might be conservative, because they require the existence of a common Lyapunov matrix.

In a recent paper, Petterson and Lennartson [2] have studied a similar problem and have come up with stability conditions that do not require a single Lyapunov matrix for the solution of the LMIs, but has some extra assumptions on the different Lyapunov functions. However, these are analysis results only. Our approach is based on the LMI formulation of the stability conditions for closed-loop T–S systems given in [3] and [7]. We extend these

Fuzzy Logic. Edited by S. Farinwata, D. Filev and R. Langari
© 2000 John Wiley & Sons, Ltd

results to the case when states are not available for measurement of feedback, by introducing the notion of a *fuzzy observer*.

This chapter is organized as follows. In Section 11.2 we give an overview of T–S fuzzy systems and sufficient stability conditions for both continuous-time and discrete-time cases. Section 11.3 deals with the LMI formulation of the stability results given in Section 11.2. In Section 11.4 we introduce the notion of a fuzzy observer for both continuous-time and discrete-time systems. We also state and prove the separation property of the closed-loop observer and controller, i.e. we show that in both the continuous-time and discrete-time cases we can design the observer and controller gains by solving two separate sets of LMI feasibility problems. We present a numerical example in Section 11.5 to illustrate our approach. Finally, we present our conclusions in Section 11.6.

11.2 TAKAGI–SUGENO MODELS

11.2.1 Continuous-Time T–S Models

A continuous-time T–S model is represented by a set of fuzzy **If–Then** rules written as follows:

> ith plant rule: **If** $x_1(t)$ is M_{i1} and ... $x_n(t)$ is M_{in} **Then** $\dot{x} = A_i x$

where $x \in \mathbb{R}^{n \times 1}$ is the state vector, $i = \{1, \ldots, r\}$, r is the number of rules, M_{ij} are input fuzzy sets, and the matrices $A_i \in \mathbb{R}^{n \times n}$.

Using a singleton fuzzifier, product inference and weighted average deffuzifier [4, 5], the aggregated fuzzy model can be written as follows:

$$\dot{x} = \frac{\sum_{i=1}^{r} w_i(x)(A_i x)}{\sum_{i=1}^{r} w_i(x)}, \tag{11.1}$$

and w_i is defined as

$$w_i(x) = \prod_{j=1}^{n} \mu_{ij}(x_j), \tag{11.2}$$

where μ_{ij} is the membership function of the jth fuzzy set in the ith rule. Now, defining

$$\alpha_i(x) = \frac{w_i(x)}{\sum_{i=1}^{r} w_i(x)} \tag{11.3}$$

we can write (11.1) as

$$\dot{x} = \sum_{i=1}^{r} \alpha_i(x) A_i x, \quad i = 1, \ldots, r, \tag{11.4}$$

where $\alpha_i(x) > 0$ and

$$\sum_{i=1}^{r} \alpha_i(x) = 1.$$

The interpretation of equation (11.4) is that the overall system is a 'fuzzy' blending of the implications. It is evident that the system (11.4) is generally nonlinear due to the non-linearity of the α_i's. In the next section we present sufficient conditions based on Lyapunov stability theory for the stability of an open-loop system (11.4). The following theorem, due to Tanaka and Sugeno, is first presented [6]:

Theorem 1 *The continuous-time T–S system (11.4) is globally asymptotically stable if there exists a common positive definite matrix $P > 0$ which satisfies the following inequalities:*

$$A_i^T P + P A_i < 0, \quad \forall i = 1, \ldots, r, \quad (11.5)$$

where r is the number of T–S rules.

11.2.2 Continuous-Time T–S Controllers and Closed-Loop Stability

In the previous section, we discussed the open-loop T–S fuzzy systems as well as sufficient conditions for the stability of the open-loop system. Now, we introduce the notion of the T–S controller in the same fashion as the T–S system. The controller consists of fuzzy **If–Then** rules. Each rule is a local state-feedback controller, and the overall controller is obtained by the aggregation of local controllers. A generic non-autonomous T–S plant rule can be written as follows:

> ith plant rule: **If** $x_1(t)$ is M_{i1} and ... $x_n(t)$ is M_{in} **Then** $\dot{x} = A_i x + B_i u$

The overall plant dynamics can be written as

$$\dot{x} = \sum_{i=1}^{r} \alpha_i(x)(A_i x + B_i u). \quad (11.6)$$

In the same fashion, a generic T–S controller rule can be written as

> ith controller rule: **If** $x_1(t)$ is M_{i1} and ... $x_n(t)$ is M_{in} **Then** $u = -K_i x$

The overall controller, using the same inference method as before, is given as

$$u(t) = -\sum_{i=1}^{r} \alpha_i(x) K_i x(t), \quad (11.7)$$

where the α_i's are defined in (11.3). Note that we are using the same fuzzy sets for the controller rules and the plant rules. Replacing (11.7) in (11.6), and keeping in mind that

$$\sum_{i=1}^{r} \alpha_i(x) = 1,$$

we can write the closed-loop equation as follows:

$$\dot{x} = \sum_{i=1}^{r} \sum_{j=1}^{r} \alpha_i(x) \alpha_j(x)(A_i - B_i K_j) x. \quad (11.8)$$

The following theorem presents sufficient conditions for closed-loop stability [7].

242 OBSERVER-BASED CONTROLLER SYNTHESIS

Theorem 2 *The closed-loop T–S fuzzy system (11.8) is globally asymptotically stable if there exists a common positive-definite matrix P which satisfies the following Lyapunov inequalities:*

$$(A_i - B_iK_i)^T P + P(A_i - B_iK_i) < 0, \quad \forall i = 1, \ldots, r,$$
$$G_{ij}^T P + PG_{ij} < 0, \quad j < i \leq r, \quad (11.9)$$

where G_{ij} is defined as

$$G_{ij} = A_i - B_iK_j + A_j - B_jK_i, \quad j < i \leq r. \quad (11.10)$$

Proof The proof can be easily obtained by multiplying the first set of inequalities in (11.9) by α_i^2 and the second set of inequalities by $\alpha_i\alpha_j$ and summing them up.

11.2.3 Discrete-Time T–S Controllers

We can define the non-autonomous discrete-time T–S system in the same fashion as the continuous-time system. The non-autonomous discrete-time T–S system can be written as

$$x(k+1) = \sum_{i=1}^{r} \alpha_i(x)(A_ix(k) + B_iu(k)) \quad (11.11)$$

We define the discrete-time T–S controller as a set of fuzzy implications. A generic implication can be written as

---
*i*th controller rule: **If** $x_i(k)$ is M_{i1} and $\ldots x_n(k)$ is M_{in} **Then** $u(k) = -K_ix(k)$
---

where $K_i \in \mathbb{R}^{m \times n}$. The overall controller will be

$$u(k) = -\sum_{i=1}^{r} \alpha_i(x)K_ix(k). \quad (11.12)$$

Replacing (11.12) in (11.11) we obtain the following closed-loop equation:

$$x(k+1) = \sum_{i=1}^{r}\sum_{j=1}^{r} \alpha_i(x)\alpha_j(x)(A_i - B_iK_j)x(k). \quad (11.13)$$

Sufficient conditions for the stability of the closed-loop system are given by the following theorem [7].

Theorem 3 *The closed-loop system (11.13) is globally asymptotically stable if there exists a common positive-definite matrix P that satisfies the following matrix inequalities:*

$$(A_i - B_iK_i)^T P(A_i - B_iK_i) - P < 0, \quad i = 1, \ldots, r,$$
$$G_{ij}^T PG_{ij} - P < 0, \quad i < j \leq r, \quad (11.14)$$

where G_{ij} is the same as in (11.10).

Proof The proof is similar to that of Theorem 2. □

11.3 LMI STABILITY CONDITIONS FOR T-S FUZZY SYSTEMS

11.3.1 The Continuous-Time Case

Sufficient stability conditions for open-loop continuous time T–S systems were derived in Theorem 1. These conditions, as discussed earlier, are LMIs in the matrix variable P. Note that equation (11.4) is the equation for a Polytopic Linear Differential Inclusion [1].

On the other hand, the closed-loop case is different. Theorem 2 provides sufficient conditions for the stability of the closed loop system. However these Lyapunov inequalities are not LMIs in P and K_i, since they contain the product of P and K_i. Using a clever change of variables due to Bernussou et al. [8], we can recast the matrix inequalities in (11.9) as LMIs. In fact, let

$$P^{-1} = Y, \quad X_i = K_i Y. \tag{11.15}$$

Then pre-multiplying and post-multiplying the inequalities in (11.9) by Y and using the above change of variable, we obtain the following LMIs [7]:

$$0 < Y,$$
$$0 < YA_i^T + A_i Y - B_i X_i - X_i^T B_i^T, \quad \forall i = 1, \ldots, r, \tag{11.16}$$
$$0 < Y(A_i + A_j)^T + (A_i + A_j)Y - (B_i X_j + B_j X_i) - (B_i X_j + B_j X_i)^T, \quad j < i \leq r.$$

If the above LMIs have a solution, stability of the closed-loop T–S system is guaranteed. We can find the T–S controller gains by reversing the transformations in (11.15), i.e.

$$K_i = X_i Y^{-1}.$$

Again, we point out the fact that the resulting T–S controller is conservative, because we are searching for a common quadratic Lyapunov function. In the next section we derive the LMI conditions for the stability of discrete-time T–S fuzzy systems.

11.3.2 The Discrete-Time Case

The closed-loop stability conditions in (11.14) can be recast as the following LMIs [3, 7]:

$$Y > 0,$$
$$\begin{bmatrix} Y & (A_i Y - B_i X_i)^T \\ (A_i Y - B_i X_i) & Y \end{bmatrix} > 0, \quad i = 1, \ldots, r, \tag{11.17}$$
$$\begin{bmatrix} Y & [(A_i + A_j)Y - M_{ij}]^T \\ (A_i + A_j)Y - M_{ij} & Y \end{bmatrix} > 0, \quad j < i \leq r,$$

where Y and X_i are defined in (11.15), and M_{ij} is given by

$$M_{ij} = B_i X_j + B_j X_i. \tag{11.18}$$

If the LMIs are feasible, then the controller gains can be obtained from

$$K_i = X_i Y^{-1}.$$

244 OBSERVER-BASED CONTROLLER SYNTHESIS

Once the controller gains are obtained, we can write the control action as (11.7) for the continuous-time case and as (11.12) in the discrete-time case. In the next section, we generalize our design methodology and present T–S output feedback controllers using an asymptotic observer methodology.

11.4 FUZZY OBSERVERS

11.4.1 Why Output Feedback?

So far we have developed a systematic framework for the design of T–S state feedback controllers. An implicit assumption in all previous sections was that the states are available for measurement. However, we know that measuring the states can be physically difficult and costly. Moreover, sensors are often subject to noise and failure. This motivates the question: 'How can we design output feedback controllers for T–S fuzzy systems?'

We already know from classical control theory that using an observer, we can estimate the states of an observable LTI system from output measurements. In fact, we even know how to estimate the states of a linear time-invariant (LTI) system in the presence of additive noise in the system, and measurement noise in the output, using a Kalman filter [9]. Our attempt is to generalize the observer methodology to the case of a PLDI instead of a single LTI system, or more specifically, to the case of T–S fuzzy systems. We present a new approach, which is to design an observer based on fuzzy implications, with fuzzy sets in the antecedents, and an asymptotic observer in the consequents. Each fuzzy rule is responsible for observing the states of a locally linear subsystem. The following section will describe the observer design in the continuous-time case [10].

11.4.2 Continuous-Time T–S Fuzzy Observers

Consider the closed-loop fuzzy system described by r plant rules and r controller rules as follows:

$$\dot{x}(t) = \sum_{i=1}^{r} \alpha_i(y)(A_i x(t) + B_i u(t)),$$

$$y(t) = \sum_{i=1}^{r} \alpha_i(y) C_i x(t). \quad (11.19)$$

We define a *fuzzy observer* as a set of T–S **If–Then** rules which estimate the states of the system (11.19). A generic observer rule can be written as

> ith rule: **If** $y_1(t)$ is M_{i1} and ... $y_p(t)$ is M_{ip} **Then** $\dot{\hat{x}} = A_i \hat{x} + B_i u + L_i(y - \hat{y})$

where p is the number of measured outputs, $y_i = C_i x$ is the output of the ith T–S plant rule, $\hat{y}$ is the global output estimate, and $L_i \in \mathbb{R}^{n \times p}$ is the local observer gain matrix. The deffuzified global output estimate can be written as

$$\hat{y}(t) = \sum_{j=1}^{r} \alpha_j(y) C_j \hat{x}(t),$$

where the α_i's are the normalized membership functions as in (11.3). The aggregation of all fuzzy implications results in the following state equations:

$$\dot{\hat{x}} = \sum_{i=1}^{r} \alpha_i(y)(A_i\hat{x} + B_i u) + \sum_{i=1}^{r}\sum_{j=1}^{r} \alpha_i(y)\alpha_j(y) L_i C_j (x - \hat{x}). \quad (11.20)$$

Since $\sum_{j=1}^{r} \alpha_j(y) = 1$, we can write equation (11.20) as

$$\dot{\hat{x}} = \sum_{i=1}^{r}\sum_{j=1}^{r} \alpha_i(y)\alpha_j(y)[(A_i - L_i C_j)\hat{x} + B_i u + L_i C_j x]. \quad (11.21)$$

Note that we wrote the normalized membership functions as a function of y instead of x since the antecedents are measured output variables, not states. The controller is also based on the estimate of the state rather than the state itself, i.e. we have

$$u(t) = -\sum_{j=1}^{r} \alpha_j(y) K_j \hat{x}(t). \quad (11.22)$$

Substituting (11.22) in (11.4) we get the following equation for the closed-loop system:

$$\dot{x} = \sum_{i=1}^{r}\sum_{j=1}^{r} \alpha_i(y)\alpha_j(y)(A_i x - B_i K_j \hat{x}). \quad (11.23)$$

Defining the state estimation error as

$$\tilde{x} = x - \hat{x}$$

and subtracting (11.23) from (11.21) we get

$$\dot{\tilde{x}} = \sum_{i=1}^{r}\sum_{j=1}^{r} \alpha_i(y)\alpha_j(y)(A_i - L_i C_j)\tilde{x}. \quad (11.24)$$

To guarantee that the estimation error goes to zero asymptotically, we can use Theorem 2. The observer dynamics is stable if a common positive-definite matrix P_2 exists such that the following matrix inequalities are satisfied:

$$(A_i - L_i C_i)^T P_2 + P_2(A_i - L_i C_i) < 0, \quad i = 1, \ldots, r,$$
$$H_{ij}^T P_2 + P_2 H_{ij} < 0, \quad j < i \leq r, \quad (11.25)$$

where H_{ij} is defined as

$$H_{ij} = A_i - L_i C_j + A_j - L_j C_i. \quad (11.26)$$

Although the inequalities in (11.25) are not LMIs, they can be recast as LMIs by the following change of variables:

$$W_i = P_2 L_i. \quad (11.27)$$

Using the above variable change in (11.25) and utilizing the LMI Lemma [12], we obtain the following LMIs in P_2 and W_i:

$$P_2 > 0,$$
$$A_i^T P_2 + P_2 A_i - W_i C_i - C_i^T W_i^T < 0, \quad i = 1, \ldots, r, \quad (11.28)$$
$$(A_i + A_j)^T P_2 + P_2(A_i + A_j) - (W_i C_j + W_j C_i) - (W_i C_j + W_j C_i)^T < 0, \quad j < i \leq r.$$

The observer gains are obtained from

$$L_i = P_2^{-1} W_i. \tag{11.29}$$

By augmenting the states of the system with the state estimation error, we obtain the following $2n$-dimensional state equations for the observer/controller closed-loop system:

$$\begin{bmatrix} \dot{x} \\ \hline \dot{\tilde{x}} \end{bmatrix} = \begin{bmatrix} \sum_{i=1}^{r}\sum_{j=1}^{r} \alpha_i\alpha_j(A_i - B_iK_j) & \sum_{i=1}^{r}\sum_{j=1}^{r} \alpha_i\alpha_j B_iK_j \\ \hline 0 & \sum_{i=1}^{r}\sum_{j=1}^{r} \alpha_i\alpha_j(A_i - L_iC_j) \end{bmatrix} \begin{bmatrix} x \\ \hline \tilde{x} \end{bmatrix},$$

$$y = \begin{bmatrix} \sum_{j=1}^{r} \alpha_j C_j & 0 \end{bmatrix} \begin{bmatrix} x \\ \hline \tilde{x} \end{bmatrix}. \tag{11.30}$$

We then have the following theorem for the stability of the closed-loop observer/controller system.

Theorem 4 *The closed-loop observer/controller system (11.30) is globally asymptotically stable if there exists a common, positive-definite matrix $\tilde{P}$ such that the following Lyapunov inequalities are satisfied:*

$$A_{ii}^T \tilde{P} + \tilde{P} A_{ii} < 0, \quad i = 1, \ldots, r,$$
$$(A_{ij} + A_{ji})^T \tilde{P} + \tilde{P}(A_{ij} + A_{ji}) < 0, \quad j < i \leq r, \tag{11.31}$$

where A_{ij} can be defined as

$$A_{ij} = \begin{bmatrix} A_i - B_iK_j & B_iK_j \\ 0 & A_i - L_iC_j \end{bmatrix}. \tag{11.32}$$

Proof The proof follows directly from Theorem 2. □

Note that the above matrix inequalities are not LMIs in $\tilde{P}$, the K_i's, and the L_i's. We would like to know if, with the same change of variables as in (11.15) and (11.27), we can rewrite the inequalities in (11.31) as LMIs. In fact, we would like to check if we can extend the separation property of the observer/controller of a single LTI system to the case of (11.30). We will show in the next section that in the case of (11.30) we indeed have the separation property, resulting in two separate sets of LMIs for the observer and the controller [10].

11.4.3 Separation Property of the Observer/Controller

To show that the separation property holds, we have to prove that $\tilde{P}$, the common positive-definite solution of the inequalities in (11.31), is a block diagonal matrix with $\lambda P = \lambda Y^{-1}$ and P_2 as diagonal elements, where P is the positive-definite solution of inequalities in (11.9), P_2 is the solution of (11.25), and $\lambda > 0$. We now express our main result as the separation property in the following theorem:

Theorem 5: Main result—separation theorem for T–S fuzzy systems *The closed-loop system (30) is globally asymptotically stable if the inequalities in (9) and (25) are satisfied independently.*

Proof We choose $\tilde{P}$ as a block diagonal matrix with λP and P_2 as the block diagonal elements, i.e. we have the following:

$$\tilde{P} = \begin{bmatrix} \lambda P & 0 \\ \hline 0 & P_2 \end{bmatrix}. \tag{11.33}$$

We show that there always exists a $\lambda > 0$ such that $\tilde{P}$ satisfies the inequalities in (11.31), provided (11.9) and (11.25) are satisfied. Substituting for $\tilde{P}$ and A_{ij} in (11.31) we obtain the following:

$$\begin{bmatrix} \lambda[(A_i - B_i K_i)^{\mathrm{T}} P + P(A_i - B_i K_i)] & \lambda P(B_i K_i) \\ \hline \lambda(B_i K_i)^{\mathrm{T}} P & (A_i - L_i C_i)^{\mathrm{T}} P_2 + P_2(A_i - L_i C_i) \end{bmatrix} < 0. \tag{11.34}$$

Using the LMI lemma [12], (11.34) is negative-definite if and only if the following conditions are satisfied:

$$\lambda[(A_i - B_i K_i)^{\mathrm{T}} P + P(A_i - B_i K_i)] < 0,$$
$$\lambda P(B_i K_i)[(A_i - L_i C_i)^{\mathrm{T}} P_2 + P_2(A_i - L_i C_i)]^{-1}(B_i K_i)^{\mathrm{T}} P \tag{11.35}$$
$$- [(A_i - B_i K_i)^{\mathrm{T}} P + P(A_i - B_i K_i)] > 0,$$

Since (11.9) is satisfied, the first inequality is already true. The second condition is satisfied for any $\lambda > 0$ such that

$$\lambda \min_{1 \le j \le r} \mu_i > \max_{1 \le j \le r} \nu_i,$$

where

$$\mu_i = \lambda_{\min}\{P(B_j K_j)[(A_j - L_j C_j)^{\mathrm{T}} P_2 + P_2(A_j - L_j C_j)]^{-1}(B_j K_j)^{\mathrm{T}} P\}$$

and

$$\nu_j = \lambda_{\max}[(A_j - B_j K_j)^{\mathrm{T}} P + P(A_j - B_j K_j)]$$

and where $\lambda_{\min}$ and $\lambda_{\max}$ are the minimum and maximum eigenvalues. Since (11.9) and (11.25) are already satisfied, such a λ always exists. Using the same argument, we can also show that the second set of inequalities in (11.31) is satisfied. Therefore, the two sets of inequalities can be solved independently, and the separation property holds. $\square$

In the next section we present the dual case of discrete-time T–S observers.

11.4.4 Discrete-Time T–S Fuzzy Observers

We can define the T–S fuzzy observer in the same fashion as the continuous-time system [11]. A generic rule for the discrete-time T–S fuzzy observer is

> ith rule: **If** $y_1(k)$ is M_{i1} and ... $y_p(k)$ is M_{ip}
> **Then** $\hat{x}(k+1) = A_i x(k) + B_i u(k) + L_i(y(k) - \hat{y}(k))$

where p is the number of measured outputs, $y_i(k) = C_i x(k)$ is the output of each T–S plant rule, $\hat{y}$ is the global output estimate, and $L_i \in \mathbb{R}^{n \times p}$ is the local observer gain matrix. The deffuzified output estimate can be written as

$$\hat{y}(k) = \sum_{j=1}^{r} \alpha_j(y) C_j \hat{x}(k),$$

where α_i are the normalized membership functions as in (11.3). The overall output can also be written in a similar manner:

$$y(k) = \sum_{j=1}^{r} \alpha_j(y) C_j x(k).$$

The aggregation of all fuzzy implications results in the following state equation:

$$\hat{x}(k=1) = \sum_{i=1}^{r} \alpha_i(y)(A_i \hat{x} + B_i u) + \sum_{i=1}^{r} \sum_{j=1}^{r} \alpha_i(y) \alpha_j(y) L_i C_j (x - \hat{x}). \quad (11.36)$$

Since

$$\sum_{j=1}^{r} \alpha_j(y) = 1,$$

we can write equation (11.36) as

$$\hat{x}(k+1) = \sum_{i=1}^{r} \sum_{j=1}^{r} \alpha_i(y) \alpha_j(y) [(A_i - L_i C_j)\hat{x} + B_i u + L_i C_j x]. \quad (11.37)$$

By defining the estimation error as before, we can write the estimation error $\tilde{x}(k)$ as follows:

$$\dot{\tilde{x}} = \sum_{i=1}^{r} \sum_{j=1}^{r} \alpha_i(y) \alpha_j(y) (A_i - L_i C_j) \tilde{x}. \quad (11.38)$$

To guarantee that the estimation error goes to zero asymptotically, we can use Theorem 3. The observer dynamics is stable if a common positive-definite matrix P_2 exists such that the following matrix inequalities are satisfied:

$$\begin{aligned} (A_i - L_i C_i)^{\mathrm{T}} P_2 (A_i - L_i C_i) - P_2 < 0, & \quad i = 1, \ldots, r, \\ H_{ij}^{\mathrm{T}} P_2 H_{ij} - P_2 < 0, & \quad j < i \leq r, \end{aligned} \quad (11.39)$$

where H_{ij} is defined as in (11.26). Although the inequalities in (39) are not LMIs, they can be recast as LMIs using the change of variables of equation (27). Using the above variable change and also utilizing the LMI lemma, we obtain the following LMIs in P_2 and W_i:

$$P_2 > 0,$$

$$\left[\begin{array}{c|c} P_2 & (P_2 A_i - W_i C_i)^{\mathrm{T}} \\ \hline P_2 A_i - W_i C_i & P_2 \end{array} \right] > 0, \quad i = 1, \ldots, r,$$

$$\left[\begin{array}{c|c} P_2 & (P_2 (A_i + A_j) - W_i C_j + W_j C_i)^{\mathrm{T}} \\ \hline (P_2 (A_i + A_j) - W_i C_j + W_j C_i) & P_2 \end{array} \right] > 0, \quad j < i \leq r.$$

$$(11.40)$$

The closed-loop observer/controller system can be written as

$$\left[\begin{array}{c} x(k+1) \\ \hline \tilde{x}(k+1) \end{array}\right] = \left[\begin{array}{c|c} \sum_{i=1}^{r}\sum_{j=1}^{r}\alpha_i\alpha_j(A_i - B_iK_j) & \sum_{i=1}^{r}\sum_{j=1}^{r}\alpha_i\alpha_j B_i K_j \\ \hline 0 & \sum_{i=1}^{r}\sum_{j=1}^{r}\alpha_i\alpha_j(A_i - L_iC_j) \end{array}\right] \left[\begin{array}{c} x(k) \\ \hline \tilde{x}(k) \end{array}\right],$$

$$y = \left[\sum_{j=1}^{r}\alpha_j C_j \Big| 0\right] \left[\begin{array}{c} x(k) \\ \tilde{x}(k) \end{array}\right]. \tag{11.41}$$

Using Theorem 3, the system (11.41) is globally asymptotically stable if there exists a positive definite matrix $\tilde{P} > 0$ such that

$$A_{ii}^T \tilde{P} A_{ii} - \tilde{P} < 0, \quad i = 1, \ldots, r,$$
$$(A_{ij} + A_{ji})^T \tilde{P} (A_{ij} + A_{ji}) - \tilde{P} < 0, \quad j < i \le r, \tag{11.42}$$

where A_{ij} is the same as in (11.32).

As in the continuous-time case, we can show that the Lyapunov matrix $\tilde{P}$ is indeed block diagonal, i.e. the discrete-time version of Theorem 5 holds, and the observer and controller gains can be found via separate LMI feasibility problems. A proof of the separation property in the discrete-time case is given in [11].

11.5 NUMERICAL EXAMPLE

We present a numerical example to illustrate the results obtained in this chapter. We use a two-rule T–S fuzzy model which approximates the motion of an inverted pendulum on a cart. This system has been studied in [7] and [13]. The T–S fuzzy rules are obtained by approximation of the nonlinear system around $0°$ and $88°$. The T–S rules can be written as:

- Plant rule 1: **If** y is *around* 0 **Then** $\dot{x} = A_1 x + B_1 u$.
- Plant rule 2: **If** y is *around* $\pm\pi/2$ **Then** $\dot{x} = A_2 x + B_2 u$.
- Controller rule 1: **If** y is *around* 0 **Then** $u = -K_1 x$.
- Controller rule 2: **If** y is *around* $\pm\pi/2$ **Then** $u = -K_2 x$.
- Observer rule 1: **If** y is *around* 0 **Then** $\dot{\hat{x}} = A_1\hat{x} + B_1 u + L_1 C(x - \hat{x})$.
- Observer rule 2: **If** y is *around* $\pm\pi/2$ **Then** $\dot{\hat{x}} = A_2\hat{x} + B_2 u + L_2 C(x - \hat{x})$.

Here $y = x_1$ is the measured angle from the vertical point and A_1, A_2, B_1, B_2, and C are given as follows:

$$A_1 = \begin{bmatrix} 0 & 1 \\ 17.3 & 0 \end{bmatrix}; \quad B_1 = \begin{bmatrix} 0 \\ -0.177 \end{bmatrix}, \tag{11.43}$$

$$A_2 = \begin{bmatrix} 0 & 1 \\ 9.45 & 0 \end{bmatrix}; \quad B_2 = \begin{bmatrix} 0 \\ -0.03 \end{bmatrix},$$

$$C = \begin{bmatrix} 1 & 0 \end{bmatrix}. \tag{11.44}$$

The membership functions μ_1 and μ_2 for the two fuzzy sets *close to zero*, and *close to* $\pm\pi/2$, are plotted in Figure 11.1. We design the observer and controller gains by local pole

250 OBSERVER-BASED CONTROLLER SYNTHESIS

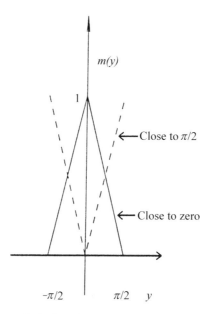

Figure 11.1 Membership functions for the angle

placement, and look for common Lyapunov matrices P and P_2. We place the closed-loop poles of the system at $-2, -2$, and the poles of the observer dynamics at $-6, -6.5$, respectively. The observer and controller gains are

$$K_1 = [-120.67 \quad -66.67]; \quad K_2 = [-2551.6 \quad -764.0],$$
$$L_1 = [12.5 \quad 57.3]^T \qquad L_2 = [12.5 \quad 50.0]^T. \qquad (11.45)$$

Fortunately, the LMIs are feasible and we can find positive-definite Lyapunov matrices P and P_2 as solutions to the LMIs in (11.16) and (11.28). The simulation results for the states of the system $x(1)$ and $x(2)$ as well as the estimation error $x(3)$ and $x(4)$ are depicted in Figures 11.2–11.5. Although we were able to solve for positive-definite Lyapunov matrices

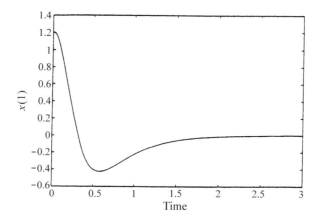

Figure 11.2 Initial condition response of the pendulum angle

NUMERICAL EXAMPLE 251

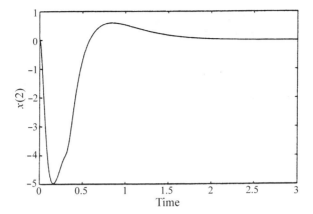

Figure 11.3 Initial condition response of the angular velocity

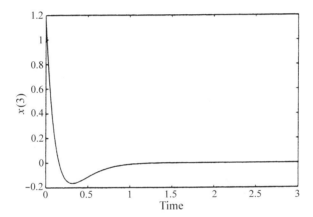

Figure 11.4 Estimation error for angle

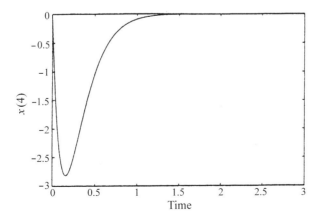

Figure 11.5 Estimation error for angular velocity

P and P_2 using local pole placement, this might not always be possible. This is the reason why we need to design for performance in addition to stability. The details are discussed in [13].

11.6 CONCLUSION

The purpose of this chapter was to extend the current methods of designing stabilizing state feedback controllers for T–S fuzzy systems to the output feedback case for continuous time and discrete time. We stated and proved a separation theorem which makes it possible to design for the observer and controller gains separately. Future research can be done in this area by extending these results to the case where performance is needed in addition to stability. Using a guaranteed-cost framework developed in [13], we can design an observer/controller system which minimizes an upper bound on a quadratic performance index. Another extension would be to include additive noise and develop a Kalman filter for T–S fuzzy systems.

REFERENCES

1. Boyd, S., L. El Ghaoui, E. Feron and V. Balakrishnan, *Linear Matrix Inequalities in System and Control Theory*, SIAM Studies in Applied Mathematics, vol. 15, SIAM, Philadelphia, 1994.
2. Petterson, S. and B. Lennartson, An LMI approach for stability analysis of nonlinear systems, *Proceedings of the European Control Conference, ECC'98*, Brussels, Belgium, 1997.
3. Zhao, J., V. Wertz and R. Gorez, Fuzzy gain scheduling controllers based on fuzzy models, *Proceedings of the 1996 FUZZ-IEEE*, New Orleans, 1996, pp. 1670–1676.
4. Jamshidi, M., N. Vadiee and T. J. Ross (eds.), *Fuzzy Logic and Control: Hardware and Software Applications*, Prentice-Hall, Englewood Cliffs, 1993.
5. Ross, T. J., *Fuzzy Logic With Engineering Applications*, McGraw-Hill, New York, 1995.
6. Tanaka, K. and M. Sugeno, Stability analysis and design of fuzzy control systems, *Fuzzy Sets and Systems*, **45**(2), 1992, 135–156.
7. Wang, H. O., K. Tanaka and M. F. Griffin, An approach to fuzzy control of nonlinear systems: stability and design issues, *IEEE Transactions on Fuzzy Systems*, **4**(1), 1996, 14–23.
8. Bernussou, J., P. L. D. Peres and J. C. Geromel, A linear programming oriented procedure for quadratic stabilization of uncertain systems, *System Control Letters*, **13**, 1989, 65–72.
9. Kalman, R. E., A new approach to filtering and prediction problems, *Transactions of the ASME: Journal of Basic Engineering* **82D**(1), 1960, 35–45.
10. Jadbabaie, A., A. Titli and M. Jamshidi, A separation property of observer/controller for continuous-time fuzzy systems, *Proceedings of the Thirty-Fifth Allerton National Conference*, Allerton House, 1997.
11. Jadbabaie, A., A. Titli and M. Jamshidi, Fuzzy observer based control of nonlinear systems, *Proceedings of the 1997 IEEE Conference on Decision and Control*, San Diego, CA., 1997.
12. Dorato, P., C. T. Abdallah and V. Cerone, *Linear Quadratic Control: An Introduction*, Prentice-Hall, Englewood Cliffs, 1995.
13. Jadbabaie, A., M. Jamshidi and A. Titli, Guaranteed cost design of continuous-time Takagi–Sugeno fuzzy systems, *Proceedings of the 1998 FUZZ-IEEE*, Anchorage, Alaska, 1998.

12

LMI-Based Fuzzy Control: Fuzzy Regulator and Fuzzy Observer Design via LMIs

Kazuo Tanaka
University of Electro-Communications, Tokyo, Japan

Hua O. Wang
Duke University, Durham, USA

12.1 INTRODUCTION

Recently, the issue of the stability of fuzzy control systems has been considered extensively in non-linear stability frameworks [1–8]. Specially, the stabilization of a feedback system containing a fuzzy regulator and a fuzzy observer was discussed in [3]. This chapter extends the work of [3] and, more importantly, new relaxed stability conditions and LMI (linear matrix inequality) based design procedures are obtained for both continuous and discrete fuzzy systems. The stability analysis and design procedures proposed here are straightforward and natural, although the non-linear regulator and observer design is difficult in general.

Linear regulators and linear observers play an important role in modern control theory and practice. We envisage that a systematic design method of fuzzy regulators and fuzzy observers would be important for fuzzy control as well. This chapter presents a model-based fuzzy control of non-linear systems. The fuzzy regulator and fuzzy observer are designed by solving LMIs (linear matrix inequalities) that represent control performance such as *decay rate, disturbance rejection, robust stability, minimization of quadratic performance function and constraints on the control input and output*. It is based on some of the newly developed systematic fuzzy control design techniques [4–8, 10, 13–16] using LMI-based technique. Details of this approach are given in [22].

This approach begins with constructing a Takagi–Sugeno (T–S) type fuzzy model [9] for a non-linear plant. Once the fuzzy model is obtained, the control design is carried out via the

Fuzzy Logic. Edited by S. Farinwata, D. Filev and R. Langari
© 2000 John Wiley & Sons, Ltd

so-called parallel distributed compensation (PDC) scheme. The idea is that for each local linear model, there is an associated linear feedback control. The resulting overall controller, which is non-linear in general, is the fuzzy blending of each individual linear controller. Hence the PDC approach is a multiple control with automatic switching via fuzzy logic rules. Our approach offers several advantages. The design procedure is conceptually simple and natural. The stability analysis and control design problems are reduced to LMI problems where powerful numerical algorithms [11] can be applied.

It has been generally assumed in the model-based fuzzy control that all the states of the T–S fuzzy system are measurable. All states of the systems are not always measurable in most practical applications. Hence we envisage that a fuzzy observer would be important in fuzzy control theory and practice. This chapter also considers a fuzzy observer design to estimate the states of the T–S fuzzy system. In particular, we guarantee the stability of the overall control system including a fuzzy regulator and a fuzzy observer.

Section 12.2 presents the T–S fuzzy model. Section 12.3 introduces the fuzzy regulator design via LMIs that represent control performance such as *decay rate, disturbance rejection, robust stability, minimization of quadratic performance function and constraints on the control input and output* [22]. New stability conditions are also presented by relaxing the previous stability results. Section 12.4 shows the fuzzy observer and an LMI-based design procedure for the augmented system containing the fuzzy regulator and the fuzzy observer.

12.2 TAKAGI–SUGENO FUZZY MODEL

The main feature of a Takagi–Sugeno (T–S) fuzzy model [9] is to express the join dynamics of each fuzzy implication (rule) by a linear system model. The ith rule of the T–S fuzzy model is of the following form:

Plant rule i:

If $z_1(t)$ is M_{i1} and ... and $z_p(t)$ is M_{ip}

Then $\begin{cases} \dot{x}(t) = A_i x(t) + B_i u(t), \\ y(t) = C_i x(t), \end{cases}$ $\quad i = 1, 2, \ldots, r,$ (12.1)

where M_{ij} is the fuzzy set and r is the number of **If–Then** rules. $x(t) \in R^n$ is the state vector, $u(t) \in R^m$ is the input vector, $y(t) \in R^q$ is the output vector, $A_i \in R^{n \times n}, B_i \in R^{n \times m}$ and $C_i \in R^{q \times n}$. $z_1(t) \sim z_p(t)$ are the premise variables. The premise variables are assumed to be independent of the input variables $u(t)$. This assumption is employed to prevent a complicated defuzzification process of the PDC fuzzy controllers [4–7]. Note that stability conditions derived in this chapter can be applied even to the case when the premise variables depend on the input variables $u(t)$. Each linear state and output equations in the consequent parts are called a 'subsystem'.

The overall fuzzy model is achieved by fuzzy 'blending' of the linear system models. Given a pair of $(x(t), u(t))$, the final output of the fuzzy system is inferred as follows:

$$\dot{x}(t) = \frac{\sum_{i=1}^{r} w_i(z(t))\{A_i x(t) + B_i u(t)\}}{\sum_{i=1}^{r} w_i(z(t))}$$

$$= \sum_{i=1}^{r} h_i(z(t))\{A_i x(t) + B_i u(t)\}, \quad (12.2)$$

$$y(t) = \frac{\sum_{i=1}^{r} w_i(z(t)) C_i x(t)}{\sum_{i=1}^{r} w_i(z(t))} = \sum_{i=1}^{r} h_i(z(t)) C_i x(t), \tag{12.3}$$

where

$$z(t) = [z_1(t)\, z_2(t) \ldots z_p(t)],$$

$$w_i(z(t)) = \prod_{j=1}^{p} M_{ij}(z_j(t)).$$

$M_{ij}(z_j(t))$ is the grade of membership of $z_j(t)$ in M_{ij}. In (12.2) and (12.3),

$$h_i(z(t)) = \frac{w_i(z(t))}{\sum_{i=1}^{r} w_i(z(t))}. \tag{12.4}$$

Since

$$\begin{cases} \sum_{i=1}^{r} w_i(z(t)) > 0, \\ w_i(z(t)) \geq 0, \quad i = 1, 2, \ldots, r, \end{cases} \tag{12.5}$$

for all t, we have

$$\begin{cases} \sum_{i=1}^{r} h_i(z(t)) = 1 \\ h_i(z(t)) \geq 0, \quad i = 1, 2, \ldots, r, \end{cases} \tag{12.6}$$

for all t. $h_i(z(t))$ can be regarded as the normalized weight of each **If–Then** rule.

The open-loop system of (12.2) is defined as follows:

$$\dot{x}(t) = \frac{\sum_{i=1}^{r} w_i(z(t)) A_i x(t)}{\sum_{i=1}^{r} w_i(z(t))} = \sum_{i=1}^{r} h_i(z(t)) A_i x(t). \tag{12.7}$$

The well-known sufficient stability condition for ensuring stability of (12.7) is derived using the Lyapunov approach.

Theorem 1 *The equilibrium of the fuzzy system* (12.7) *is asymptotically stable in general if there exists a common positive definite matrix P such that*

$$A_i^T P + P A_i < 0, \tag{12.8}$$

for $i = 1, 2, \ldots, r$, i.e. for all the subsystems.

This theorem is reduced to the Lyapunov stability theorem when $r=1$. Theorem 1 gives a sufficient condition for ensuring the stability of (12.7). For the system (12.7), the following question naturally arises: Are the systems stable if all its subsystems are stable, i.e. if all A_i's are stable? The answer is 'no' in general [4, 7].

12.3 FUZZY REGULATOR DESIGN VIA LMIs

12.3.1 Parallel Distributed Compensation

The parallel distributed compensation (PDC) introduced in [4]–[7] is employed to design a fuzzy controller. The main idea of the PDC is to design each local control rule so as to compensate each local rule of a fuzzy system. For instance, control rule i has the same structure as rule i of the fuzzy model. The consequent parts are replaced with the feedback laws, i.e. the state feedback laws in fuzzy regulators.

Control rule i:

$$\text{If } z_1(t) \text{ is } M_{i1} \text{ and } \ldots \text{ and } z_p(t) \text{ is } M_{ip} \text{ Then } u(t) = -F_i x(t). \quad (12.9)$$

The fuzzy controller is inferred as follows:

$$u(t) = -\sum_{i=1}^{r} h_i(z(t)) F_i x(t). \quad (12.10)$$

The fuzzy regulator design is to determine the local feedback gains F_i in the consequent parts. The feedback gains F_i are determined by an LMI-based design technique presented in the next subsection. Thus, the PDC is simple and natural. Other non-linear control techniques require special and rather involved knowledge.

12.3.2 Control Performance Represented via LMIs

Control performance represented via LMIs is derived and employed to obtain the feedback gains F_i. All the proof are omitted due to lack of space.

12.3.2.1 Stability

By substituting (12.10) into (12.2), the fuzzy control system can be represented as

$$\dot{x}(t) = \sum_{i=1}^{r} \sum_{j=1}^{r} h_i(z(t)) h_j(z(t)) \{A_i - B_i F_j\} x(t)$$

$$= \sum_{i=1}^{r} h_i(z(t)) h_i(z(t)) G_{ii} x(t) + 2 \sum_{i<j}^{r} h_i(z(t)) h_j(z(t)) \left\{ \frac{G_{ij} + G_{ji}}{2} \right\} x(t), \quad (12.11)$$

where

$$G_{ij} = A_i - B_i F_j.$$

For the fuzzy control system (12.11), we obtain LMI stability conditions with respect to $X = P^{-1}$ and $M_i = F_i X$.

Theorem 2 *The fuzzy model (12.2) can be stabilized via the PDC fuzzy controller (12.10) if there exists a common positive definite matrix X such that*

$$-XA_i^T - A_i X + M_i^T B_i^T + B_i M_i > 0 \quad (12.12)$$

for all i and

$$-XA_i^T - A_i X - XA_j^T - A_j X + M_j^T B_i^T + B_i M_j + M_i^T B_j^T + B_j M_i \geq 0 \quad (12.13)$$

for $i < j$ except all the pairs (i, j) such that $h_i(z(t)) h_j(z(t)) = 0$, $\forall t$, where

$$X = P^{-1} > 0, \quad M_i = F_i X.$$

The fuzzy controller design is to determine $F_i (i = 1, 2, \ldots, r)$ satisfying the conditions of Theorem 2. The feedback gains are obtained as

$$F_i = M_i X^{-1}$$

from the solutions X and M_i of the above LMIs.

A trial-and-error type of procedure was first used in [4]. In [12], a procedure to construct a common $P = X^{-1}$ is given for second-order fuzzy systems, i.e. the dimension of the state is two. We first stated in [5]–[8] that the common P problem for fuzzy controller design can be solved numerically, i.e. the stability conditions of Theorems 1 and 2 can be expressed in LMIs. Note that equations (12.12) and (12.13) are LMIs with respect to X and M_i. The solution X and M_i satisfying (12.12) and (12.13) can be obtained by convex optimization techniques or by determining that no such $X = P^{-1}$ and M_i exist. This is a convex feasibility problem. Numerically, this feasibility problem can be solved very efficiently by means of the most powerful tools available to date in the mathematical programming literature. For instance, recently developed interior-point methods are extremely efficient in practice.

Next, consider the common B case, i.e. $B_1 = B_2 = \cdots = B_r$. In this case the stability conditions of Theorem 2 can be simplified.

Corollary 1 *Assume that $B_1 = B_2 = \cdots = B_r$. The fuzzy model (12.2) can be stabilized via the PDC fuzzy controller (12.10) if there exists a common positive definite matrix X satisfying* 12.

Corollary 1 means that (12.13) implies (12.12) in the common B case.

12.3.2.2 Relaxed stability

We have shown that the stability analysis of the fuzzy control system is reduced to a problem of finding a common $X = P^{-1}$. If r is large, then it might be difficult to find a common X satisfying the conditions of Theorem 2. Next, new stability conditions are derived by relaxing the conditions of Theorem 2. Theorem 3 gives the new relaxed stability conditions.

Theorem 3 *Assume that the number of rules that fire for all t is less than or equal to s, where $1 < s \leq r$. The fuzzy model (12.2) can be stabilized via the PDC fuzzy controller (12.10) if there exists a common positive definite matrix X and a common positive semidefinite matrix Y_0 such that*

$$-XA_i^T - A_i X + M_i^T B_i^T + B_i M_i - (s-1)Y_0 > 0 \qquad (12.14)$$

for all i and

$$2Y_0 - XA_i^T - A_i X - XA_j^T - A_j X + M_j^T B_i^T + B_i M_j + M_i^T B_j^T + B_j M_i \geq 0 \qquad (12.15)$$

for $i < j$ except all the pairs (i, j) such that, $h_i(z(t))h_j(z(t)) = 0$, $\forall t$, and $s > 1$, where.

$$M_i = F_i X.$$

Y_0 should be a positive semidefinite matrix. The above relaxed conditions (12.14) and (12.15) are reduced to (12.12) and (12.13), respectively, when $Y_0 = \mathbf{0}$.

It is confirmed that a bigger difference between the previous stability conditions and the relaxed stability conditions generally appears when simultaneously solving them with other

LMI conditions, e.g. the pole assignment condition,

$$\begin{bmatrix} r^2 X & A_i X - B_i M_j - cX \\ XA_i^T - M_j B_i^T - cX & X \end{bmatrix} > 0,$$

etc. The above condition satisfies

$$|c - \lambda_k(A_i - B_i F_j)| < r, \quad k = 1, \ldots, n,$$

where $c < 0$ and $r > 0$.

12.3.2.3 Decay rate

Theorem 4 [8, 18, 19] *The largest lower bound on the decay rate that we can find using a quadratic Lyapunov function can be found by solving the following GEVP (generalized eigenvalue minimization problem) in X, Y_0, M_i and α:*

$$\underset{M_1 \ldots M_r}{\text{maximize}} \; \alpha$$

subject to

$$X > 0,$$
$$Y_0 \geq 0,$$
$$-XA_i^T - A_i X + M_i^T B_i^T + B_i M_i - (s-1)Y - 2\alpha X > 0, \quad (12.16)$$
$$2Y_0 - XA_i^T - A_i X - XA_j^T - A_j X$$
$$+ M_j^T B_i^T + B_i M_j + M_i^T B_j^T + B_j M_i \geq 0, \quad i < j, \quad (12.17)$$

where $M_i = F_i X$.

From the solution, the feedback gains can be obtained as $F_i = M_i X^{-1}$. The above design problem is reduced to a stable controller design problem (12.14) and (12.15) if we solve (12.16) and (12.17) for X, Y_0 and M_i with $\alpha = 0$ [8, 18, 19]. The complete version, including the discrete case, is presented in [18].

12.3.2.4 Constraints on control input and output (8, 18, 19)

Assume that the initial condition $x(0)$ is known. The constraint $\|u(t)\|_2 \leq \mu$ is enforced at all times $t \geq 0$ if the LMIs (12.18) hold, where $X > 0, M_i = F_i X$:

$$\begin{bmatrix} 1 & x(0)^T \\ x(0) & X \end{bmatrix} \geq 0, \quad \begin{bmatrix} X & M_i^T \\ M_i & \mu^2 I \end{bmatrix} \geq 0. \quad (12.18)$$

Assume that the initial condition $x(0)$ is known. The constraint $\|y(t)\|_2 \leq \lambda$ is enforced at all times $t \geq 0$ if the LMIs (12.19) hold, where $X > 0$:

$$\begin{bmatrix} 1 & x(0)^T \\ x(0) & X \end{bmatrix} \geq 0, \quad \begin{bmatrix} X & XC_i^T \\ C_i X & \lambda^2 I \end{bmatrix} \geq 0. \quad (12.19)$$

Note that the LMIs for the constraints on the control input and output do not imply a guarantee of stability [8, 18, 19].

12.3.2.5 Disturbance rejection

Consider the following T–S fuzzy model (12.20) instead of (12.2):

$$\dot{x}(t) = \sum_{i=1}^{r} h_i(t)\{A_i x(t) + B_i u(t) + E_i v(t)\}, \qquad (12.20)$$

where $v(t)$ is the disturbance.

Theorem 5 [17] *The PDC fuzzy controller* (12.10) *that minimizes*

$$\sup_{\|v(t)\|_2 \neq 0} \frac{\|y(t)\|_2}{\|v(t)\|_2} \qquad (12.21)$$

can be designed by solving the following LMIs:

$$\underset{X, M_i}{\text{minimize}} \; \gamma$$

$$\text{subject to } X > 0,$$

$$\begin{bmatrix} T_{ij} & XC_i^T & XC_j^T \\ C_i X & -I & 0 \\ C_j X & 0 & -I \end{bmatrix} \leq 0, \qquad (12.22)$$

where

$$T_{ij} = XA_i^T - M_j^T B_i + A_i X - B_i M_j$$
$$+ XA_j^T - M_i^T B_j + A_j X - B_j M_i + \frac{1}{2\gamma^2}(E_i + E_j)(E_i^T + E_j^T),$$

$$M_i = F_i X$$

Then, the PDC fuzzy controller (10) *satisfies*

$$\sup_{\|v(t)\|_2 \neq 0} \frac{\|y(t)\|_2}{\|v(t)\|_2} \leq \gamma.$$

The constraints on the output and disturbance rejection in the discrete case are also discussed in [20] from other points of view.

12.3.2.6 Robust stability

Consider the following fuzzy model with uncertainty:

Plant rule i:

If $z_1(t)$ is M_{i1} and ... and $z_p(t)$ is M_{ip} Then $\dot{x}(t) = (A_i + D_{ai}\Delta_{ai}(t)E_{ai})x(t)$
$+ (B_i + D_{bi}\Delta_{bi}(t)E_{bi})u(t), \quad i = 1, \ldots, r,$ (12.23)

where the uncertain blocks satisfy

$$\|\Delta_{ai}(t)\| \leq 1/\gamma_{ai},$$
$$\Delta_{ai}(t) = \Delta_{ai}^T(t),$$
$$\|\Delta_{bi}(t)\| \leq 1/\gamma_{bi},$$
$$\Delta_{bi}(t) = \Delta_{bi}^T(t)$$

LMI-BASED FUZZY CONTROL

for all i. The fuzzy model is represented as

$$\dot{x}(t) = \sum_{i=1}^{r} h_i(t)\{(A_i + D_{ai}\Delta_{ai}(t)E_{ai})x(t) + (B_i + D_{bi}\Delta_{bi}(t)E_{bi})u(t)\}. \tag{12.24}$$

A robust fuzzy controller is designed to maximize the norm of the uncertainty blocks, i.e. minimize γ_{ai} and γ_{bi} in the class of the PDC controller (12.10) satisfying the robust stability conditions (12.25) and (12.26).

Theorem 6 [15, 21] *The feedback gains F_i that stabilize the fuzzy model (12.23) and maximize the norms of the uncertain blocks (i.e. minimize $\sum_{i=1}^{r}\{\alpha_i \cdot \gamma_{ai}^2 + \beta_i \cdot \gamma_{bi}^2\}$) can be obtained by solving the following LMIs, where $\alpha_i, \beta_i > 0$ are design parameters:*

$$\underset{X, M_i}{\text{minimize}} \sum_{i=1}^{r}\{\alpha_i \cdot \gamma_{ai}^2 + \beta_i \cdot \gamma_{bi}^2\}$$

subject to

$$X > 0, \quad Y_1, Y_2 \geq 0,$$
$$\hat{S}_{ii} + (s-1)Y_1 < 0 \tag{12.25}$$

for all i and

$$\hat{T}_{ij} - 2Y_2 < 0, \tag{12.26}$$

for $i < j$ except the pairs (i, j) such that $h_i(z(t))h_j(z(t)) = 0$, $\forall t$ and $s > 1$, where

$$\hat{S}_{ii} = \begin{bmatrix} XA_i^T + A_iX - B_iM_j - M_j^TB_i^T & D_{ai} & D_{bi} & XE_{ai}^T & -M_j^TE_{bi}^T \\ D_{ai}^T & -I & 0 & 0 & 0 \\ D_{bi}^T & 0 & -I & 0 & 0 \\ E_{ai}X & 0 & 0 & -\gamma_{ai}^2 I & 0 \\ -E_{bi}M_j & 0 & 0 & 0 & -\gamma_{bi}^2 I \end{bmatrix},$$

$$\hat{T}_{ij} = \begin{bmatrix} \begin{pmatrix} XA_i^T + A_iX - B_iM_j - M_j^TB_i^T \\ +XA_j^T + A_jX - B_jM_i - M_i^TB_j^T \end{pmatrix} & D_{ai} & D_{bi} & D_{aj} & D_{bj} & XE_{ai}^T & -M_j^TE_{bi}^T & XE_{aj}^T & -M_i^TE_{bj}^T \\ D_{ai}^T & -I & 0 & 0 & 0 & 0 & 0 & 0 & 0 \\ D_{bi}^T & 0 & -I & 0 & 0 & 0 & 0 & 0 & 0 \\ D_{aj}^T & 0 & 0 & -I & 0 & 0 & 0 & 0 & 0 \\ D_{bj}^T & 0 & 0 & 0 & -I & 0 & 0 & 0 & 0 \\ E_{ai}X & 0 & 0 & 0 & 0 & -\gamma_{ai}^2 I & 0 & 0 & 0 \\ -E_{bi}M_j & 0 & 0 & 0 & 0 & 0 & -\gamma_{bi}^2 I & 0 & 0 \\ E_{aj}X & 0 & 0 & 0 & 0 & 0 & 0 & -\gamma_{aj}^2 I & 0 \\ -E_{bj}M_i & 0 & 0 & 0 & 0 & 0 & 0 & 0 & -\gamma_{bj}^2 I \end{bmatrix},$$

$$Y_1 = \text{diag}(Y_0 \ 0 \ 0 \ 0 \ 0),$$
$$Y_2 = \text{diag}(Y_0 \ 0 \ 0 \ 0 \ 0 \ 0 \ 0 \ 0 \ 0).$$

12.3.2.7 Minimization of the quadratic performance function

We define a design problem to minimize the upper bound of the performance function

$$J = \int_0^\infty \{y^T(t)Wy(t) + u^T(t)Ru(t)\}\,dt. \qquad (12.27)$$

Theorem 7 [15, 21] *The feedback gains F_i to minimize the upper bound of the performance function can be obtained by solving the following LMIs:*

$$\underset{X, M_1, \dots, M_r}{\text{minimize}} \; \lambda$$

subject to

$$X > 0,$$

$$\begin{bmatrix} \lambda & x^T(0) \\ x(0) & X \end{bmatrix} > 0, \qquad (12.28)$$

$$\hat{U}_{ii} + (s-1)Y_3 < 0 \qquad (12.29)$$

for all i and

$$\hat{V}_{ii} - 2Y_4 < 0, \qquad (12.30)$$

for $i < j$ except the pairs (i, j) such that $h_i(z(t))h_j(z(t)) = 0$, $\forall t$ and $s > 1$, where

$$\hat{U}_{ii} = \begin{bmatrix} XA_i^T + A_iX - B_iM_i - M_i^TB_i^T & XC_i^T & -M_i^T \\ C_iX & -W^{-1} & 0 \\ -M_i & 0 & -R^{-1} \end{bmatrix},$$

$$\hat{V}_{ij} = \begin{bmatrix} \begin{pmatrix} XA_i^T + XA_i - B_iM_j - M_j^TB_i^T \\ +XA_j^T + XA_j - B_jM_i - M_i^TB_j^T \end{pmatrix} & XC_i^T & -F_j^T & XC_j^T & -M_i^T \\ C_iX & -W^{-1} & 0 & 0 & 0 \\ -M_j & 0 & -R^{-1} & 0 & 0 \\ C_jX & 0 & 0 & -W^{-1} & 0 \\ -M_i & 0 & 0 & 0 & -R^{-1} \end{bmatrix},$$

$$Y_3 = \text{diag}(Y_0 \quad 0 \quad 0),$$
$$Y_4 = \text{diag}(Y_0 \quad 0 \quad 0 \quad 0 \quad 0).$$

From the solution of the LMIs, $F_i = M_i X^{-1}$ for all i. Then, the performance index satisfies

$$J < x^T(0)X^{-1}x(0) < \lambda.$$

Equations (12.29) and (12.30) are reduced to the ordinary conditions when $Y_3 = 0$ and $Y_4 = 0$.

12.3.2.8 Mixed control problem of the robust-optimal fuzzy control

We define a mixed control problem that simultaneously considers the robust stability (Theorem 6) and the minimization of quadratic performance function (Theorem 7).

Theorem 8 [15, 21] *The PDC controller (12.10) that simultaneously considers both the robust fuzzy controller design (Theorem 6) and the optimal fuzzy control design (Theorem 7) can be designed by solving the following LMIs:*

$$\underset{X, M_i}{\text{minimize}} \; \lambda + \sum_{i=1}^{r} \{\alpha_i \cdot \gamma_{ai}^2 + \beta_i \cdot \gamma_{bi}^2\}$$

subject to

$$X > 0, \; Y_1, \; Y_2 \geq 0, \quad (12.25), (12.28) \text{ and } (12.29)$$

for all i and (12.26) and (12.30) for $i < j$ except the pairs (i, j) such that $h_i(z(t))h_j(z(t)) = 0$, $\forall t$ and $s > 1$.

The same idea can be easily applied to other LMIs. That is, by utilizing other different combinations among the above LMIs, other mixed control problems can be defined and are efficiently solved via LMIs. For more details, see [15].

12.4 FUZZY OBSERVER DESIGN

A fuzzy observer is designed using the PDC, where $\hat{x}(t)$ denotes the state vector estimated by a fuzzy observer.

Observer rule i:

    If $z_1(t)$ is M_{i1} and ... and $z_p(t)$ is M_{ip}

    Then $\begin{cases} \dot{\hat{x}}(t) = A_i \hat{x}(t) + B_i u(t) + K_i(y(t) - \hat{y}(t)) \\ \hat{y}(t) = C_i \hat{x}(t), \end{cases} \quad i = 1, 2, \ldots, r.$

For simplicity, assume that $z_1(t) \sim z_p(t)$ do not depend on the state variables $x(t)$. The assumption will realize the separation principle of the fuzzy regulator design and the fuzzy observer design. We have also considered the case when $z_1(t) \sim z_p(t)$ depend on the state variables $x(t)$. For more details, see [18] and [19]. The complete version, including both the continuous and discrete cases, is given in [18].

The fuzzy observer has the linear state observer's laws in its consequent parts. The overall fuzzy observer is represented as

$$\dot{\hat{x}}(t) = \sum_{i=1}^{r} h_i(z(t))\{A_i \hat{x}(t) + B_i u(t) + K_i(y(t) - \hat{y}(t))\},$$

$$\hat{y}(t) = \sum_{i=1}^{r} h_i(z(t)) C_i \hat{x}(t).$$
(12.31)

The fuzzy observer design is to determine the local gains K_i in the consequent parts. We should use the PDC controller (12.32) instead of (12.10) in the use of the fuzzy observer

$$u(t) = -\sum_{i=1}^{r} h_i(z(t)) F_i \hat{x}(t). \tag{12.32}$$

The design problem for the augmented control system containing the fuzzy regulator and the fuzzy observer is to determine the feedback gains F_i and the observer gains K_i. The design problem can be efficiently solved by using Theorem 9. It should be emphasized that the feedback gains F_i and the observer gains K_i can be separately determined under the assumption, i.e. the separation principle is satisfied even in observer-based fuzzy control under the assumption.

Theorem 9 [18, 19] *The largest lower bound on the decay rate that we can find using a quadratic Lyapunov function can be found by solving the following GEVP (generalized eigenvalue minimization problem) in $P, Q, \hat{Y}, M_i, N_i$ and α:*

$$\max \alpha$$

subject to

$$P_1 > 0, \quad P_2 > 0, \quad \hat{Y} \geq 0, \quad Q_{22} \geq 0,$$

$$A_i P_1 + P_1 A_i^T - M_i^T B_i^T - B_i M_i + (s-1)\hat{Y} + 2\alpha P_1 < 0, \tag{12.33}$$

$$P_2 A_i + A_i^T P_2 - C_i^T N_i^T - N_i C_i + (s-1)Q_{22} + 2\alpha P_2 < 0, \tag{12.34}$$

for all i and

$$A_i P_1 + P_1 A_i^T - M_j^T B_i^T - B_i M_j$$
$$+ A_j P_1 + P_1 A_j^T - M_i^T B_j^T - B_j M_i + 2\hat{Y} + 4\alpha P_1 < 0, \tag{12.35}$$

$$P_2 A_i + A_i^T P_2 - C_j^T N_i^T - N_i C_j$$
$$+ P_2 A_j + A_j^T P_2 - C_i^T N_j^T - N_j C_i - 2Q_{22} + 4\alpha P_2 < 0, \tag{12.36}$$

for $i < j$ except the pairs (i, j) such that $h_i(z(t))h_j(z(t)) = 0, \forall t$ and $s > 1$, where

$$M_i = F_i P_1, \quad N_i = P_2 K_i.$$

Equations (12.33)–(12.36) are reduced to the ordinary stability conditions of the overall control system including (12.2), (12.31) and (12.32) when $\hat{Y} = 0, Q_{22} = 0$ and $\alpha = 0$.

Equations (12.33) and (12.35) are LMIs with respect to $\hat{Y}, M_i$ and P_1. Equations (12.34) and (12.36) are LMIs with respect to Q_{22}, N_i and P_2 as well. As mentioned above, the feedback gains F_i and the observer gains K_i can be separately determined by solving each set of the above LMIs. The feedback gains can be obtained as $F_i = M_i P_1^{-1}$ from solutions of (12.33) and (12.35). The observer gains can be obtained as $K_i = P_2^{-1} N_i$ from the solutions of (12.34) and (12.36). By solving (12.33)–(12.36) with other LMIs presented here, multi-objective fuzzy control [15, 17, 18, 21] can be easily and effectively realized.

12.5 CONCLUSIONS

This chapter has presented a model-based fuzzy control of non-linear systems. The fuzzy regulator and fuzzy observer have been designed by solving LMIs (linear matrix inequalities) that represent control performance such as *decay rate, disturbance rejection, robust stability, minimization of quadratic performance function and constraints on the control input and output*. In addition, this chapter has presented new stability conditions by relaxing the previous stability results. The relaxed stability conditions have been also utilized in the LMI-based design procedure. For more details on this topic, see [22].

Recently, the LMI-based fuzzy control technique developed in our previous papers (e.g. [4]–[8], [13]–[16]) is applied to multi-objective control of a high-rise/high-speed next generation elevator [17] which is an industrial complicated (non-linear) process. The LMI-based design technique effectively achieves the multi-objective control.

REFERENCES

1. Langari, R. and M. Tomizuka, Analysis and synthesis of fuzzy linguistic control systems, *1990 ASME Winter Annual Meeting*, 1990, pp. 35–42.
2. Farinwata, S. S. *et al.*, Stability analysis of the fuzzy logic controller designed by the phase portrait assignment algorithm, *Proceedings of the Second IEEE International Conference on Fuzzy Systems*, 1993, pp. 1377–1382.
3. Tanaka, K. and M. Sano, On the concept of fuzzy regulators and fuzzy observers, *Proceedings of the Third IEEE International Conference on Fuzzy Systems*, vol. 2, 1994, pp. 767–772.
4. Tanaka, K. and M. Sugeno, Stability analysis and design of fuzzy control systems, *Fuzzy Sets and Systems*, **45**(2), 1992, 135–156.
5. Wang, H., K. Tanaka and M. Griffin, Parallel distributed compensation of nonlinear systems by Takagi and Sugeno's fuzzy model., *Proceedings of the 1995 FUZZ-IEEE* 1995, pp. 531–538.
6. Tanaka, K., T. Ikeda and H. Wang, Robust stabilization of a class of uncertain nonlinear systems via fuzzy control, *IEEE Transactions on Fuzzy Systems*, **4**(1), 1996, 1–13.
7. Wang, H., K. Tanaka and M. Griffin, An approach to fuzzy control of nonlinear systems: stability and design issues, *IEEE Transactions on Fuzzy Systems*, **4**(1), 1996, 14–23.
8. Tanaka, K. *et al.* Fuzzy control system designs via LMIs, *Proceedings of the 1997 American Control Conference*.
9. Takagi, T. and M. Sugeno, Fuzzy identification of systems and its applications to modeling and control, *IEEE Transactions on Systems, Man and Cybernetics*, **15**, 1985, 116–132.
10. Tanaka, K. and Y. Yamaichi, Model-based fuzzy control approach to a nonlinear control benchmark problem, *Proceedings of the Fortieth Joint Automatic Control Conference*, 1997, 405–506.
11. Boyd, S. *et al.*, *Linear Matrix Inequalities in Systems and Control Theory*, SIAM, Philadelphia, 1994.
12. Kawamoto, S. *et al.*, An approach to stability analysis of second order fuzzy systems, *Proceedings of the First IEEE International Conference on Fuzzy Systems*, vol. 1, 1992, pp. 1427–1434.
13. Tanaka, K., T. Ikeda and H. O. Wang, Design of Fuzzy Control Systems Based on Relaxed LMI Stability Conditions, *Proceedings of the Thirty-fifth IEEE Conference on Decision and Control*, 1996, pp. 598–603.
14. Tanaka, K., T. Ikeda and H. O. Wang, A unified approach to controlling chaos via an LMI-based fuzzy control system design, *IEEE Transactions on Circuits and Systems*, **45**(10), 1998, 1021–1040.
15. Tanaka, K., T. Taniguchi and H. O. Wang, A mixed control problem of robust fuzzy control and optimal fuzzy control and its solution via linear matrix inequalities, *IEEE Transactions on Fuzzy Systems*, submitted.
16. Li, J., H. O. Wang and K. Tanaka, Stable fuzzy control of the benchmark nonlinear control problem: A system-theoretic approach, *Proceedings of the Third Joint Conference of Information Sciences*, Durham, vol. 1, 1997, pp. 263–266.
17. Tanaka, K. and M. Nishimura, Multi-objective fuzzy control of high rise / high speed elevators using LMIs, *Proceedings of the 1998 American Control Conference*, 1998, pp. 3450–3454.
18. Tanaka, K., T. Ikeda and H. O. Wang, Fuzzy regulators and fuzzy observers: Relaxed stability conditions and LMI based designs, *IEEE Transactions on Fuzzy Systems*, **6**(2), 1998, 250–265.

19. Tanaka, K., T. Ikeda and H. O. Wang, Fuzzy regulators and fuzzy observers: A linear matrix inequality approach, *Proceedings of the Thirty-sixth IEEE Conference on Decision and Control*, 1997.
20. Zhao, J. *et al.*, Synthesis of fuzzy control system with desired performances, *Proceedings of the 1996 IEEE International Symposium on Intelligent Control*, 1996, pp. 115–120.
21. Tanaka, K., T. Taniguchi and H. O. Wang, Model-based fuzzy control of TORA system: Fuzzy regulator and fuzzy observer design via LMIs that represent decay rate, disturbance rejection, robustness, optimality, *Proceedings of the 1998 FUZZ-IEEE*, Alaska, 1998.
22. Tanaka, K. and H. O. Wang, *Fuzzy Control Systems Analysis and Design: A Linear Matrix Inequality Approach*, John Wiley and Sons Publishers, New York, 2000.

13
A Framework for the Synthesis of PDC-Type Takagi–Sugeno Fuzzy Control Systems: An LMI Approach

Joongseon Joh
Changwon National University, Changwon, Korea

Reza Langari
Texas A&M University, Texas, USA

13.1 INTRODUCTION

The application of fuzzy logic to control problems has been an important issue for the past two decades since Zadeh's seminal paper [1]. The main stream of research has focused on its application to industrial systems and a number of successful results have been reported. There have been, however, few works on the development of systematic design methods for fuzzy controllers. This has been a major obstacle to the wide acceptance of fuzzy control in the field automatic control.

It is common to categorize fuzzy controllers by the so-called Mamdani and Takagi–Sugeno (T–S) types. In general, a Mamdani-type fuzzy controller is designed empirically. However, a T–S type fuzzy controller can be designed from a T–S fuzzy model which can adopt locally linearized mathematical models of the plant as the consequent part of each rule. Wang et al.'s PDC (parallel distributed compensator) is an example of such efforts [2]. In this chapter we seek systematic synthesis methods for fuzzy controllers based on a PDC-type T–S fuzzy control scheme.

13.1.1 Brief Historical Overview

In the literature, there was an important breakthrough in 1992 in the field of fuzzy control. Tanaka and Sugeno [3] proposed a theorem on the stability analysis of a T–S fuzzy model.

Fuzzy Logic. Edited by S. Farinwata, D. Filev and R. Langari
© 2000 John Wiley & Sons, Ltd

It states the sufficient conditions for global asymptotic stability of T–S fuzzy models. To establish these conditions requires 'finding a common symmetric positive definite matrix P which satisfies n simultaneous Lyapunov inequalities'. Wang et al. [2] proposed the so-called PDC as a design framework and also modified Tanaka's stability theorem to include a control algorithm. An important observation is that the stability problem is a standard feasibility problem with several linear matrix inequalities (LMIs) when the feedback gains are pre-determined and can be solved numerically using an algorithm named the *interior-point* method. It is, however, not a design method since it needs pre-determined feedback gains. Wang et al.'s method can be considered as a stability checking method for pre-designed system and requires trial-and-error for control design. Therefore, it is still necessary to develop systematic design methods that guarantee stability and desired performance.

Basically, Wang et al.'s stability criterion is adopted in this chapter. However, it cannot be used directly because it involves non-linear matrix inequalities (NMIs) when P and feedback gains, the K_i's, are both matrix variables. Therefore the stability LMIs are derived for both of continuous and discrete PDC-type T–S fuzzy control systems by applying the Schur complements [4, p. 7] to Wang et al.'s stability criterion. A systematic synthesis method for PDC-type T–S fuzzy control systems which guarantees global asymptotic stability and satisfies the desired performance of the closed-loop system is proposed based on the stability LMIs. This is accomplished by including pole placement constraints (in form of LMIs). The desired performance is represented as a region in the complex plane where the desired closed-loop poles lie [5, 6]. The proposed synthesis method is extended to include parameteric uncertainties. The corresponding stability LMIs are derived. A PDC-type T–S fuzzy controller for an inverted pendulum with a cart is designed using the proposed method as an illustrative example.

13.2 BACKGROUND MATERIALS

13.2.1 T–S Fuzzy Model of Non-linear Dynamic Systems and its Stability

A T–S Fuzzy model [7] is composed for r plant rules that can be represented as

$$\text{Plant rule } i : \textbf{If } x_1(t) \text{ is } M_1^i \text{ and } x_2(t) \text{ is } M_2^i \text{ and} \ldots \text{and } x_n(t) \text{ is } M_n^i \tag{13.1}$$
$$\textbf{Then } \delta x(t) = A_i x(t) + B_i u(t), \quad i = 1, 2, \ldots, r,$$

where

$x_j = j$th state (or linguistic) variable,

$M_j = $ fuzzy term set of x_j,

$M_j^i = $ a fuzzy term of M_j selected for plant rule i,

$x(t) = $ state vector and $x(t) = [x_1(t) \ldots x_n(t)]^T \in R^n$,

$u(t) = $ input vector and $u(t) = [u_1(t) \ldots u_m(t)]^T \in R^m$,

$A_i \in R^{n \times n}$,

$B_i \in R^{n \times m}$,

It should be noted that $\delta x(t) = \dot{x}(t)$ for the continuous-time T–S fuzzy model and $\delta x(t) = x(t+1)$ for the discrete-time T–S fuzzy model.

For any current state vector $x(t)$ and input vector $u(t)$, the T–S fuzzy model infers $\delta x(t)$ as the output of the fuzzy model as follows:

$$\delta x(t) = \frac{\sum_{i=1}^{r} w_i [A_i x(t) + B_i u(t)]}{\sum_{i=1}^{r} w_i}, \qquad (13.2)$$

where

$$w_i = \prod_{k=1}^{r} M_k^i(x_k(t)). \qquad (13.3)$$

For a free system (i.e. $u(t) \equiv 0$), (13.2) can be written as

$$\delta x(t) = \frac{\sum_{i=1}^{r} w_i A_i x(t)}{\sum_{i=1}^{r} w_i}. \qquad (13.4)$$

It is assumed, from now on, that a proper T–S fuzzy model is available.

Tanaka and Sugeno [3] suggested an important criterion for the stability of the discrete T–S fuzzy model.

Theorem 1: Stability criterion for the continuous-time T–S fuzzy model *The equilibrium state of the continuous-time T–S fuzzy model (13.4) (namely, $x = 0$) is globally asymptotically stable if there exists a common symmetric positive definite matrix P such that*

$$A_i^T P + P A_i < 0 \quad \text{for all } i = 1, 2 \ldots, r. \qquad (13.5)$$

□

Proof See [3].

Theorem 2: Stability criterion for the discrete-time T–S fuzzy model *The equilibrium of the discrete-time T–S fuzzy model (namely, $x = 0$) is globally asymptotically stable if there exists a common symmetric positive definite matrix P such that*

$$A_i^T P A_i - P < 0 \quad \text{for all } i = 1, 2, \ldots, r. \qquad (13.6)$$

□

Proof See [3].

Remark 1 It is well known that the total system may not be stable even if every subsystem is stable [3].

Remark 2 It should be noted that (13.5) and (13.6) are *sufficient conditions* for stability but are not necessary conditions. Therefore there may exist better criterion for the stability to T–S fuzzy models.

13.2.2 PDC-type T–S Fuzzy Control System and its Stability

Wang et al. [2] proposed a framework which can be used as a guideline to design a T–S fuzzy controller using a T–S fuzzy model. It can be summarized as

A T–S fuzzy controller can be designed using a T–S fuzzy model by using the antecedent part of the T–S fuzzy model as that of the T–S fuzzy controller.

270 SYNTHESIS OF PDC-TYPE T–S FUZZY CONTROL SYSTEMS

In this case we can use a proper linear control method for each pair of plant and controller rules. Wang et al. [2] called it PDC (parallel distributed compensation).

A PDC-type T–S fuzzy controller which uses full state feedback is composed of r control rules that can be represented as

$$\text{Control rule } i : \textbf{If } x_1(t) \text{ is } M_1^i \text{ and } x_2(t) \text{ is } M_2^i \text{ and} \ldots \text{and } x_n(t) \text{ is } M_n^i$$
$$\textbf{Then } u(t) = K_i x(t), \quad i = 1, 2, \ldots, r. \tag{13.7}$$

For any current state vector $x(t)$, the T–S fuzzy controller infers $u(t)$ as the output of the fuzzy controller as follows:

$$u(t) = \frac{\sum_{j=1}^{r} w_j K_j x(t)}{\sum_{j=1}^{r} w_j}. \tag{13.8}$$

It has a very important advantage because it makes it easy (or manageable) to apply (13.8) to (13.2). Therefore the closed-loop behavior of the T–S fuzzy model (13.1) with the T–S fuzzy controller (13.7) using PDC can be obtained by substituting (13.8) into (13.2) as follows:

$$\delta x(t) = \frac{\sum_{i=1}^{r} \sum_{j=1}^{r} w_i w_j (A_i + B_i K_j) x(t)}{\sum_{i=1}^{r} \sum_{j=1}^{r} w_i w_j}. \tag{13.9}$$

The sufficient conditions for the stability of (13.9) are obtained as follows.

Theorem 3: Stability condition for the PDC-type continuous-time T–S fuzzy control system *The equilibrium of the fuzzy system (13.9) (namely, $x = 0$) is globally asymptotically stable if there exists a common symmetric positive definite matrix P such that*

$$(A_i + B_i K_j)^T P + P(A_i + B_i K_j) < 0 \quad \text{for all } i, j = 1, 2, \ldots, r. \tag{13.10}$$
□

Proof See [2].

Theorem 4: Stability condition for the PDC-type discrete-time T–S fuzzy control system *The equilibrium of the discrete-time T–S fuzzy control system (namely, $x = 0$) is globally asymptotically stable if there exists a common symmetric positive definite matrix P such that*

$$(A_i + B_i K_j)^T P(A_i + B_i K_j) - P < 0 \quad \text{for all } i, j = 1, 2, \ldots, r. \tag{13.11}$$
□

Proof See [2].

It is suggested that P can be determined numerically by solving the LMIs in (13.10) when the K_i's, $i = 1, 2, \ldots, r$, are pre-determined. It should be noted that (13.10) and (13.11) have r^2 LMIs. Wang et al. [2] rewrote (13.9) by grouping the same terms as follows:

$$\delta x(t) = \frac{\sum_{i=1}^{r} w_i w_i (A_i + B_i K_i) x(t) + 2 \sum_{i<j}^{r} w_i w_j G_{ij} x(t)}{\sum_{i=1}^{r} \sum_{j=1}^{r} w_i w_j}, \tag{13.12}$$

where

$$G_{ij} = \frac{(A_i + B_i K_j) + (A_j + B_j K_i)}{2}, \quad i < J \leq r, \tag{13.13}$$

to yield a less conservative stability condition. The corresponding sufficient conditions for the stability of (13.12) is summarized in the Theorems 5 and 6.

Theorem 5: **Less conservative stability condition for the PDC-type continuous-time T–S fuzzy control system** *The equilibrium state of the continuous-time fuzzy control system in (13.12) (namely, $x = 0$) is globally asymptotically stable if there exists a common symmetric positive definite matrix P such that*

$$(A_i + B_i K_i)^T P + P(A_i + B_i K_i) < 0, \quad i = 1, 2, \ldots, r,$$
$$G_{ij}^T P + P G_{ij} < 0, \quad i < j \leq r. \tag{13.14}$$

$\square$

Proof See [2]

Theorem 6: **Less conservative stability condition for the PDC-type discrete-time T–S fuzzy control system** *The equilibrium of the discrete-time T–S fuzzy control system in (13.12) (namely, $x = 0$) is globally asymptotically stable if there exists a common symmetric positive definite matrix P such that*

$$(A_i + B_i K_i)^T P(A_i + B_i K_i) - P < 0, \quad i = 1, 2, \ldots, r,$$
$$G_{ij}^T P + P G_{ij} < 0, \quad i < j \leq r. \tag{13.15}$$

$\square$

Proof See [2].

The number of LMIs for (13.14) and (13.15) is $r(r+1)/2$. Therefore the number of LMIs to be solved is reduced from r^2 of (13.10) and (13.11) to $r(r+1)/2$.

Remark 3 It should be noted that the advantage to (13.14) and (13.15) over (13.10) and (13.11) is not only the reduction of the number of LMIs to be solved but also the relaxation of the stability criterion of (13.14) and (13.15). In the authors' experience some standard feasibility problem that are infeasible from (13.10) and (13.11) can be solved from (13.14) and (13.15).

Remark 4 The sufficient condition for the stability of (13.14) and (13.15) can be used only for the purpose of checking of the stability of the T–S fuzzy control system in which the feedback gains, the K_i's, $i = 1, 2, \ldots, r$, are pre-determined by a proper linear controller design method.

13.3 STABILITY LMIs AS A FRAMEWORK FOR THE SYNTHESIS OF PDC-TYPE T–S FUZZY CONTROL SYSTEMS

The goal of the research on the synthesis of PDC-type T–S fuzzy control systems is to find proper feedback gains which simultaneously guarantee stability and a system's performance. Therefore, the stability conditions in Section 13.2 can play a central role for this purpose. However, it should be emphasized that the stability criteria (13.14) and (13.15) are LMIs *if and only if the feedback gains, the K_i's are pre-determined* using the proper design method in Wang et al. [2]. It means that if the K_i's and P are treated as matrix variables then (13.14) and (13.15) are no longer LMIs since they contain terms like $K_i^T B_i^T P$ and $K_i^T B_i^T P B_i K_i$, i.e. they are NMIs (non-linear matrix inequalties). An analytic or numercial method that solves such NMI problems does not yet exist. So, it is definitely necessary to convert the NMIs (13.14) and (13.15) to the LMIs (13.10) and (13.11). They are called the *stability LMIs* in this chapter.

Theorem 7: Stability LMIs for PDC-type continuous-time T–S fuzzy control systems The equilibrium state of the continuous-time T–S fuzzy system (13.9) (namely, $x = 0$) where the K_i's are unknown is globally asymptotically stable if there exists a common symmetric positive definite matrix $Q = P^{-1} > 0$ which satisfies

$$QA_i^T + A_iQ + V_i^T B_i^T + B_iV_i < 0, \quad i = 1, 2, \ldots, r,$$
$$QA_i^T + A_iQ + QA_j^T + A_jQ + V_j^T B_i^T + B_iV_j + V_i^T B_j^T + B_jV_i < 0, \quad i < j \leq r, \quad (13.16)$$

where Q and $V_i = K_iQ$, $i = 1, 2, \ldots, r$, are the new matrix variables of the LMIs. □

Proof Substitution of (13.13) into (13.14) yields

$$(A_i + B_iK_i)^T P + P(A_i + B_iK_i) < 0, \quad i = 1, 2, \ldots, r,$$
$$(A_i + B_iK_j)^T P + (A_j + B_jK_i)^T P + P(A_i + B_iK_j) + P(A_j + B_jK_i) < 0, \quad i < j \leq r. \quad (13.17)$$

If we let $Q = P^{-1}$ and pre- and post-multiply (13.17) by $Q > 0$, then the negative definiteness of (13.17) will not be changed due to Sylvester's law of inertia since Q is symmetric and positive definite. Equation (13.17) becomes

$$QA_i^T + A_iQ + QK_i^T B_i^T + B_iK_iQ < 0, \quad i = 1, 2, \ldots, r,$$
$$QA_i^T + A_iQ + QA_j^T + A_jQ + QK_j^T B_i^T + B_iK_jQ + QK_i^T B_j^T + B_jK_iQ < 0, \quad i < j \leq r. \quad (13.18)$$

Now we can obtain the LMIs by letting $K_iQ = V_i$ as follows:

$$QA_i^T + A_iQ + V_i^T B_i^T + B_iV_i < 0, \quad i = 1, 2, \ldots, r,$$
$$QA_i^T + A_iQ + QA_j^T + A_jQ + V_j^T B_i^T + B_iV_j + V_i^T B_j^T + B_jV_i < 0, \quad i < j \leq r, \quad (13.19)$$

where Q and V_i, $i = 1, 2, \ldots, r$, are the new matrix variables of the LMIs. Equation (13.19) is certainly the LMIs for matrix variables Q, V_i, and V_j.

Theorem 8: Stability LMIs for PDC-type discrete-time T–S fuzzy control systems The equilibrium of the discrete-time T–S fuzzy control system (namely, $x = 0$) where the K_i's are unknown is globally asymptotically stable if there exists a common symmetric positive definite matrix $Q = P^{-1} > 0$ which satisfies

$$\begin{bmatrix} -Q + A_iQA_i^T + A_iV_i^T B_i^T + B_iV_iA_i^T & B_iV_i \\ V_i^T B_i^T & -Q \end{bmatrix} < 0, \quad i = 1, 2, \ldots, r, \quad (13.20)$$

and

$$\begin{bmatrix} -4Q + S_{ij} & B_iV_j + B_jV_i \\ V_j^T B_i^T + V_i^T B_j^T & -Q \end{bmatrix} < 0, \quad i < j \leq r, \quad (13.21)$$

where

$$S_{ij} = A_iQA_i^T + A_iV_j^T B_i^T + B_iV_jA_i^T + A_iQA_j^T + A_iV_i^T B_j^T + B_iV_iA_j^T \\ + A_jQA_i^T + A_jV_j^T B_i^T + B_jV_iA_i^T + A_jQA_j^T + A_jV_i^T B_j^T + B_jV_iA_j^T \quad (13.22)$$

and the Q and V_i's are matrix variables. □

Proof The proof is composed of two parts. The first part is the derivations of (13.20) from

$$(A_i + B_iK_i)^T P(A_i + B_iK_i) - P < 0, \quad i = 1, 2, \ldots, r, \quad (13.23)$$

and (13.21) from
$$G_{ij}^T P + P G_{ij} < 0, \quad i < j \leq r, \tag{13.24}$$

where (13.23) and (13.24) are both from Theorem 6. Recall that, for any metrices Φ_{11}, Φ_{12}, and Φ_{22}, where Φ_{11} and Φ_{22} are symmetric, the following are equivalent:

(i) $\begin{bmatrix} \Phi_{11} & \Phi_{12} \\ \Phi_{12}^T & \Phi_{22} \end{bmatrix} < 0,$ \hfill (13.25)

(ii) $\Phi_{11} < 0, \quad \Phi_{22} < \Phi_{12}^T \Phi_{11}^{-1} \Phi_{12},$ \hfill (13.26)

(iii) $\Phi_{22} < 0, \quad \Phi_{11} < \Phi_{12} \Phi_{22}^{-1} \Phi_{12}^T.$ \hfill (13.27)

They are called Schur complements in [4, p. 7]. Using (13.25) and (13.27), (13.23) is equivalent to

$$\begin{bmatrix} -P & (A+BK)^T \\ (A+BK) & -P^{-1} \end{bmatrix} < 0. \tag{13.28}$$

Note that the subscript i is omitted for simplicity. From (13.25) and (13.26), (13.28) is equivalent to

$$-P^{-1} + (A+BK)P^{-1}(A+BK)^T < 0. \tag{13.29}$$

Let $Q = P^{-1} > 0$, then (13.29) becomes

$$-Q + (A+BK)Q(A+BK)^T < 0. \tag{13.30}$$

Equation (13.30) can be expanded as

$$-Q + AQA^T + AQK^T B^T + BKQA^T + BKQK^T B^T < 0. \tag{13.31}$$

From (13.25) and (13.27), (13.31) is equivalent to

$$\begin{bmatrix} -Q + AQA^T + AQK^T B^T + BKQA^T & BKQ \\ QK^T B^T & -Q \end{bmatrix} < 0. \tag{13.32}$$

Let $V = KQ$, then (13.32) becomes

$$\begin{bmatrix} -Q + AQA^T + AV^T B^T + BVA^T & BV \\ V^T B^T & -Q \end{bmatrix} < 0. \tag{13.33}$$

Equation (13.33) is certainly a LMI for the matrix variables Q and V and it completes the proof of the first part of the Theorem 8.

Expansion of (13.24) becomes

$$\tfrac{1}{4}\{(A_i + B_i K_j) + (A_j + B_j K_i)\}^T P\{(A_i + B_i K_j) + (A_j + B_j K_i)\} - P < 0. \tag{13.34}$$

From (13.25) and (13.27), (13.34) is equivalent to

$$\begin{bmatrix} -P & \tfrac{1}{2}\{(A_i + B_i K_j) + (A_j + B_j K_i)\}^T \\ \tfrac{1}{2}\{(A_i + B_i K_j) + (A_j + B_j K_i)\} & -P^{-1} \end{bmatrix} < 0. \tag{13.35}$$

Using (13.25) and (13.26), (13.35) is also equivalent to

$$-P^{-1} + \tfrac{1}{4}\{(A_i + B_i K_j) + (A_j + B_j K_i)\} P^{-1} \{(A_i + B_i K_j) + (A_j + B_j K_i)\}^\mathrm{T} < 0. \quad (13.36)$$

Let $Q = P^{-1} > 0$, then (13.36) becomes

$$-Q + \tfrac{1}{4}\{(A_i + B_i K_j) + (A_j + B_j K_i)\} Q \{(A_i + B_i K_j) + (A_j + B_j K_i)\}^\mathrm{T} < 0 \quad (13.37)$$

or

$$\begin{aligned} -4Q + \{&A_i Q A_i^\mathrm{T} + A_i Q K_j^\mathrm{T} B_i^\mathrm{T} + B_i K_j Q A_i^\mathrm{T} + A_i Q A_j^\mathrm{T} + A_i Q K_i^\mathrm{T} B_j^\mathrm{T} + B_i K_j Q A_j^\mathrm{T} \\ &+ A_j Q A_i^\mathrm{T} + A_j Q K_j^\mathrm{T} B_i^\mathrm{T} + B_j K_i Q A_i^\mathrm{T} + A_j Q A_j^\mathrm{T} + A_j Q K_i^\mathrm{T} B_j^\mathrm{T} + B_j K_i Q A_j^\mathrm{T} \\ &+ (B_i K_j + B_j K_i) Q (K_j^\mathrm{T} B_i^\mathrm{T} + K_i^\mathrm{T} B_j^\mathrm{T}) \} < 0. \end{aligned} \quad (13.38)$$

From (13.25) and (13.27), (13.38) is equivalent to

$$\begin{bmatrix} -4Q + S_{ij} & B_i V_j + B_j V_i \\ V_j^\mathrm{T} B_i^\mathrm{T} + V_i^\mathrm{T} B_j^\mathrm{T} & -Q \end{bmatrix} < 0. \quad (13.39)$$

Equation (13.39) is certainly a LMI for the matrix variables Q, V_i, and V_j and it completes the proof of the second part of the Theorem 8.

13.4 POLE PLACEMENT CONSTRAINT LMIs AS PERFORMANCE SPECIFICATIONS FOR THE SYNTHESIS OF PDC-TYPE T–S FUZZY CONTROL SYSTEMS

Stability LMIs play very important roles in the synthesis of PDC-type T–S fuzzy control systems since they are able to treat the feedback gains as unknowns. It is, however, still necessary to find an effective way of including the performance specifications of the system to be designed. Chilali and Gahinet [5] and Gahinet et al. [6] proposed that a convex region that represents the desired closed-loop pole placement constraints can be represented as LMIs.

Theorem 9: LMIs for pole placement constraints *The closed-loop poles lie in the LMI region*

$$D = \{z \in \mathbb{C} \mid f_D(z) := L + Mz + M^\mathrm{T} \bar{z} < 0\} \quad (13.40)$$

iff there exists a symmetric positive definite matrix X_pol satisfying

$$[\lambda_{ij} X_\mathrm{pol} + \mu_{ij}(A + BK) X_\mathrm{pol} + \mu_{ji} X_\mathrm{pol} (A + BK)^\mathrm{T}]_{1 \leq i,j \leq m} < 0. \quad (13.41)$$

Here, z is a complex variable, A, B, and K are the system, and feedback gain matrices of a linear system, respectively, and $L = L^\mathrm{T} = [\lambda_{ij}]_{1 \leq i,j \leq m}$ and $M = [\mu_{ij}]_{1 \leq i,j \leq m}$ are known real matrices which can be determined by specifying the desired closed-loop pole region in s-plane. □

Proof See [5].

The notation in (13.41) represents an $m \times m$ matrix whose (i,j)th element is given in the bracket. The size of the matrix (13.41), i.e. m, is determined by the subjective way of representing the closed-loop pole regions in the s-plane. Three useful LMI regions are introduced as follows [6]:

- a disk with center at $(-q, 0)$ and radius r

$$f_D(z) = \begin{pmatrix} -r & \bar{z}+q \\ z+q & -r \end{pmatrix};\qquad(13.42)$$

- a conic sector with center at the origin and with inner angle θ

$$f_D(z) = \begin{pmatrix} \sin\frac{\theta}{2}(z+\bar{z}) & -\cos\frac{\theta}{2}(z-\bar{z}) \\ \cos\frac{\theta}{2}(z-\bar{z}) & \sin\frac{\theta}{2}(z+\bar{z}) \end{pmatrix};\qquad(13.43)$$

- a vertical strip $h_1 < x < h_2$

$$f_D(z) = \begin{pmatrix} 2h_1 - (z+\bar{z}) & 0 \\ 0 & (z+\bar{z}) - 2h_2 \end{pmatrix}.\qquad(13.44)$$

Therefore, we can obtain L and M using (13.42)–(13.44).

Now we propose to apply the Chilali and Gahinet LMI regions [5, 6] to each local T–S fuzzy controller in order to specify the desired closed-loop performance. Therefore, we have r LMI regions corresponding to r local T–S fuzzy controllers as follows:

$$[\lambda_{kl}X_{\text{pol}} + \mu_{kl}(A_i + B_iK_i)X_{\text{pol}} + \mu_{lk}X_{\text{pol}}(A_i + B_iK_i)^{\text{T}}]_{1\leq k,l\leq m} < 0, \quad i = 1, 2, \ldots, r. \quad(13.45)$$

By letting $X_{\text{pol}} = Q$ since Q is symmetric positive definite and $K_iQ = V_i$, we obtain

$$[\lambda_{kl}Q + \mu_{kl}A_iQ + \mu_{kl}B_iV_i + \mu_{lk}QA_i^{\text{T}} + \mu_{lk}V_i^{\text{T}}B_i^{\text{T}}]_{1\leq k,l\leq m} < 0, \quad i = 1, 2, \ldots, r. \quad(13.46)$$

The r LMIs in (13.46) represent the desired closed-loop pole placement constraints for the PDC-type T–S fuzzy control systems. We can combine (13.46) with the stability LMIs to yield a synthesis method for the PDC-type T–S fuzzy control systems.

Theorem 10: Synthesis of PDC-type continuous-time T–S fuzzy control systems *A PDC-type continuous-time T–S fuzzy controller that guarantees global asymptotic stability and satisfies the desired performance by placing closed-loop poles for each local model within the desired region can be designed by solving*

$$\begin{aligned} QA_i^{\text{T}} + A_iQ + V_i^{\text{T}}B_i^{\text{T}} + B_iV_i &< 0, \quad i = 1, 2, \ldots, r, \\ QA_i^{\text{T}} + A_iQ + QA_j^{\text{T}} + A_jQ + V_j^{\text{T}}B_i^{\text{T}} + B_iV_j + V_i^{\text{T}}B_j^{\text{T}} + B_jV_i &< 0, \quad i < j \leq r, \\ [\lambda_{kl}Q + \mu_{kl}A_iQ + \mu_{kl}B_iV_i + \mu_{lk}QA_i^{\text{T}} + \mu_{lk}V_i^{\text{T}}B_i^{\text{T}}]_{1\leq k,l\leq m} &< 0, \quad i = 1, 2, \ldots, r, \\ Q > \alpha I, \alpha &= \text{positive constant}. \end{aligned}\qquad(13.47)$$

□

Proof Combining Theorem 7 and (13.46) yields (13.47).

Theorem 11: Synthesis of PDC-type discrete-time T–S fuzzy control systems *A PDC-type discrete-time T–S fuzzy controller which guarantees global asymptotic stability* and satisfies the desired performance by placing closed-loop poles for each local model within the desired region can be designed by solving

$$\begin{bmatrix} -Q + A_i Q A_i^T + A_i V_i^T B_i^T + B_i V_i A_i^T & B_i V_i \\ V_i^T B_i^T & -Q \end{bmatrix} < 0, \quad i = 1, 2, \ldots, r,$$

$$\begin{bmatrix} -4Q + S_{ij} & B_i V_j + B_j V_i \\ V_j^T B_i^T + V_i^T B_j^T & -Q \end{bmatrix} < 0, \quad i < j \le r, \quad (13.48)$$

$$[\lambda_{kl} Q + \mu_{kl} A_i Q + \mu_{kl} B_i V_i + \mu_{lk} Q A_i^T + \mu_{lk} V_i^T B_i^T]_{1 \le k, l \le m} < 0, \quad i = 1, 2, \ldots, r,$$

$$Q > \alpha I, \quad \alpha = \text{positive constant}.$$

$\square$

Proof Combining Theorem 8 and (13.46) yields (13.48)

Remark 5 In equations (13.47) and (13.48), $Q > \alpha I$ is included in orded to avoid a trivial solution and to easily differentiate between feasibility and infeasibility. If $\alpha = 0$, then we have the original constraint on Q, i.e. positive definiteness.

Remark 6 We can obtain Q and $V_i, i = 1, 2, \ldots, r$, by solving (13.47) or (13.48). Then the common symmetric positive definite matrix P and the feedback gains $K_i, i = 1, 2, \ldots, r$, can be determined as follows:

$$P = Q^{-1},$$
$$K_i = V_i Q^{-1} = V_i P, \quad i = 1, 2, \ldots, r. \quad (13.49)$$

13.5 AN EXTENSION TO PDC-TYPE T–S FUZZY CONTROL SYSTEMS WITH PARAMETER UNCERTAINTIES

The framework that was proposed in the Sections (13.3) and (13.4) can be extended to include parametric uncertainties. The corresponding T–S fuzzy model can be represented as

Plant rule i: **If** $x_1(t)$ is M_1^i and $x_2(t)$ is M_2^i and ... and $x_n(t)$ is M_n^i
Then $\delta x(t) = (A_i + \Delta A_i(t))x(t)$
$+ (B_i + \Delta B_i(t))u(t), \quad i = 1, 2, \ldots, r, \quad (13.50)$

where $\Delta A_i(t)$ and $\Delta B_i(t)$ are system and input uncertainty matrices, respectively. It is assumed that

$$[\Delta A_i(t) \quad \Delta B_i(t)] = H_i F_i(t)[E_{A_i} \quad E_{B_i}], \quad i = 1, 2, \ldots, r, \quad (13.51)$$

where H_i, E_{A_i}, and E_{B_i} are known constant matrices and $F_i(t), i = 1, 2, \ldots, r$, are unknown matrice defined as

$$F_i(t) \in \Omega := \{F(t) | F(t)^T F(t) \le I, \text{elements of } F(t) \text{ are Lebesgue measurable}\}. \quad (13.52)$$

For any current state vector $x(t)$ and input vector $u(t)$, the T–S fuzzy model with parameter uncertainties infers $\delta x(t)$ as the output of the fuzzy model as follows:

$$\delta x(t) = \frac{\sum_{i=1}^{r} w_i [\hat{A}_i x(t) + \hat{B}_i u(t)]}{\sum_{i=1}^{r} w_i}, \quad (13.53)$$

where $\hat{A}_i = (A_i + \Delta A_i(t))$ and $\hat{B}_i = (B_i + \Delta B_i(t))$. The corresponding PDC-type T–S fuzzy controller is the same as (13.7).

We can obtain sufficient conditions for stability of the resulting system in a manner similar to Theorems 3 and 4 as follows.

Theorem 12: Stability condition for the PDC-type continuous-time T–S fuzzy control system with parameteric uncertainties *The equilibrium state of the PDC-type continuous-time T–S fuzzy control system with parameter uncertainties (namely, $x = 0$) is globally asymptotically stable if there exists a common symmetric positive definite matrix P such that*

$$(\hat{A}_i + \hat{B}_i K_j)^T P + P(\hat{A}_i + \hat{B}_i K_j) < 0 \quad \text{for all } i,j = 1,2,\ldots,r. \tag{13.54}$$

Theorem 13: Stability condition for the PDC-type discrete-time T–S fuzzy control system with parameter uncertainties *The equilibrium of the PDC-type discrete-time T–S fuzzy control system with parameter uncertainties (namely, $x = 0$) is globally asymptotically stable if there exists a common symmetric positive definite matrix P such that*

$$(\hat{A}_i + \hat{B}_i K_j)^T P(\hat{A}_i + \hat{B}_i K_j) - P < 0, \quad \text{for all } i,j = 1,2,\ldots,r. \tag{13.55}$$

□

Stability LMIs corresponding to the Theorems 12 and 13 can be obtained using similar approaches as in Section 13.3. We need, however, an additional theorem as follows.

Theorem 14. *For constant matrices A and B with proper dimensions, it holds that*

$$A^* F^*(t) B + B^* F(t) A \leq \lambda A^* A + \lambda^{-1} B^* B, \quad \text{for all } F(t) \in \Omega, \lambda > 0. \tag{13.56}$$

□

Proof See [9]

Theorem 15: Stability LMIs for the PDC-type continuous-time T–S fuzzy control system with parameteric uncertainties *The equilibrium of the PDC-type continuous-time T–S fuzzy control system with parameter uncertainties in (13.50)–(13.52) (namely, $x = 0$) is globally asymptotically stable if there exists a common symmetric positive definite matrix Q, matrices V_j, $j = 1,2,\ldots,r$, and positive constants λ_{ij}, $i,j = 1, 2,\ldots,r$, such that*

$$\begin{bmatrix} \hat{S}_{ij} & QE_{A_i}^T + V_j^T E_{B_i}^T \\ E_{A_i} Q + E_{B_i} V_j & -\lambda_{ij} I \end{bmatrix} < 0, \quad \text{for all } i,j = 1,2,\ldots,r, \tag{13.57}$$

where

$$\hat{S}_{ij} = QA_i^T + A_i Q + V_j^T B_i^T + B_i V_j + \lambda_{ij} H_i H_i^T, \quad i,j = 1,2,\ldots,r, \tag{13.58}$$

$$Q = P^{-1}, \tag{13.59}$$

$$V_j = K_j Q, \quad j = 1,2,\ldots,r. \tag{13.60}$$

Furthermore, this is an LMI representation of Theorem 12. □

Proof Using the Schur complements, (13.57) can be written as

$$\hat{S}_{ij} + \lambda_{ij}^{-1} Q(E_{A_i} + E_{B_i} K_j)^T (E_{A_i} + E_{B_i} K_j) Q < 0. \tag{13.61}$$

Pre- and post-multiplication by $P > 0$ of (13.61) yields

$$(A_i + B_i K_j)^T P + P(A_i + B_i K_j)$$
$$+ \underbrace{\lambda_{ij} P H_i H_i^T P + \lambda_{ij}^{-1}(E_{A_i} + E_{B_i} K_j)^T (E_{A_i} + E_{B_i} K_j)}_{(**)} < 0. \quad (13.62)$$

Using Theorem 14, (**) can be written as

$$\lambda_{ij} P H_i H_i^T P + \lambda_{ij}^{-1}(E_{A_i} + E_{B_i} K_j)^T (E_{A_i} + E_{B_i} K_j)$$
$$\geq P H_i F_i(t)(E_{A_i} + E_{B_i} K_j) + (E_{A_i} + E_{B_i} K_j)^T F_i^T(t) H_i^T P, \quad \forall F_i(t) \in \Omega. \quad (13.63)$$

Substitution of (13.63) into (13.62) gives

$$[(A_i + \Delta A_i(t)) + (B_i + \Delta B_i(t))K_j]^T P + P[(A_i + \Delta A_i(t)) + (B_i + \Delta B_i(t))K_j] < 0 \quad (13.64)$$

or

$$(\hat{A}_i + \hat{B}_i K_j)^T P + P(\hat{A}_i + \hat{B}_i K_j) < 0 \quad \text{for all } i, j = 1, 2, \ldots, r. \quad (13.65)$$

This completes the proof.

Theorem 16: Stability LMIs for the PDC-type discrete-time T–S fuzzy control system with parameter uncertainties *The equilibrium of the PDC-type discrete-time T–S fuzzy control system with parameteric uncertainties in (13.50)–(13.52) (namely, $x = 0$) is globally asymptotically stable if there exists a common symmetric positive definite matrix Q, matrices $V_j, j = 1, 2, \ldots, r$, and positive constants $\lambda_{ij}, i, j = 1, 2, \ldots, r$, such that*

$$\begin{bmatrix} -Q & QA_i^T + V_j^T B_i^T & QE_{A_i}^T + V_j^T E_{B_i}^T \\ A_i Q + B_i V_j & -Q + \lambda_{ij} H_i H_i^T & 0 \\ E_{A_i} Q + E_{B_i} V_j & 0 & -\lambda_{ij} I \end{bmatrix} < 0, \quad \text{for all } i, j = 1, 2, \ldots, r, \quad (13.66)$$

where $Q = P^{-1}$ and $V_j = K_j Q$, $j = 1, 2, \ldots, r$. Furthermore, this is an LMI representation of Theorem 13. ☐

Proof Using the Schur complements, (13.66) can be written as

$$-Q + Q(A_i + B_i K_j)^T (Q - \lambda_{ij} H_i H_i^T)^{-1}(A_i + B_i K_j)Q$$
$$+ \lambda_{ij}^{-1} Q(E_{A_i} + E_{B_i} K_j)^T (E_{A_i} + E_{B_i} K_j)Q < 0. \quad (13.67)$$

Pre- and post-multiplication of $P > 0$ by (13.67) and application of Theorem 14 gives

$$0 > -P + (A_i + B_i K_j)^T P(I - \lambda_{ij} H_i H_i^T P)^{-1}(A_i + B_i K_j)$$
$$+ \lambda_{ij}^{-1}(E_{A_i} + E_{B_i} K_j)^T (E_{A_i} + E_{B_i} K_j)$$
$$= -P + (A_i + B_i K_j)^T P(A_i + B_i K_j) + \lambda_{ij}^{-1}(E_{A_i} + E_{B_i} K_j)^T (E_{A_i} + E_{B_i} K_j)$$
$$(A_i + B_i K_j)^T P H_i(\lambda_{ij}^{-1} I - H_i^T P H_i)^{-1} H_i^T P(A_i + B_i K_j)$$
$$\geq -P + (A_i + B_i K_j)^T P(A_i + B_i K_j) + \lambda_{ij}^{-1}(E_{A_i} + E_{B_i} K_j)^T F_i^T(t) F_i(t)(E_{A_i} + E_{B_i} K_j)$$
$$+ (A_i + B_i K_j)^T P H_i(\lambda_{ij}^{-1} I - H_i^T P H_i)^{-1} H_i^T P(A_i + B_i K_j)$$

$$- [H_i^T P(A_i + B_i K_j) - (\lambda_{ij}^{-1} I - H_i^T P H_i) F_i(t)(E_{A_i} + E_{B_i} K_j)]^T$$
$$\times (\lambda_{ij}^{-1} I - H_i^T P H_i)^{-1} [H_i^T P(A_i + B_i K_j) - (\lambda_{ij}^{-1} I - H_i^T P H_i) F_i(t)(E_{A_i} + E_{B_i} K_j)]$$
$$= [(A_i + \Delta A_i(t)) + (B_i + \Delta B_i(t)) K_j]^T P[(A_i + \Delta A_i(t)) + (B_i + \Delta B_i(t)) K_j]^T - P$$
$$= (\hat{A}_i + \hat{B}_i K_j)^T P(\hat{A}_i + \hat{B}_i K_j) - P < 0 \quad \text{for all } i, j = 1, 2, \ldots, r. \tag{13.68}$$

This completes the proof.

13.6 A SIMULATED EXAMPLE

The proposed design method is verified by designing a controller for an inverted pendulum on a cart which is adopted from Wang et al. [2]. The equations of motion for the pendulum are

$$\dot{x}_1 = x_2,$$
$$\dot{x}_2 = \frac{g \sin(x_1) - amlx_2^2 \sin(2x_1)/2 - a\cos(x_1)u}{4l/3 - aml\cos^2(x_1)}, \tag{13.69}$$

where x_1 is the angle (in radians) of the pendulum from the vertical, x_2 is the angular velocity, and u is the control force (in Newtons) applied to the cart. The other parameters are as follows:

g = the gravity constant (9.8 m/s^2),
m = mass of the pendulum (2.0 kg),
M = mass of the cart (8.0 kg),
$2l$ = length of the pendulum (1.0 m),
a = $1/(m+M)$.

The T–S fuzzy model of (13.69) in Wang et al. [2] is adopted in this chapter. It is composed of two plant rules:

Plant rule 1: **If** x_1 is about 0 **Then** $\dot{x} = A_1 x + B_1 u$;
Plant rule 2: **If** x_1 is about $\pm (\pi/2)(|x_1| < \pi/2)$ **Then** $\dot{x} = A_2 x + B_2 u$, (13.70)

where

$$A_1 = \begin{bmatrix} 0 & 1 \\ \frac{g}{4l/3 - aml} & 0 \end{bmatrix}, \quad B_1 = \begin{bmatrix} 0 \\ -\frac{a}{4l/3 - aml} \end{bmatrix},$$

$$A_2 = \begin{bmatrix} 0 & 1 \\ \frac{2g}{\pi(4l/3 - aml\beta^2)} & 0 \end{bmatrix}, \quad B_2 = \begin{bmatrix} 0 \\ -\frac{a\beta}{4l/3 - aml\beta^2} \end{bmatrix},$$

and where $\beta = \cos(88°)$. Refer to Wang et al. [2] for a detailed description. Membership functions of 'about 0' and 'about $\pm\pi/2$' are shown in Figure 13.1

A T–S controller has the following structure according to the PDC framework:

Control rule 1: **If** x_1 is about 0 **Then** $u = K_1 x$;
Control rule 2: **If** x_1 is about $\pm (\pi/2)(|x_1| < \pi/2)$ **Then** $u = K_2 x$, (13.71)

280 SYNTHESIS OF PDC-TYPE T–S FUZZY CONTROL SYSTEMS

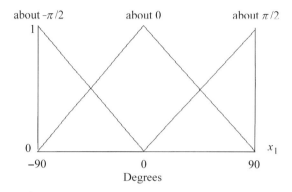

Figure 13.1 Membership functions of 'about 0' and 'about $\pm\pi/2$'

and the resulting output of the controller is

$$u = w_1 K_1 x + w_2 K_2 x \tag{13.72}$$

since $w_1 + w_2 = 1$ from Figure 13.1. Here, w_1 and w_2 are membership grades of the antecedent parts of the control rules 1 and 2, respectively. It should be noted that K_i, $i = 1, 2$, are not determined in advance, i.e. they are unknown feedback gains to be designed. Wang et al. [2] and Tanaka et al. [8] suggested predetermining the $K_i, i = 1, 2$, using a proper design method and then checking its stability using (13.14).

The design purpose of this example is to place the closed-loop poles of each local model, i.e. plant rules 1 and 2, within the desired region as shown in Figure 13.2 as a shaded polygon. It corresponds to restricting damping and response times to within a certain range. Since the plant is the second-order system, the response should be

$$\zeta > 0.995 \text{ or } \%OS < 2.55 \times 10^{-12}\% \text{ and } T_s < 2.67(s). \tag{13.73}$$

From (13.40), (13.43) and (13.44) the LMI region is defined by the L and M matrices as,

$$L = \begin{bmatrix} 3 & 0 & 0 \\ 0 & 0 & 0 \\ 0 & 0 & 0 \end{bmatrix}, \quad M = \begin{bmatrix} 1 & 0 & 0 \\ 0 & 0.0998 & -0.9950 \\ 0 & 0.9950 & 0.0998 \end{bmatrix}, \tag{13.74}$$

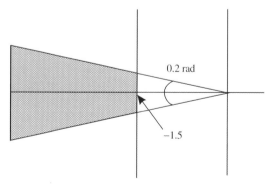

Figure 13.2 Desired pole placement constraint region

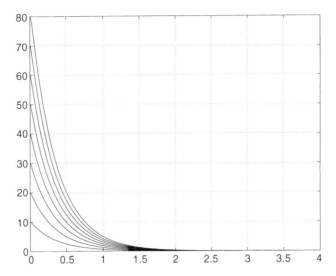

Figure 13.3 Response of the example system with eight initial conditions

where L and M are constructed by arranging two constraints as block diagonal matrices. Therefore, $m = 3$ in this case and the LMI in (13.46) is 3×3.

Now, we have to solve six LMIs in (13.47) since $r = 2$. The interior-point method yield Q, V_1, and V_2. The desired solution can be determined from (13.49) as

$$P = \begin{bmatrix} 1.129 \times 10^{-1} & 4.005 \times 10^{-2} \\ 4.005 \times 10^{-2} & 1.465 \times 10^{-2} \end{bmatrix}, \tag{13.75}$$

$$K_1 = [1029.9 \quad 356.7], \qquad K_2 = [6978.9 \quad 2437.4]. \tag{13.76}$$

The P in (13.75) is obviously symmetric and positive definite since the eigenvalues are 1.271×10^{-1} and 3.885×10^{-4}. Furthermore, it can be easily checked that the stability criterion (13.14) is satisfied. Therefore, the feedback control system (13.70) and (13.71) with feedback gains (13.76) is globally asymptotically stable. The stability of the proposed design method is verified.

The performance of the controller (13.71) with feedback gains (13.76) is checked by simulation. The simulation is performed with various initial conditions (i.e. $10, 20, \ldots, 80$ (deg.)) to see the performance of the controlling non-linear system. Figure 13.3 shows the resulting response of the system, and also shows that the performance specifications (13.73) are satisfied. Therefore, the performance of the proposed method is verified.

13.7 CONCLUDING REMARKS

The stability LMIs of continuous and discrete T–S fuzzy control systems are derived from Wang et al.'s stability criterion. The stability criterion of Wang et al. [2] involves non-linear MIs when P and $K_i, i = 1, 2, \ldots, r$, are considered as matrix variables since P and K_i, $i = 1, 2, \ldots, r$, are coupled. Therefore the LMI approach cannot be applied directly to solve for P and $K_i, i = 1, 2, \ldots, r$, simultaneously. The derived stability LMIs treat the feedback gains as unknowns and can be used as a framework for the synthesis of the PDC-type T–S fuzzy control systems. The desired performance is represented as r pole placement constraint LMIs which place the closed-loop poles of r local subsystems within the desired region in the s-plane. By solving stability LMIs and pole placement constraint LMIs simultaneously, the feedback gains that guarantee global asymptotic stability and satisfy the desired performance can be determined. The proposed method is extended to include parameter uncertainties by deriving the corresponding stability LMIs. The design method is verified by designing a T–S fuzzy controller for an inverted pendulum with a cart using the proposed method.

REFERENCES

1. Zadeh, L. A., Fuzzy sets, *Information and Control*, **8**(3), 1965, 338–353.
2. Wang, H. O., K. Tanaka and M. Griffin, An analytical framework of fuzzy modeling and control of nonlinear systems: Stability and design issues, *Proceedings of the American Control Conference*, Seattle, Washington, 1995, pp. 2272–2276.
3. Tanaka, K. and M. Sugeno, Stability analysis and design of fuzzy control system, *Fuzzy Sets and Systems*, **45**, 1992, 135–156.
4. Boyd, S., L. E. Ghaoui, E. Feron and V. Balakrishnan, *Linear Matrix Inequalities in System and Control Theory*, SIAM Studies in Applied Mathematics, SIAM Philadelphia, 1994.
5. Chilali, M. and P. Gahinet, H_∞ design with pole placement constraints: An LMI approach *Proceedings of the Thirty-third IEEE Conference on Decision and Control*, Lake Buena Vista, 1994, pp. 553–558.
6. Gahinet, P., A. Nemirovski, A. J. Laub and M. Chilali, *LMI Control Toolbox: For Use with MATLAB*, The Math Works, 1995.
7. Takagi, T. and M. Sugeno, Fuzzy identification of systems and its applications to modeling and control, *IEEE Transactions on Systems, Man and Cybernetics*, **SMC-15**(1), 1985, 116–132.
8. Tanaka, K., T. Ikeda and H. Wang, Design of fuzzy control systems based on relaxed LMI stability conditions, *Proceedings of the Thirty-Fifth IEEE Conference on Decision and Control*, Kobe, Japan, 1996, pp. 598–603.
9. Shi, G., Y. Zou and C. Yang, An algebraic approach to robust H_{infty} control via state feedback, *Systems Control Letters*, **18**, 1992, 365–370.

14

On Adaptive Fuzzy Logic Control of Non-linear Systems—Synthesis and Analysis

J. X. Lee and **G. Vukovich**
Canadian Space Agency, St. Hubert, Canada

14.1 INTRODUCTION

For most engineering systems, there are two important information sources: sensors, which provide numerical measurements of variables, and human experts, who provide linguistic instructions and descriptions of the systems. The information of the first type is called numerical information, and the second is called linguistic information, and is usually represented in fuzzy terms. Conventional engineering control approaches can make use of only the numerical information and have difficulty incorporating linguistic information. In contrast to this, the fuzzy logic approach provides a systematic and efficient framework for incorporating linguistic descriptions of human expert knowledge into the design of automatic controllers.

Classically, fuzzy logic controllers have been designed by directly utilizing expert knowledge which describes control actions from given system states, such as '**If** the speed of the car is slow, **Then** give more gas', and a vast body of literature deals with a form of non-linear fuzzy proportional-derivative-integral control for set point regulation problems, such as the applications presented in [7]–[11], [16], [17], [21], [23]. But this kind of fuzzy logic control approach generally lacks formal synthesis techniques which guarantee the basic requirements of global stability and acceptable performance. Their design is generally *ad hoc*, and controller parameters must often be manually adjusted by trial and error. There is thus a need to develop a systematic design and analysis approach for fuzzy logic control systems such that the resulting control systems are adaptive in the sense that they not only incorporate liguistic fuzzy information from human experts, but also maintain system stability and consistent performance in the presence of system uncertainties and variations. Since the latter part of this objective generally coincides with the basic objective of adaptive

Fuzzy Logic. Edited by S. Farinwata, D. Filev and R. Langari
© 2000 John Wiley & Sons, Ltd

control, which is to maintain consistent performance in the presence of variations of system parameters, the design and analysis tools from the realm of adaptive control are useful in the development of adaptive fuzzy logic systems.

Pioneering work of this kind was reported in [15], [18] and [20]. The former developed a direct adaptive fuzzy logic control scheme in which fuzzy logic systems are used as controllers whose parameters are directly adjusted to reduce some norm of the output error between the plant and the reference model. Linguistic **If–Then** rules describing control actions from given system states can be directly incorporated into controllers of this kind. The latter developed an indirect adaptive fuzzy logic control scheme where fuzzy logic systems are used to model the plant whose parameters are estimated, and the controller is chosen assuming that the estimated parameters represent the true values of the plant parameters. The fuzzy **If–Then** rules describing the behavior of the plant can then be incorporated into the indirect adaptive fuzzy controllers.

The fuzzy logic systems of the publications just cited are static in nature. Since the physical systems of interest in control engineering are generally dynamic, this suggests that one might include dynamic elements into fuzzy logic systems to take advantage of the intrinsic dynamics. This idea was implemented in [5] where the so-called *dynamic fuzzy logic system* (DFLS) was introduced. Based on this dynamic fuzzy logic system, a stable identification algorithm was developed in [5] and an indirect adaptive control algorithm was presented in [6].

In this chapter we synthesize a DFLS-based indirect adaptive control scheme which generalizes the adaptive control algorithm presented in [6] in such a way that the new control scheme here is applicable to a larger class of non-linear systems and is more useful in practical applications. The closed-loop system performance and stability properties are also analyzed theoretically and presented in the form of Theorem–Proof. The application of this DFLS-based adaptive control algorithm to non-linear systems is illustrated in a simulation example.

The basics of the theory of fuzzy sets and fuzzy logic are not repeated here, and can be found in many text books, such as [13], [21], [24]. In this regard, the authors would like to mention specifically the seminal paper by Zadeh [22] in which the approach to analysis of complex systems upon which fuzzy logic controllers would be later built was formulated, and an excellent review paper by Lee [4] which has been extensively quoted in the literature.

14.2 CONTROL OBJECTIVE

Consider the following class of non-linear dynamic systems:

$$x^{(N)} = f(\boldsymbol{x}) + g(\boldsymbol{x})u, \tag{14.1}$$

where $u \in R$ and $x \in R$ are the input and output of the system, respectively, R is the set of real numbers, $\boldsymbol{x} \triangleq \{x, \dot{x}, \ldots, x^{(N-1)}\}^T \in R^N$ is the state vector, which is assumed to be available for measurement, and $x^{(N)}$ represents the Nth order derivative of x with respect to time. $f : R^N \to R$ and $g : R^N \to R$ are unknown continuous non-linear functions defined in a certain controllability region $\mathcal{X} \subset R^N$ (whereas in [6] the gain function, g, was assumed known). Let

$$x_1 \triangleq x, \quad x_2 \triangleq \dot{x}, \ldots, x_N \triangleq x^{(N-1)}, \tag{14.2}$$

Then $\mathbf{x} = \{x_1 \ldots, x_N\}^T$, and equation (14.1) can be written in state space form as

$$\begin{cases} \dot{x}_1 &= x_2, \\ \vdots & \\ \dot{x}_{N-1} &= x_N, \\ \dot{x}_N &= f(\mathbf{x}) + g(\mathbf{x})u. \end{cases} \quad (14.3)$$

For the system to be controllable, we impose mild restrictions on the gain function g.

Assumption 2.1 $0 < \varepsilon_g \leq |g(\mathbf{x})| \leq M_g, \forall\, \mathbf{x} \in \mathcal{X}$ and $0 \leq t < \infty$, where M_g is a large positive constant, ε_g is a small positive constant, and the sign of $g(\mathbf{x})$ is also known.

Let q be a desired trajectory for the system output, x, and denote

$$q_1 \triangleq q, \quad q_2 \triangleq \dot{q}, \ldots, q_N \triangleq q^{(N-1)}. \quad (14.4)$$

Let e be the difference between x and q, i.e.

$$e \triangleq x - q. \quad (14.5)$$

Denote

$$e_1 \triangleq e, \quad e_2 \triangleq \dot{e}, \ldots, e_N \triangleq e^{(N-1)}. \quad (14.6)$$

In vector form,

$$\mathbf{q} \triangleq \{q_1, \ldots, q_N\}^T, \quad \mathbf{e} \triangleq \mathbf{x} - \mathbf{q} = \{e_1, \ldots, e_N\}^T. \quad (14.7)$$

Our objective is to develop a DFLS-based stable adaptive control system such that the plant output $x(t)$ follows the predefined bounded reference trajectory $q(t)$ under the constraint that all quantities involved must be bounded.

14.3 DFLS IDENTIFIER

Consider the dynamic fuzzy logic system (DFLS) as shown in Figure 14.1 [5], where for $p = 1, \ldots, P, w_p \in \mathcal{W}_p \subset R$ is an input variable, $\mathcal{W}_p$ is the *universe of discourse* [4] of the variable w_p '$1/S$' represents an integral operator, α is a positive constant, and $y \in R$ is the scalar output of the DFLS. The block 'FLS' represents a static fuzzy logic system illustrated in Figure 14.2 [20], where $W_p \in \mathcal{W}_p$ is the *fuzzy set* [4] generated from the crisp input w_p by a corresponding fuzzifier, $Z \in \mathcal{Z}$ is the fuzzy set induced by the fuzzy inference engine, $z \in \mathcal{Z} \subset R$ is a scalar variable defuzzified from Z, and $\mathcal{Z}$ is the universe of discourse of the variable z. Because a multi-input–multi-output (MIMO) system can often be decomposed into a group of multi-input–single-output (MISO) systems, only these latter will be considered in this chapter.

The dynamic fuzzy logic system of Figure 14.1 can be described by a differential equation of the form

$$\dot{y} = -\alpha y + f_{\text{FLS}}(\mathbf{w}). \quad (14.8)$$

The second term on the right-hand side of the above equation, $f_{\text{FLS}}(\mathbf{w})$, represents the output of an ordinary fuzzy logic system as in Figure 14.2, i.e. $z = f_{\text{FLS}}(\mathbf{w})$ and the vector, $\mathbf{w} \triangleq \{w_1, \ldots, w_P\}^T$, is its input.

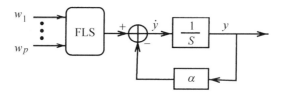

Figure 14.1 Dynamic fuzzy logic system

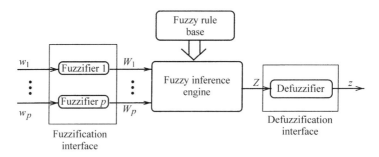

Figure 14.2 Basic structure of a fuzzy logic system

In the rest of this chapter we restrict the function $f_{\text{FLS}}(w)$, i.e., the ordinary fuzzy logic system, to be of the form

$$f_{\text{FLS}}(w) \triangleq \Theta^{\text{T}}(w)\bar{Y}, \tag{14.9}$$

where $\Theta(w)$ and $\bar{Y}$ are defined as

$$\begin{cases} \Theta(w) & \triangleq \{\theta_1(w), \ldots, \theta_I(w)\}^{\text{T}}, \\ \bar{Y} & \triangleq \{\bar{y}_1, \ldots, \bar{y}_I\}^{\text{T}}, \end{cases} \tag{14.10}$$

where I represents the size of the fuzzy rule base, and the function $\theta_i(w)$ is given by

$$\theta_i(w) \triangleq \frac{\prod_{p=1}^{P} \exp\left[-\frac{1}{2}\left(\frac{w_p - \bar{w}_p^i}{\sigma_p^i}\right)^2\right]}{\sum_{i=1}^{I} \prod_{p=1}^{P} \exp\left[-\frac{1}{2}\left(\frac{w_p - \bar{w}_p^i}{\sigma_p^i}\right)^2\right]}. \tag{14.11}$$

Equations (14.9)–(14.11) represent a static fuzzy logic system characterized by a singleton fuzzifier, centroid defuzzifier, the product implication rule, the sup–product compositional operator, Gaussian-type membership functions for primary fuzzy sets, and the algebraic product operation for t-norms [4, 20]. This specific static FLS has been theoretically demonstrated to be a universal approximator [20]. Details can be found in [4], [20], [22].

Substituting (14.9) into (14.8) yields

$$\dot{y} = -\alpha y + \Theta^{\text{T}}(w)\bar{Y}. \tag{14.12}$$

This DFLS will be used as an on-line identifier for the unknown state x_N of the nonlinear system in (14.3), and

$$w \triangleq \{w_1, \ldots, w_{N+1}\}^{\text{T}} \triangleq \{x_1, \ldots, x_N, u\}^{\text{T}}. \tag{14.13}$$

Note that $P \triangleq N + 1$. In this work the parameters α and Θ are to be defined in an off-line design process, leaving only $\bar{Y}$ as a free parameter vector to be adaptively tuned on-line.

Following a Lyapunov synthesis approach, the control and adaptive laws were derived. For compactness of presentation we present the control and the adaptive laws only without showing the derivation procedure, which is, in a sense, a process inverse to that of the stability proofs, which we do show. We verify the stability properties of the closed-loop system in the form of theorems, and give proofs of the theorems in detail in the appendix at the end of this chapter.

14.4 CONTROL LAW OF THE SYSTEM

For the system states to have the desired tracking performance and at the same time always be bounded, we specify a control law consisting of two components, u_c and u_s,

$$u \triangleq u_c - S_u u_s, \tag{14.14}$$

where

$$S_u \triangleq \begin{cases} 0, & \text{if } \|e\| < M_e, \\ 1, & \text{if } \|e\| \geq M_e. \end{cases} \tag{14.15}$$

M_e is a finite positive design parameter that bounds $\|e\|$.

The first component, u_c, is the usual control term for tracking performance when the system states are within a desired bound:

$$u_c \triangleq \frac{1}{\hat{g}} \left[\dot{q}N + \alpha q_N - \Theta^T(x, 0)\bar{Y} - \sum_{k=1}^{N-1} \alpha_k e_k \right], \tag{14.16}$$

where $\alpha_1, \ldots, \alpha_{N-1}$, and α are design parameters to be determined, $\hat{g}$ is the estimated value of the unknown function $g(x)$, and

$$\Theta(x, 0) \triangleq \Theta(x, u)|_{u=0}. \tag{14.17}$$

Let $\alpha_N \triangleq \alpha$, and

$$A \triangleq \begin{bmatrix} 0 & 1 & 0 & \cdots & 0 \\ 0 & 0 & 1 & \cdots & 0 \\ \vdots & \vdots & \vdots & & \vdots \\ 0 & 0 & 0 & \cdots & 1 \\ -\alpha_1 & -\alpha_2 & -\alpha_3 & \cdots & -\alpha_N \end{bmatrix}.$$

The parameters $\{\alpha_1, \ldots, \alpha_N\}$ are chosen in such a way that A is a stable matrix, i.e. the roots of the polynomial $\det[sI - A] = s^N + \alpha_N s^{N-1} + \cdots + \alpha_1$ are all located in the open left half of the complex plane.

The second component, u_s, is a supervisory control term that restrains the system states from drifting beuond a desired bound, and is only activated when system states actually reach this bound. This term follows the idea presented in [19]. Here, it is

$$u_s(t) \triangleq \text{sgn}(B^T Pe) \cdot \text{sgn}(g) \cdot \frac{1}{\varepsilon_g} \cdot [\alpha|x_N| + |\Theta^T(x, 0)\bar{Y}| + M_f + (M_g + |\hat{g}|) \cdot |u_c|], \tag{14.18}$$

where $B^T \triangleq [0, \ldots, 0, 1]_{1 \times N}$, M_f is a large positive constant that bounds $f(x)$, i.e. $\forall x \in \mathcal{X}, |f(x)| \leq M_f$, and the sign function, sgn, is defined as

$$\text{sgn}(x) \triangleq \begin{cases} 1, & \text{if } x \geq 0, \\ -1, & \text{if } x < 0, \end{cases} \tag{14.19}$$

and P is a positive definite symmetric matrix that satisfies the Lyapunov equation

$$A^T P + P^T A = -Q, \tag{14.20}$$

where Q is some give positive definite symmetric matrix.

14.5 ADAPTIVE LAW FOR THE PARAMETER VECTOR $\bar{Y}$

The expression for the state variable $\dot{x}_N$ in (14.3) is now rewritten as

$$\dot{x}_N = -\alpha x_N + \Theta^T(x, u)\bar{Y} - r(x, u, \Theta, \bar{Y}), \tag{14.21}$$

where

$$r(x, u, \Theta, \bar{Y}) \triangleq \Theta^T(x, u)\bar{Y} - \alpha x_N - f(x) - g(x)u, \tag{14.22}$$

which can be considered the static modeling error of the DFLS identifier. For bounded inputs, u, and states, x, there exists an optimal parameter vector, $\bar{Y}^*$, which minimizes this static modeling error denoted here $r(x, u, \Theta, \bar{Y}^*)$ [5],

$$\bar{Y}^* \triangleq \left\{ \bar{Y}' \mid \min_{\bar{Y}'} \left\{ \sup_{\{x,u\} \in \mathcal{W}} |r(x, u, \Theta, \bar{Y}')| \right\} \right\}, \tag{14.23}$$

where $\mathcal{W} \subset R^{N+1}$ is a compact set, and $\bar{Y}' \triangleq \{\bar{Y} : \|\bar{Y}\| \leq M_{\bar{Y}}\}$, for which

$$\sup_{\{x,u\} \in \mathcal{W}} |r(x, u, \Theta, \bar{Y}')| \leq M_r. \tag{14.24}$$

$M_{\bar{Y}}$ and M_r are positive constants that bound the norms of $\bar{Y}$ and $r(x, u, \Theta, \bar{Y}^*)$, respectively. Here, we know that such an optimal parameter vector exists for bounded inputs and states [5, Lemma 1], but we have no guidance on how to find it. We wish to specify an adaptive law for $\bar{Y}$, such that $\bar{Y}^*$ can be satisfactorily approximated.

Replacing $\bar{Y}$ by $\bar{Y}^*$ in (14.21) results in

$$\dot{x}_N = -\alpha x_N + \Theta^T(x, u)\bar{Y}^* - r(x, u, \Theta, \bar{Y}^*). \tag{14.25}$$

Subtracting (14.25) from (14.12) yields

$$\dot{\xi} = -\alpha \xi + \Theta^T(x, u)\Delta_{\bar{Y}} + r(x, u, \Theta, \bar{Y}^*), \tag{14.26}$$

where ξ is the identification error for the state variable x_N,

$$\xi \triangleq y - x_N, \tag{14.27}$$

and the vector $\Delta_{\bar{Y}}$ is the parameter estimation error

$$\Delta_{\bar{Y}} \triangleq \bar{Y} - \bar{Y}^*. \tag{14.28}$$

Using (14.14) in (14.3) yields

$$\begin{cases} \dot{x}_1 = x_2, \\ \vdots \\ \dot{x}_{N-1} = x_N, \\ \dot{x}_N = f(\boldsymbol{x}) + \hat{g}u_c - \Delta_g u_c - S_u g u_s, \end{cases} \quad (14.29)$$

where Δ_g is the estimation error of the function $g(\boldsymbol{x})$,

$$\Delta_g \triangleq \hat{g} - g. \quad (14.30)$$

Substituting (14.16) into (14.29) and considering (14.6) gives

$$\begin{cases} \dot{e}_1 = e_2, \\ \vdots \\ \dot{e}_{N-1} = e_N, \\ \dot{e}_N = -\alpha e_N - \sum_{k=1}^{N-1} \alpha_k e_k - \Theta^T(\boldsymbol{x},0)\Delta_{\bar{Y}} - \Delta_g u_c - r(\boldsymbol{x},0,\Theta,\bar{Y}^*) - S_u g u_s, \end{cases} \quad (14.31)$$

where

$$r(\boldsymbol{x},0,\Theta,\bar{Y}) \triangleq r(\boldsymbol{x},u,\Theta,\bar{Y})|_{u=0} = \Theta^T(\boldsymbol{x},0)\bar{Y} - \alpha x_N - f(\boldsymbol{x}). \quad (14.32)$$

Noting that $\alpha = \alpha_N$ equation (14.31) can be written in matrix form as

$$\dot{\boldsymbol{e}} = A\boldsymbol{e} + B[-\Theta^T(\boldsymbol{x},0)\Delta_{\bar{Y}} - \Delta_g u_c - r(\boldsymbol{x},0,\Theta,\bar{Y}^*) - S_u g u_s]. \quad (14.33)$$

The adaptive law for $\bar{Y}$ is obtained via the 'inverse' of the stability demonstration procedure as mentioned earlier,

$$\dot{\bar{Y}} = -H[\Theta(\boldsymbol{x},u)h\xi - \Theta(\boldsymbol{x},0)(B^T Pe)] - S\beta H\bar{Y}, \quad (14.34)$$

where H is a constant positive definite symmetric matrix, and h is a positive constant that weights the identification error. The last term on the right-hand side of (14.34) represents a projection algorithm modification [1, 2, 14], to avoid circumstances under which the parameter vector, $\bar{Y}$, becomes too large or even drifts to infinity [14]. The 'on' or 'off' status of this term is controlled by the switch S, which is defined as

$$S = \begin{cases} 0, & \text{if} \quad \|\bar{Y}\| < M_{\bar{Y}}, \\ & \text{or} \quad \|\bar{Y}\| = M_{\bar{Y}} \text{ and } \bar{Y}^T H[\Theta(\boldsymbol{x},u)h\xi - \Theta(\boldsymbol{x},0)(B^T Pe)] \geq 0; \\ & \text{or} \quad \|\bar{Y}\| > M_{\bar{Y}} \text{ and } \bar{Y}^T H[\Theta(\boldsymbol{x},u)h\xi - \Theta(\boldsymbol{x},0)(B^T Pe)] > 0; \\ 1, & \text{otherwise;} \end{cases} \quad (14.35)$$

β is a positive design parameter which satisfies

$$\begin{cases} \beta \geq \dfrac{\bar{Y}^T H[\Theta(\boldsymbol{x},0)(B^T Pe) - \Theta(\boldsymbol{x},u)h\xi]}{\bar{Y}^T H\bar{Y}}, & \text{if } \|\bar{Y}\| = M_{\bar{Y}} \text{ and} \\ & \bar{Y}^T H[\Theta(\boldsymbol{x},u)h\xi - \Theta(\boldsymbol{x},0)(B^T Pe)] < 0; \\ \beta > \dfrac{\bar{Y}^T H[\Theta(\boldsymbol{x},0)(B^T Pe) - \Theta(\boldsymbol{x},u)h\xi]}{\bar{Y}^T H\bar{Y}}, & \text{if } \|\bar{Y}\| > M_{\bar{Y}} \text{ and} \\ & \bar{Y}^T H[\Theta(\boldsymbol{x},u)h\xi - \Theta(\boldsymbol{x},0)(B^T Pe)] \leq 0. \end{cases} \quad (14.36)$$

14.6 ADAPTIVE LAW FOR $\hat{g}$

Since the function $g(x)$ is of a static nature, a static FLS is used here as its estimator. Consider a FLS similar to the one defined in equations (14.9)–(14.11). Let

$$\hat{g} = \Psi^{T}(x)\bar{G}, \tag{14.37}$$

where

$$\begin{cases} \Psi(x) \triangleq \{\psi_1(x), \ldots, \psi_J(x)\}^{T}, \\ \bar{G} \triangleq \{\bar{g}_1, \ldots, \bar{g}_J\}^{T}, \end{cases} \tag{14.38}$$

J is the size of the fuzzy rule base, and the membership function, ψ_j, is defined as

$$\psi_j(x) \triangleq \frac{\prod_{n=1}^{N} \exp\left[-\left(\frac{x_n - \bar{x}_{gn}^j}{\sigma_{gn}^j}\right)^2\right]}{\sum_{j=1}^{J} \prod_{n=1}^{N} \exp\left[-\left(\frac{x_n - \bar{x}_{gn}^j}{\sigma_{gn}^j}\right)^2\right]}. \tag{14.39}$$

Equations (14.37)–(14.39) represent a static FLS characterized by a singleton fuzzifier, centroid defuzzifier, sup-product compositional rule of inference, algebraic product operation for fuzzy implication and t-norm, and Gaussian-type membership functions for primary fuzzy sets.

In (14.37), $\Psi(x)$ is designed off-line and $\bar{G}$ is the parameter vector adjusted on-line. If expert knowledge of $g(x)$ is available, it can be incorporated by choosing the parameter values of $\Psi(x)$ and the initial values of $\bar{G}$ accordingly. If there is no expert knowledge of $g(x)$ at all, we can determine the parameter values of $\Psi(x)$ through some other means, e.g. trial and error, and set the initial values of $\bar{G}$ randomly. In any situation $\bar{G}$ is further adaptively tuned in such a way that both the control objective and the stability properties are satisfied, except that less training may be required with some expert knowledge available [5]. In what follows we present such an adaptive law. But we first impose an assumption on $\bar{G}$, which is necessary for the system to be stable.

Assumption 6.1 $\forall x \in \mathcal{X}$, if $\|\bar{G}\| \geq M_{\bar{G}}$, then $|\hat{g}| \geq \varepsilon_g$, where $M_{\bar{G}}$ is some large positive constant.

Again, by the Lyapunov synthesis approach, the adaptive law for the parameter vector, $\bar{G}$, is

$$\dot{\bar{G}} = \gamma \Psi(x)(B^{T}Pe)u_c - S_g \beta_g \gamma \bar{G}, \tag{14.40}$$

where γ is a positive gain, S_g is a switch to control the 'on' or 'off' status of the projection modification term, and β_g is a positive design parameter:

$$\{S_g, \beta_g\}^{T} \triangleq \begin{cases} \{S_g^{U}, \beta_g^{U}\}^{T}, & \text{if } |\hat{g}| > \varepsilon_g, \\ \{-S_g^{L}, \beta_g^{L}\}^{T}, & \text{if } |\hat{g}| = \varepsilon_g \text{ and } \|\bar{G}\| < M_{\bar{G}}, \end{cases} \tag{14.41}$$

where

$$S_g^U \triangleq \begin{cases} 0, & \text{if } \|\bar{G}\| < M_{\bar{G}}, \\ & \text{or } \|\bar{G}\| = M_{\bar{G}} \text{ and } \bar{G}^T \Psi(x)(B^T Pe)u_c \leq 0, \\ & \text{or } \|\bar{G}\| > M_{\bar{G}} \text{ and } \bar{G}^T \Psi)(x)(B^T Pe)u_c < 0, \\ 1, & \text{otherwise}; \end{cases} \quad (14.42)$$

$$S_g^L \triangleq \begin{cases} 0, & \text{if } \hat{g}(B^T Pe)u_c \geq 0, \\ 1, & \text{if } \hat{g}(B^T Pe)u_c < 0; \end{cases} \quad (14.43)$$

$$\begin{cases} \beta_g^U \geq \dfrac{\bar{G}^T \Psi(x)(B^T Pe)u_c}{\bar{G}^T \bar{G}}, & \text{if } \|\bar{G}\| = M_{\bar{G}} \text{ and } \bar{G}^T \Psi(x)(B^T Pe)u_c > 0, \\ \beta_g^U > \dfrac{\bar{G}^T \Psi(x)(B^T Pe)u_c}{\bar{G}^T \bar{G}}, & \text{if } \|\bar{G}\| > M_{\bar{G}} \text{ and } \bar{G}^T \Psi(x)(B^T Pe)u_c \geq 0; \end{cases} \quad (14.44)$$

$$\beta_g^L \geq -\dfrac{\hat{g}(\Psi^T \Psi)(B^T Pe)u_c}{\hat{g}^2}. \quad (14.45)$$

$M_{\bar{G}}$ is a positive design constant that bounds $\|\bar{G}\|$ and satisfies

$$M_{\bar{G}} \leq \dfrac{1}{\sqrt{J}} M_g. \quad (14.46)$$

where J is the size of the vectors Ψ and $\bar{G}$ as in (14.38).

It is possible that $|\hat{g}|$ exceeds its lower bound, ε_g, during training, as the training is implemented computationally as

$$\hat{g}(t + dt) = \hat{g}(t) + \dot{\hat{g}}(t)dt = \hat{g}(t) + \Psi^T \dot{\bar{G}}(t)dt \quad (14.47)$$

(where $\dot{\bar{G}}(t)$ is given by the training law, (14.40), so that a $|\hat{g}(t)|$ very close to but still within its lower bound, ε_g, produces a $|\hat{g}(t + dt)|$ which can possibly exceed this prespecified lower bound. In this situation, we can change the value of the gain, γ and obtain a new value of $\dot{\bar{G}}(t)$, which results in a new value of $\hat{g}(t + dt)$ (such that $|\hat{g}(t + dt)| \geq \varepsilon_g$). Another choice could simply be to let $\hat{g}(t + dt) \triangleq \text{sgn}[\hat{g}(t)]\varepsilon_g$ to avoid this situation, and carry on the adaptive calculation with new samples.

14.7 STABILITY PROPERTIES OF THE DFLS CONTROL ALGORITHM

The stability properties of the DFLS based adaptive control system described in the previous sections are summarized as follows.

Theorem 7.1 *Consider an unknown non-linear dynamic system described by (14.1), which is subject to the control law, (14.14). If the DFLS of (14.12) is used to identify the state, x_N, using adaptive law, (14.34), for tuning its parameter vector, $\bar{Y}$, and if furthermore the unknown function, $g(x)$, is estimated by a FLS, with adaptive law (14.40), then the closed loop system has the following stability properties:*

7.1.1 $\|\bar{Y}\| \leq M_{\bar{Y}}$, where $M_{\bar{Y}}$ is defined in Section 14.5;
7.1.2 $\varepsilon_g \leq |\hat{g}| \leq M_g$, where ε_g, and M_g are defined in Section 14.2;

7.1.3 $\|\boldsymbol{e}\| \leq M_e, \|\boldsymbol{x}\| \leq M_e + \|\boldsymbol{q}\|,$ and $|\xi| \leq M_\xi,$ where $\boldsymbol{q}$ and $\boldsymbol{e}$ are defined in (14.7), and M_ξ is a positive constant;

7.1.4

$$|u(t)| \leq \frac{1}{\varepsilon_g} \cdot \left[\left(2 + \frac{M_g + |\hat{g}(t)|}{\varepsilon_g} \right) \cdot (\sqrt{I} M_{\bar{Y}} + \alpha |x_N(t)|) \right.$$
$$\left. + \left(1 + \frac{M_g + |\hat{g}(t)|}{\varepsilon_g} \right) \cdot (|\dot{q}_N(t)| + \|\boldsymbol{a}\| M_e) + M_f \right],$$

where $\boldsymbol{a} \triangleq \{\alpha_1, \ldots, \alpha_N\}^{\mathrm{T}}$, I is the size of the vector Θ and M_f is a positive constant that bounds $|f(\boldsymbol{x})|$;

7.1.5

$$\int_0^t \|\boldsymbol{e}(\tau)\|^2 \mathrm{d}\tau \leq b + \eta \int_0^t [r^2(\boldsymbol{x}, 0, \Theta, \bar{\boldsymbol{Y}}^*) + r^2(\boldsymbol{x}, u, \Theta, \bar{\boldsymbol{Y}}^*) + \Delta_g^2] \mathrm{d}\tau,$$

$$\int_0^t \xi^2(\tau) \mathrm{d}\tau \leq b + \eta \int_0^t [r^2(\boldsymbol{x}, 0, \Theta, \bar{\boldsymbol{Y}}^*) + r^2(\boldsymbol{x}, u, \Theta, \boldsymbol{Y}^*) + \Delta_g^2] \mathrm{d}\tau,$$

where b and η are positive constants;

7.1.6 if $r(t) \in \mathcal{L}_2[0, \infty)$ and $\Delta_g \in \mathcal{L}_2[0, \infty)$, i.e. $(\int_0^\infty \|r(\tau)\|^2 \mathrm{d}\tau)^{1/2} < \infty$ and $(\int_0^\infty \Delta_g^2 \mathrm{d}\tau)^{1/2} < \infty$, then

$$\lim_{t \to \infty} \|\boldsymbol{e}(t)\| = 0, \qquad \lim_{t \to \infty} |\xi(t)| = 0.$$

The proofs are given in the appendix.

14.8 ILLUSTRATIVE APPLICATION

Consider the following non-linear system:

$$\dot{x} = \frac{\cos(x)}{1+x^2} + \frac{2}{1+\cos^2(x)} u. \tag{14.48}$$

Our objective is to command the state x to track a desired trajectory, $q(t)$, by using the DFLS-based adaptive control scheme developed in previous sections, where this trajectory is

$$q \triangleq 2\sin(t). \tag{14.49}$$

A DFLS identifier, labeled D, is used to identify x, i.e.

$$y = D(x, u), \tag{14.50}$$

where y represents the identifier's output and is the estimate of x. The tracking error, e, and identification error, ξ, are

$$\begin{cases} e \triangleq x - q, \\ \xi \triangleq y - x. \end{cases} \tag{14.51}$$

The identifier, D, has two inputs, $x \in \mathcal{X} \subset R$ and $u \in \mathcal{U} \subset R$, and one output, $y \in \mathcal{Y} \subset R$, where $\mathcal{X}, \mathcal{U}$ and $\mathcal{Y}$ are universes of discourse of the lingusitic variables, x, u and y, respectively. For conciseness of expression, let $\boldsymbol{w} \triangleq \{x, u\}^{\mathrm{T}} \triangleq \{w_1, w_2\}^{\mathrm{T}}$. In both $\mathcal{X}$ and $\mathcal{U}$, five primary fuzzy sets, A_{1j_1} and A_{2j_2}, $j_1, j_2 = 1, \ldots, 5$, are defined. Gaussian-type membership functions are used for all the primary fuzzy sets,

$$\mu A_{p1}(w'_p) \triangleq \begin{cases} 1, & \text{if } w'_p < \bar{w}_{p1}, \\ \exp\left[-\frac{1}{2}\left(\frac{w'_p - \bar{w}_{p1}}{\sigma_{p1}}\right)^2\right], & \text{if } w'_p \geq \bar{w}_{p1}; \end{cases}$$

$$\mu A_{pj_p}(w'_p) \triangleq \exp\left[-\frac{1}{2}\left(\frac{w'_p - \bar{w}_{pj_p}}{\sigma_{pj_p}}\right)^2\right], \quad j_1, j_2 = 2, 3, 4;$$

$$\mu A_{p5}(w'_p) \triangleq \begin{cases} \exp\left[-\frac{1}{2}\left(\frac{w'_p - \bar{w}_{p5}}{\sigma_{p5}}\right)^2\right], & \text{if } w'_p < \bar{w}_{p5}, \\ 1, & \text{if } w'_p \geq \bar{w}_{p5}. \end{cases} \quad (14.52)$$

Based on experience, we set the shape parameters of all the primary fuzzy sets to 0.45, i.e.

$$\sigma_{pj_p} \triangleq 0.45, \quad (14.53)$$

and define the position parameters, $\bar{w}_{pj_p}$, as

$$\{\bar{w}_{p1}, \bar{w}_{p2}, \bar{w}_{p3}, \bar{w}_{p4}, \bar{w}_{p5}\}^T \triangleq \{-2, -1, 0, 1, 2\}^T. \quad (14.54)$$

In (14.52), $w'_p, p = 1, 2$, are filtered values of inputs, w_p. The purpose of the prefilters is to transform the input values of the system states into desired regions. Here, they are defined as

$$\begin{cases} w'_1 = f_1(w_1) \triangleq w_1 \cdot \frac{1}{2}, \\ w'_2 = f_2(w_2) \triangleq w_2 \cdot \frac{1}{25}. \end{cases} \quad (14.55)$$

The output of D, y, is defined via the DFLS of (14.12), i.e.

$$D(w') : \dot{y} = -\alpha y + \Theta^T(w')\bar{Y}. \quad (14.56)$$

In this equation,

$$\begin{cases} \Theta(w') \triangleq \{\theta_1(w'), \theta_2(w'), \ldots, \theta_{25}(w')\}^T, \\ \bar{Y} \triangleq \{\bar{y}_1, \bar{y}_2, \ldots, \bar{y}_{25}\}^T, \end{cases} \quad (14.57)$$

and

$$\theta_i(w') = \frac{\prod_{p=1}^{2} \mu_{A_p^i}(w'_p)}{\sum_{i=1}^{25} \prod_{p=1}^{2} \mu_{A_p^i}(w'_p)}, \quad i = 1, \ldots, 25, \quad (14.58)$$

where

$$A_1^i \in \{A_{1j_1}; j_1 = 1, \ldots, 5\}, \qquad A_2^i \in \{A_{2j_2}; j_2 = 1, \ldots, 5\}. \quad (14.59)$$

$\bar{Y}$ is the free parameter vector to be adaptively adjusted using the training law, (14.34),

$$\dot{\bar{Y}} = -H[\Theta(x', u')h\xi - \Theta(x', 0)(B^T Pe)] - S\beta H\bar{Y}. \quad (14.60)$$

Setting $A = [-\alpha]$ and $B = [1]$, let $Q \triangleq [2\alpha]$, then $P = [1]$. Furthermore, let $h \triangleq 1$, then

$$\dot{\bar{Y}} = -H[\Theta(x', u')\xi - \Theta(x', 0)e] - S\beta H\bar{Y}. \quad (14.61)$$

294 FUZZY LOGIC CONTROL OF NON-LINEAR SYSTEMS

In computer implementation, the derivative is approximated by a difference, i.e.

$$\bar{Y} \triangleq \frac{\bar{Y}(kT+T) - \bar{Y}(kT)}{T}, \quad k = 0, 1, 2, \ldots, \tag{14.62}$$

where T is the time incremental step. Using (14.62) in (14.61) yields

$$\bar{Y}(kT+T) = \bar{Y}(kT) - T\mathbf{H}[\Theta(x'(kT), u'(kT))\xi(kT) - \Theta(x'(kT), 0)e(kT) + S\beta\bar{Y}(kT)]. \tag{14.63}$$

Based on experience, we define the gain matrix $\mathbf{H}$ to be a diagonal matrix with all diagonal elements equal to 4. We set α to 10, the time increment, T, to 0.05 s, the bound of the parameter vector $M_{\bar{Y}}$ to 10^3, and the bound of the function f to $M_f = 1$. The initial values of the elements of the parameter vector, $\bar{Y}(0)$, are random numbers uniformly distributed in $(-1, 1)$, and we assume the system to be controlled is initially at rest, i.e. $x(0) = 0$.

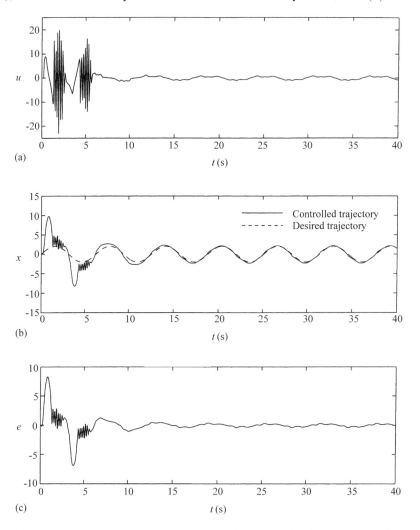

Figure 14.3 Trajectory control. (a) Controller output; (b) controlled trajectory, x; (c) trajectory error, e. *Note*: Adaptation terminates at $t = 20$ s

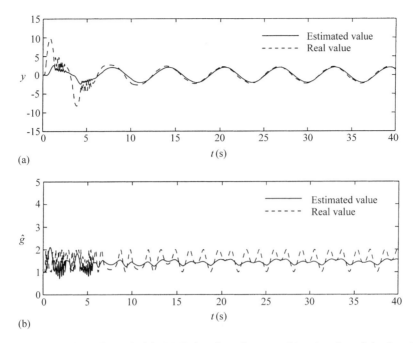

Figure 14.4 Estimation of x and $g(x)$. (a) Estimation of state x; (b) estimation of the function $g(x)$

The unknown gain function, $g(x)$, is estimated with a FLS, i.e.

$$\hat{g} \triangleq \Psi^T(x')\bar{G}, \tag{14.64}$$

where $\hat{g}$ represents the identifier output and is the estimate of g. The FLS has one input, $w'_1 = x'$, and one output, $\hat{g}$. Five primary fuzzy sets, $B_j, j = 1, \ldots, 5$, are defined in the input universe of discourse of the FLS. Gaussian-type membership functions are used for all the primary fuzzy sets, which are the same as those given in (14.52)–(14.54). The adaptive law of this FLS is given in (14.40),

$$\dot{\bar{G}} = \gamma e u_c \Psi(w'_1) - S_g \beta_g \gamma \bar{G}, \tag{14.65}$$

where $\gamma \triangleq 0.1$. Let $M_g \triangleq 100, \varepsilon_g \triangleq 0.1, M_{\bar{G}} \triangleq \frac{1}{\sqrt{5}} \cdot M_g$, and all the elements of the parameter vector, $\bar{G}(0)$, be set to 1. The controller output, u, the controlled trajectory of system output, x, and the tracking error, e, are shown in Figure 14.3(a)–(c), respectively, where the adaptation process stops at $t = 20$ s. The estimations for the system state, x, and the unknown function, g, are shown in Figure 14.4(a), (b), respectively. It is observed that the system performance is very good.

14.9 CONCLUDING REMARKS

1. A DFLS-based indirect adaptive control scheme was developed in this chapter, via a Lyapunov synthesis approach for a class of non-linear systems in companion form, equation (14.1). Closed-loop system performance and stability properties are analysed, with the results supported by proofs. The application of this DFLS-based adaptive control

algorithm to a non-linear system was illustrated in a simulation example, where the detailed design procedure of the controller was presented and satisfactory system performance observed.

2. The development of the DFLS-based indirect control scheme consists of off-line design (for Θ, etc.), and on-line adaptation (for $\bar{Y}$). In off-line design, *ad hoc* analysis based on physical intuition is required. This currently is an active but immature area of research; that is, explicit procedures for determining the optimal values of the parameters have yet to be developed. After off-line design, the parameter vector $\bar{Y}$ is adjusted on-line with the adaptive law given in (14.34). If satisfactory results are not obtained by adjusting $\bar{Y}$, then one must return to the off-line design process to modify the off-line designed parameters and repeat the entire procedure.

3. If there is no expert knowledge about the system state, x_N, and the gain function, g, available, then the initial values of the parameter vectors, $\bar{Y}$ of (14.12), and $\bar{G}$ of (14.37), must be assigned randomly. With some expert knowledge, these initial values can be determined accordingly, resulting in improved performance and requiring less training [5]. In any situation, the closed-loop system performance is at least as good as that specified in Theorem 7.1.

4. In the DFLS-based adaptive control algorithm a measurable state variable, x_N, was identified, rather than identifying the unknown function, $f(x)$, as in the case of static FLS-based control algorithms [15, 20]. It is easier to acquire knowledge of the dynamic behavior of a measurable state than to acquire knowledge of the unknown function, $f(x)$, which is not directly accessible to us. In addition, in the DFLS-based control algorithm, the system control input, u, is explicitly included as an input of the DFLS, where any expert knowledge about the relationship between the system input, u, and the system state, x_N, can be explicitly incorporated. This is in contrast to the situation of the static FLS based control algorithm [15, 20], where u is not explicitly included as an input of the FLS.

5. Assumption 6.1 is required to ensure the system stability properties specified in Theorem 7.1. Since ε_g is a very small number and $M_{\bar{G}}$ is a very large number, this assumption is not unreasonable. Here, the elements of the parameter vector, $\bar{G}$, are not restricted to have the same sign as $g(x)$, as was the case in [20], which has the effect of reducing the search region of the elements of $\bar{G}$ by half and can produce a poorer set of parameters. If Assumption 6.1 is violated in practice, then it is necessary to change the initial values of $\bar{G}$ as well as those of the off-line determined parameters, and repeat the training process.

APPENDIX: PROOF OF THEOREM 7.1

A.1 Proof of Theorem 7.1.1

This proof is similar to that in [6, Proof of Theorem 1.1], and therefore is omitted here.

A.2 Proof of Theorem 7.1.2

We consider two cases:

Case 1: $|\hat{g}| > \varepsilon_g$. Consider the Lyapunov function candidate

$$V_{\bar{G}} = 1/2 \bar{G}^T \bar{G}. \tag{14.66}$$

APPENDIX: PROOF OF THEOREM 7.1

Differentiating both sides of (14.66) yields

$$\dot{V}_{\bar{G}} = \bar{G}^{\mathrm{T}}\dot{\bar{G}}. \tag{14.67}$$

Using (14.40) in (14.67), and considering (14.41) gives

$$\dot{V}_{\bar{G}} = \gamma \bar{G}^{\mathrm{T}}\Psi(x)(B^{\mathrm{T}}Pe)u_c - S_g^U \beta_g^U \gamma \bar{G}^{\mathrm{T}}\bar{G}. \tag{14.68}$$

Let $0 < M_{\bar{G}} < \infty$ be the nominal bound of the norm of $\bar{G}$. We now consider different situations.

- $\|\bar{G}\| < M_{\bar{G}}$: This is the desired situation, and the norm of the parameter vector, $\bar{G}$, is of no concern to us.
- $\|\bar{G}\| = M_{\bar{G}}$: Considering (14.42),
  — for $\bar{G}^{\mathrm{T}}\Psi(x)(B^{\mathrm{T}}Pe)u_c \leq 0$, $S_g^U = 0$,

$$\dot{V}_{\bar{G}} = \gamma \bar{G}^{\mathrm{T}}\Psi(x)(B^{\mathrm{T}}Pe)u_c \leq 0; \tag{14.69}$$

  — for $\bar{G}^{\mathrm{T}}\Psi(x)(B^{\mathrm{T}}Pe)u_c > 0$, $S_g^U = 1$, and

$$\dot{V}_{\bar{G}} = \gamma \bar{G}^{\mathrm{T}}\Psi(x)(B^{\mathrm{T}}Pe)u_c - \beta_g^U \gamma \bar{G}^{\mathrm{T}}\bar{G}, \tag{14.70}$$

$$\beta_g^U \geq \frac{\bar{G}^{\mathrm{T}}\Psi(x)(B^{\mathrm{T}}Pe)u_c}{\bar{G}^{\mathrm{T}}\bar{G}}, \tag{14.71}$$

or

$$\dot{V}_{\bar{G}} \leq 0. \tag{14.72}$$

Equations (14.69) and (14.72) indicate that $\|\bar{G}\|$ does not increase, and therefore the boundedness of $\|\bar{G}\|$ can be concluded.

- $\|\bar{G}\| > M_{\bar{G}}$: In this case, $\|\bar{G}\|$ exceeds its nominal bound, $M_{\bar{G}}$, which may occur during the training process, and as the training is implemented computationally as

$$\bar{G}(t+\mathrm{d}t) = \bar{G}(t) + \dot{\bar{G}}(t)\mathrm{d}t \tag{14.73}$$

(where $\dot{\bar{G}}(t)$ is given by the training law (14.40), then in the situation in which $\|\bar{G}(t)\|$ is very close to the nominal bound, $M_{\bar{G}}$, but still within this bound, the resulting $\|\bar{G}(t+\mathrm{d}t)\|$ can exceed the prespecified nominal bound by the norm of the training increment, $\dot{\bar{G}}(t)\mathrm{d}t$, at most. Considering (14.42), then,

  — for $\bar{G}^{\mathrm{T}}\Psi(x)(B^{\mathrm{T}}Pe)u_c < 0$, $S_g^U = 0$,

$$\dot{V}_{\bar{G}} = \gamma \bar{G}^{\mathrm{T}}\Psi(x)(B^{\mathrm{T}}Pe)u_c < 0; \tag{14.74}$$

  — for $\bar{G}^{\mathrm{T}}\Psi(x)(B^{\mathrm{T}}Pe)u_c \geq 0$, $S_g^U = 1$, and

$$\dot{V}_{\bar{G}} = \gamma \bar{G}^{\mathrm{T}}\Psi(x)(B^{\mathrm{T}}Pe)u_c - \beta_g^U \gamma \bar{G}^{\mathrm{T}}\bar{G}, \tag{14.75}$$

$$\beta_g^U > \frac{\bar{G}^{\mathrm{T}}\Psi(x)(B^{\mathrm{T}}Pe)u_c}{\bar{G}^{\mathrm{T}}\bar{G}}, \tag{14.76}$$

or
$$\dot{V}_{\bar{G}} < 0. \tag{14.77}$$

Equations (14.74) and (14.77) indicate that as soon as $\|\bar{G}\|$ exceeds its nominal bound, $M_{\bar{G}}$, it immediately returns within this bound. Therefore, this excess is *small* and *temporary*. Formally, denote

$$\delta M_{\bar{G}} \triangleq \sup_{\mathbf{x} \in \mathcal{X}} \|\dot{\bar{G}}(t) \mathrm{d}t\|, \tag{14.78}$$

and let

$$M'_{\bar{G}} \triangleq M_{\bar{G}} + \delta M_{\bar{G}}, \tag{14.79}$$

then we have

$$\|\bar{G}(t + \mathrm{d}t)\| \leq M'_{\bar{G}}, \tag{14.80}$$

which again indicates the boundedness of $\|\bar{G}\|$.

For simplicity but without loss of generality, we shall continue to use $M_{\bar{G}}$ rather than $M'_{\bar{G}}$, which is equivalent to implicitly specifying that $M_{\bar{G}} \triangleq \max\{M_{\bar{G}}, M'_{\bar{G}}\}$.

Therefore, in the situation in Case 1, i.e. $|\hat{g}| > \varepsilon_g$, it is concluded that

$$\|\bar{G}\| \leq M_{\bar{G}}. \tag{14.81}$$

Considering now (14.37)

$$|\hat{g}| = |\Psi^T \bar{G}| \leq \|\Psi\| \cdot \|\bar{G}\|, \tag{14.82}$$

and from (14.39) it is clear that of all the elements of the vector Ψ, none is larger than one, i.e. $\psi_j \leq 1$, for $j = 1, \ldots, J$. Therefore,

$$\|\Psi\| = \sqrt{\sum_{j=1}^{J} \psi_j^2} \leq \sqrt{J}. \tag{14.83}$$

Using (14.81) and (14.83) in (14.82) yields

$$|\hat{g}| \leq \sqrt{J} M_{\bar{G}}, \tag{14.84}$$

and by considering (14.46), i.e.

$$M_{\bar{G}} \leq \frac{1}{\sqrt{J}} M_g, \tag{14.85}$$

we have

$$|\hat{g}| \leq M_g. \tag{14.86}$$

Case 2: $|\hat{g}| = \varepsilon_g$ and $\|\bar{G}\| < M_{\bar{G}}$. In this situation, $\|\bar{G}\|$ is of no concern to us. Consider the Lyapunov function candidate

$$V_{\hat{g}} = \tfrac{1}{2} \hat{g}^2. \tag{14.87}$$

Differentiating both sides of (14.87) yields

$$\dot{V}_{\hat{g}} = \hat{g} \dot{\hat{g}} = \hat{g} \Psi^T \dot{\bar{G}}. \tag{14.88}$$

Referring now to (14.40) and (14.41), we have

$$\dot{V}_{\hat{g}} = \gamma \hat{g} \Psi^T \Psi (B^T P e) u_c + S_g^L \beta_g^L \gamma \hat{g} (\Psi^T \bar{G})$$
$$= \gamma \hat{g} \Psi^T \Psi (B^T P e) u_c + S_g^L \beta_g^L \gamma \hat{g}^2. \tag{14.89}$$

Furthermore, referring to (14.43) and (14.45),

- for $\hat{g}(B^T P e) u_c \geq 0$, $S_g^L = 0$,

$$\dot{V}_{\hat{g}} = \gamma \hat{g} \Psi^T \Psi (B^T P e) u_c \geq 0; \tag{14.90}$$

- for $\hat{g}(B^T P e) u_c < 0$, $s_g^L = 1$, and

$$\dot{V}_{\hat{g}} = \gamma \hat{g} \Psi^T \Psi (B^T P e) u_c + \gamma \beta_g^L \hat{g}^2, \tag{14.91}$$

$$\beta_g^L \geq -\frac{(\Psi^T \Psi) \hat{g} (B^T P e) u_c}{\hat{g}^2}, \tag{14.92}$$

or

$$\dot{V}_{\hat{g}} \geq 0. \tag{14.93}$$

Equations (14.90) and (14.93) indicate that $|\hat{g}|$ does not decrease, and therefore, the lower boundedness of $|\hat{g}|$ is concluded.

This concludes the proof of Theorem 7.1.2.

A.3 Proof of Theorem 7.1.3

Using (14.32) in (14.33) yields

$$\dot{e} = Ae + B[-\Theta^T(x, 0)\bar{Y} + \alpha x_N + f(x) + g(x)u_c - \hat{g}u_c - S_u g u_s]. \tag{14.94}$$

Consider the Lyapunov function candidate

$$V_e = \tfrac{1}{2} e^T P e. \tag{14.95}$$

Differentiating both sides of (14.95) yields

$$\dot{V}_e = \tfrac{1}{2} \dot{e}^T P e + \tfrac{1}{2} e^T P \dot{e}. \tag{14.96}$$

Using (14.94) in (14.96) yields

$$\dot{V}_e = \tfrac{1}{2} e^T (A^T P + P^T A) e + (B^T P e)[\alpha x_N + f(x) + g(x) u_c - \Theta^T(x, 0) \bar{Y} - \hat{g} u_c]$$
$$\quad - S_u g (B^T P e) u_s$$
$$= -\tfrac{1}{2} e^T Q e + (B^T P e)[\alpha x_N + f(x) + g(x) u_c - \Theta^T(x, 0) \bar{Y} - \hat{g} u_c]$$
$$\quad - S_u g (B^T P e) u_s. \tag{14.97}$$

Then

$$\dot{V}_e \leq -\tfrac{1}{2} e^T Q e + |B^T P e| \cdot [\alpha |x_N| + |\Theta^T(x,0) \bar{Y}| + M_f + (M_g + |\hat{g}|) \cdot |u_c|]$$
$$\quad - S_u g (B^T P e) u_s \tag{14.98}$$

where M_f is the upper bound for $|f(x)|$. If $\|e\| \geq M_e$, $S_u = 1$. Using (14.18) with (14.98) yields

$$\dot{V}_e \leq -\tfrac{1}{2} e^T Q e + \left(1 - \frac{|g|}{\varepsilon_g}\right) \cdot |B^T P e| \\ \times [\alpha |x_N| + |\Theta^T(x,0)\bar{Y}| + M_f + (M_g + |\hat{g}|) \cdot |u_c|]. \quad (14.99)$$

Since $|g| \geq \varepsilon_g$, then for $\|e\| \geq M_e$,

$$\dot{V}_e \leq -\tfrac{1}{2} e^T Q e \leq 0. \quad (14.100)$$

Therefore, when $\|e\| \geq M_e$, $\dot{V}_e \leq 0$, which indicates the boundedness of $\|e\|$, i.e.

$$\|e\| \leq M_e. \quad (14.101)$$

Furthermore, since $e = x - q$,

$$\|x\| \leq \|e\| + \|q\| \leq M_e + \|q\|. \quad (14.102)$$

Now, consider (14.12)

$$\dot{y} = -\alpha y + \Theta^T(x,u)\bar{Y} \quad (14.103)$$

The solution to (14.103) is

$$y(t) = e^{-\alpha t} y(0) + \int_0^t e^{-\alpha(t-\tau)} \Theta^T(x,u)\bar{Y} d\tau, \quad (14.104)$$

so that

$$|y(t)| \leq e^{-\alpha t}|y(0)| + \int_0^t e^{-\alpha(t-\tau)} |\Theta^T(x,u)\bar{Y}| d\tau, \\ \leq e^{-\alpha t}|y(0)| + e^{-\alpha t} \int_0^t e^{\alpha \tau} \|\Theta^T(x,u)\| \cdot \|\bar{Y}\| d\tau. \quad (14.105)$$

Note that

$$\begin{cases} \|\Theta^T(x,u)\| = \sqrt{\sum_{i=1}^I \theta_i^2(x,u)} \leq \sqrt{I}, \\ \|\bar{Y}\| \leq M_{\bar{Y}}, \end{cases} \quad (14.106)$$

then

$$|y(t)| \leq e^{-\alpha t}|y(0)| + e^{-\alpha t} \sqrt{I} M_{\bar{Y}} \cdot \frac{1}{\alpha} e^{\alpha \tau}\big|_0^t \\ = e^{-\alpha t}|y(0)| + \frac{1}{\alpha} \sqrt{I} M_{\bar{Y}} (1 - e^{-\alpha t}) \quad (14.107)$$

For $t \geq 0$, $0 < e^{-\alpha t} \leq 1$, therefore

$$|y(t)| \leq |y(0)| + \frac{1}{\alpha} \sqrt{I} M_{\bar{Y}}. \quad (14.108)$$

Since $\xi = y - x$,

$$|\xi| \leq |y(t)| + |x| \leq |y(0)| + \frac{1}{\alpha} \sqrt{I} M_{\bar{Y}} + M_e + \|q\|. \quad (14.109)$$

Let

$$M_q \triangleq \sup_{t \geq 0} \{\|q(t)\|\} \quad (14.110)$$

and

$$M_\xi \triangleq |y(0)| + \frac{1}{\alpha}\sqrt{I}M_{\bar{Y}} + M_e + M_q. \tag{14.111}$$

Then M_ξ is a positive constant that bounds $|\xi|$, i.e.

$$|\xi| \le M_\xi \tag{14.112}$$

This completes the proof of Theorem 7.1.3

A.4 Proof of Theorem 7.1.4

Equations (14.14) and (14.18) together give

$$u = u_c - S_u\text{sgn}(\boldsymbol{B}^T\boldsymbol{Pe})\cdot\text{sgn}(g)\cdot\frac{1}{\varepsilon_g}\cdot[\alpha|x_N| + |\Theta^T(\boldsymbol{x},0)\bar{\boldsymbol{Y}}| + M_f + (M_g + |\hat{g}|)\cdot|u_c|] \tag{14.113}$$

or

$$|u| \le \left(1 + \frac{M_g + |\hat{g}|}{\varepsilon_g}\right)\cdot|u_c| + \frac{1}{\varepsilon_g}\cdot[\alpha|x_N| + |\Theta^T(\boldsymbol{x},0)\bar{\boldsymbol{Y}}| + M_f]. \tag{14.114}$$

Using (14.16) in (14.114) with the added definition $\alpha \triangleq \alpha_N$ gives

$$\begin{aligned}|u| &\le \left(1 + \frac{M_g + |\hat{g}|}{\varepsilon_g}\right)\cdot\frac{1}{|\hat{g}|}\cdot\left|\left[\dot{q}_N + \alpha x_N - \Theta^T(\boldsymbol{x},0)\bar{\boldsymbol{Y}} - \sum_{k=1}^N \alpha_k e_k\right]\right| \\ &+ \frac{1}{\varepsilon_g}\cdot[\alpha|x_N| + |\Theta^T(\boldsymbol{x},0)\bar{\boldsymbol{Y}}| + M_f] \\ &\le \left(1 + \frac{M_g + |\hat{g}|}{\varepsilon_g}\right)\cdot\frac{1}{\varepsilon_g}\cdot\left[|\dot{q}_N| + \alpha|x_N| + |\Theta^T(\boldsymbol{x},0)\bar{\boldsymbol{Y}}| + \left|\sum_{k=1}^N \alpha_k e_k\right|\right] \\ &+ \frac{1}{\varepsilon_g}\cdot[\alpha|x_N| + |\Theta^T(\boldsymbol{x},0)\bar{\boldsymbol{Y}}| + M_f].\end{aligned} \tag{14.115}$$

Now

$$\begin{cases}|\Theta^T(\boldsymbol{x},0)\bar{\boldsymbol{Y}}| &\le \|\Theta(\boldsymbol{x},0)\|\cdot\|\bar{\boldsymbol{Y}}\| \le \sqrt{I}M_{\bar{Y}}, \\ |\sum_{k=1}^N \alpha_k e_k| &= |\boldsymbol{a}^T\boldsymbol{e}| \le \|\boldsymbol{a}\|\cdot\|\boldsymbol{e}\| \le \|\boldsymbol{a}\|\cdot M_e\end{cases} \tag{14.116}$$

where I is the size of vector Θ. Using (14.116) in (14.115) yields

$$\begin{aligned}|u| \le \frac{1}{\varepsilon_g}\bigg[&\left(2 + \frac{M_g + |\hat{g}|}{\varepsilon_g}\right)\cdot(\alpha|x_N| + \sqrt{I}M_{\bar{Y}}) \\ &+ \left(1 + \frac{M_g + |\hat{g}|}{\varepsilon_g}\right)\cdot(|\dot{q}_N| + \|\boldsymbol{a}\|\cdot M_e) + M_f\bigg].\end{aligned} \tag{14.117}$$

This completes the proof of Theorem 7.1.4.

A.5 Proof of Theorem 7.1.5

Case 1: $|\hat{g}| > \varepsilon_g$. Let Δ_g be the modeling error of the FLS, i.e.

$$\Delta_g(x, \Psi, \bar{G}) \triangleq \hat{g} - g(x) = \Psi^T \bar{G} - g(x), \tag{14.118}$$

and let $\bar{G}^*$ be the optimal parameter vector that minimizes $|\Delta_g(x, \Psi, \bar{G}^*)|$. Formally,

$$\bar{G}^* \triangleq \arg \min_{\bar{G}'} \left\{ \sup_{x \in \mathcal{X}} |\Delta_g(x, \Psi, \bar{G}')| \right\}, \tag{14.119}$$

where $\bar{G}' \triangleq \{\bar{G} : \|\bar{G}\| \leq M_{\bar{G}}\}$. The existence of such an optimal parameter vector, $\bar{G}^*$, is justified by Lemma 1 of [5]. $M_{\bar{G}}$ is a positive constant that bounds $\|\bar{G}\|$. Let

$$\hat{g}^* \triangleq \Psi^T \bar{G}^*. \tag{14.120}$$

Consider (14.33):

$$\dot{e} = Ae + B[-\Theta^T(x,0)\Delta_{\bar{Y}} - \Delta_g u_c - r(x,0,\Theta,\bar{Y}^*) - S_u g u_s]. \tag{14.121}$$

Since

$$\Delta_g u_c = (\hat{g} - g)u_c = (\hat{g} - \hat{g}^*)u_c + (\hat{g}^* - g)u_c = \Psi^T(\bar{G} - \bar{G}^*)u_c + (\Psi)^T \bar{G}^* - g)u_c, \tag{14.122}$$

and referring to (14.118) and letting

$$\Delta_{\bar{G}} \triangleq \bar{G} - \bar{G}^*, \tag{14.123}$$

we have

$$\Delta_g u_c = \Psi^T \Delta_{\bar{G}} u_c + \Delta_g(x, \Psi, \bar{G}^*) u_c. \tag{14.124}$$

Equation (14.124) in (14.121) yields

$$\dot{e} = Ae + B[-\Theta^T(x,0)\Delta_{\bar{Y}} - \Psi^T \Delta_{\bar{G}} u_c - r(x,0,\Theta,\bar{Y}^*) - \Delta_g(x, \Psi, \bar{G}^*)u_c - S_u g u_s]. \tag{14.125}$$

Consider the Lyapunov function candidate

$$V = \tfrac{1}{2}\left[e^T Pe + h\xi^2 + \Delta_{\bar{Y}}^T H^{-1} \Delta_{\bar{Y}} + \gamma^{-1} \Delta_{\bar{G}}^T \Delta_{\bar{G}}\right]. \tag{14.126}$$

Differentiating both sides of (14.126) yields

$$\dot{V} = \tfrac{1}{2}\dot{e}^T Pe + \tfrac{1}{2}e^T P\dot{e} + h\xi\dot{\xi} + \Delta_{\bar{Y}}^T H^{-1} \dot{\Delta}_{\bar{Y}} + \gamma^{-1} \Delta_{\bar{G}}^T \dot{\Delta}_{\bar{G}}. \tag{14.127}$$

In the current situation, $\{S_g, \beta_g\}^T = \{S_g^U, \beta_g^U\}^T$. Using (14.26) and (14.125) in (14.127) yields

$$\begin{aligned}\dot{V} &= \tfrac{1}{2}e^T(A^T P + P^T A)e - \alpha h \xi^2 + \Delta_{\bar{Y}}^T[H^{-1}\dot{\Delta}_{\bar{Y}} - \Theta(x,0)(B^T Pe) + \Theta(x,u)h\xi]\\ &+ \Delta_{\bar{G}}^T[\gamma^{-1}\dot{\Delta}_{\bar{G}} - \Psi(B^T Pe)u_c] - S_u g(B^T Pe)u_s\\ &- r(x,0,\Theta,\bar{Y}^*)(B^T Pe) + r(x,u,\Theta,\bar{Y}^*)h\xi - \Delta_g(x,\Psi,\bar{G}^*)(B^T Pe)u_c.\end{aligned} \tag{14.128}$$

Now using (14.20) in (14.128), and noting that

$$\dot{\Delta}_{\bar{Y}} = \dot{\bar{Y}}, \quad \dot{\Delta}_{\bar{G}} = \dot{\bar{G}}, \quad S_u g(B^T Pe)u_s \geq 0, \tag{14.129}$$

we have

$$\dot{V} \leq -\tfrac{1}{2}e^{\mathrm{T}}Qe - \alpha h \xi^2 + \Delta_{\bar{Y}}^{\mathrm{T}}[H^{-1}\dot{\bar{Y}} - \Theta(x,0)(B^{\mathrm{T}}Pe) + \Theta(x,u)h\xi]$$
$$+ \Delta_{\bar{G}}^{\mathrm{T}}[\gamma^{-1}\dot{\bar{G}} - \Psi(B^{\mathrm{T}}Pe)u_c] \qquad (14.130)$$
$$- r(x,0,\Theta,\bar{Y}^*)(B^{\mathrm{T}}Pe) + r(x,u,\Theta,\bar{Y}^*)h\xi - \Delta_g(x,\Psi,\bar{G}^*)(B^{\mathrm{T}}Pe)u_c.$$

Substitution of (14.34) and (14.40) into (14.130) yields

$$\dot{V} \leq -\tfrac{1}{2}e^{\mathrm{T}}Qe - \alpha h \xi^2 - S\beta \Delta_{\bar{Y}}^{\mathrm{T}}\bar{Y} - S_g^{\mathrm{U}}\beta_g^{\mathrm{U}}\Delta_{\bar{G}}^{\mathrm{T}}\bar{G}$$
$$- r(x,0,\Theta,\bar{Y}^*)(B^{\mathrm{T}}Pe) + r(x,u,\Theta,\bar{Y}^*)h\xi - \Delta_g(x,\Psi,\bar{G}^*)(B^{\mathrm{T}}Pe)u_c. \qquad (14.131)$$

After some manipulations, we obtain

$$\Delta_{\bar{Y}}^{\mathrm{T}}\bar{Y} = \tfrac{1}{2}\|\Delta_{\bar{Y}}\|^2 + \tfrac{1}{2}\|\bar{Y}\|^2 - \tfrac{1}{2}\|\bar{Y}^*\|^2. \qquad (14.132)$$

We know that $\|\bar{Y}^*\| \leq M_{\bar{Y}}$, and for $S = 1$, $\|\bar{Y}\| \geq M_{\bar{Y}}$, therefore

$$S\beta \Delta_{\bar{Y}}^{\mathrm{T}}\bar{Y} \geq \tfrac{1}{2}S\beta\|\Delta_{\bar{Y}}\|^2 \geq 0. \qquad (14.133)$$

Since the eigenvalues of a symmetric positive definite matrix are positive [3], the eigenvalues of Q are positive. Denoting $\lambda_{\min}(Q)$ as the minimum eigenvalue of Q, then

$$e^{\mathrm{T}}Qe \geq \lambda_{\min}(Q)\|e\|^2 \geq 0. \qquad (14.134)$$

Also,

$$\Delta_{\bar{G}}^{\mathrm{T}}\bar{G} = \tfrac{1}{2}\|\Delta_{\bar{G}}\|^2 + \tfrac{1}{2}\|\bar{G}\|^2 - \tfrac{1}{2}\|\bar{G}^*\|^2. \qquad (14.135)$$

- For $S_g^{\mathrm{U}} = 0$,

$$-S_g^{\mathrm{U}}\beta_g^{\mathrm{U}}\Delta_{\bar{G}}^{\mathrm{T}}\bar{G} = 0. \qquad (14.136)$$

- For $S_g^{\mathrm{U}} = 1$, $\|\bar{G}\| \geq M_{\bar{G}}$, but $\|\bar{G}^*\| \leq M_{\bar{G}}$. Therefore

$$-S_g^{\mathrm{U}}\beta_g^{\mathrm{U}}\Delta_{\bar{G}}^{\mathrm{T}}\bar{G} \leq -1/2\beta_g^{\mathrm{U}}\|\Delta_{\bar{G}}\|^2. \qquad (14.137)$$

From (14.136) and (14.137) we conclude that

$$-S_g^{\mathrm{U}}\beta_g^{\mathrm{U}}\Delta_{\bar{G}}^{\mathrm{T}}\bar{G} \leq -\tfrac{1}{2}S_g^{\mathrm{U}}\beta_g^{\mathrm{U}}\|\Delta_{\bar{G}}\|^2. \qquad (14.138)$$

Using (14.138) and (14.133), (14.134) in (14.131) yields

$$\dot{V} \leq -\tfrac{1}{2}\lambda_{\min}(Q)\|e\|^2 - \alpha h \xi^2 - \tfrac{1}{2}S\beta\|\Delta\bar{Y}\|^2 - \tfrac{1}{2}S_g^{\mathrm{U}}\beta_g^{\mathrm{U}}\|\Delta_{\bar{G}}\|^2$$
$$- r(x,0,\Theta,\bar{Y}^*)(B^{\mathrm{T}}Pe) + r(x,u,\Theta,\bar{Y}^*)h\xi - \Delta_g(x,\Psi,\bar{G}^*)(B^{\mathrm{T}}Pe)u_c. \qquad (14.139)$$

Individual terms in (14.139) can be reorganized as

$$-\alpha h \xi^2 + r(x,u,\Theta,\bar{Y}^*)h\xi = -\tfrac{1}{2}\alpha h \xi^2 - \tfrac{1}{2}\left[(\alpha h)^{1/2}\xi - \left(\tfrac{h}{\alpha}\right)^{1/2}r(x,u,\Theta,\bar{Y}^*)\right]^2$$
$$+ \tfrac{h}{2\alpha}r^2(x,u,\Theta,\bar{Y}^*) \leq -\tfrac{1}{2}\alpha h \xi^2 + \tfrac{h}{2\alpha}r^2(x,u,\Theta,\bar{Y}^*);$$

$$(14.140)$$

304 FUZZY LOGIC CONTROL OF NON-LINEAR SYSTEMS

$$
\begin{aligned}
&-\tfrac{1}{2}\lambda_{\min}(Q)\|e\|^2 - r(x,0,\Theta,\bar{Y}^*)(B^{\mathrm{T}}Pe) - \Delta_g(x,\Psi,\bar{G}^*)(B^{\mathrm{T}}Pe)u_c \\
&= -\tfrac{1}{4}\lambda_{\min}(Q)\|e\|^2 - \tfrac{1}{8}\lambda_{\min}(Q)\left\|\left[e + 4\frac{r(x,0,\Theta,\bar{Y}^*)}{\lambda_{\min}(Q)}(P^{\mathrm{T}}B)\right]\right\|^2 \\
&\quad + 2\frac{r^2(x,0,\Theta,\bar{Y}^*)}{\lambda_{\min}(Q)}\|(P^{\mathrm{T}}B)\|^2 + 2\frac{\Delta_g^2(x,\Psi,\bar{G}^*)u_c^2}{\lambda_{\min}(Q)}\|(P^{\mathrm{T}}B)\|^2 \\
&\quad - \tfrac{1}{8}\lambda_{\min}(Q)\left\|\left[e + 4\frac{\Delta_g(x,\Psi,\bar{G}^*)u_c}{\lambda_{\min}(Q)}(P^{\mathrm{T}}B)\right]\right\|^2 \\
&\leq -\tfrac{1}{4}\lambda_{\min}(Q)\|e\|^2 + 2\frac{r^2(x,0,\Theta,\bar{Y}^*)}{\lambda_{\min}(Q)}\|P^{\mathrm{T}}B\|^2 \\
&\quad + 2\frac{\Delta_g^2(x,\Psi,\bar{G}^*)u_c^2}{\lambda_{\min}(Q)}\|P^{\mathrm{T}}B\|^2.
\end{aligned}
\tag{14.141}
$$

Using (14.140) and (14.141) in (14.139), and noting that

$$-\tfrac{1}{2}S\beta\|\Delta_{\bar{Y}}\|^2 \leq 0, \qquad -\tfrac{1}{2}S_g^{\mathrm{U}}\beta_g^{\mathrm{U}}\|\Delta_{\bar{G}}\|^2 \leq 0, \tag{14.142}$$

yields

$$
\begin{aligned}
\dot{V} &\leq -\tfrac{1}{4}\lambda_{\min}(Q)\|e\|^2 - \tfrac{1}{2}\alpha h\xi^2 \\
&\quad + 2\frac{\|P^{\mathrm{T}}B\|^2}{\lambda_{\min}(Q)}[r^2(x,0,\Theta,\bar{Y}^*) + \Delta_g^2(x,\Psi,\bar{G}^*)u_c^2] + \frac{h}{2\alpha}r^2(x,u,\Theta,\bar{G}^*).
\end{aligned}
\tag{14.143}
$$

Since

$$u_c \triangleq \frac{1}{\hat{g}}\left[\dot{q}_N + \alpha q_N - \Theta^{\mathrm{T}}(x,0)\bar{Y} - \sum_{k=1}^{N-1}\alpha_k e_k\right], \tag{14.144}$$

then

$$
\begin{aligned}
|u_c| &\leq \frac{1}{\varepsilon_g}\left[|\dot{q}_N| + \alpha|x_N| + |\Theta^{\mathrm{T}}(x,0)\bar{Y}| + \left|\sum_{k=1}^{N}\alpha_k e_k\right|\right] \\
&\leq \frac{1}{\varepsilon_g}[M_{\dot{q}_N} + \alpha M_{x_N} + \|\Theta^{\mathrm{T}}(x,0)\| \cdot \|\bar{Y}\| + \|a\| \cdot \|e\|] \\
&\leq \frac{1}{\varepsilon_g}[M_{\dot{q}_N} + \alpha M_{x_N} + \sqrt{I}M_{\bar{Y}} + \|a\|M_e],
\end{aligned}
\tag{14.145}
$$

where

$$M_{\dot{q}_N} \triangleq \sup_{t\geq 0}\{|\dot{q}_N|\}, \qquad M_{x_N} \triangleq \sup_{t\geq 0}\{|x_N|\}. \tag{14.146}$$

Now letting

$$M_{u_c} \triangleq \frac{1}{\varepsilon_g}[M_{\dot{q}_N} + \alpha M_{x_N} + \sqrt{I}M_{\bar{Y}} + \|a\|M_e], \tag{14.147}$$

we have

$$|u_c| \leq M_{u_c}, \tag{14.148}$$

and if we define

$$\zeta \triangleq \min\left\{\frac{\lambda_{\min}(\boldsymbol{Q})}{4}, \frac{\alpha h}{2}\right\}, \tag{14.149}$$

$$\eta \triangleq \max\left\{2\frac{\|\boldsymbol{P}^{\mathrm{T}}\boldsymbol{B}\|^2}{\zeta\lambda_{\min}(\boldsymbol{Q})}, 2\frac{\|\boldsymbol{P}^{\mathrm{T}}\boldsymbol{B}\|^2}{\zeta\lambda_{\min}(\boldsymbol{Q})}M_{u_c}^2, \frac{h}{2\zeta\alpha}\right\}. \tag{14.150}$$

Then

$$\dot{V} \leq -\zeta(\|\boldsymbol{e}\|^2 + \xi^2) + \zeta\eta[r^2(\boldsymbol{x}, 0, \Theta, \bar{\boldsymbol{Y}}^*) + r^2(\boldsymbol{x}, u, \Theta, \bar{\boldsymbol{Y}}^*) + \Delta_g^2(\boldsymbol{x}, \Psi, \bar{\boldsymbol{G}}^*)]. \tag{14.151}$$

Integrating both sides of (14.151) yields

$$V(t) - V(0) \leq -\zeta \int_0^t \|\boldsymbol{e}^2(\tau)\|\mathrm{d}\tau - \zeta \int_0^t \xi^2(\tau)\mathrm{d}\tau$$
$$+ \zeta\eta \int_0^t [r^2(\boldsymbol{x}, 0, \Theta, \bar{\boldsymbol{Y}}^*) + r^2(\boldsymbol{x}, u, \Theta, \bar{\boldsymbol{Y}}^*) + \Delta_g^2(\boldsymbol{x}, \Psi, \bar{\boldsymbol{G}}^*)]\mathrm{d}\tau. \tag{14.152}$$

Let

$$b \triangleq \frac{1}{\zeta}\sup_{t \geq 0}\{V(0) - V(t)\}, \tag{14.153}$$

then we have

$$\int_0^t \|\boldsymbol{e}(\tau)\|^2\mathrm{d}\tau + \int_0^t \xi^2(\tau)\mathrm{d}\tau$$
$$\leq b + \eta \int_0^t [r^2(\boldsymbol{x}, 0, \Theta, \bar{\boldsymbol{Y}}^*) + r^2(\boldsymbol{x}, u, \Theta, \bar{\boldsymbol{Y}}^*) + \Delta_g^2(\boldsymbol{x}, \Psi, \bar{\boldsymbol{G}}^*)]\mathrm{d}\tau, \tag{14.154}$$

or

$$\int_0^t \|\boldsymbol{e}(\tau)\|^2\mathrm{d}\tau \leq b + \eta \int_0^t [r^2(\boldsymbol{x}, 0, \Theta, \bar{\boldsymbol{Y}}^*) + r^2(\boldsymbol{x}, u, \Theta, \bar{\boldsymbol{Y}}^*) + \Delta_g^2(\boldsymbol{x}, \Psi, \bar{\boldsymbol{G}}^*)]\mathrm{d}\tau, \tag{14.155}$$

$$\int_0^t \xi^2(\tau)\mathrm{d}\tau \leq b + \eta \int_0^t [r^2(\boldsymbol{x}, 0, \Theta, \bar{\boldsymbol{Y}}^*) + r^2(\boldsymbol{x}, u, \Theta, \bar{\boldsymbol{Y}}^*) + \Delta_g^2(\boldsymbol{x}, \Psi, \bar{\boldsymbol{G}}^*)]\mathrm{d}\tau. \tag{14.156}$$

Case 2: $|\hat{g}| = \varepsilon_g$. Consider the Lyapunov function candidate

$$V = \tfrac{1}{2}[\boldsymbol{e}^{\mathrm{T}}\boldsymbol{P}\boldsymbol{e} + h\xi^2 + \Delta_{\bar{Y}}^{\mathrm{T}}\boldsymbol{H}^{-1}\Delta_{\bar{Y}} + \gamma^{-1}(\Psi^{\mathrm{T}}\Psi)^{-1}\Delta_g^2]. \tag{14.157}$$

By mathematical manipulations similar to those of (14.127)–(14.143) we have

$$\dot{V} \leq -\tfrac{1}{4}\lambda_{\min}(\boldsymbol{Q})\|\boldsymbol{e}\|^2 - \tfrac{1}{2}\alpha h\xi^2 + \Delta_g[\gamma^{-1}(\Psi^{\mathrm{T}}\Psi)^{-1}\dot{\hat{g}} - (\boldsymbol{B}^{\mathrm{T}}\boldsymbol{P}\boldsymbol{e})u_c]$$
$$+ \frac{\|\boldsymbol{P}^{\mathrm{T}}\boldsymbol{B}\|^2}{\lambda_{\min}(\boldsymbol{Q})}r^2(\boldsymbol{x}, 0, \Theta, \bar{\boldsymbol{Y}}^*) + \frac{h}{2\alpha}r^2(\boldsymbol{x}, u, \Theta, \bar{\boldsymbol{Y}}^*). \tag{14.158}$$

Considering $\dot{\hat{g}} = \Psi^{\mathrm{T}}\dot{\hat{\boldsymbol{G}}}$, and using (14.40) and (14.41) gives

$$\Delta_g[\gamma^{-1}(\Psi^{\mathrm{T}}\Psi)^{-1}\dot{\hat{g}} - (\boldsymbol{B}^{\mathrm{T}}\boldsymbol{P}\boldsymbol{e})u_c] = S_g^{\mathrm{L}}\beta_g^{\mathrm{L}}(\Psi^{\mathrm{T}}\Psi)^{-1}\Delta_g\Psi^{\mathrm{T}}\boldsymbol{G} = S_g^{\mathrm{L}}\beta_g^{\mathrm{L}}(\Psi^{\mathrm{T}}\Psi)^{-1}\Delta_g\hat{g}. \tag{14.159}$$

Since
$$\Delta_g \hat{g} = \tfrac{1}{2}\Delta_g^2 + \tfrac{1}{2}\hat{g}^2 - \tfrac{1}{2}g^2, \quad |\hat{g}| = \varepsilon_g, \quad |g| \geq \varepsilon_g, \tag{14.160}$$

we have
$$\Delta_g \hat{g} \leq \tfrac{1}{2}\Delta_g^2. \tag{14.161}$$

Using (14.161) in (14.159), and considering both $S_g^L = 0$ and $S_g^L = 1$, we have
$$\Delta_g[\gamma^{-1}(\mathbf{\Psi}^T\mathbf{\Psi})^{-1}\dot{\hat{g}} - (\mathbf{B}^T\mathbf{P}\mathbf{e})u_c] \leq \tfrac{1}{2}\beta_g^L(\mathbf{\Psi}^T\mathbf{\Psi})^{-1}\Delta_g^2. \tag{14.162}$$

Equation (14.162) in (14.158) gives
$$\dot{V} \leq -\tfrac{1}{4}\lambda_{\min}(\mathbf{Q})\|\mathbf{e}\|^2 - \tfrac{1}{2}\alpha h \xi^2 + \tfrac{1}{2}\beta_g^L(\mathbf{\Psi}^T\mathbf{\Psi})^{-1}\Delta_g^2 \\ + \frac{\|\mathbf{P}^T\mathbf{B}\|^2}{\lambda_{\min}(\mathbf{Q})} r^2(\mathbf{x}, 0, \Theta, \bar{\mathbf{Y}}^*) + \frac{h}{2\alpha} r^2(\mathbf{x}, u, \Theta, \bar{\mathbf{Y}}^*). \tag{14.163}$$

Let
$$\zeta \triangleq \min\left\{\frac{\lambda_{\min}(\mathbf{Q})}{4}, \frac{\alpha h}{2}\right\}, \tag{14.164}$$

$$\eta \triangleq \max\left\{\frac{\|\mathbf{P}^T\mathbf{B}\|^2}{\zeta\lambda_{\min}(\mathbf{Q})}, \frac{h}{2\zeta\alpha}, \frac{\beta_g^L(\mathbf{\Psi}^T\mathbf{\Psi})^{-1}}{2\zeta}\right\}. \tag{14.165}$$

Then
$$\dot{V} \leq -\zeta(\|\mathbf{e}\|^2 + \xi^2) + \zeta\eta[r^2(\mathbf{x}, 0, \Theta, \bar{\mathbf{Y}}^*) + r^2(\mathbf{x}, u, \Theta, \bar{\mathbf{Y}}^*) + \Delta_g^2]. \tag{14.166}$$

Integrating both sides of (14.166) yields
$$V(t) - V(0) \leq -\zeta\int_0^t \|\mathbf{e}^2(\tau)\|\mathrm{d}\tau - \zeta\int_0^t \xi^2(\tau)\mathrm{d}\tau \\ + \zeta\eta\int_0^t [r^2(\mathbf{x}, 0, \Theta, \bar{\mathbf{Y}}^*) + r^2(\mathbf{x}, u, \Theta, \bar{\mathbf{Y}}^*) + \Delta_g^2]\mathrm{d}\tau. \tag{14.167}$$

Let
$$b \triangleq \frac{1}{\zeta}\sup_{t\geq 0}\{V(0) - V(t)\}, \tag{14.168}$$

then
$$\int_0^t \|\mathbf{e}(\tau)\|^2\mathrm{d}\tau + \int_0^t \xi^2(\tau)\mathrm{d}\tau \leq b + \eta\int_0^t [r^2(\mathbf{x}, 0, \Theta, \bar{\mathbf{Y}}^*) + r^2(\mathbf{x}, u, \Theta, \bar{\mathbf{Y}}^*) + \Delta_g^2]\mathrm{d}\tau, \tag{14.169}$$

or
$$\int_0^t \|\mathbf{e}(\tau)\|^2\mathrm{d}\tau \leq b + \eta\int_0^t [r^2(\mathbf{x}, 0, \Theta, \bar{\mathbf{Y}}^*) + r^2(\mathbf{x}, u, \Theta, \bar{\mathbf{Y}}^*) + \Delta_g^2]\mathrm{d}\tau, \tag{14.170}$$

$$\int_0^t \xi^2(\tau)\mathrm{d}\tau \leq b + \eta\int_0^t [r^2(\mathbf{x}, 0, \Theta, \bar{\mathbf{Y}}^*) + r^2(\mathbf{x}, u, \Theta, \bar{\mathbf{Y}}^*) + \Delta_g^2]\mathrm{d}\tau, \tag{14.171}$$

This completes the proof of Theorem 7.1.5.

A.6 Proof of Theorem 7.1.6

For $r(x, u, \Theta, \bar{Y}^*) \in \mathcal{L}_2[0, \infty)$ and $\Delta_g \in \mathcal{L}_2[0, \infty)$, we have

$$\left[\int_0^\infty r^2(x, u, \Theta, \bar{Y}^*) d\tau\right]^{1/2} < \infty, \quad \left[\int_0^\infty r^2(x, 0, \Theta, \bar{Y}^*) d\tau\right]^{1/2} < \infty, \quad \left[\int_0^\infty \Delta_g^2 d\tau\right]^{1/2} < \infty \tag{14.172}$$

By Theorem 7.1.5,

$$\left[\int_0^\infty \|e(\tau)\|^2 d\tau\right]^{1/2} < \infty, \quad \left[\int_0^\infty \xi^2(\tau) d\tau\right]^{1/2} < \infty. \tag{14.173}$$

Therefore, $e \in \mathcal{L}_2$ and $\xi \in \mathcal{L}_2$. From Theorem 7.1.3, it is known that $e \in \mathcal{L}_\infty$ and $\xi \in \mathcal{L}_\infty$. Observing the expressions for $\dot{\xi}$ and $\dot{e}$, (14.26) and (14.33) reveals that all the components on the right-hand sides of the expressions are bounded. Thefore, $\dot{\xi}$ and $\dot{e}$ are bounded, i.e. $\dot{e} \in \mathcal{L}_\infty$ and $\dot{\xi} \in \mathcal{L}_\infty$. From Corollary 2.9 in [12], we conclude that

$$\lim_{t \to \infty} \|e\| = 0, \quad \lim_{t \to \infty} |\xi| = 0. \tag{14.174}$$

This completes the proof of Theorem 7.1.6.

REFERENCES

1. Goodwin, G. C. and D. Q. Mayne, A parameter estimation perspective of continuous time model reference adaptive control, *Automatica*, **23**, 1987, 57–70.
2. Ioannou, P.A. and A. Datta, Robust adaptive control: Design, analysis and robustness bounds, In: P. V. Kokotovic (ed.), *Foundations of Adaptive Control*, Springer-Verlag, Berlin, 1991, pp. 71–152.
3. Kailath, T., *Linear Systems*, Prentice-Hall, Englewood Cliffs, 1980.
4. Lee, C. C., Fuzzy logic in control systems, fuzzy logic controller—Parts I and II, *IEEE Transactions on Systems, Man and Cybernetics*, **20**(2), 1990, 404–435.
5. Lee, J. X. and G. Vukovich, The dynamic fuzzy logic system: Nonlinear system identification and application to robotic manipulators, *Journal of Robotic Systems*, **14**(6), 1997, 391–405.
6. Vukovich, G. and J. X. Lee, Stable identification and adaptive control—A dynamic fuzzy logic system approach, In: W. Pedrycz (ed.), *Fuzzy Evolutionary Computation*, Kluwer, Boston, London, Dordrecht, 1997, pp. 223–248.
7. Lee, J., On methods for improving performance of PI-type FLCs, *IEEE Transactions on Fuzzy Systems*, **1**(4), 1993, 298–301.
8. Malki, H. A., H. Li and G. Chen, New design and stability analysis of fuzzy proportional-derivative control systems, *IEEE Transactions on Fuzzy Systems*, **2**(4), 1994, 245–254.
9. Mamdani, E. H., Application of fuzzy Algorithm for control of simple dynamic plant, *Proceedings of the IEEE*, **121**, 1974, 1585–1588.
10. Mamdani, E. H. and S. Assilian, A case study on the application of fuzzy set theory to automatic control, *Proceedings of the IFAC Stochastic Control Symposium*, Budapest, Hungary, 1974.
11. Mamdani, E. H., Advances in the linguistic synthesis of Fuzzy controllers, In: E. H. Mamdani and B. R. Gaines (eds.), *Fuzzy Reasoning and its Applications*, Academic Press, New York, 1981, pp. 352–334.
12. Narendra, K. S. and A. M. Annaswamy, *Stable Adaptive Systems*, Prentice-Hall, Englewood Cliffs, 1989.

13. Pedrycz, W., *Fuzzy Control and Fuzzy Systems*, 2nd extended edition, Research Studies Press, Taunton/Toronto, John Wiley, Chichester, 1993.
14. Polycarpou, M. M. and P. A. Ioannou, Stable nonlinear system identification using neural network models, In: G. Bekey and K. Goldberg (eds.), *Neural Networks in Robotics*, Kluwer, Norwell, 1993, pp. 147–164.
15. Su, C.-Y. and Y. Stepanenko, Adaptive control of a class of non-linear systems with fuzzy logic, *IEEE Transactions on Fuzzy Systems*, **2**(4), 1994, 285–294.
16. Sugeno, M. (ed.), *Industrial Applications of Fuzzy Control*, Elsevier Science Publishers, Amsterdam, 1985.
17. Tong, R. M., A control engineering review of fuzzy systems, *Automatica*, **13**, 1977, 559–569.
18. Wang, L. X., Stable adaptive control of nonlinear systems, *IEEE Transactions on Fuzzy Systems*, **1**(2), 1993, 146–155.
19. Wang, L. X., A supervisory controller for fuzzy control systems that guarantees stability, *IEEE Transactions on Automatic Control*, **39**(9), 1994, 1845–1847.
20. Wang, L. X., *Adaptive Fuzzy Systems and Control*, Prentice-Hall, Englewood Cliffs, 1994.
21. Yager, R. R. and D. P. Filev, *Essentials of Fuzzy Modeling and Control*, John Wiley, New York, 1994.
22. Zadeh, L. A., Outline of a new approach to the analysis of complex systems and decision processes, *IEEE Transactions on Systems, Man and Cybernetics*, **SMC-3**, 1973, 28–44.
23. Zheng, L., A practical guide to tune of proportional and integral (PI) like fuzzy controller, *Proceedings of the First IEEE International Conference on Fuzzy Systems*, 1992, pp. 663–640.
24. Zimmermann, H.-J., *Fuzzy Set Theory and its Applications*, 3rd edn., Kluwer, Dordrecht, 1996.

15
Stabilization of Direct Adaptive Fuzzy Control Systems: Two Approaches

Hwan-Chun Myung, Zenn. Z. Bien
Korea Advanced Institute of Science and Technology, Yusong-gu, Korea

Yong-Tae Kim
SAMSUNG Electronics Co. Ltd, Suwon, Korea

15.1 INTRODUCTION

The fuzzy logic controller (FLC) is considered applicable for plants that are poorly understood or very complex in their dynamic models. But, the inherent non-linearity in the FLC structure and heuristic aspects in its design are known as major disadvantages of the approach. For example, the membership function of fuzzy sets representing values of variables or parameters are often heuristically designed by trial and error. In particular, the proof of stability of any FLC based on system remains a difficult task. Since 1990 there have been a number of papers that deal with the above problems. Wang [1] proposed using an adaptive law to tune the membership function automatically with a supervisory controller appended to stabilize the total system. Chen *et al.* [2] presented a modified adaptive law to advance the performance of the controller suggested by [1], using a Riccati-like equation. Su [3] proposed a design method similar to [1]. In the latter case, a sliding mode scheme is inserted into the controller and this modification renders the possibility to prove the stability of the FLC. In [1] and [3] it is assumed that the upper bound of the plant under consideration is known. Lu [4] proposed a method to use an FLC and a sliding mode controller (SMC) simultaneously in the additive form for stability of the FLC, where the minimum approximation error (MAE) was compensated with a SMC.

Compared with the previous research, in the first approach, the MAE is handled by the SMC instead of the upper bound of the plant and the switching gain is newly designed, because, when the SMC is used with the FLC for stabilization, the switching gain does not need to be dependent on the upper bound of the plant any more. In addition, the reaching phase problem is also considered by designing the sliding surface of the SMC with the integral part and it is shown that this integral action can always guarantee the performance of

Fuzzy Logic. Edited by S. Farinwata, D. Filev and R. Langari
© 2000 John Wiley & Sons, Ltd

the control system with the known bound of the approximation error, under the switching condition. As a second approach, a new adaptive law is proposed which can efficiently eliminate the influence of perturbations on the parameter adaptation. The proposed adaptation scheme is that the parameter modification in the adaptive process should occur in consideration of both the system's current performance and the system's behavioral tendency as well. Also, the concept of persistent excitation in the adaptive fuzzy control systems is first introduced to guarantee the convergence and bound of the adaptation parameters in the fuzzy system.

In Section 15.2 the first approach is described and the second approach is given in Section 15.3. Some simulations and experiments are provided to illustrate the performance of the proposed methods in Section 15.4. Finally, the conclusion is given in Section 15.5.

15.2 INTEGRAL SLIDING-MODE ADAPTIVE FLC: APPROACH I

15.2.1 Structure of an Integral Sliding-Mode Adaptive FLC

Consider a plant of the form

$$x^{(n)} = f(x, \dot{x}, \ldots, x^{(n-1)}) + bu, \qquad y = x. \tag{15.1}$$

If the function f and the constant b are known, then the ideal control law u^* is given by

$$u^* = \frac{1}{b}[-f(\bar{x}) + y_m^{(n)} + \bar{k}^T \bar{e}], \tag{15.2}$$

where $\bar{x} = [x, \dot{x}, \ldots, x^{(n-1)}]^T$ is a state vector, y_m is a reference input, $\bar{e} = [y_m - x, \dot{y}_m - \dot{x}, \ldots, y_m^{(n-1)} - x_m^{(n-1)}]^T$ is an error vector and $\bar{k} = [k_1, k_2, \ldots, k_n]^T$ is a design parameter vector. Applying (15.2) to (15.1), results in

$$e^{(n)} + k_1 e^{(n-1)} + \cdots + k_n e = 0. \tag{15.3}$$

The proposed controller is as follows:

$$u = u_c(\bar{x}|\bar{\theta}) + K\,\text{sgn}(s), \tag{15.4}$$

where s stands for a variable specified in (15.11). The control law (15.4) can be considered as the summation of the FLC and the SMC. If (15.1), (15.2) and (15.4) are combined, the error equation

$$e^{(n)} = -\bar{k}^T \bar{e} + b[u^* - u_c(\bar{x}|\bar{\theta}) - K\,\text{sgn}(s)], \tag{15.5}$$

or, equivalently,

$$\frac{d\bar{e}}{dt} = A\bar{e} + \bar{b}_c[u^* - u_c(\bar{x}|\bar{\theta}) - K\,\text{sgn}(s)], \tag{15.6}$$

where

$$\bar{b}_c^T = [0, \ldots, 0, b], \qquad A = \begin{bmatrix} 0 & 1 & 0 & 0 & \cdots & 0 \\ 0 & 0 & 1 & 0 & \cdots & 0 \\ \vdots & \vdots & \vdots & \vdots & \vdots & \vdots \\ 0 & 0 & 0 & 0 & \cdots & 1 \\ -k_n & -k_{n-1} & -k_{n-2} & -k_{n-3} & \cdots & -k_1 \end{bmatrix}$$

is obtained. Equation (15.6) will be used in analyzing the proposed FLC.

15.2.2 Stabilization of the Integral Sliding-Mode Adaptive FLC

Two definitions are stated for the proof of the stability. One is the Fuzzy Basis Function (FBF) [1], which is defined as

$$u_c(\bar{x}|\bar{\theta}) = \bar{\theta}^T \xi(\bar{x}), \tag{15.7}$$

where

$$\bar{\theta}^T = [c^1, \ldots, c^m], \xi_l = \frac{\Pi_{i=1}^n \mu_{Fil}(x_i)}{\sum_{l=1}^m (\Pi_{i=1}^n \mu_{Fil}(x_i))} : \text{FBF},$$

m is the number of rules, n is the number of states, and μ_{Fil} is the membership function. $\bar{\theta}$ means a center-value vector of the membership functions in consequent parts of the **If–Then** rules and ξ_l means the FBF. Another is the MAE [1]:

$$\omega(\bar{x}) \equiv u_c(\bar{x}|\bar{\theta}^*) - u^* : \text{MAE}, \tag{15.8}$$

where

$$\bar{\theta}^* = \arg \min_{|\bar{\theta}|_2 \leq M_\theta} \left[\sup_{|\bar{x}|_2 \leq M_x} |u_c(\bar{x}|\bar{\theta}) - u^*| \right].$$

M_θ and M_x are the design parameters based on practical constraints. According to (15.8), the MAE is the difference between the ideal conventional controller and an optimal fuzzy controller that has the same number of rules as the designed FLC. Using (15.7) and (15.8), equation (15.6) can be modified as

$$\begin{aligned}\frac{d\bar{e}}{dt} &= A\bar{e} + \bar{b}_c[u_c(\bar{x}|\bar{\theta}^*) - u_c(\bar{x}|\bar{\theta})] - \bar{b}_c K \operatorname{sgn}(s) - \bar{b}_c \omega, \\ &= A\bar{e} + \bar{b}_c \bar{\phi}^T \xi - \bar{b}_c (K \operatorname{sgn}(s) + \omega).\end{aligned} \tag{15.9}$$

To prove the stability of the FLC, suppose that

- the sign of b is known,
- $|\omega|_\infty < \infty$ is known, and
- $|\bar{e}(t)|_2 \leq M_e, |\bar{\theta}^*(t)|_2 \leq M_\theta$ for $\forall t \leq 0$.

Define the Lyapunov function candidate as

$$V = \tfrac{1}{2}\bar{e}^T P \bar{e} + \tfrac{b}{2\gamma}\bar{\phi}^T\bar{\phi} + \tfrac{1}{2}\bar{\psi}^T\bar{\psi}, \tag{15.10}$$

where

$$A^T P + PA = -Q : \text{Lyapunov equation},$$

$$s = \bar{b}_c^T P^2 \left[(I + P^{-1})\bar{e} - A \int_0^t \bar{e}\,dt \right] = \bar{b}_c^T P^2 \bar{h}, \quad \forall t \geq 0, \tag{15.11}$$

and

$$\bar{\phi} = \bar{\theta}^* - \bar{\theta}, \quad \bar{\psi} = -\bar{e} + P\bar{h}, \quad \bar{h} = (I + P^{-1})\bar{e} - A\int_0^t \bar{e}\,dt.$$

Then, differentiating the Lyapunov function is given by

$$\frac{dV}{dt} = -\tfrac{1}{2}\bar{e}^T Q\bar{e} + \tfrac{b}{\gamma}\bar{\phi}^T\left[\gamma\bar{e}^T\bar{p}_n\bar{\xi}(\bar{x}) + \tfrac{d\bar{\phi}}{dt}\right] - \bar{e}P\bar{b}_c^T(K\operatorname{sgn}(s)+w)$$
$$+ (-\bar{e}^T + \bar{h}^T P^T)\left(-\frac{d\bar{e}}{dt} + P\left[(I+P^{-1})\frac{d\bar{e}}{dt} - A\bar{e}\right]\right)$$
$$= -\tfrac{1}{2}\bar{e}^T Q\bar{e} + \tfrac{b}{\gamma}\bar{\phi}^T\left[\gamma\bar{e}^T\bar{p}_n\bar{\xi}(\bar{x}) + \tfrac{d\bar{\phi}}{dt}\right] - \bar{e}P\bar{b}_c^T(K\operatorname{sgn}(s)+w)$$
$$+ (-\bar{e}^T + \bar{h}^T P^T)P[\bar{b}_c\bar{\phi}^T\bar{\xi}(\bar{x}) - \bar{b}_c(K\operatorname{sgn}(s)+w)],$$
$$= -\tfrac{1}{2}\bar{e}^T Q\bar{e} + b\bar{\phi}^T\left[\bar{e}^T\bar{p}_n\bar{\xi}(\bar{x}) + (-\bar{e}^T\bar{p}_n + \bar{h}^T P\bar{p}_n)\bar{\xi} + \tfrac{1}{\gamma}\tfrac{d\bar{\phi}}{dt}\right]$$
$$- \bar{h}^T P^2 \bar{b}_c(K\operatorname{sgn}(s)+w),$$
$$= -\tfrac{1}{2}\bar{e}^T Q\bar{e} - s(K\operatorname{sgn}(s)+w) \le 0,$$

where $\bar{e}^T P \bar{b}_c^T = \bar{e}^T \bar{p}_n b$, $\bar{p}_n$: nth column of matrix P. By the projection algorithm [1], the adaptive law is obtained as,

if $|\bar{\theta}|_2 < M_\theta$ or $s\bar{\theta}^T\bar{\xi}(\bar{x}) \le 0$ for $|\bar{\theta}|_2 = M_\theta$,

$$\frac{d\bar{\theta}}{dt} = \gamma \bar{h}^T P\bar{p}_n\bar{\xi}(\bar{x}),$$

if $s\bar{\theta}^T\bar{\xi}(\bar{x}) \ge 0$ for $|\bar{\theta}|_2 = M_\theta$,

$$\frac{d\bar{\theta}}{dt} = P\{\gamma s\bar{\xi}(\bar{x})\},$$

where the projection operator $P\{*\}$ is defined as

$$P\{\gamma s\bar{\xi}(\bar{x})\} = \gamma s\bar{\xi}(\bar{e}) - \gamma s\frac{\bar{\theta}\bar{\theta}^T\bar{\xi}(\bar{x})}{|\bar{\theta}|_2^2}.$$

The above adaptive law plays the role of tuning M.F. and, simultaneously, keeping the maximum of $|\bar{\theta}|_2$ under the given assumption. Accordingly, by the second assumption,

$$\frac{dV}{dt} \le -\frac{1}{2}\bar{e}^T Q\bar{e} < 0,$$

which sequentially results in

$$V \in L_\infty,$$

and

$$|\bar{e}|_2 \in L_\infty \cap L_2,$$

because $V = \int \frac{dV}{dt} dt < \infty$. Since $w \in L_\infty$ by the second assumption and $|\bar{\phi}|_2 \in L_\infty$ by the third assumption and the projection algorithm,

$$\left|\frac{d\bar{e}}{dt}\right|_2 \in L_\infty,$$

by (15.6). Therefore, by the above relations and Barbalat's lemma, an asymptotic stability of the fuzzy logic control system in (15.6),

$$\lim_{t\to\infty} |\bar{e}|_2 = 0,$$

is obtained.

15.2.3 Properties of the Integral Sliding-Mode Adaptive FLC

Integrating both sides of the relation

$$\frac{dV}{dt} \le -\tfrac{1}{2}\bar{e}^T Q \bar{e} < -\frac{\lambda_{Q_{min}}}{2}|\bar{e}|_2^2,$$

where $\lambda_{Q_{min}}$ is a minimum eigenvalue of a matrix Q, yields

$$\int_0^t |\bar{e}(\tau)|_2^2\, d\tau \le \frac{2}{\lambda_{Q_{min}}}[|V(0)| + |V(t)|].$$

Since $V(0)$ includes $\underline{\phi}^T(0)\underline{\phi}(0)$, which gives the exactness of the prior knowledge used, the above relation means that the more exact the prior knowledge, the better the performance. The characteristics are clarified in Section 15.4. In the previous case, there is no need to satisfy a switching condition, because the proof of stability is independent of the switching condition. But, if the sliding surface is modified to resolve a reaching phase problem and the switching gain $K(t)$ is designed to satisfy the switching condition, then the performance can be made better. In this case, the sliding mode controller plays the main role of both the performance and stability based on the switching condition, while the FLC provides the MAE and allows room to reduce the switching gain by the exact prior knowledge that makes M_θ designed with less value, which will be shown in simulation experiments. Suppose the modified first assumption

- b_L, a lower bound of b, is known: $0 < b_L \le b$.

Equations (15.12) and (15.13) explain the modification of the sliding surface and the new design of switching gain $K(t)$, respectively. Let

$$s = \bar{b}_c^T P^2[(I + P^{-1})\bar{e} - A\int \bar{e}\, dt - (I + P^{-1})\bar{e}(0)] = \bar{b}_c^T P^2 \bar{h}, \qquad (15.12)$$

where

$$\bar{h} = (I + P^{-1})\bar{e} - A\int \bar{e}\, dt - (I + P^{-1})\bar{e}(0).$$

To satisfy the switching condition

$$s\frac{ds}{dt} \le -\eta|s|, \quad \eta > 0,$$

define the switching gain function as

$$K(t) = \max\left(0, \frac{\bar{p}_n^T A\bar{e}(t)}{b_L(|\bar{p}_n|_2^2 + p_{nn})} + |\omega|_\infty + 2M_\theta|\bar{\xi}(\bar{e})|_2\right) + \eta_1, \qquad (15.13)$$

where $\bar{p}_n = [p_{n1}, p_{n2}, \ldots, p_{nn}]^T$, $|\bar{\phi}|_2 = |\bar{\theta}^* - \bar{\theta}|_2 \le |\bar{\theta}^*|_2 + |\bar{\theta}|_2 < 2M_\theta$, and $\eta_1 > \eta > 0$. The designed switching gain gives the relation

$$s\frac{ds}{dt} = \bar{b}_c^T P^2\left[\frac{d\bar{e}}{dt} - A\bar{e} + P^{-1}\frac{d\bar{e}}{dt}\right],$$

$$= -\bar{b}_c^T P^2 \bar{b}_c K \operatorname{sgn}(s) - \bar{b}_c^T P^2 \bar{b}_c \omega + \bar{b}_c^T P^2 \bar{b}_c \bar{\phi}^T \bar{\xi}(\bar{x}) + \bar{b}_c^T P \frac{d\bar{e}}{dt},$$

$$\le -\eta_1|s|.$$

By (15.12) and (15.13), the sliding surface can become zero from the initial time, because (15.12) makes it zero with the initial state part independent of any initial states. Therefore, without the chattering problem, the state response will be described by:

$$\bar{b}_c^T P^2 [(I + P^{-1})\bar{e} - A \int_0^t \bar{e} dt - (I + P^{-1})\bar{e}(0)] = 0 \text{ for } t \geq 0. \quad (15.14)$$

In (15.14), an appropriate state response can be designed by modifying P and A.

15.3 NEW FUZZY LOGIC BASED LEARNING CONTROL: APPROACH II

15.3.1 Structure of the New Fuzzy Logic Based Learning Control

Consider the plant

$$x^{(n)} = f(x, \dot{x}, \ldots, x^{(n-1)}) + u + d(t), \quad y = x, \quad (15.15)$$

where $d(t)$ means an external disturbance. The ideal controller, the error equation, and sliding surface in this second approach are defined respectively as follows:

$$u^* = -f(\bar{x}) + y_m^{(n)} + k_d s + \sum_{i=1}^{n-1} c_i e_{i+1}, \quad (15.16)$$

$$e^{(n)} = -\bar{k}^T \bar{e} + [u_c(\bar{x}|\bar{\theta}^*) - u_c(\bar{x}|\bar{\theta})] + [u^* - u_c(\bar{x}|\bar{\theta}^*)] - d,$$
$$= -\bar{k}^T \bar{e} + \bar{\phi}^T(\bar{x})\bar{\xi} - (w + d), \quad (15.17)$$

$$s = \bar{c}^T \bar{e}, \frac{ds}{dt} = -k_d s + \bar{\phi}^T \bar{\xi}(\bar{x}) - (w + d). \quad (15.18)$$

To overcome deterioration caused by continual improper modification of the parameters of fuzzy systems, a new learning law is defined as

$$\frac{d\bar{\theta}}{dt} = \gamma \left(\frac{ds}{dt} + \lambda ds \right) \bar{\xi}(x). \quad (15.19)$$

To prove the stability of the FLC, suppose that

- $|d(t)| + |w(\bar{x})| < \varepsilon_d$,
- $\lambda_d = k_d + (1/k_d)$,
- $|\bar{x}(t)| \leq M_x$, $|\bar{\theta}(t)| \leq M_\theta$ for $\forall t \geq 0$.

For clarification of the remaining proof, the simplified notations $u_c \equiv u_c(\bar{x}|\bar{\theta})$ and $u_c^* \equiv u_c(\bar{x}|\bar{\theta}^*)$ are used.

15.3.2 Stabilization of the New Fuzzy Logic Based Learning Control

Consider the Lyapunov function candidate as

$$V(s, \bar{\phi}) = \frac{1}{2} \left(\frac{1}{k_d} s^2 + \frac{1}{\gamma} \bar{\phi}^T \bar{\phi} \right). \quad (15.20)$$

Differentiating the above Lyapunov function with respect to time, we obtain:

$$\frac{dV}{dt} = \frac{1}{k_d} s \frac{ds}{dt} + \frac{1}{\gamma} \frac{d\bar{\phi}^T}{dt} \bar{\phi}, \tag{15.21}$$

$$= \frac{1}{k_d} s \frac{ds}{dt} - \left(\frac{ds}{dt} + \lambda_d s\right)(\bar{\theta}^* - \bar{\theta})^T \bar{\xi}(\bar{x}),$$

$$= -s^2 + \frac{1}{k_d} s[(u_c^* - u_c) - (\omega + d)] - [(\lambda_d - k_d)s + (u_c^* - u_c) - (\omega + d)](u_c^* - u_c),$$

$$= -s^2 - \frac{1}{k_d}(\omega + d)s - (u_c^* - u_c)^2 + (\omega + d)(u_c^* - u_c), \tag{15.22}$$

$$< -\left[s^2 - \frac{1}{k_d}\varepsilon_d |s| + (u_c^* - u_c)^2 - \varepsilon_d |u_c^* - u_c|\right],$$

$$< 0,$$

where

$$\forall |s| > \left(\frac{1}{2} - \frac{1}{k_d}\right)\varepsilon_d \quad \text{or} \quad |u_c^* - u_c| > \left(1 + \frac{1}{2k_d}\right)\varepsilon_d. \tag{15.23}$$

From (15.23), $dV/dt < 0$ outside a compact region Ω, where the set Ω is defined as

$$\Omega \equiv \left\{(s, \bar{\phi}), \|s\| \leq \left(\frac{1}{2} - \frac{1}{k_d}\right)\varepsilon_d, |u_c^* - u_c| \leq \left(1 + \frac{1}{2k_d}\right)\varepsilon_d\right\}. \tag{15.24}$$

Since $V(s, \bar{\phi})$ is a scalar function with continuous partial derivatives in Ω^c and satisfies

1. $V(s, \bar{\phi}) > 0$ for $\forall s, \bar{\phi} \in \Omega^c$,
2. $\frac{dV}{dt}(s, \bar{\phi}) \leq 0$ for $\forall s, \bar{\phi} \in \Omega^c$,
3. $V(s, \bar{\phi})$ belongs to the class L_∞, s is uniformly bounded, i.e. $|s| \leq \Delta_d$.

Also, (15.18) can be rewritten as

$$e^{(n-1)} + c_{n-1} e^{(n-2)} + \cdots + c_2 e^2 + c_1 e = s. \tag{15.25}$$

From the input-to-state stability theory and the design assumption about the vector $\bar{c}$, if the input s of the system (15.25) is bounded, then the tracking errors are bounded. Also, the bound of all tracking errors means the bound of all states. The learning law (15.19) can be rewritten as

$$\frac{d\bar{\phi}}{dt} = -\gamma \bar{\xi}(\bar{x}) \bar{\xi}(\bar{x})^T \bar{\phi}^T + \gamma((\lambda_d - k_d)s - \omega - d)\bar{\xi}(\bar{x}). \tag{15.26}$$

Since $\bar{\xi}(\bar{x})$ is piecewise continuous and bounded, when $\bar{\xi}(\bar{x})$ is persistently exciting, i.e.

$$\int_t^{t+T_0} \bar{\xi}(\bar{x}(\tau))\bar{\xi}^T(\bar{x}(\tau))d\tau \geq \alpha I, \quad \forall t \geq t_0, \tag{15.27}$$

it can be shown that the nominal system $d\bar{\phi}/dt = -\gamma \bar{\xi}(\bar{x})\bar{\xi}^T(\bar{x})\bar{\phi}$ has an exponentially stable equilibrium state $\bar{\phi} = 0$. Since $\bar{\xi}(\bar{x})$, s, ω and d are bounded, if s is assumed to be bounded, the perturbation term $\bar{\rho} = \gamma((\lambda_d - k_d)s - \omega - d)\bar{\xi}(\bar{x})$ is also bounded. Since the system (15.26) has the exponentially stable nominal dynamics and the perturbation $\bar{\rho}$ is bounded, the parameter error vector $\bar{\phi}$ is bounded. Therefore, the parameter vector $\bar{\theta}$ is also bounded. From equation (15.22),

$$\frac{dV}{dt} \leq -\frac{1}{2}s^2 - \frac{1}{2}(u_c^* - u_c)^2 + \frac{1}{2}\left(1 + \frac{1}{k_d^2}\right)\varepsilon_d^2. \tag{15.28}$$

Integrating both sides of (15.28), we obtain

$$\int_0^\infty [s^2 + (u_c^* - u_c)^2] dt \leq 2V_0 - 2V_\infty + \int_0^\infty \left(1 + \frac{1}{k_d^2}\right)\varepsilon_d^2 dt. \tag{15.29}$$

Because V_0 and V_∞ are bounded, if w, d, $\varepsilon_d \in L_2$, then it follows that s, $\bar{e}$, $u_c^* - u_c \in L_2$. Also, from (15.17) and (15.18) we can find that s, $\bar{e}$, $u_c^* - u_c \in L_\infty$. Therefore, from Barbalat's lemma, s, $\bar{e}$, and $u_c^* - u_c$ converge to zero as $t \to \infty$. Since w converges to zero, the controller output u_c also converges to the ideal controller output u_c^*.

15.3.3 Discussion of the New Fuzzy Logic Based Learning Control

The typical learning component of the control law dictates that the parameter vector $\bar{\theta}$ is updated to be proportional to a linear combination of the error state as

$$\frac{d\bar{\theta}}{dt} = \gamma s \bar{\xi}(\bar{x}), \tag{15.30}$$

where γ is the learning rate. The update law (15.30) means that the parameters of the fuzzy system are modified in proportion to the magnitude of the distance from the hyperplane $s = \bar{c}^T \bar{e} = 0$ in the error space. This kind of direct fuzzy logic based learning control method may modify the parameters of the fuzzy system whenever the perturbations such as approximation error and external disturbance force the error state to move outside the hyperplane $s = 0$. Thus, it is not difficult to infer that the fuzzy logic based learning control method (15.30) is sensitive to both approximation error and external disturbance, and the control performance can be deteriorated by continual improper modification of the parameters of the fuzzy system. To overcome such problematic behavior, a learning law is designed on the basis of the proposed learning scheme (15.19). Equation (15.19) means that, even though the error state is outside the good trajectory hyperplane, i.e. $s \neq 0$, as long as the state exponentially approaches the hyperplane, i.e. $\frac{ds}{dt} = -\lambda_d s$ is satisfied, the parameters are not modified.

15.4 SIMULATION

15.4.1 Approach I

Consider the following simple non-linear plant:

$$\dot{x}_1 = x_2,$$
$$\dot{x}_2 = x_1 + \sin(x_2) + u,$$
$$y = x_1.$$

The parameters are set as follows:

$$A = \begin{bmatrix} 0 & 1 \\ -2 & -3 \end{bmatrix}, \quad P = \begin{bmatrix} 1.25 & 0.25 \\ 0.25 & 0.25 \end{bmatrix}, \quad b = \begin{bmatrix} 0 \\ 1 \end{bmatrix},$$

$$K = 4, \quad \gamma = 10, \quad \bar{x}(0) = \begin{bmatrix} 1 \\ 0 \end{bmatrix}, \quad \bar{x}_{\text{ref}} = \begin{bmatrix} 5 \\ 0 \end{bmatrix}, \quad M_x = 19, \quad M_\theta = 100,$$

Membership function type: Gaussian; Rule interval: 2; Variance: 1.

The simulations are divided into two parts. One is the case when the switching condition is not satisfied and the other is the case when such a condition is satisfied. Figure 15.1(a) shows the comparison of the state responses in the first case. The prior knowledge used is described in Figure 15.1(b). When no prior knowledge is used, all the elements of $\bar{\theta}$ are initialized to be one. The same condition of prior knowledge is also applied to the second case. As shown in Figure 15.1(a), using prior knowledge enhances the state performance compared with when no prior knowledge is applied. Figure 15.2(a) shows that the performance is fixed (15.14) under the switching condition. In Figure 15.2(b), (c), the different control inputs are shown, depending on whether or not prior knowledge is applied. In the second case, it is shown that prior knowledge plays the role of reducing the control input under the same limitation of the sliding surface value, which is fixed to be 0.05 in this simulation.

15.4.2 Approach II

Consider the Duffing forced oscillation system [1]:

$$\dot{x}_1 = x_2,$$
$$\dot{x}_2 = -0.1x_2 - x_1^3 + 12\cos(t) + u + d,$$
$$y = x_1.$$

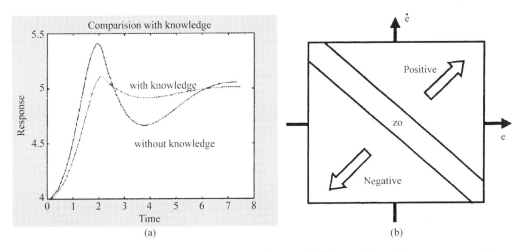

Figure 15.1 Comparison with prior knowledge under no switching condition: (a) state responses; (b) prior knowledge

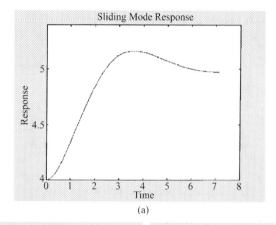

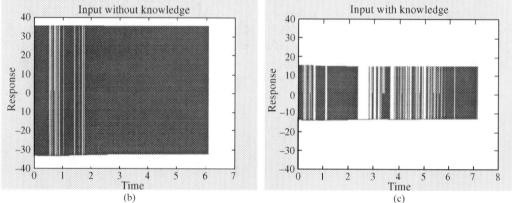

Figure 15.2 Comparison with prior knowledge under the switching condition: (a) state response under the switching condition; (b) control input without prior knowledge; (c) control input with prior knowledge

The system is chaotic if $u(t) = 0$ and $d(t) = 0$. The reference signal is assumed to be $y_m(t) = 1.5\sin(t)$ and the external disturbance $d(t) = \sin(6t)$ is present. Let the initial state $\bar{x}(0) = (2, 0)^T$ and the parameter vectors $\bar{\theta}(0) = \bar{0}$. We simply choose $k_1 = 1$, $k_2 = 2$, $\gamma = 2.5$ and $\lambda_d = 2$. Also, the membership functions for x_1 and x_2 are selected as follows:

$$\mu_{F_{1,i}}(x_i) = \exp[-((x_i + 2.5)/0.4)^2], \quad \mu_{F_{2,i}}(x_i) = \exp[-((x_i + 1.5)/0.4)^2],$$
$$\mu_{F_{3,i}}(x_i) = \exp[-((x_i + 0.5)/0.4)^2], \quad \mu_{F_{4,i}}(x_i) = \exp[-(x_i/0.4)^2],$$
$$\mu_{F_{5,i}}(x_i) = \exp[-((x_i - 0.5)/0.4)^2], \quad \mu_{F_{6,i}}(x_i) = \exp[-((x_i - 1.5)/0.4)^2],$$
$$\mu_{F_{7,i}}(x_i) = \exp[-((x_i - 2.5)/0.4)^2],$$

which cover the interval $[-2.5\ \ 2.5]$. We directly integrate the difference equations of the closed-loop system and the learning law with step size 0.02. Figures 15.3 and 15.4 show the simulation results by the proposed method (15.19) and the conventional method using the adaptive law (15.30). We can see that, in the conventional method, even though the tracking error is bounded, the parameter vector $\bar{\theta}$ diverges as $t \to \infty$. However, in the proposed method, we can find that the effects of both approximation error and external

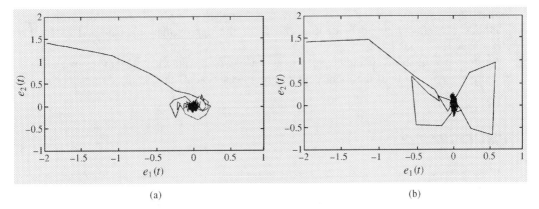

Figure 15.3 Error state trajectory in the phase plane: (a) using the proposed adaptive law (15.19); (b) using the adaptive law (15.30)

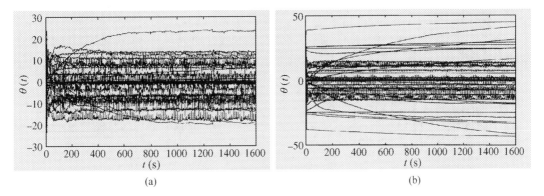

Figure 15.4 Trajectories of elements of parameter vector when the disturbance is present: (a) using the proposed adaptive law (15.19); (b) using the adaptive law (15.30)

disturbances are attenuated efficiently, and all the signals in the system are bounded (Figure 15.4).

15.5 CONCLUDING REMARKS

In this work sliding-mode adaptive fuzzy logic controllers are given to stabilize a non-linear control system in a prespecified bounded state region. In the first approach it is shown that even if there is no requirement of the switching condition, stability can guaranteed. But if we revise a sliding surface with (15.11) and select a time-varying switching gain $K(t)$ according to (15.12), then the performance and the robustness of the given system can be made better, which mainly comes from the integral action of the designed sliding surface (15.10). A direct adaptive fuzzy control method with a novel adaptation law is proposed to improve the robustness of the adaptive systems in the second approach. It is shown that the proposed method can guarantee the bound of all the signals in the system on the assumption of persistent excitation of the fuzzy basis vector. Simulation results show that the proposed methods can efficiently attenuate the effects of the approximation error and external disturbances.

Further investigations should be performed to deal with the practical systems by extending the proposed methods to the more general non-linear systems and to the output feedback control case.

BIBLIOGRAPHY

1. Wang, L. X., Stable adaptive fuzzy control of nonlinear systems, *IEEE Transactions on Fuzzy Systems*, **1**(2), 1993, 146–155.
2. Chen, B. S., C. H. Lee and Y. C. Chang, H^∞ tracking design of uncertain nonlinear SISO systems: Adaptive fuzzy approach, *IEEE Transactions on Fuzzy Systems*, **4**(1), 1996, 32–43.
3. Su, C. Y. and Y. Stephanenko, Adaptive control of a class of nonlinear system with fuzzy logic, *IEEE Transactions on Fuzzy System*, **2**(4), 1994, 285–294.
4. Lu, Y. S. and J. S. Chen, A self-organizing fuzzy sliding-mode controller design for a class of nonlinear servo systems, *IEEE Transactions on Industrial Electronics*, **41**(5), 1994, 492–502.
5. Kim, Y.-T., A study on the Design of Robust Learing Controller Based on a New Learning Paradigm, Ph.D. dissertation in KAIST, 1998.
6. Kim, Y.-T. and Z. Z. Bien, Robust self-learning fuzzy logic controller design for a class of non-linear MIMO systems, *Fuzzy Sets and Systems*, accepted.
7. Bien, Z. Z., *Fuzzy Logic Control*, Hong Reng, 1997.
8. Lee, J. H., J. S. Ko, S. K. Chung, D. S. Lee, J. J. Lee and M. J. Yoon, Continuous variable structure controller for BLDDM position control with prescribed tracking performance, *IEEE Transactions on Industrial Electronics*, **41**(5), 1994, 483–490.
9. Wu, J. C. and T. S. Liu, Fuzzy control stabilization with application to motorcycle control, *IEEE Transactions on Systems, Man and Cybernetics*, **26**(6), 1996, 836–847.
10. Cao, S. G., N. W. Rees and G. Feng, Stability analysis of fuzzy control systems, *IEEE Transactions on Systems, Man and Cybernetics*, **26**(1), 1996, 201–204.
11. Chen, J. Q., J. H. Lu and L. J. Chen, Analysis and synthesis of fuzzy closed-loop control systems, *IEEE Transactions on Systems, Man and Cybernetics*, **25**(5), 1995, 881–888.
12. Driankov, D., H. Hellendoorn and M. Reinfrank, *An Introduction of Fuzzy Control*, Springer-Verlag, Berlin, 1993.
13. Slotine, J. J. E. and W. Li, *Applied Nonlinear Control*, Prentice-Hall, Englewood Cliffs, 1991.
14. Itkis, U., *Control Systems of Variable Structure*, Israel University Press, 1976.
15. Narendra, K. S. and A. M. Annaswamy, *Stable Adaptive Systems*, Prentice-Hall, Englewood Cliffs, 1989.
16. Wang, L. X., *Adaptive Fuzzy Systems And Control*, Prentice-Hall, Englewood Cliffs, 1994.

16

Gain Scheduling Based Control of a Class of TSK Systems[1]

Dimitar P. Filev
Ford Motor Company, Detroit, USA

16.1 INTRODUCTION

It is well-established engineering practice to design controllers for non-linear plants by linearizing the non-linear dynamics at each operating point. The synthesis of a non-linear control system is substituted by synthesis of a family of linear controllers. This practical methodology for non-linear control system design by utilizing the powerful toolbox of linear control methods is known as gain scheduling control. The gain scheduling paradigm is based on the assumption that the non-linear system is represented by a collection of multiple linear models; these models locally approximate the non-linear system in the vicinity of the operating points. Associated with each linear subsystem is a linear controller. The linear controllers are parameterized for all operating points so a transition from one operating point to another results in a transition between the respective linear controllers. In its classical form the gain scheduling approach assumes parameterization by a slow changing external variable. In a broader sense the parameterization can be guided by a state dependent variable.

The Takagi–Sugeno–Kang (TSK) fuzzy model can be considered as a generalization of the gain scheduling concept [1]. Instead of linearizing strictly at an operating point it utilizes the idea of linearization in a fuzzily defined region of the state space. The fuzzy regions are parameterized and each region is associated with a linear subsystem. Owing to the fuzzily defined regions the non-linear system is decomposed into a multimodel structure consisting of linear models that are not necessarily independent. The TSK paradigm naturally defines the mechanism for interpolation between the linear structures based on a soft switching mechanism that is known as the Takagi–Sugeno reasoning method. The increased flexibility

---

[1] A short version of this paper was presented at the Fifth IEEE International Conference on Fuzzy Systems, New Orleans, 1996. Portions reprinted, with permission, from *Proceedings of the Fifth International Conference on Fuzzy Systems*, New Orleans, 1996, pp. 687–693, © 1996 IEEE.

Fuzzy Logic. Edited by S. Farinwata, D. Filev and R. Langari
© 2000 John Wiley & Sons, Ltd

of the TSK representation, however, introduces design problems that do not explicitly exist in conventional gain scheduling. These problems follow from the interpolative properties of the TSK representation. In the TSK model based gain scheduling we deal with a combination of linear subsystems that are associated with a given region of the state space. Subsequently, we consider a combination of linear controllers that are assigned to the linear subsystems of a particular fuzzy region.

The problem of analysis and synthesis based on TSK models have been under extensive investigation since the early 1990s. Primarily, work on TSK control systems have dealt with a non-linear interpretation of the TSK model as a matrix polytope. The polytopic representation establishes sufficient stability conditions for a TSK system using a common Lyapunov function for a set of Lyapunov inequalities. The problem of synthesis of a state feeback that stabilizes a TSK system is formulated in a similar way. The problem of TSK stability is transformed to a convex program [2–4]. The polytopic approach, however, does not utilize to the full extent the fact that the TSK system is characterized with known and well-defined non-linearities that result from the partitioning of the state space; this partitioning is defined by the antecedents of the rules of the TSK model. In addition, the polytopic approach as a rigorous tool for dealing with non-linear systems with a special structure does not completely take advantage of the knowledge content of the TSK fuzzy model. It provides a solution that does not take into account the partitioning of the space and the nature of the parameterization [5]. In general, this approach may lead to conservative solutions.

In this chapter we discuss the issue of stabilization of a TSK fuzzy system from the perspective of gain scheduling. We look at the TSK model as a mechanism for scheduling the model parameters. We develop a non-linear compensator that stabilizes one class of TSK systems by changing according to the schedule of changing model parameters. We derive the conditions for such a compensator to exist.

16.2 TSK MODEL AS A GAIN SCHEDULED SYSTEM

We consider a single-input–single-output (SISO) TSK fuzzy system P that is described by a rule-base of the following format:

If μ_1 is A_{i1} and ... and μ_n is A_{in}
Then $y(k) = b_{i1}u(k-1) + \cdots + b_{in}u(k-n) - a_{i1}y(k-1) - \cdots - a_{in}y(k-n)$, $\quad i = [1, m]$,

$$(16.1)$$

where u and y are the plant input and output, and $\mu_i, i = [1, n]$, are called the markers of the TSK system. These variables may coincide with (some of) the state variables $y(k-1), y(k-2), \ldots, y(k-n)$. They determine regions in the state space (not necessarily disjunct) where the non-linear system dynamics is approximated by the linear models $y(k) = b_{i1}u(k-1) + \cdots + b_{in}u(k-n) - a_{i1}y(k-1) - \cdots - a_{in}y(k-n)$. The fuzzy sets $A_{i1}, A_{i2}, \ldots, A_{in}$ represent linguistic labels that determine a fuzzy partitioning of the space of marker variables into rectangular fuzzy regions $R_i = A_{i1} \times A_{i2} \times \cdots \times A_{in}, i = [1, m]$. The output $y(k)$ of the overall model is inferred from the individual linear subsystems according to the TSK reasoning mechanism:

$$y(k) = \sum_{i=1}^{m} \frac{\tau_i}{\sum_{j=1}^{m} \tau_j} [b_{i1}u(k-1) + \cdots + b_{in}u(k-n) - a_{i1}y(k-1) - \cdots - a_{in}y(k-n)]$$

$$= \sum_{i=1}^{m} v_i [b_{i1}u(k-1) + \cdots + b_{in}u(k-n) - a_{i1}y(k-1) - \cdots - a_{in}y(k-n)], \quad (16.2)$$

where the weights $v_i = \tau_i / \sum_{j=1}^{m} \tau_j$, $i = [1, m]$, are normalized degrees of firing the rules $\tau_i = A_{i1}(\mu_{i1}) \wedge \cdots \wedge A_{in}(\mu_{in})$. For systems of higher order and in the case of multiple inputs and outputs it can be difficult to develop a TSK model of type (16.1). Instead the conjunctive rules can be replaced by a set of rules that assign a linear model to each fuzzy clusters R_i:

If $[\mu_1 \quad \mu_1 \ldots \mu_n]'$ is R_i
Then $y(k) = b_{i1}u(k-1) + \cdots + b_{in}u(k-n) - a_{i1}y(k-1) - \cdots - a_{in}y(k-n), \quad i = [1, m]$.
(16.3)

The output of the overall model is formally the same as in expression (16.2) but the weights v_i are the normalized distances to the cluster centers. The clusters and the models can be obtained by learning from the data [6–8] or by linearization around the operating points.

The markers can be also associated with auxiliary variables that are correlated with changes in the operating conditions [13]. In this case each set of markers determines an operating point and the antecedents of the rules define a measure of closeness to these reference operating points; the marker variables are not dependent on the state vector. The linear models in the consequents of the rules are obtained by linearization at the operating points.

The elements of vector $v = [v_1 \quad v_2 \ldots v_m]$ are non-negative and sum to one. They preserve the ordering of the firing levels of the rules, i.e. $\tau_i \geq \tau_j$ implies $v_i \geq v_j$, but they are not proportional to the firing levels because of the normalization. Henceforth, the v_i's represent the relative rather than the actual degrees of membership to the R_i regions. Because of this reason two completely different types of partitioning may result in the same weight vector v; due to normalization the vector v does not provide a relevant measure of the level of interaction between two regions (and between the linear models that are associated with them).

Overall, the TSK model is a non-linear system with parameters that are scheduled based on the weight vector $v = [v_1 \quad v_2 \ldots v_m]'$; the v_i's are non-linear functions of the markers, i.e. the past output values $y(k-1), y(k-2), \ldots, y(k-n)$, through the degrees of firing, the τ_i's. Only in the case of markers that are associated with operating conditions rather than with the state variables is the TSK model linear. In a broad sense it can be considered a gain scheduling system with the firing levels as scheduling variables. If the marker vector $[\mu_1 \quad \mu_2 \ldots \mu_n]'$ resides in only one R_i region, then the TSK model coincides with the ith linear ARMA model. Because of fuzzy partitioning of the R_i regions the state vector $[y(k-1) \ldots y(k-n)]'$ may belong to more than one R_i region and in this case the output of the TSK is formed by the convex combination of the subsystems that are associated with these regions. The idea behind using the TSK models is to provide a smooth approximation of a non-linear system by combining a set of piecewise linear models. The role of the TSK reasoning mechanism is to switch between the linear subsystems and to provide a smooth approximation in the areas that are covered by more than one region. It is reasonable to expect that the partitioning of the space is dominated by areas that belong to only one region where the TSK model exhibits linear behavior. The main objective of the TSK model is to represent the switching linear structures associated with each of the regions rather than the system dynamics in the areas covered by more than one region. Otherwise, if the TSK model is introduced to approximate a non-linear system that lacks linear areas, then this model does not have significant advantages since it replaces one non-linear model with another (i.e. a TSK model that is designed as a non-linear model).

For a fixed vector of weights $v = [v_1^* \ldots v_m^*]'$ the TSK model transforms into a linear ARMA model:

$$y(k) = b_1^* u(k-1) + \cdots + b_n^* u(k-n) - a_1^* y(k-1) - \cdots - a_n^* y(k-n)],$$

where $b_j^* = \sum_{i=1}^m v_i^* b_{ij}$, and $a_j^* = \sum_{i=1}^m v_i^* a_{ij}$, $j = [1, n]$. It can be viewed as the ratio of polynomials that are parameterized by the weight vector v^* and can be represented (for zero initial conditions) by the transfer function:

$$P^*(z) = \frac{B^*(z)}{A^*(z)} = \frac{b_1^* z^{n-1} + b_2^* z^{n-2} + \cdots + b_n^*}{z^n + a_1^* z^{n-1} + \cdots + a_n^*}. \quad (16.4)$$

One manifestation of the set of all transfer functions (16.4) for all possible vectors v is the polytopic transfer function TSK model:

$$P(z, v) = \frac{\sum_{i=1}^m v_i (b_{i1} z^{n-1} + \cdots + b_{in})}{\sum_{i=1}^m v_i (z^n + a_1 z^{n-1} + \cdots + a_n)}. \quad (16.5)$$

In the following section we discuss the necessary conditions for the stability of the TSK systems represented by the polytopic model (16.5).

16.3 STABILITY CONDITIONS FOR TSK FUZZY SYSTEMS

An evident necessary condition for the stability of a polytopic TSK fuzzy system is the condition for its stability in any region of the state space. This necessary condition translates into a requirement for stability of the characteristic polytope $A(z, v)$ for any vector v. We distinguish two main cases:

1. The marker vector $[\mu_1 \ \mu_2 \ldots \mu_n]'$ resides in one of the regions $R_i = A_{i1} \times A_{i2} \times \cdots \times A_{in}$, $i = [1, m]$; in other words, this is the special case $v_i = 1$ and $v_j = 0$, $j \neq i$, $i, j = [1, m]$;

2. The marker vector $[\mu_1 \ \mu_2 \ldots \mu_n]'$ resides in more than one region R_i, $i = [1, m]$; this is the general case where $v = [v_1 \ v_2 \ldots v_n]'$, i.e. v is any vector with non-negative elements that sum to one.

For the special cases $v_i = 1$ and $v_j = 0$, $j \neq i$, $i, j = [1, m]$, stability of the TSK fuzzy system is implied by the stability of the linear subsystems. This case reflects situations in which the TSK system dynamics is identical to the dynamics of the ith linear subsystem. Stability of these linear subsystems is defined by the Schur stability of denominator polynomials $A_i(z), i = (1, m)$. To deal with the general case we apply the edge theorem [9],

according to which the zeroes of a polytope of polynomials,
$$A(z,v) = \sum_{i=1}^{m} v_i A_i(z) = \sum_{i=1}^{m} v_i(z^n + a_{i1}z^{n-1} + \cdots + a_{in}), \tag{16.6}$$
lie in the open unit disc if and only if the edges $e_{ij}(z)$ of a polytope $A(z,v)$,
$$E_{ij}(z) = tA_i(z) + (1-t)A_j(z), \quad i,j = [1,m], t \in [0,1], \tag{16.7}$$
are Schur stable. Thus the conditions for the stability of a characteristic polytope $A(z,v)$ can be transformed to conditions for the stability of edges $E_{ij}(z), i,j = [1,m]$. An easy check for stability of the convex edge polynomials $E_{ij}(z)$'s follows from the Schur-Cohn stability criterion [10]: stability of a pair of convex monic polynomials $A_i(z)$ and $A_j(z)$ is implied by the lack of negative real eigenvalues of the matrix $T(A_i)T^{-1}(A_j), i,j = [1,m]$, where

$$T(A_i) = \begin{bmatrix} 1 & a_{i1} & \cdots & a_{i,n-3} & a_{i,n-2} - a_{in} \\ 0 & 1 & \cdots & a_{i,n-4} - a_{in} & a_{i,n-3} - a_{i,n-1} \\ \cdots & & & & \\ 0 & -a_{in} & \cdots & 1 - a_{i4} & a_{i1} - a_{i3} \\ -a_{in} & -a_{i,n-1} & \cdots & -a_{i3} & 1 - a_{i2} \end{bmatrix} \tag{16.8}$$

We summarize the above discussion in the following theorem covering the necessary conditions for the stability of a polytopic TSK fuzzy system.

Theorem 1. *The polytopic TSK system (16.5) is stable if the linear subsystem polynomials $A_i(z)$ are stable and the matrices $T(A_i)T^{-1}(A_j), i,j = [1,m]$, have no negative eigenvalues.*

The example below illustrates the application of Theorem 1 for analyzing the necessary conditions for the stability of a polytopic TSK system.

Example 1 Consider the TSK fuzzy system:

If $y(k-2)$ is [shape] $y(k-2)$ Then $y(k) = u(k) + 1.29\ y(k-1) - 1.19\ y(k-2) + 0.41\ y(k-3);$

If $y(k-2)$ is [shape] $y(k-2)$ Then $y(k) = u(k) - 1.32\ y(k-1) - 1.22\ y(k-2) - 0.44\ y(k-3).$

The non-linear transfer function associated with this model is
$$P(z,v) = \frac{B(z)}{A(z,v)} = \frac{1}{z^3 + (-1.29v_1 + 1.32v_2)z^2 + (1.19v_1 + 1.22v_2)z + (-0.41v_1 + 0.44v_2)}.$$

The denominator polytope $A(z,v)$ is formed by the convex sum of the stable polynomials:
$$A_1(z) = z^3 - 1.29z^2 + 1.19z - 0.41, \qquad A_2(z) = z^3 + 1.32z^2 + 1.22z + 0.44.$$

Furthermore, we check the conditions for stability of the polytope $A(z,v)$. The matrices $T(A_1)$ and $T(A_2)$ are
$$T(A_1) = \begin{bmatrix} 1 & -0.88 \\ 0.41 & -0.19 \end{bmatrix}, \qquad T(A_2) = \begin{bmatrix} 1 & -0.88 \\ -0.44 & -0.22 \end{bmatrix}.$$

We form the matrix product:
$$T(A_1)T^{-1}(A_2) = \begin{bmatrix} -3.54 & -10.19 \\ -1 & -3.16 \end{bmatrix}.$$

Its eigenvalues are $\lambda_1 = -0.152$ and $\lambda_2 = -6.533$; henceforth the polytope $A(z, v)$ is unstable and so is the TSK system. We can easily verify that for different values of the parameter a this TSK system is unstable.

If the v_i's are independent variables (as they are considered in robust control) then the necessary conditions of Theorem 1 are sufficient conditions as well. This is the case when the markers are not dependent on the state variables (operating points are used as markers) and the TSK model is a linear system. In general, as we pointed out above, the weights v_i's are non-linear functions of the state variables rather than independent variables and the conditions of Theorem 1 are only necessary conditions for the stability of a TSK system. Common sufficient conditions for the stability of an autonomous state TSK model

If μ_1 is A_{i1} and ... and μ_n is A_{in} **Then** $x(k + 1) = F_i\, x(k)$, $\quad i = [1, m]$, $\qquad$ (16.9)

follow from its matrix polytopic structure:

$$x(k+1) = \sum_{i=1}^{m} F_i x(k) \qquad (16.10)$$

and were originally derived in [2].

Theorem 2 [2] *The equilibrium of a discrete fuzzy system (16.10) is asymptotically stable in the large if there exists a common positive definite matrix P such that*

$$F_i^T P F_i - P < 0$$

for $i = [1, m]$.

In the next sections we focus on one special class of TSK systems that are described by TSK models of the type

If μ_1 is A_{i1} and ... and μ_n is A_{in}
Then $y(k) = b_1 u(k-1) + \cdots + b_n u(k-n) - a_{i1} y(k-1)$
$\qquad - \cdots - a_{in} y(k-n)$, $\quad i = [1, m]$, $\qquad$ (16.11)

or alternatively for constant vector v^* by the ratio of polynomials:

$$P^*(z) = \frac{B^*(z)}{A^*(z)} = \frac{b_1 z^{n-1} + b_2 z^{n-2} + \cdots + b_n}{z^n + a_1^* z^{n-1} + \cdots + a_n^*} \qquad (16.12)$$

and for all vectors v by the polytopic ratio:

$$P(z, v) = \frac{b_1 z^{n-1} + \cdots + b_n}{\sum_{i=1}^{m} v_i(z^n + a_1 z^{n-1} + \cdots + a_n)}. \qquad (16.13)$$

The denominator polynomial of the model (16.13) is a convex sum of polynomials (polytope). We deal with the subsystems $(B(z), A_i(z, v))$ that have different dynamics in the individual R_i regions (different characteristic polynomials $A_i(z), i = [1, m]$) but the same zeroes (the same numerator polynomials $B_i(z) = B(z)$). The assumption of the same zeroes and variable poles is a reasonable simplification of the TSK model structure. According to it the subsystems associated with individual fuzzy R_i regions are characterized by different dynamics (different characteristic polynomials $A_i(z), i = [1, m]$) but the input–output connections (system zeroes) remain fixed (the same numerator polynomial $B(z)$ for all linear subsystems).

16.4 SYNTHESIS OF TSK COMPENSATORS

Our final goal is to design a compensator that stabilizes a plant P which is described by the TSK model (16.12). It is reasonable to expect the compensator to have a model that is similar to the plant model. The polytopic TSK compensator C consists of a feedforward compensator C_1 and a feedback compensator C_2. The output of the compensator C is a combination of the outputs of the feedforward and feedback compensators:

$$u(z) = \frac{T^*(z)}{R^*(z)} w(z) - \frac{S^*(z)}{R^*(z)} y(z). \tag{16.14}$$

For some particular vector v^* the compensator is described by its transfer function:

$$C^*(z) = [C_1^*(z) \ C_2^*(z)] = \left[\frac{T^*(z)}{R^*(z)} \ \frac{S^*(z)}{R^*(z)} \right], \tag{16.15}$$

where

$$C_1^*(z) = \frac{T^*(z)}{R^*(z)} = \frac{\sum_{j=0}^{c} t_j^* z^{c-j}}{z^c + \sum_{j=1}^{c} r_j^* z^{c-j}}, \tag{16.16}$$

$$C_2^*(z) = \frac{S^*(z)}{R^*(z)} = \frac{\sum_{j=0}^{c} s_j^* z^{c-j}}{z^c + \sum_{j=1}^{c} r_j^* z^{c-j}}, \tag{16.17}$$

in which $r_j^* = \sum_{i=1}^{m} v_i^* r_{ij}, j = [1, c], t_j^* = \sum_{i=1}^{m} v_i^* t_{ij}$, and $s_j^* = \sum_{i=1}^{m} v_i^* s_{ij}, j = [0, c]$. The set of compensators C is manifested by the ratio of polytopes:

$$[C_1(z, v) C_2(z, v)] = \left[\frac{\sum_{i=1}^{m} v_i(t_{i0} z^c + t_{i1} z^{c-1} + \cdots + t_{ic})}{\sum_{i=1}^{m} v_i(z^c + r_{i1} z^{c-1} + \cdots + r_{ic})} \ \frac{\sum_{i=1}^{m} v_i(s_{i0} z^c + s_{i1} z^{c-1} + \cdots + s_{ic})}{\sum_{i=1}^{m} v_i(z^c + r_{i1} z^{c-1} + \cdots + r_{ic})} \right]. \tag{16.18}$$

We call expression (16.18) the *non-parameteric TSK compensator*. The compensator and the plant have the same weights. Any change in the plant model invokes a change in the compensator. Compensator parameters are scheduled based on the plant model parameters following the vector of weights v. One possible approach to the design of the compensator that guarantees stability of the closed-loop system is to assign the same characteristic polynomial to all linear subsystems [12]. In addition, we require that the numerator polynomial of the closed-loop transfer function be independent of the vector v.

If we determine a set of compensators of the type (16.12) that assign a stable characteristic polynomial $D(z)$ and a given numerator polynomial $N(z)$ to the closed-loop system for any vector v, then the closed-loop characteristic polynomial will be $D(z)$ independently of v. In addition, the structure of the compensator (16.14) guarantees that the numerator polynomial of the transfer function of the closed-loop system does not depend on v, i.e. the closed-loop system will be stable. Lemma 1 determines the conditions under which we can calculate the polynomials $T^*(z), S^*(z)$ and $R^*(z)$ that assign the desired polynomials $D(z)$ and $N(z)$ to the closed-loop system for any vector v.

Lemma 1 *Given a plant P with a TSK model (16.13) and desired polynomials $D(z)$ and $N(z)$, there exists a TSK compensator (16.14) that assigns given polynomials $D(z)$ and $N(z)$ to the closed-loop system transfer function if and only if the numerator polynomial $B(z)$ and* the denominator polytope $A(z, v)$ of the TSK polytopic representation are coprime for any weight vector v and polynomial $B(z)$ is a stable polynomial. For some v^* the polynomials $T^*(z), S^*(z)$ and $R^*(z)$ of the compensator C are the solutions of the Diophantine equations:

$$A^*(z)R^*(z) + B^*(z)S^*(z) = D(z)B(z)Q(z), \quad (16.19)$$

$$T^*(z) = N(z)Q(z), \quad (16.20)$$

where $Q(z)$ is an arbitrary stable polynomial that is chosen to guarantee proper transfer functions $C_1^*(z) = T^*(z)/R^*(z)$ and $C_2^*(z) = S^*(z)/R^*(z)$ of the compensator subsystems C_1 and C_2.

Proof By substitution into the transfer function of the closed loop system under the control law (16.19)–(16.20) for some v^* we get $y(z)/w(z) = N(z)/D(z)$. Diophantine equation (16.19) follows from the known algebraic pole assignment theorem of linear control. The requirement for stability of the polynomial $B(z)$ reflects the fact that the unstable zeroes of a linear system (the zeroes of the polynomial $B(z)$) cannot be eliminated from closed-loop system dynamics by the compensator.

In the above lemma we assumed that $B(z)$ and $A(z, v)$ are coprime for any $v \in V$. While the condition of coprimeness of the numerator and denominator polynomials in linear system is equivalent to controllability and observability of the system, in the case of a polytopic model it has a different formulation. The necessary and sufficient conditions for coprimeness of the polynomial $B(z)$ and the polytope $A(z, v)$ are given by the following theorem.

Theorem 3 [12] Let Z_b be the set of zeroes of the numerator polynomial $B(z)$. The polynomial $B(z)$ and the polytope $A(z, v) = \sum_{i=1}^{m} v_i A_i(z)$ are coprime for any $v \in V$ if and only if $A_i(z_0), i = [1, m]$, are of the same sign for any $z_0 \in Z_b$.

Proof $B(z)$ and $A(z, v)$ have a common factor z_0 if and only if $B(z_0) = 0$ and $A(z_0, v) = 0$ for some $v \in V$. But $A(z_0, v) = \sum_{i=1}^{m} v_i A_i(z_0) = 0$ for some $v \in V$ if and only if $0 \in [\min(A_i(z_0)), \max(A_i(z_0))]$, because of $v_i \geq 0$ and $\sum_{i=1}^{m} v_i = 1$.

Example 2 We consider a plant described by the following TSK model:

If $y(k-2)$ is ⟋ $y(k-2)$ Then $y(k) = u(k-2) - y(k-1) - 0.5y(k-2)$,

If $y(k-2)$ is ⟍ $y(k-2)$ Then $y(k) = u(k-2) + y(k-1) - 0.5y(k-2)$.

Our objective is to design a compensator that stabilizes this plant by assigning a characteristic polynomial

$$D(z) = z^2 - 0.2z - 0.05.$$

The desired numerator polynomial of the closed-loop transfer function is $N(z) = 1$. The TSK model of this plant in polytopic form is

$$P(z, v) = \frac{1}{v_1(z^2 + z + 0.5) + v_2(z^2 - z + 0.5)} = \frac{1}{z^2 + (v_1 - v_2)z + 0.5}.$$

We are looking for a TSK compensator C by (16.19)–(16.20) that assigns the characteristic polynomial $D(z)$ to the closed-loop system. Such a compensator can be obtained by application of Lemma 1. Evidently, $B(z)$ and $A(z, v)$ are coprime for any $v \in V$.

Furthermore, we chose $Q(z) = z + 0.1$. We can calculate $S^*(z)$ and $R^*(z)$ from the Diophantine equation (16.19) for any weight vector $v \in V$:

$$(z^2 + (v_1 - v_2)z + 0.5) R^*(z) + 1S(z) = (z^2 - 0.2z - 0.05)(z + 0.1)$$
$$= z^3 - 0.1z^2 + 0.03z - 0.005.$$

Because of the required $N(z) = 1$ polynomial, $T^*(z)$ equals the polynomial $Q(z)$, i.e. $T^*(z) = z + 0.1$.

Five consecutive values of the antecedent state variable $y(k - 2)$, vector v, and respective polynomials $S^*(z)$ and $R^*(z)$ are listed in Table 16.1.

Example 3 Given a plant with a TSK model as follows:

If $y(k - 3)$ is ⟋⟍ Then $y(k) = u(k - 1) - 0.25 u(k - 3) + 3.37 y(k - 1)$
$\quad\quad\quad -2\quad 2$
$- 8.42 y(k - 2) + 5.44 y(k - 3),$

If $y(k - 3)$ is ⟋ Then $y(k) = u(k - 1) - 0.25 u(k - 3) + 4.1 y(k - 1)$
$\quad\quad\quad -2\quad 2$
$- 0.25 y(k - 2) + 0.98 y(k - 3).$

The goal is to stabilize this system. The polytopic TSK model of the system is

$$P(z, v) = \frac{z^2 - 0.25}{v_1(z^3 - 3.37z^2 + 8.42z - 5.4375) + v_2(z^3 - 4.1z^2 + 0.25z - 0.98)}.$$

In order to apply Lemma 1 we check the coprimeness of the numerator polynomial $B(z) = z^2 - 0.25$ and the denominator polytope

$$A(z, v) = v_1(z^3 - 3.37z^2 + 8.42z - 5.4375) + v_2(z^3 - 4.1z^2 + 0.25z - 0.98)$$

by using Theorem 3. The set of zeroes of the numerator polynomial is $Z_b = [-0.5, 0.5]$. By substituting for z into the polynomials $A_1 = z^3 - 3.37z^2 + 8.42z - 5.4375$ and $A_2(z) = z^3 - 4.1z^2 + 0.25z - 0.98$ we calculate $A_1(-0.5) = -10.6175$ and $A_1(0.5) = -1.9425$; $A_2(-0.5) = -2.255$ and $A_2(0.5) = -1.755$. Therefore, according to Theorem 3, the numerator and denominator of $P(z, v)$ are coprime for any vector $v \in V$. $B(z)$ is a stable polynomial and by Lemma 1 there exists a compensator assigning the desired characteristic and numerator polynomials to the transfer function of the closed-loop system. We chose the

Table 16.1 Five consecutive values of $y(k - 2)$, vector v, and polynomials $S^*(z)$ and $R^*(z)$

k	$y(k - 2)$	$v(1)$	$v(2)$	$S^*(z)$	$R^*(z)$
3	1	0	1	$0.430z - 0.445$	$z + 0.900$
4	0.200	0.400	0.600	$-0.450z - 0.045$	$z + 0.100$
5	-0.010	0.505	0.495	$-0.469z + 0.060$	$z - 0.110$
6	-0.012	0.506	0.494	$-0.469z + 0.061$	$z - 0.112$
7	-0.002	0.501	0.499	$-0.470z - 0.056$	$z - 0.102$

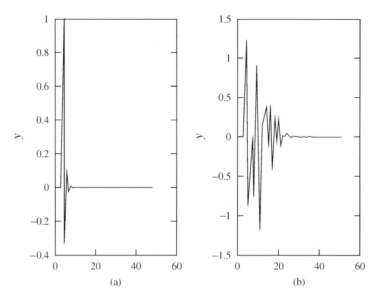

Figure 16.1 Example 3: a. Impulse response of the closed loop system. Controller parameters are calculated from the nominal TSK model; b. Impulse response of the closed loop system. Controller parameters are calculated from a perturbed TSK model

desired characteristic polynomial to be the stable polynomial:

$$D(z) = z^3 - 0.3093\, z^2 - 0.002\, z - 0.0024$$

and the desired numerator polynomial $N(z) = 1$. We further augment the polynomial $D(z)$ with the numerator polynomial $B(z)$. We chose $Q(z) = 1$ and calculated $S^*(z)$ and $R^*(z)$ from the Diophantine equation (16.19) for any weight vector $v \in V$:

$$(v_1(z^3 - 3.37z^2 + 8.42z - 5.4375) + v_2(z^3 - 4.1z^2 + 0.25z - 0.98))R^*(z) + (z^2 - 0.25)S^*(z)$$
$$= (z^3 - 0.3093\, z^2 - 0.002\, z - 0.0024)(z^2 - 0.25)$$
$$= z^5 - 0.3093\, z^4 - 0.2520\, z^3 - 0.0797\, z^2 + 0.0005\, z + 0.0006.$$

This equation is the solution for any vector v. According to equation (16.20) we get for the polynomial $T^*(z) = 1$. The impulse response of the closed-loop system is presented in Figure 16.1(a). The same characteristic is shown in Figure 16.1(b) under the assumption of parametric disturbances within the range of 30% of the nominal values of the TSK model parameters.

16.5 ANALYTIC FORM OF THE POLYTOPIC TSK COMPENSATOR

Lemma 1 does not provide an analytic form of the non-parametric TSK compensator C that assigns the desired transfer function to the closed-loop system. In what follows we explicitly specify the parameters of a proper TSK compensator that stabilizes the closed-loop system.

From equation (16.19) it follows that for a given TSK model (16.13) and the desired closed-loop system of order n i.e. $D(z) = d_0 z^n + d_1 z^{n-1} + \cdots + d_n$, there exists a proper non-parametric TSK compensator of order $c = n - 1$ for any v if and only if the conditions of the Theorem 3 are satisfied. The order of the arbitrary stable polynomial $Q(z)$ is chosen based on the order of the numerator polynomial $B(z)$:

$$\text{order}(Q(z)) = n - 1 - \text{order}(B(z)), \tag{16.21}$$

assuming that $\text{order}(B(z)) = 0$ implies $Q(z) = 1$. The polynomial $\bar{D}(z) = D(z)B(z)Q(z)$ is of order $(2n - 1)$. The order of $\bar{D}(z)$ is $(2n - 1)$, i.e. $\bar{D}(z) = z^{2n-1} + \bar{d}_1 z^{2n-2} + \cdots + \bar{d}_{2n-1}$. For any $v \in V$ the vector $\bar{C}_2 = [s_0^* \ s_1^* \ldots s_{n-1}^* \ r_0^* \ r_1^* \ldots r_{n-1}^*]$ formed by the coefficients of the polynomials $S^*(z)$ and $R^*(z)$ of the compensator C_2 is calculated as a solution to the matrix equation:

$$\bar{C}_2 = \tilde{S}^{-1}\bar{D}', \tag{16.22}$$

where the vector $\bar{D}$ is formed by the coefficients of the polynomial $\bar{D}(z) = [\bar{d}_0 \ \bar{d}_1 \ \bar{d}_2 \ldots \bar{d}_{2n-1}]$ and the matrix $\tilde{S}$ is the modified Sylvester matrix of the polytopic TSK model (16.13):

$$\tilde{S} = \sum_{i=1}^{m} v_i \begin{bmatrix} 0 & 0 & \cdots & 0 & 1 & 0 & \cdots & 0 \\ b_1 & 0 & \cdots & 0 & a_{i1} & 1 & \cdots & 0 \\ \cdot & b_1 & \cdots & 0 & \cdot & a_{i1} & \cdots & 0 \\ b_n & \cdot & \cdots & 0 & a_{in} & \cdot & \cdots & 0 \\ 0 & b_n & \cdots & 0 & 0 & a_{in} & \cdots & 1 \\ 0 & 0 & \cdots & b_1 & 0 & 0 & \cdots & a_{i1} \\ 0 & 0 & \cdots & \cdot & 0 & 0 & \cdots & \cdot \\ 0 & 0 & \cdots & b_n & 0 & 0 & \cdots & a_{in} \end{bmatrix}. \tag{16.23}$$

The order n of the desired polynomial $D(z)$ and the requirement for a strictly proper polytopic TSK system P follow from the consideration of a proper TSK compensator. Inversion of the modified Sylvester matrix $\tilde{S}$ exists for any weight vector v due to the coprimeness of $B(z)$ and $A(z, v)$ which is guaranteed according to the Theorem. The polynomial $T(z)$ of the compensator C_1 is obtained by multiplying $Q(z)$ and $N(z)$ according to equation (16.20). This guarantees that for every v the transfer function of the closed-loop system is the ratio of polynomials $N(z)$ and $D(z)$.

Example 4 Given a TSK model:

If $y(k-2)$ is ⟋‾ $y(k-2)$ **Then** $y(k) = u(k-2) - 1.17u(k-3) + 0.371u(k-4)$
$-0.029u(k-5) - 0.376y(k-1) - 1.411y(k-2) - 2.324y(k-3) - 0.471y(k-4)$
$-1.564y(k-5),$

If $y(k-2)$ is ⟋‾ $y(k-2)$ **Then** $y(k) = u(k-2) - 1.17u(k-3) + 0.371u(k-4)$
$-0.029u(k-5) - 2.528y(k-1) - 3.793y(k-2) - 0.694y(k-3) + 1.055y(k-4)$
$-2.213y(k-5),$

and a desired transfer function of the closed-loop system

$$P_0(z) = \frac{1}{z^9 - 0.91z^8 - 0.38z^7 + 1.09z^6 - 0.29z^5 - 0.34z^4 + 0.13z^3 + 0.008z^2 - 0.005z + 0.0004}.$$

We calculate a non-parametric compensator by (16.14) that assigns the above transfer function to the closed-loop system. The polytopic TSK model associated with this plant is

$$P(z, v) = \frac{B(z)}{v_1 A_1(z) + v_2 A_2(z)},$$

where

$$B(z) = z^3 - 1.17 z^2 + 0.371 z - 0.029,$$
$$A_1(z) = z^5 + 0.376 z^4 + 1.411 z^3 + 2.324 z^2 + 0.471 z + 1.564,$$
$$A_2(z) = z^5 + 2.528 z^4 + 3.793 z^3 + 0.694 z^2 - 1.055 z + 2.213.$$

It is easy to verify that individual subsystems are in minimal realization. The numerator has three zeros, i.e. $Z_b = \{0.7, 0.35, 0.12\}$. By substituting for z into the polynomials $A_1(z)$, and $A_2(z)$ we calculate $A_1(0.7) = 3.775, A_2(0.7) = 3.891; A_1(0.35) = 2.085, A_2(0.35) = 2.135; A_1(0.12) = 1.657, A_2(0.12) = 2.103$. Therefore, according to Theorem 3 the numerator and denominator of $P(z, v)$ are coprime for any vector $v \in V$. The polynomial $B(z)$ is stable and $P(z, v)$ is strictly proper, i.e. the conditions of Lemma 1 are satisfied. According to (16.21) the polynomial $Q(z)$ should be of order 1 so we chose $Q(z) = z + 0.71$. Thus for the polynomial $\bar{D}(z)$ we get:

$$\bar{D}(z) = (z^9 - 0.91 z^8 - 0.38 z^7 + 1.09 z^6 - 0.29 z^5 - 0.34 z^4 + 0.13 z^3 + 0.008 z^2 - 0.005 z + 0.0004)$$
$$\times (z^3 - 1.17 z^2 + 0.371 z - 0.029)(z + 0.71)$$

and for the vector of coefficients $\bar{D}$:

$$\bar{D} = [1 \quad -0.91 \quad -0.385 \quad 1.097 \quad -0.289 \quad -0.336 \quad 0.127 \quad 0.008 \quad -0.005 \quad 0.0004]$$

By substituting in (16.22) we obtain an analytic expression for the joint vector $\bar{C}_2$ containing the coefficients of the compensator numerator and denominator polytopes $S^*(z)$

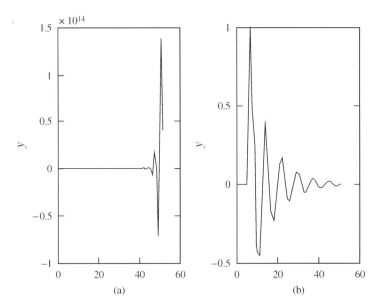

Figure 16.2 Example 4: a. Impulse response of the TSK model; b. Impulse response of the closed loop system

and $R^*(z)$:

$$\bar{C}_2 = [s_0^* \ldots s_4^* \quad r_0^* \ldots r_4^*] = [v_1 S_1 + v_2 S_2]^{-1} \bar{D}'.$$

The square matrices S_1 and S_2 are the Sylvester matrices of the subsystems (A_1, B) and (A_2, B). Following expression (16.20) the polynomial $T^*(z)$ coincides with $Q(z)$ since

$$T^*(z) = N(z)Q(z) = z + 0.71.$$

The impulse responses of the given TSK model and the closed-loop system are shown in Figure 16.2(a) and (b).

16.6 PARAMETERIZATION OF NON-PARAMETERIC TSK COMPENSATORS

In the previous sections we considered assigning same characteristic polynomial $D(z)$ to all subsystems. We shall further discuss the case of assigning different characteristic polynomials by guiding the selection based on the partitioning of the marker variables. The mechanism of selection of alternative characteristic polynomials is realized by considering the compensator of variable parameters:

$$\bar{C} = \tilde{S}^{-1} D_t', \qquad (16.24)$$

where the index t is selected according to the rule:

$$t = \begin{cases} j & \tau_j = \max_i \tau_i; \tau_j \geq \tau_{\min}, i, j = [1, m], \\ 0, & \text{otherwise}, \end{cases} \qquad (16.25)$$

and the vectors D_t' are formed by the coefficients of the desired $(m+1)$ stable characteristic polnomials of the closed-loop system.

The TSK compensator as defined by expressions (16.24) and (16.25), switches between alternative control laws. It picks the jth linear subsystem that is dominating the system dynamics, i.e. the subsystem that is associated with the rule of maximal firing level $\tau_j > \tau_{\min}$ and assigns to it the jth characteristic polynomial. If all the firing levels are below $\tau_{\min}$, i.e. the system is in transition from one operating point to another, then it assigns a default characteristic polynomial D_0. The reason for introducing a TSK compensator of this type is the fact that the normalization of the firing levels effectively modifies the membership functions, although the ordering between the firing levels of the rules is preserved in their normalized counterparts. One important feature of this compensator is that the linear compensator subsystems do not interact. In a sense, it can be thought of as a TSK model with a max type defuzzifier rather than the conventionally used COA type defuzzifier in (16.2).

In order to deal with the problem that the switching from one compensator to another may result in instability we shall use Theorem 2 to guarantee that the autonomous closed-loop model is stable. The autonomous state space model of the closed-loop system under the control law (16.24)–(16.25) is

$$\bar{x}(k+1) = \bar{F}_t \bar{x}(k), \qquad (16.26)$$

where t changes according to the switching rule (16.25). This system is of order $(2n-1)$ and the matrices $\bar{F}_i$ contain the coefficients of the characteristic polynomials $D_i, i = [0, m]$.

Thus, by selecting the matrices $\bar{F}_i$ that satisfy the sufficient condition of Theorem 2 we can effectively determine such characteristic polynomials D_i that guarantee stability of the free system (16.26). Henceforth, a sufficient condition for stability of the system (16.26) is the existence of a positive definite matrix $\bar{P}$ such that the set of matrix inequalities

$$\bar{F}_i^{\mathrm{T}} \bar{P} \bar{F}_i - \bar{P} < 0, \quad i = [0, m]. \tag{16.27}$$

holds. Then the associated characteristic polynomials $D_i, i = [0, m]$, respectively, of the compensator (16.24) guarantee stability of (16.26).

16.7 CONCLUSION

We have demonstrated that the conditions for synthesis of stable TSK fuzzy systems can be derived by considering the TSK model as a gain scheduling system. This leads to the design of non-parameteric compensators that change with the TSK model. By using results from linear control theory we can eliminate the effects of non-linearity of the TSK model and achieve stability of the closed-loop system for one class of TSK systems.

REFERENCES

1. Takagi, T. and M. Sugeno, Fuzzy identification of systems and its application to modeling and control, *IEEE Transactions on Systems, Man and Cybernetics*, **15**, 1985, 116–132.
2. Tanaka, K. and M. Sugeno, Stability analysis of fuzzy systems using Lyapunov's direct method, *Proceedings of NAFIPS'90*, 1990, pp. 133–136.
3. Tanaka, K. and M. Sano, A robust stabilization problem of fuzzy control systems and its application to backing up control of a truck-trailer, *IEEE Transactions on Fuzzy Systems*, **2**, 1994, 119–134.
4. Wang, H. O., K. Tanaka and M. Griffin, An approach to fuzzy control of nonlinear systems: Stability and design issues, *IEEE Transactions on Fuzzy Systems*, **4**(1), 1996, 14–23.
5. Sugeno, M., Toward stability analysis of TSK fuzzy control systems, Plenary lecture at IFSA '95, Sao Paulo, 1995.
6. Filev, D., Control of nonlinear systems described by quasilinear fuzzy models, In: H.-J. Sebastian and K. Tamer (eds.), *Lecture Notes in Control and Information Sciences*, vol.143: *System Modelling and Optimization*. Springer-Verlag, New York, 1989, pp. 591–598.
7. Filev, D., Fuzzy modelling of complex systems, *International Journal of Approximate Reasoning*, **4**, 1991, 281–290.
8. Yager, R. and D. Filev, *Essentials of Fuzzy Modelling and Control*, John Wiley, New York, 1994.
9. Bartlet, A. C., C. V. Hollot and L. Huang, Root locations of an entire polytope of polynomials: it suffices to check the edges, *Math. Control Signals Systems* **1**, 1988, 67–71.
10. Ackermann, J. E. and B. R. Barmish, Robust Schur stability of a polytope of polynomials, *IEEE Transactions on Automatic Control*, **AC-33**, 1988, 984–986.
11. Barmish, B. R., *New Tools for Robustness of Linear Systems*, Macmillan, New York, 1994.
12. Filev, D., Polytopic TSK fuzzy systems: Analysis and synthesis, *Proceedings of the fifth IEEE International Conference on Fuzzy Systems*, 1996, pp. 687–693.
13. Palm, R., Design of fuzzy controllers, In: H. Nguyen and M. Sugeno (eds.), *Fuzzy Systems Modeling and Control*, Kluwer, Boston, 1998, pp. 227–272.

17
Output Tracking Using Fuzzy Neural Networks

Dimitrios K. Pirovolou
Anadrill-Schlumberger, Sugar Land, Texas, USA

George Vachtsevanos
Georgia Institute of Technology, Atlanta, USA

17.1 INTRODUCTION

Manipulators used for welding, painting, etc. are required to track specified trajectories in the presence of uncertainties in their dynamics. In process control very often temperature or concentration are required to track a specific profile. In many of these cases, a detailed model of the system to be controlled is not available. Therefore, there is a need to develop control methodologies that do not rely on the existence of a mathematical model and that can take advantage of qualitative information about the system (heuristics) or any available data.

The tracking problem is one that, by its nature, is more complicated than set-point control. If state feedback is used, then the states of the system must be available directly or through an observer. If output feedback is employed, then the controller itself turns out to be in the form of a dynamic system.

The problem of controlling a system so that its output follows some desired path was first addressed by Brockett and Mesarovic [1] in 1965 for time-invariant linear control systems. The authors used an inverse system to generate the required control. Since then, a number of approaches to the linear output tracking have been explored [2–4].

For non-linear control systems, the output tracking problem is more difficult, but a number of well-known results from the theory of linear systems have been generalized to the non-linear case [5–9]. To a large extent, this has been possible because the methods used by Silverman [3] in the linear case can be generalized, provided that one removes from the state space a co-dimension, i.e. one sub-manifold of 'singular points', which do not exist in the linear case. At a singular point, the control effort becomes unbounded.

Hirschorn and Davis [10, 11], identify a class of output functions for which global output tracking results can be obtained. These are output functions that have certain 'observability'

Fuzzy Logic. Edited by S. Farinwata, D. Filev and R. Langari
© 2000 John Wiley & Sons, Ltd

properties that permit one to accomodate transversals of the set of singular points by use of the available output data. The authors develop conditions that a reference trajectory must fulfill, using the dynamics of the system under consideration, so that it can be tracked by the output of the system.

Behtash [12] studies the output tracking problem in the presence of uncertainties. First, a control law is designed that makes the output of a minimum-phase non-linear system track a bounded reference signal with bounded derivatives. Next, this control law is modified to compensate for the effect of uncertainties. Two types of uncertainties are considered: those satisfying a generalized matching condition but otherwise unstructured, and linear parametric uncertainties. For the first kind of uncertainties it is proved that high gain control can result in an error of $O(\varepsilon)$, where ε is a small design parameter. For the second kind of uncertainties, the author uses adaptive control techniques. This scheme is considerably simpler than the augmented error scheme proposed by Narendra et al. [13] for linear systems, and by Sastry and Isodori [14] for non-linear systems. The tracking error becomes small, but it does not become zero, as opposed to the augmented error methods.

The above-mentioned methodologies are attempting to solve the output tracking problem for non-linear systems by extending the results derived in linear systems theory. One basic concept is the inevitability of the input–output mapping. In the non-linear case, however, there are problems that do not exist in linear systems, such as singular points in the state space and lack of simple tests for the observability and controllability properties. More recently, a considerable amount of work in these problem areas has been reported. Several results have been produced and many techniques, such as the Internal Model Principle, have been developed that can be applied to non-linear systems.

The fundamental weakness in these approaches is that a model of the system is required for the design of a control law. In the adaptive case, the value of a parameter vector is unknown, but the structure of the model is known. Unfortunately, a model good for control purposes is not always available. In addition, the cost of deriving a model from first principles or from data might be prohibitively high. Therefore, non-model-based techniques, among other techniques, are currently being investigated for the output tracking problem.

References to the tracking problem using non-conventional techniques, like fuzzy logic or neural networks, have surfaced recently. Most of the designs presented in these articles refer to specific systems. The results seem to be satisfactory, but, unfortunately, there is no theoretical background to justify their widespread use. Narendra and Parthasarathy [15, 16] use feedforward neural networks and learning by back-propagation for the solution of the tracking problem. There are basically two drawbacks with this approach. First, the learning mechanism using back-propagation may result in local minima of the error surface. Second, the appropriate size of a network that will solve the problem is not known a priori. Therefore, this approach might give good results, but such results cannot be guaranteed in the design phase. If the size of the network is not adequate, for example, then a new network must be trained, the size of which is decided by trial and error.

Isidori's work on this problem [17, 18] is based on the availability of a model. However, a number of concepts developed by the author can be used for qualitative analysis.

Schneider et al. [19] design a fuzzy controller for tracking a non-maneuvering target. They train the controller off-line using a set of already observed trajectories. The disadvantages are that this approach runs off-line, and that there is basically no theoretical background on the capabilities of this controller. Nothing can be said about the performance of the controller before simulation or experimental results are made available.

Wang and Mendel [20] designed a rule-based system to represent a mapping from an input space to R. A fuzzy controller is of this variety. In this architecture, the position and the shape of the membership functions, as well as the number of rules, are free parameters, which are adjusted so that a performance index is minimized. This approach, however, requires the availability of output training data. Also, it is an off-line methodology. In the examples presented in [20], the authors use the beam and ball problem from [21], and they design a fuzzy controller as a substitute for the controller proposed in [21]. The output of the original controller is used as training data for the fuzzy controller. As it is to be expected, the original controller performs better. This design has two limitations. First, it cannot run on-line because it is computationally expensive, and second, it requires the availability (not just the existence) of some control. Its advantage is that it can integrate very naturally into the same framework qualitative knowledge about the system and analytical information about the control law.

Nedungadi [22] uses a fuzzy system to translate from desired Cartesian coordinates to appropriate joint variables for a robotic manipulator. In other words, he avoids solving the inverse kinematic equations. The author is presenting some satisfactory results in the sense that the response is faster, but the error is larger compared with the case where the inverse kinematics are used to find the joint variables. The drawback is that he is relying heavily on heuristics. For example, there is no justification for the number of membership functions used. This is typical in many applications of fuzzy logic and/or neural networks.

In [23] the authors are training a neural network, which is then used as a computed torque controller. The inputs to the network are the desired joint variables and their derivatives, and the output is a vector of the joint torques. The idea here is similar to the concept pursued by Narendra and his students.

Johnson and Leahy [24] suggest an adaptation mechanism based on neural networks in order to estimate the payload of a manipulator, which is then used to/adjust the feedforward compensator. The training of the network is done off-line. The adaptation concept appears in the adjustment of the compensator. A model of the manipulator is still required.

Most of the published results on fuzzy logic control [25], and more particularly on the tracking problem, depend heavily on heuristics. There are some good results but they are application-specific. There is no systematic way to decide on the size of the rule base, to assess the performance of the system without going to experiments, or to make claims about stability.

This chapter is organized as follows. In Section 17.2 the problem statement and the underlying assumptions are given. Section 17.3 describes the structure of the controller. Section 17.4 presents a result, which falls in best approximation theory, and which provides a tool for determining the size of the rule base. In Section 17.5 the learning algorithm is explained and in Section 17.6 some examples are given to illustrate the effectiveness of the scheme.

17.2 PROBLEM STATEMENT—ASSUMPTIONS

Consider a system that has input $u(t) \in R$, state $x(t) \in R^n$ and output $y(t) \in R$. The system is described by equation (17.1):

$$\dot{x} = f(x) + g(x)u,$$
$$y = h(x), \tag{17.1}$$

where f and g are smooth vector fields, and h is also smooth. Many industrial processes fit this description: furnaces, robotic manipulators, chemical processes, etc. It is assumed that this system equation is not known explicitly, and that the states of the system are available, either by direct measurements or indirectly through other measurements. The objective is to design a controller so that for some $\varepsilon > 0$ there exists $t^* > 0$ such that the output of the system approaches a desired output (reference signal), $y_R(t)$, within $\pm \varepsilon$ at $t = t^*$ and stays within this envelope ($\pm \varepsilon$) for all $t \geq t^*$. Expressed mathematically, a controller has to be designed so that

$$\text{for } \varepsilon > 0, \exists\, t^*, 0 < t^* < \infty, \text{ such that } |y_R(t) - y(t)| < \varepsilon, \quad \forall t \geq t^*. \tag{17.2}$$

From now on this will be referred to as the *control objective*. It is also assumed that there exists an (unknown) ideal control $u^*(t)$, within the available control authority, that will drive the tracking error asymptotically to zero. The idea is to design a controller that will be able to generate a control $u(\mathbf{x}, \mathbf{e})$, which is sufficiently close to u^*. There are two questions that have to be answered. First, how close should $u(\mathbf{x}, \mathbf{e})$ be to u^* so that the control objective is achieved? Second, what is the complexity of the controller (i.e. the number of rules) that can produce such a control u? The first question is being addressed next. The second question will be addressed in Section 17.4.

Perturbation theory will be used to answer the first question. Let the nominal system (with control $u^*(t)$ as input) be described by equation (17.3) as

$$\dot{x} = f(x) + g(x)u^*(t) = F(t, x). \tag{17.3}$$

The perturbed system (with control $u(t)$ as input) is given by equation (17.4) as

$$\dot{x} = F(t, x) + G(t, x), \tag{17.4}$$

where $G(t, x) = g(x)(u(t) - u^*(t))$. Let $z_1(t)$ and $z_2(t)$ be the solutions of the nominal (equation (17.3)) and the perturbed (equation (17.4)) system, respectively. By assumption, $\lim_{t \to \infty}[h(z_1(t)) - y_R(t)] = 0$. This means that for some $c \in (0, 1)$, there exists $t' > 0$ so that $\forall t > t'$ the following holds:

$$|h(z_1(t)) - y_R(t)| < c\varepsilon \Rightarrow \tag{17.5}$$

$$-c\varepsilon < h(z_1(t))y_R(t) < c\varepsilon. \tag{17.6}$$

Also assume (for now) that $\exists\, t'' > 0$ so that $\forall t > t''$, $z_1(t)$ and $z_2(t)$ are close enough so that

$$|h(z_2(t)) - h(z_1(t))| < (1 - c)\varepsilon \Rightarrow \tag{17.7}$$

$$-(1 - c)\varepsilon < h(z_2(t)) - h(z_1(t)) < (1 - c)\varepsilon. \tag{17.8}$$

Then, adding equations (17.6) and (17.8) yields

$$|h(z_2(t)) - y_R(t)| < \varepsilon, \quad \forall t > t^* = \max(t', t''). \tag{17.9}$$

This is the control objective. The question that naturally arises at this point is: How close does the state of the perturbed system have to be to the state of the nominal system so that $|h(z_2) - h(z_1)| < (1 - c)\varepsilon$? This question can be answered using the reasoning that follows:

$$\begin{aligned} h(z_2) &= h(z_1) + \nabla h(\xi)(z_2 - z_1) \Rightarrow \\ h(z_2) - h(z_1) &= \nabla h(\xi)(z_2 - z_1) \Rightarrow \\ |h(z_2) - h(z_1)| &= |\nabla h(\xi)(z_2 - z_1)| \Rightarrow \\ |h(z_2) - h(z_1)| &< Z \, \| z_2 - z_1 \|, \end{aligned} \tag{17.10}$$

where $Z = \max \| \nabla h(x) \|_2$ for all x in the area of interest.

So, if $\| z_2 - z_1 \|_2 < (1-c)\varepsilon/Z$, as assumed under closeness of solutions, then equation (17.7) will hold. The next question is : How close should u be to u^* so that $\| z_2 - z_1 \|_2 < (1-c)\varepsilon/Z$? In answering this question, the following theorem will be used, which is taken from [26].

Theorem 1 *Let $D = \{x \in R^n | \| x \|_2 < r\}$ and suppose that the following assumptions are satisfied for all $(t, x) \in [0, \infty) \times D$:*

- $F(t, x)$ is continuously differentiable and the Jacobian matrix $[\partial F/\partial x]$ is bounded and Lipschitz in x, and uniformly in t.
- The origin $x = 0$ is an exponentially stable equilibrium point of the nominal system (17.3).

The perturbation term $G(t, x)$ is piecewise continuous in t and locally Lipschitz in x, and satisfies

$$\| G(t, x) \|_2 \leq \delta, \quad \forall t \geq t_0 \geq 0, \quad \forall x \in D. \tag{17.11}$$

Let $z_1(t)$ and $z_2(t)$ denote solutions of the nominal system (17.3) and the perturbed system (17.4), respectively. Then, there exist positive constants $\beta, \gamma, \eta, \mu, \lambda$ and κ, independent of δ, such that if $\delta < \eta$, $\| y(t_0) \|_2 < \lambda$, and $\| z_2(t_0) - z_1(t_0) \|_2 < \mu$, then the solutions $z_1(t)$ and $z_2(t)$ will be uniformly bounded for all $t \geq t_0 \geq 0$, and

$$\| z_1(t) - z_2(t) \|_2 \leq k e^{-\gamma(t-t_0)} \| z_1(t_0) - z_2(t_0) \|_2 + \beta\delta. \tag{17.12}$$

It was mentioned earlier that $G(t, x) = g(x)(u(t) - u^*(t))$. If $\max_{x \in D} \| g(x) \|_2$ is denoted by C, then equation (17.11) can be written as

$$\| u(t) - u^*(t) \|_2 \leq (1/C)\delta, \quad \forall t \geq t_0. \tag{17.13}$$

A value (even approximate) of δ is required in order to derive the condition (17.13). If the system satisfies the conditions mentioned in Theorem 1 and the control satisfies the condition (17.13) for all t greater than some t_1, then the control objective will be achieved. The condition in (17.13) will drive the design of the fuzzy controller in Section 17.4.

17.3 THE STRUCTURE OF THE CONTROLLER

The controller is a fuzzy controller. The rules have the reference signal as inputs and its first $(n-1)$, time derivatives, as well as the tracking error and its first $(n-1)$ time derivatives. A typical rule for a second-order system will appear as follows:

ith rule : **If** x_1 is NB and x_2 is PM and e_1 is NM and e_2 is PM **then** u is w_i,

where NB means Negative Big, PM is Positive Medium, and NM is Negative Medium. The universe of discourse of the input variables is normalized to $[-1, 1]$. Also, the rule base covers exhaustively the input space. The degree of fulfillment of a rule is equal to the product of the degrees of fulfillment of the constituent linguistic statements. The output of each rule, w_i, is a crisp number which is initialized as a small random number, and which will be adjusted during normal operation by a learning algorithm. All input variables are assigned the same number of membership functions, N (i.e. N^{2n} total number of rules), which are uniformly distributed on the interval $[-1, 1]$, and which are of the form:

$$g_k(x) = \begin{cases} \cos^2\left(\frac{\pi}{2}\frac{x - \mu_k}{\sigma}\right), & \text{for } x \in [\mu_k - \sigma, \mu_k + \sigma], \\ 0, & \text{elsewhere}, \end{cases} \quad (17.14)$$

where $\sigma = 2/(N-1)$ and $\mu_k = -1 + (k-1) \cdot \sigma, k = 1, \ldots, N$. The defuzzification is done according to equation (17.15):

$$u = \frac{\sum_{r=1}^{N^{2n}} w_r \cdot \mu_r(\tilde{x})}{\sum_{r=1}^{N^{2n}} \mu_r(\tilde{x})} = \frac{\sum_{r=1}^{N^{2n}} w_r \cdot \prod_{p=1}^{2n} \mu_r^p(x_p)}{\sum_{r=1}^{N^{2n}} \prod_{p=1}^{2n} \mu_r^p(x_p)}, \quad (17.15)$$

where w_r is the consequence of the rth rule, and $\mu_r^p(x_p)$ is the degree of certainty for the pth linguistic statement (premise) of the rth rule. The membership functions defined in equation (17.14) make the denominator of equation (17.15) equal to 1, for all input vectors to the controller. Therefore, equation (17.15) is reduced to

$$u = \sum_{r=1}^{N^{2n}} w_r \cdot \prod_{p}^{2n} \mu_r^p(x_p) = W^T * \Phi, \quad (17.16)$$

where W is a column vector that contains the output weights of all the rules, and Φ is a column vector with the degrees of fulfillment of each rule. The degree of fulfillment of each rule is a 'bell' function on the $2n$-dimensional imput space. So, the control $u(x, e)$ will be approximated by a linear combination of the functions, instances of which constitute the vector $\Phi(x, e)$.

17.4 THE MAIN RESULTS

Consider that the controller explained in the previous section is used as an approximator of the ideal control $u^*(x, e)$. Define the approximation error as

$$e(x, e) = u^*(x, e) - W^T * \Phi(x, e). \quad (17.17)$$

It can be proven [27] that there exists a weight vector W^* which makes the integral of the squared error minimal. This vector W^* is given by

$$W^* = A^{-1}Y, \quad (17.18)$$

where $W^* = [w_1, w_2 \ldots w_N^{2n}]^T$, $Y = [(u^*, g_1)(u^*, g_2) \ldots (u^*, g_N^{2n})]^T$ and

$$A = \begin{bmatrix} (g_1, g_1) & (g_1, g_2) & \cdots & (g_1, g_{N^{2n}}) \\ (g_2, g_1) & (g_2, g_2) & \cdots & (g_2, g_{N^{2n}}) \\ \vdots & \vdots & \ddots & \vdots \\ (g_{N^{2n}}, g_1) & (g_{N^{2n}}, g_2) & \cdots & (g_{N^{2n}}, g_{N^{2n}}) \end{bmatrix}. \quad (17.19)$$

The inner product (g_i, g_j) is defined as

$$(g_i, g_j) = \int_{-1}^{1}\int_{-1}^{1} \cdots \int_{-1}^{1} g_i(x, e)g_j(x, e)\,dx_1dx_2 \cdots dx_n de_1 \cdots de_n. \quad (17.20)$$

Moreover, if $M = \sup\{|u^*(x,e)| : (x,e) \in [-1,1]^{2n}\}$, $M_1 = \sup\{|u^{*\prime}(x,e)| : (x,e) \in [-1,1]^{2n}\}$ and $M_2 = \sup\{|u^{*\prime\prime}(x,e)| : (x,e) \in [-1,1]^{2n}\}$, then the integral of the squared error is bounded by

$$(e,e) = \| u^*(x,e) - W^{*T}\Phi(x,e) \|^2 \leq \frac{2n\sigma^2 M M_2}{3} + \frac{M_1^2 + M M_2}{4N^2}. \quad (17.21)$$

The proof is given in [27]. The proposed design methodology does not require the availability of a state model of the system, but it does require an estimate (even approximate) of the quantities M, M_1, and M_2. This result can be used for the determination of N once the estimates for M, M_1, and M_2 are available.

17.5 THE LEARNING ALGORITHM

In this section the learning algorithm that will update the output weights of the rules will be presented. Consider the instantaneous error between the ideal control $u^*(t)$ and the output of the controller $u(t)$. The derivative of the square of the instantaneous error is given by equation (17.22):

$$\frac{\partial e_k^2}{\partial w_i} = 2 * e_k * \frac{\partial e_k}{\partial w_i} = -2 * \Phi_k^i * e_k, \quad (17.22)$$

where k is the discrete time. Therefore, the output weight of the ith rule should be adjusted in the opposite direction:

$$w_{k+1}^i = w_k^i + \mu * \Phi_k^i * e_k \quad (17.23)$$

or in vector form:

$$W_{k+1} = W_k + \mu * \Phi_k * e. \quad (17.24)$$

It can be proved [1] that the update law (17.24) will keep the error $\tilde{W} = W^* - W_k$ bounded, provided that the training sequence $((x_k, e_k), u_k^*)$ is persistently exciting. Moreover, the error $e_k = u_k^* - u_k$ will become as small as possible, always under the limitations imposed by the number of rules used. The problem is that the ideal control u_k^* is not directly available. Therefore another way must be found to train the neuro-fuzzy controller. In solving this problem, consider the normal form of the state equation of the system, which is

$$\begin{aligned} \dot{z}_1 &= z_2, \\ \dot{z}_2 &= z_3, \\ &\ldots \\ \dot{z}_n &= b(z(t)) + \alpha(z(t))u(t), \\ y &= z_1. \end{aligned} \quad (17.25)$$

Consider the control law given by equation (17.26):

$$u^*(t) = \frac{1}{\alpha(z(t))} \left[-b(z(t)) + y_d^{(n)} + \sum_{i=0}^{n-1} c_i e^{(i)} \right], \quad (17.26)$$

where $e^{(i)} = y_d^{(i)} - z_{i+1}$, i.e. $e_0 = y_d - z_1$, $e_1 = y_d^{(1)} - z_2$, etc. The term $\alpha(z(t))$ in the denominator of equation (17.26) does not become zero, by the assumption that the relative degree of the system is $r - n$ for all points of interest. If this control law is applied to the system described by equation (17.25), then the equation for z_n can be rewritten as

$$\dot{z}_n = y_d^{(n)} + \sum_{i=0}^{n-1} c_i e^{(i)} \Rightarrow e^{(n)} + c_{n-1} e^{(n-1)} + \cdots + c_0 e = 0. \tag{17.27}$$

In other words, the tracking error dynamics is governed by the ordinary differential equation (17.27). The coefficients c_i can be chosen so that the error converges to zero arbitrarily fast. There is, however, one more issue that has to be addressed here: The expressions $\alpha(z(t))$ and $\beta(z(t))$ used in equation (17.26) are unknown. Therefore, their values have to be estimated.

Let us address the first issue first. The equation for z_n can be written in discrete time for the last two sampling instants as

$$\hat{\beta}(z(k)) + \hat{\alpha}(z(k))u(k) = \frac{z_n(k) - z_n(k-1)}{T} \tag{17.28}$$

and

$$\hat{\beta}(z(k-1)) + \hat{\alpha}(z(k-1))u(k-1) = \frac{z_n(k-1) - z_n(k-2)}{T}, \tag{17.29}$$

where $\hat{\alpha}(z(k))$ and $\hat{\beta}(z(k))$ are estimates of the unknown $\alpha(z(t))$ and $\beta(z(t))$. Under the assumption that the system is sufficiently smooth, and that the sampling period is sufficiently short, one can assume that $\hat{\beta}(z(k)) \cong \hat{\beta}(z(k-1))$ and $\hat{\alpha}(z(k)) \cong \hat{\alpha}(z(k-1))$. Under this approximation, one can solve the system of equations (17.28) and (17.29) for $\hat{\alpha}(z(k))$ and $\hat{\beta}(z(k))$. These two estimates can be substituted in equation (17.26) to derive the ideal control u_k^* which will be used to train the controller.

17.6 ILLUSTRATIVE EXAMPLES

The first example is a disk-and-pendulum system, as shown in Figure 17.1. The output is the angle of the pendulum from the perpendicular. The control variable is the torque applied on

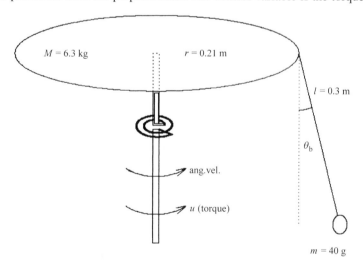

Figure 17.1 The disk and pendulum system

the axle. The objective is to have the output track a desired trajectory, within $\pm 1°$. The angle of the pendulum from the perpendicular is denoted by θ_b, the angle that the disk has rotated from the starting position is denoted by θ_d, and the angle that the axle has rotated from the starting position is denoted by θ_a. This means that there is some flexibility at the joint between the axle and the disk. This is denoted by K. The generalized force acting on each angle is denoted by $p_{\theta_b}, p_{\theta_d}, p_{\theta_a}$, respectively. The complete space model of this system is given as

$$\dot{\theta}_b = \frac{1}{I_b} p_{\theta_b},$$

$$\dot{\theta}_d = \frac{1}{I_M} p_{\theta_d},$$

$$\dot{\theta}_a = \frac{1}{I_M} p_{\theta_a},$$

$$\dot{p}_{\theta_b} = -1/2 mgl \cdot \sin(\theta_b) + \frac{I_C}{2I_M^2} p_{\theta_d}^2 - \frac{C_{\theta_b}}{I_b} p_{\theta_b}, \qquad (17.30)$$

$$\dot{p}_{\theta_d} = -\frac{C_\theta}{I_M} p_{\theta_d} - K(\theta_d - \theta_a),$$

$$\dot{p}_{\theta_a} = u - \frac{C_\theta}{I_M} p_{\theta_a} + K(\theta_d - \theta_a),$$

where

$$I_b = 1/3 ml^2,$$
$$I_M = 1/2 Mr^2 + m(r + 1/2l \cdot \sin(\theta_b))^2,$$
$$I_C = ml(r + 1/2l \cdot \sin(\theta_b))\cos(\theta_b).$$

Some (rather tedious) calculations show that the relative degree of this system is $r = 5$. A new variable θ is introduced, which is equal to $(\theta_d - \theta_a)$. Therefore, by subtracting the third state equation from the second one of the above model, the new (equivalent) model is derived:

$$\dot{\theta}_b = \frac{1}{I_b} p_{\theta_b},$$

$$\dot{\theta} = \frac{1}{I_M}(p_{\theta_d} - p_{\theta_a}),$$

$$\dot{p}_{\theta_b} = -\frac{1}{2} mgl \cdot \sin(\theta_b) + \frac{I_C}{2I_M^2} p_{\theta_d}^2 - \frac{C_{\theta_b}}{I_b} p_{\theta_b}, \qquad (17.31)$$

$$\dot{p}_{\theta_d} = -\frac{C_\theta}{I_M} p_{\theta_d} - K\theta,$$

$$\dot{p}_{\theta_a} = u - \frac{C_\theta}{I_M} p_{\theta_a} + K\theta.$$

This model has order $n = 5$ and relative degree $r = 5$. This model serves as the 'actual system' to test the control laws. It is assumed that it describes completely the dynamics of the system under study. However, it is not used for designing the control laws. Instead, for the model-based approaches, a simplified version of the above model is used. This way, the various designs will be evaluated based on their performance in the presence of unmodeled dynamics.

For all simulations the reference is the same. Equation (17.32) describes this signal:

$$y_r(t) = 0.25 + \sum_{i=1}^{30} A_i \cdot \cos\left(2\pi \cdot \frac{i}{200} \cdot t + \phi_i\right). \quad (17.32)$$

The parameters A_i and ϕ_i are random numbers. This signal is periodic with period 200 seconds. However, for the simulations, only the first 100 seconds are used. This means that the controller does not see the same protion of the signal twice. Therefore, it is a virtually non-periodic signal. This prohibits the use of the approach suggested in [28], where a set of basis functions with time as their sole argument was used for the calculation of the control law. The proposed approach is more general than the approach presented in [28], and can cover a wider range of problems involving non-periodic signals.

In addition, for most of the simulations the actual state of the system is not available. Only the output of the system is available, but it is corrupted by measurement noise. The noise is additive in nature, and its amplitude follows a normal distribution with 0 mean and 0.1 degree deviation. This noise corresponds to a 40 dB signal-to-noise ratio, compared with the reference signal.

For designs that require an estimate of the state vector, a polynomial-based estimator is used. For all simulations that use the complete model of the system, the joint flexibility is $K = 10$, while the friction coefficients are $C_\theta = 0.002$ and $C_{\theta_b} = 0.05$. Finally, the sampling period T is 0.01, for all simulations.

The proposed method was implemented and tested. The structure of the controller, i.e. the number of basis functions, f_r, or, correspondingly, the number of membership functions per variable, N, was determined first. The update law was programmed next with the components of the vector C chosen so that the learning is stable. The controller was first tested on the simplified model. The response and the tracking error are shown in Figures 17.2 and 17.3.

The same controller and estimation scheme were used next to simulate the behavior of the complete system. The results are depicted in Figures 17.4 and 17.5.

The second example is an industrial furnace used for the heat treatment of carbon parts. Heat is transferred to the part mainly through radiation (difference of the fourth power of the temperatures), which makes the system highly non-linear. Figure 17.6(a) illustrates the reference and the actual angle, while Figure 17.6(b) illustrates the error.

There is no means to cool down the part faster than the way it cools when it is allowed to cool down by itself. This is the reason for the large error at the second downhill ramp of the

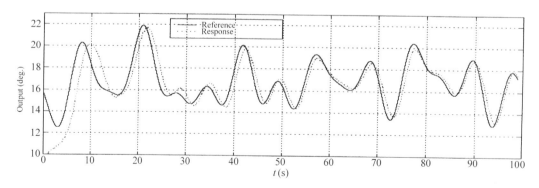

Figure 17.2 The response of the simplified system

ILLUSTRATIVE EXAMPLES **345**

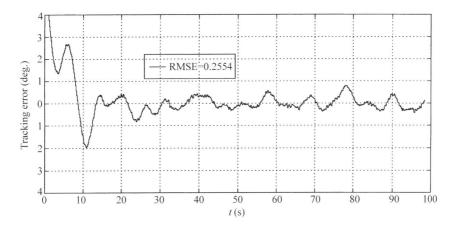

Figure 17.3 The tracking error

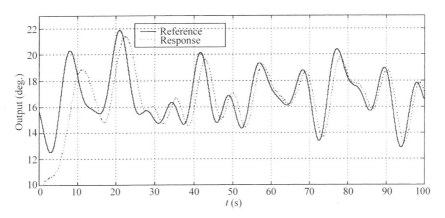

Figure 17.4 The response of the complete system

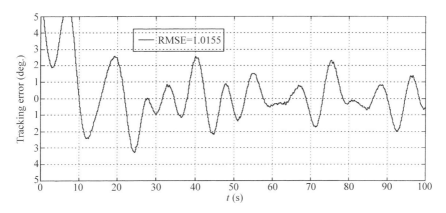

Figure 17.5 The tracking error

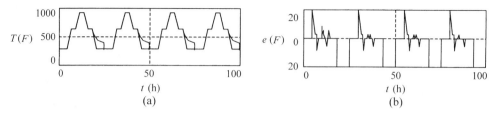

Figure 17.6 The furnace results

profile. In other words, there is not enough control authority to follow the profile within ±10° throughout. For the same reason, at the beginning of the first uphill ramp of each cycle the error is almost 20°. The control action achieves its maximum value in two sampling periods (36 s), but, put simply, there is not enough control to keep the error within limits.

17.7 COMPREHENSIVE RESULTS AND CONCLUSIONS

The disk-and-pendulum example was simulated and tested with two well-known methods. A hybrid, PD and neural network based, controller suggested by Sanner and Slotine [29] was implemented and tested against unmodeled dynamics. As a second case, the controller proposed by Yesildirek and Lewis [30] was also implemented. This methodology targets SISO systems, with relative degree $r = n$. Tables 17.1 and 17.2 summarize the results. The entry 'simple' in the tables refers to a third-order model of the system derived from the complete model ignoring friction and joint flexibility.

Table 17.2 summarizes the results of the simulations. For completeness, an input–output linearized version of the model and a conventional PID controller were also included in the study. The results suggest that the proposed methodology performed better than the other approaches under the conditions imposed on the simulations employed. The proposed approach maintains the error within $\pm 1°$ after $t = 13$ when the control algorithm is applied to the 'simple' model. When, however, the complete system is employed, involving unmodeled dynamics, none of the approaches satisfied the error requirement. The proposed approach, though, is more robust in the presence of noise and unmodeled dynamics. The fact that the proposed method outperformed others does not necessarily imply that these

Table 17.1 Conditions under which the results were produced

Control algorithm	System	Noise	Which state is fed back to the controller
Input–Output linearized	Simple	No	Actual
	Complete	Yes	Estimated
PID	Complete	Yes	Estimated
Sanner and Slotine	Simple	No	Actual
	Complete	Yes	Estimated
	Complete	Yes	Estimated
Lewis	Simple	Yes	Estimated
	Complete	Yes	Estimated
Proposed approach	Simple	Yes	Estimated
	Complete	Yes	Estimated

Table 17.2 How the various methodologies performed

Control Architecture	RMSE (simple model)	RMSE (complete model)
Input–output linearization	0.071	39.993
PID		2.217
Design of Sanner and Slotine	0.398	1.233
Design of Lewis	0.779	1.461
Proposed approach	0.255	1.016

results can be generalized. Every approach has some advantages and disadvantages and an appropriate methodology should be selected after careful consideration of the problem at hand. For example, if a relatively accurate model of the system is available, a version of the input–output linearization method perhaps would be the most appropriate. On the other hand, when unmodeled dynamics are present and the uncertainty is high, probably the proposed approach would perform better.

For all three non-model-based methodologies (Sanner–Slotine, Lewis, and the proposed one) there is no analytical proof of stability when the control is discretized. It can be argued, however, that for a sampling period T that is small enough, the deviation from the continuous case is small, and therefore results about boundedness derived for the proposed method would be valid.

REFERENCES

1. Brockett, R. W. and M. D. Mesarovic, The reproducibility of multivariable systems, *Journal of Mathematical Analysis and Applications*, **11**, 1965, 548–563.
2. Sain, M. K. and J. L. Massey, Invertibility of linear time-invariant dynamical systems, *IEEE Transactions on Automatic Control*, **14**, 1969, 141–149.
3. Silverman, L. M., Inversion of multivariable linear systems, *IEEE Transactions on Automatic Control*, **14**, 1969, 270–276.
4. Willsky, A. S., On the invertibility of linear systems, *IEEE Transactions on Automatic Control*, **19**, 1974, 272–274.
5. Hirschorn, R. M., Invertibility of nonlinear control systems, *SIAM Journal of Control and Optimization*, **17**, 1979, 289–297.
6. Hirschorn, R. M., Output tracking in multivariable nonlinear systems, *IEEE Transactions on Automatic Control*, **26**, 1981, 593–595.
7. Singh, S. N., Reproducibility in nonlinear systems using dynamic compensation and output feedback, *IEEE Transactions on Automatic Control*, **27**, 1982, 958–960.
8. Hirschorn, R. M., Genralized functional reproductibility condition for non-linear systems, *IEEE Transactions on Automatic Control*, **27**, 1982, 958–960.
9. Nijmeijer, H., Invertibility of affine nonlinear control systems: A geometric approach, *Systems and Control Letters*, **2**, 1982, 163–168.
10. Hirschorn, R. M. and J., Davis, Output tracking for nonlinear systems with singular points, *SIAM Journal of Control and Optimization*, **25**(3), 1987, 547–557.
11. Hirschorn, R. M. and J. Davis, Global output tracking for nonlinear systems, *SIAM Journal of Control and Optimization*, **26**(6), 1988, 1321–1330.
12. Behtash, S., Robust output tracking for nonlinear systems, *International Journal of Control*, **51**(6), 1990, 1381–1407.

13. Narendra, K. S., Y. H. Lin and I. S. Valavani, Stable adaptive controller design–direct control, *IEEE Transactions on Automatic Control*, **23**, 1978, 570–583.
14. Sastry, S. S. and A. Isidori, Adaptive control of linearizable systems, *UCB/ERL Memorandum*, No. M87, 1987, p. 53.
15. Narendra, K. S. and K. Parthasarathy, Identification and control of dynamical systems using Neural networks, *IEEE Transactions of Neural Networks*, **1**(1), 1990, 4–27.
16. Narendara, K. S. and K. Parthasarathy, Gradient methods for the optimization of dynamical systems containing neural networks, *IEEE Transactions on Neural Networks*, **2**(2), 1991, 252–262.
17. Isidori, A., *Nonlinear Control Systems*, Springer-Verlag, New York, 1989.
18. Isidori, A. and C. Byrnes, Output regulation of nonlinear systems, *IEEE Transactions on Automatic Control*, **35**(2), 1990, 131–140.
19. Schneider, D., P. Wang and M. Togai, Design of a fuzzy logic controller for a target tracking system, *Proceedings of the First IEEE International Conference on Fuzzy Systems*, 1992, pp. 1131–1138.
20. Wang, Li-Xin and J. M. Mendel, Fuzzy basis functions, universal approximation, and orthogonal least-squares learning, *IEEE Transactions on Neural Networks*, **3**(5), 1992, 807–814.
21. Hauser, J., S. Sastry and P. Kokotovic, Nonlinear control via approximate input–output linearization: The ball and beam example, *IEEE Transactions on Automatic Control*, **37**(3), 1992, 392–398.
22. Nedungadi, A., A fuzzy robot controller—hardware implementation, *Proceedings of the First IEEE International Conference on Fuzzy Systems*, 1992, 1125–1131.
23. Miller, W. T., R. P. Hewes, F. H. Glanz and L. G. Kraft, Real-time dynamic control of an industrial manipulator using a neural-network based learning controller, *IEEE Transactions on Robotics and Automation*, **6**(1), 1990, 1–9.
24. Johnson, M. A. and M. B. Leahy, Adaptive model-based neural-network control, *Proceedings of the IEEE International Conference on Robotics and Automation*, Los Alamitos, 1990, pp. 1704–1709.
25. Mamdani, E. H. and S. Assilian, An experiment in linguistic synthesis with a fuzzy logic controller, *International Journal of Man-Machine Studies*, **7**, 1975, 1–13.
26. Khalil, H. K., *Nonlinear Systems*, Macmillan, New York, 1992.
27. Pirovolu, D., The Tracking Problem Using Fuzzy Neural Networks, Ph.D. Thesis, Georgia Institute of Technology, 1996.
28. Sadegh, N. and K. Guglielmo, Design and implementation of adaptive and repetitive controllers for mechanical manipulators, *IEEE Transactions on Robotics and Automation*, **8**, 1992, 395–400.
29. Sanner, R. M. and J. J. E. Slotine, Gaussian networks for direct adaptive control, *IEEE Transactions on Neural Networks*, **3**, 1992, 837–863.
30. Yesildirek, A. and F. L. Lewis, A neural network controller for feedback linearization, *Proceedings of the Thirty-Third IEEE Conference on Decision and Control*, 1994, 2494–2499.

18
Fuzzy Life-Extending Control of Mechanical Systems

P. Kallappa, Asok Ray and **Michael S. Holmes**
The Pennsylvania State University, University Park, USA

18.1 INTRODUCTION

This chapter presents a methodology for the synthesis of fuzzy-logic-based Life-Extending Control Systems (LECS) with the objectives of performance enhancement, and durability and life extension of mechanical structures. A major role of LECS is to reduce structural damage to mechanical system components, thereby extending the safety and reliability of plant operation, without significant loss of performance. Damage is a structural degradation process (e.g. cracks, corrosion, creep and plasticity) which leads to a reduction in the functional life of the component. Usually, plant dynamic performance is defined in terms of reference signal tracking, disturbance rejection, and/or control effort minimization. Often performance specifications do not explicitly address the dynamics of material damage (e.g. fatigue cracking) in critical plant components. The key idea of LECS is that a significant improvement in service life can be achieved, especially during transient operations, by a small reduction in the dynamic performance of the system. A well-designed LECS should be able to achieve, in some sense, an optimal solution to this performance–damage trade-off problem. Such an LECS increases the service life of mechanical system components, thereby increasing system availability and the mean time between maintenance. By keeping damage rates low, the risk of unscheduled shutdowns and catastrophic accidents is also reduced.

Lorenzo and Merril [12] proposed a concept of *damage mitigation* in the context of reusable rocket engines. Noll *et al.* [13] addressed the issue of a trade-off between flight maneuverability and durability of critical components in advanced aircraft. Dai and Ray [5] have developed a damage mitigating control system for a reusable rocket engine where a computationally fast model of structural damage in the coolant channel ligament is used for decision and control system synthesis. Ray *et al.* [16] have developed a damage prediction model for turbine blades of a reusable rocket engine. In both cases these models were used to create an optimal feedforward (open loop) control policy for life extension of engine components. The control policy was developed using constrained non-linear optimization of

Fuzzy Logic. Edited by S. Farinwata, D. Filev and R. Langari
© 2000 John Wiley & Sons, Ltd

a cost functional of damage and performance specifications with appropriate constraints. Kallappa et al. [9] have proposed a feedforward–feedback methodology for the synthesis of robust linear control systems for life extension of fossil-fueled power plants. Holmes and Ray [7] have used a similar approach for life-extending control of reusable rocket engines via output feedback. Tangirala et al. [22] have demonstrated this control concept on a laboratory testbed to achieve a trade-off between dynamic performance and structural durability. While these model-based control systems have achieved significant life extension, they do not possess the ability to modify operational strategies and performance criteria on-line. A good LECS should have this ability. For example, if an LECS observes low structural damage rates, it could make the performance criteria more stringent. On the other hand, a high damage rate might require relaxation of performance criteria to reduce the current damage rate. To achieve such on-line adaptive capabilities, a knowledge-based system in the setting of approximate reasoning is a viable option. Approximate reasoning is important in order to emulate some of the characteristics of a human operator. Yen et al. [26] and Jamshidi et al. [8] have demonstrated, through various industrial applications, the ability and versatility of fuzzy logic to emulate approximate reasoning. Thus fuzzy logic is introduced into the LECS to improve its performance.

Model-based control synthesis requires dynamic mathematical model(s) of the process (e.g. the rocket engine model of Dai and Ray [4] and the power plant model of Weng and Ray [25]). In addition, structural damage models are needed for control synthesis and simulation testing. Damage models for power plant components by Kallappa et al. [9] and Lele et al. [11] and for rocket engine components by Dai and Ray [4] and Ray et al. [16]. Medium-order and large-order mechanical systems are usually non-linear. On the other hand, advanced techniques of feedback control are mostly linear because they allow the designer to take advantage of the many versatile and easily available mathematical tools applicable to linear systems. A common practice in controller design for non-linear systems is to linearize these systems at various operating points, design a family of linear controllers and schedule their gains over the entire operating range. This practice of gain scheduling [19] is commonly used in aircraft, rotorcraft, power plants, and chemical processes. Even an elaborate procedure like gain scheduling does not guarantee global stability of the non-linear control systems except (possibly) for slow variation of the gain scheduling variable (Shamma and Athans [20]). Fuzzy logic plays a key role in the LECS to ensure stability of the gain scheduled, non-linear mechanical systems. It also assures smooth dynamics and bumpless controller switching (Graebe and Ahlen [6]) under gain scheduling.

This chapter explains the advantages of supplementing model-based control systems with appropriate fuzzy logic. The concepts and applications presented here are partly based on earlier publications by Ray et al. [16], Kallappa et al. [9], Holmes and Ray [7], and Kallappa and Ray [10]. A key feature of all control systems discussed in this chapter is the incorporation of expert knowledge of the dynamics of mechanical systems and material damage into the control synthesis. The proposed fuzzy-logic-based life-extending control system has a two-tier architecture. The controller at the upper tier makes decisions based on a trade-off between performance enhancement and life extension. This control algorithm is synthesized based on approximate reasoning embedded with rule-based expert knowledge of process dynamics and the mechanical behavior of structural materials. Using the fuzzy logic concept, the plant operation and control strategy is modified on-line for a trade-off between plant dynamic performance and structural damage in critical components. The lower tier consists of a feedforward control policy and one or more linear multivariable robust feedback controllers. The optimal feedforward policy is formulated on the principle of non-linear programming. The sampled-data feedback control laws are synthesized based on an induced L_2-norm technique which minimizes the worst case gain between the energy of the exogenous inputs and the energy of the regulated outputs. If there are more than one linear controller, then the fuzzy algorithm facilitates bumpless controller switching for gain scheduling, under wide range operation and control. The theory of fuzzy-logic-based life-extending control (LECS) is supported by examples of two specific applications in mechanical systems, namely a rocket engine and a fossil-fuel power plant that are susceptible to very high damage during transient operations. For each case, the design and simulation of the LECS focus on damage reduction and performance enhancement during transient conditions.

This chapter is organized in five sections and two appendices. Section 18.2 presents the architecture of LECS and control systems syntheses for two applications, a rocket engine and a power plant, that are briefly described in Appendix A and Appendix B, respectively. Section 18.3 discusses the details of the first LECS as applied to a rocket engine. Section 18.4 describes the second LECS and its application to power plants. Finally, the chapter is concluded in Section 18.5.

18.2 ARCHITECTURE OF LIFE-EXTENDING CONTROL SYSTEMS

Two different configurations of Life-Extending Control Systems (LECS) are presented in this section. Both configurations have two tiers and an inner loop consisting of one or more linear feedback controllers. The main difference between the two configurations is that the first one uses actual on-line damage information for control purposes while the second one uses plant process variables as an indicator of structural damage. The first configuration, referred to as LECS-1, requires either on-line measurement of actual structural damage or its on-line estimation. In the absence of on-line damage information, the plant process variables can be used as indicators of structural damage and damage rate. For example, a rapid variation in temperature in certain components is an indicator of high damage rate. This information could be built into the fuzzy logic to obtain a computationally fast and relatively simple fuzzy-logic-based controller although it may not yield as good performance. The second configuration, referred to as LECS-2, is formulated on this principle. LECS-2 is ideally suited for large systems with many critical components, operating under extreme temperature and pressure conditions that make it difficult to measure structural damage and the large number of components results in on-line execution of too many damage models.

As mentioned in the Section 18.1, gain scheduling of linear feedback controllers provides better performance for non-linear systems. Fuzzy logic can be used to implement smooth gain scheduling of these controllers. Smooth gain scheduling ensures stability, good dynamic performance and low damage rates. In this chapter the gain scheduling feature is added on to only LECS-2. It can be implemented with equal ease with LECS-1. For simulation and testing, LECS-1 is used with the model of a rocket engine (similar to the Space Shuttle main engine) and LECS-2 with gain scheduling is used with the model of a commercial-scale power plant. The power plant dynamics are relatively more non-linear and have a larger number of critical components.

18.3 LIFE-EXTENDING CONTROL OF A ROCKET ENGINE

A brief description of the rocket engine under consideration is give in Appendix A. The schematic diagram in Figure 18.1 shows a general architecture for the first configuration, LECS-1, where the plant has three types of sensor outputs: $y^{\text{dam}}(t)$, $y^{\text{dyn}}(t)$, and $y^{\text{reg}}(t)$. The vector signal $y^{\text{dam}}(t)$ contains those plant outputs that are necessary for calculation of damage (e.g. torque and shaft speed for calculating turbine blade damage in a rocket engine). The vector signal $y^{\text{dyn}}(t)$ consists of the plant outputs whole reference trajectory vector is determined by the fuzzy based controller (e.g. main thrust chamber pressure in a rocket engine). The vector signal $y^{\text{reg}}(t)$ contains regulated plant outputs whole reference signals are either constants or unaltered by the fuzzy-based controller. For example, the oxygen/hydrogen mixture ratio for the Space Shuttle main engine should ideally be kept at a constant value of 6.02 at all times (Ray and Dai [17]).

The purpose of the Linear Tracking Controller in Figure 18.1 is to keep the error signals as close to zero as possible, i.e. to track $y^{\text{set}}(k)$ and $y^{\text{ref}}(k)$, and to provide robust stability in the inner control loop. The Feedforward Signal Generator is selected as $u^{\text{ff}}(k) = c_1 y^{\text{ref}}(k) + c_2$, where the two constants, c_1 and c_2, are identified a priori, based on a linear interpolation of steady-state inputs.

The Structural Model in Figure 18.1 uses the vector $y^{\text{dam}}(t)$ as an input to generate damage-causing variables such as stresses, $y^{\text{str}}(k)$. The signal $y^{\text{str}}(k)$ excites the Damage Model whole output is both damage rate and accumulation. The purpose of the damage model in the outer control loop is to capture the dynamic characteristics of material degradation under stress. The critical components being considered in this chapter are the H_2 (fuel) turbine blades and the O_2 (oxidizer) turbine blades. The blades in these turbines are susceptible to fatigue cracks. The damage model is highly non-linear and is normalized to

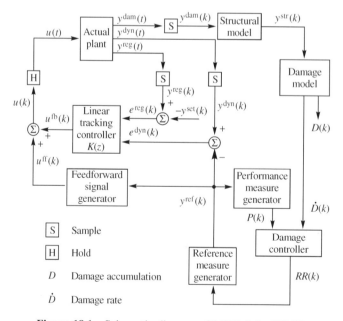

Figure 18.1 Schematic diagram of LECS-1 for SSME

have an output in the range [0, 1], where a value of 0 can be interpreted as zero damage and a value of 1 implies that the service life of the component has ended. The Damage Model used in the Damage Feedback Loop provides a measure of damage rate to the Damage Controller. Since damage is not reversible, the damage rate is always a non-negative quantity. The details of the damage model are reported by Ray and Wu [15].

The Performance Measure Generator provides the fuzzy-based controller with a measure of plant dynamic performance, P, which is function of the reference trajectory, y^{ref}, as seen in Figure 18.1. As an example, for the rocket engine under consideration in this chapter, the measure of performance is chosen to be the average ramp rate of the reference trajectory for the main thrust chamber pressure from the beginning of the maneuver to the present time:

$$P(k) = f(y^{\text{ref}}(k), k) = \frac{y^{\text{ref}}(k) - y_i^{\text{ref}}}{kT - t_0}, \qquad (18.1)$$

where t_0 is the starting time of the transient; y_i^{ref} is the value of the reference trajectory at time $t = t_0$; k indicates the kth sampling instant; $y^{\text{ref}}(k)$ is the value of the reference trajectory at the kth sampling instant; and T is the sampling time of the A/D sampler and D/A zero-order hold.

The Reference Signal Generator takes the ramp rate specified by the fuzzy-based controller and integrates it to obtain the reference signal. It is in the form of an integrator:

$$y^{\text{ref}}(k+1) = y^{\text{ref}}(k) + RR(k)T. \qquad (18.2)$$

The purpose of the fuzzy-based controller is to dynamically alter the ramp rate of the reference trajectory during the transient. The output of the fuzzy-based controller is zero before and after the transient and is constrained to lie within a pre-specified range, $RR_{\min} \leq RR(k) \leq RR_{\max}$, during the transient.

Subsection 18.3.1 gives a brief description of the inner loop control, which is based on a linear feedback control technique. Subsection 18.3.2 discusses in detail the outer loop fuzzy-based controller. Both descriptions are specific to the SSME.

18.3.1 Inner Loop Feedback Controller for LECS-1

The synthesis of LECS-1 is demonstrated for a reusable rocket engine following the control system architecture in Figure 18.1. The sampled-data tracking controller in the inner loop is designed by using the H_∞ or induced L_2-norm technique which minimizes the worst case gain between the energy of the exogenous inputs and the energy of the regulated outputs. Bamieh and Pearson [2] proposed a solution to the induced L_2-norm control synthesis problem for application to sampled-data systems. This design procedure has subsequently been incorporated as the function *sdhfsyn* in the MATLAB *mutools* toolbox (Balas, *et al*. [1]).

Figure 18.2 shows the setup used for the synthesis of the induced L_2-norm controller for the rocket engine. This is based on a plant model with two inputs, i.e. a fuel preburner oxidizer valve position and an oxidizer preburner oxidizer valve position, and two outputs, i.e. a main thrust chamber hot-gas pressure and an O_2/H_2 mixture ratio. The plant model is obtained by first linearizing an 18-state non-linear model of the rocket engine which does not contain actuator dynamics (Ray and Dai [17]) at a combustion pressure of 17.58 MPa (2550 psi) and an O_2/H_2 ratio of 6.02. The pressure 17.58 MPa (2550 psi) is chosen for

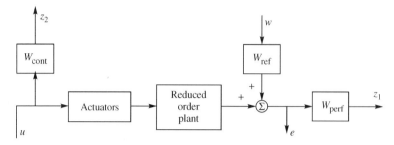

Figure 18.2 Linear controller synthesis

linearization because the controller is required to operate in the range 14.48 MPa (2100 psi) to 20.69 MPa (3000 psi). After linearization, the 18-state linear model is reduced to a 15-state linear model for the controller design via Hankel model order reduction (Zhou *et al.* [27]). A comparison of Bode plots reveals that the fifteenth-order model does not significantly alter the input-output characteristics of the original eighteenth-order model.

The frequency-dependent performance weight, W_{perf}, consists of two components: W_{press}, which penalizes the tracking error of the combustion chamber pressure, and W_{O_2/H_2}, which penalizes the tracking error of the O_2/H_2 ratio. The gain of the performance weights is related to the steady-state error at low frequencies and to the transient part of the tracking error at high frequencies. The objectives of the performance weights in this application are to keep the steady-state error and the overshoot/undershoot small while, at the same time, allowing a reasonably fast rise time. The frequency-dependent control signal weight, W_{cont}, consists of two components: W_{H_2}, which penalizes the fuel preburner oxidizer valve position, and W_{O_2}, which penalizes the oxidizer preburner oxidizer valve position. The objectives of these control signal weights are:

- prevention of large oscillations in the feedback control signal that may cause valve saturation; and
- reduction of valve wear and tear due to high-frequency movements.

In the present design the performance weights are

$$W_{\text{press}}(s) = 2\left(\frac{s+0.75}{s+3}\right); \quad W_{O_2/H_2}(s) = 1760\left(\frac{s+0.045}{s+0.8}\right), \quad (18.3)$$

and the control weights are chosen to be the same for both valves:

$$W_{H_2} = W_{O_2} = 4000\left(\frac{s+10}{s+100}\right) \times \frac{1 \times 10^5}{s+1 \times 10^5}, \quad (18.4)$$

where the additional high-frequency dynamics in equation (18.4) ensure that the generalized plant model is strictly proper, which is necessary for reasons discussed earlier. Each of the two components of the frequency-dependent disturbance weight, W_{ref}, in Figure 18.2 is chosen to be

$$W_{\text{ref}}(s) = \frac{0.5}{s+0.5}. \quad (18.5)$$

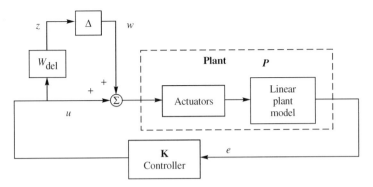

Figure 18.3 Robust stability analysis

The rationale for selecting the above transfer function is that the bandwidth of the trajectory reference signals is expected to be less than 0.5 rad/s.

The input multiplicative configuration is chosen to represent the plant model uncertainties due to actuator errors and neglected high-frequency dynamics. The sampler and zero-order hold associated with the controller are implicit in the setup used for robust stability as shown in Figure 18.3. Each of the two components of the frequency-dependent disturbance weight, W_{del}, in Figure 18.3 is chosen to be

$$W_{del}(s) = \frac{s+1}{s+10}, \quad (18.6)$$

which implies that the amount of plant uncertainty is estimated as being approximately 10% at low frequencies and 100% at high frequencies. The uncertainty model is constructed based on the assumptions of the rocket engine design and operation (Ray and Dai [17]) and can be updated as additional analytical or experimental data become available. Since the plant model is validated with steady-state design data, it is more accurate in the low-frequency range. The eighteenth-order plant model is a finite-dimensional lumped-parameter model, which may not adequately represent the dynamics of high-frequency modes. This leads to the presence of a larger amount of uncertainty in the high-frequency region of the model as compared with the uncertainty at low frequencies. Robust stability is analyzed based on the original linearized model, i.e. the model before order reduction. A sufficient condition for induced L_2-norm-based robust stability for sampled-data systems is that the induced L_2-norm of the transfer matrix from w to z in Figure 18.3 is less than unity (Sivashankar and Khargonekar [21]). The stability conditions are guaranteed for only the linearized plant if it stays within the bounds of uncertainty. A discussion on non-linear stability is included in the final section.

18.3.2 Outer Loop Fuzzy Controller for LECS-1

In this subsection we demonstrate how fuzzy-logic-based decisions can be used as a possible solution to the damage-mitigating control problem in rocket engines. It is cognitively intuitive than if an excessive damage rate or damage accumulation is detected during a ramp-up transient, reduction of the combustion chamber pressure ramp rate is likely to decrease the damage rate in turbine blades; this knowledge needs to be precisely quantified for fuzzy controller synthesis. Since it is difficult to acquire the actual experience of operating rocket engines, an intuitive knowledge base is generated from a variety of computer simulation experiments to identify the trends of the rocket engine. The standard components of a fuzzy controller (Pedrycz [14]) are described below from the perspectives of damage mitigation in the blades of a rocket engine turbine.

18.3.2.1 Fuzzy rule base

The measure of performance is chosen in this application as the average ramp rate of the pressure reference signal from the starting time of the maneuver, t_0, to the current time, $t > t_0$. Three universes of discourse (X_j, $j = 1, 2, 3$) are defined for this application: one for damage rate, one for performance, and one for the ramp rate of the thrust chamber pressure reference trajectory. The three universes of discourse (X_j, $j = 1, 2, 3$) and the associated sets of membership functions are defined as follows.

- *Natural logarithm of damage rate*:

$$\text{universe of discourse:} \quad \dot{d} \in X_1 = (-\infty, \infty),$$
$$\text{membership functions:} \quad \dot{\mathbf{D}} = \{\dot{D}_1, \dot{D}_2, \ldots, \dot{D}_{n_1}\},$$
$$\text{where } \dot{D}_i: \quad X_1 \to [0, 1], \quad i = 1, \ldots, n_1.$$

- *Performance*:

$$\text{universe of discourse:} \quad p \in X_2 = [RR_{\min}, RR_{\max}],$$
$$\text{membership functions:} \quad \mathbf{P} = \{P_1, P_2, \ldots, P_{n_2}\},$$
$$\text{where } P_i: \quad X_2 \to [0, 1], \quad i = 1, \ldots, n_2.$$

- *Ramp rate*:

$$\text{universe of discourse:} \quad rr \in X_3 = [RR_{\min}, RR_{\max}],$$
$$\text{membership functions:} \quad \mathbf{RR} = \{RR_1, RR_2, \ldots, RR_{n_3}\},$$
$$\text{where } RR_i: \quad X_3 \to [0, 1], i = 1, \ldots, n_3$$

The extreme points $RR_{\min}$ and $RR_{\max}$ are design variables which represent the minimum and maximum allowable ramp rates, and the positive integers n_1, n_2, and n_3 are design variables representing the cardinalities of the membership function sets. The lower bound of the ramp rate is chosen to be $RR_{\min} = 300\,\text{psi}/\text{s}$ and the upper bound is chosen to be $RR_{\max} = 5000\,\text{psi}/\text{s}$ based on the operating procedure of the rocket engine (Ray and Dai [17]). For the current design, $n_1 = 5$ with the $\dot{D}_i$'s representing:

- $\dot{D}_1 = $ very low damage rate,
- $\dot{D}_2 = $ low damage rate,
- $\dot{D}_3 = $ moderate damage rate,
- $\dot{D}_4 = $ high damage rate,
- $\dot{D}_5 = $ very high damage rate.

A plot of the five damage rate membership functions is shown in Figure 18.4. Similarly, Figure 18.5 shows the performance and ramp rate membership functions with $n_2 = 5$ and

LIFE-EXTENDING CONTROL OF A ROCKET ENGINE 357

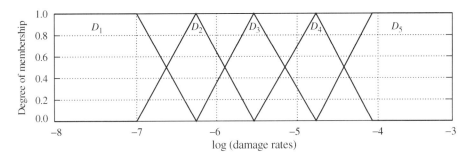

Figure 18.4 Damage rate membership functions

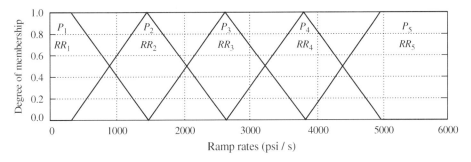

Figure 18.5 Ramp rate and performance membership functions

$n_3 = 5$. For this specific application, the performance, **P**, and ramp rate, **RR**, membership functions are made identical to each other for simplification of the control law without compromising the controller effectiveness. However, in general, **P** and **RR** may be different from each other. Furthermore, all membership functions are chosen to be triangle functions that cross each other at the membership value of 0.5. In general, any typical membership function shapes (e.g. Gaussian) can be used.

Once the membership functions are defined, they are combined into a set of $N = n_1 * n_2$ fuzzy control rules:

    Rule 1:   If $\dot{D}_1$ and P_1 **Then** RR_1.

    Rule 2:   If $\dot{D}_2$ and P_2 **Then** RR_2.

    Rule N: If $\dot{D}_N$ and P_N **Then** RR_N,

where $\dot{D}_i \in \dot{\mathbf{D}}, P_i \in \mathbf{P}$, and $RR_i \in \mathbf{RR}$. For example, the ith rule could be chosen as:

    If $\dot{D}_5$ and P_3 **Then** RR_1

which, with appropriate definitions of $\dot{D}_i = \dot{D}_5, P_i = P_3$, and $RR_i = RR_1$, represent:

> *If the damage rate is very high and performance is moderate, then set the ramp rate to a very small value.*

The experience of a human operator is captured and stored in the above rules. Table 18.1 lists a set of If–Then rules for the fuzzy controller of the rocket engine under consideration.

Table 18.1 If–Then rules for a fuzzy controller for LECS-1

	D_1	D_2	D_3	D_4	D_5
P_1	RR_5	RR_4	RR_3	RR_2	RR_1
P_2	RR_5	RR_4	RR_3	RR_2	RR_1
P_3	RR_5	RR_3	RR_2	RR_2	RR_1
P_4	RR_5	RR_3	RR_2	RR_1	RR_1
P_5	RR_5	RR_3	RR_1	RR_1	RR_1

Note that in this case the first two rows of Table 18.1 are identical. Therefore, for this particular choice of If–Then rules, it is possible to combine the first two rows into a single row without changing the input–output mapping of the fuzzy controller.

18.3.2.2 Fuzzy inputs and fuzzifier

The inputs to the inference mechanism of the fuzzy controller are fuzzy sets representing the current measure of damage rate and performance. There are two possible formats for the damage rate input to the fuzzy controller. If information about the current measure of damage is known in the form of a probability density function (pdf) (Ray and Tangirala [18]), then the fuzzy input mapping $\dot{D}: X_1 \to [0, 1]$ is simply this pdf where $\dot{D}(\dot{d})$ is large at those values of $\dot{d} \in X_1$ that are expected to be the true value of the current damage rate. If the damage input is a deterministic quantity, then a fuzzy singleton should be used:

$$\dot{D}(\dot{d}) = \begin{cases} 1, & \text{when, } \dot{d} = \text{E [current damage rate]} \equiv \bar{\dot{d}}, \\ 0, & \text{otherwise.} \end{cases} \quad (18.7)$$

The simulation experiments in this chapter are based on a deterministic damage model because the stochastic model parameters for the material (i.e. nickel-based superalloys) are not available; the deterministic rule in equation (18.7) can be easily replaced by the damage pdf as it becomes available. Furthermore, since there are two damage rates, one each for the O_2 turbine and the H_2 turbine blades, $\dot{d}$ is taken to be the maximum of the two damage rates. Since the performance measure is deterministic, a fuzzy singleton $P: X_2 \to [0, 1]$ is used as the fuzzy input:

$$P(p) = \begin{cases} 1, & \text{when } p \text{ equals the current performance measure} \equiv \bar{p}, \\ 0, & \text{otherwise.} \end{cases} \quad (18.8)$$

18.3.2.3 Inference mechanism and defuzzifier for LECS-1

The first phase of the decision process in the inference mechanism is the matching stage whose role is to determine the applicability of each fuzzy control rule to the present set of fuzzy inputs. To this end, the function $\lambda_i(D, P)$ is defined for each of the N fuzzy control rules:

$$\lambda_i = \min \left\{ \max_{\dot{d} \in X_1} \left[\min \left(\dot{D}(\dot{d}), \dot{D}_i(\dot{d}) \right) \right], \max_{p \in X_2} \left[\min \left(P(p), P_i(p) \right) \right] \right\}, \quad i = 1, \ldots, N. \quad (18.9)$$

Note the equation (18.9) is completely general in the sense that the fuzzy inputs, $\dot{D}$ and P, are not necessarily fuzzy singletons. If the two fuzzy inputs are in the form of fuzzy singletons, the equation (18.9) reduces to

$$\lambda_i = \min\{\dot{D}_i(\bar{\dot{d}}), P_i(\bar{p})\}, \tag{18.10}$$

where $\bar{\dot{d}}$ and $\bar{p}$ are the current deterministic values of damage rate and performance. The λ_i's have a simple interpretation: $\lambda_i \in [0, 1]$ represents to what extent the current damage rate and performance inputs satisfy the 'If' part (or antecedent) of the ith fuzzy control rule. If λ_i is large, then the ith rule should have a large role in determining the fuzzy controller output.

The second phase of the decision process in the inference mechanism is the summarizing stage whose role is to combine the λ_i's defined above with the RR_i's to produce a fuzzy controller output that is a deterministic quantity. A computationally efficient method to accomplish this task is presented below.

The first step is off-line identification of the centers of gravity, $\bar{rr}_i$, of each of the RR_i's. For the set of ramp rate membership functions chosen in Figure 18.5, if $RR_i = RR_j$, where $j = 2, \ldots, n_3 - 1$, then $\bar{rr}_i$ equals the $rr \in X_3$ for which $RR_j = 1$ (i.e. $RR_j(\bar{rr}_i) = 1$). If $RR_i = RR_1$, then $\bar{rr}_i = RR_{\min}$. If $RR_i = RR_{n_3}$, then $\bar{rr}_i = RR_{\max}$. The deterministic controller output is then chosen to be

$$rr = \frac{\lambda_1 \bar{rr}_1 + \lambda_2 \bar{rr}_2 + \cdots + \lambda_N \bar{rr}_N}{\lambda_1 + \lambda_2 + \cdots + \lambda_N} \tag{18.11}$$

In the summarizing stage, the controller output is created from a 'linear combination' of all N fuzzy control rules. Note, however, that only the centers of gravity of the RR_i's are used but not their shapes.

18.3.3 Results and Discussion for LECS-1

This subsection compares and discusses the results in Figures 18.6–18.13 for three different simulation experiments of the reusable rocket engine under consideration. In each of the three cases the main thrust chamber pressure is increased from 14.48 MPa (2100 psi) to 20.69 MPa (3000 psi). Also, the O_2/H_2 ratio set point is set to a constant value of 6.02 for all cases following the engine performance specifications (Ray and Dai [17]). The three case are briefly described below.

Case 1: High constant reference ramp rate. The rocket engine performance is specified in terms of chamber pressure and O_2/H_2 ratio tracking errors without any penalty on damage rate and accumulation. Therefore, the reference signal is not a function of damage accumulation or damage rate. The reference pressure ramp rate is set to a constant value of 34.48 MPa/s (5000 psi/s) during the upthrust transients of the rocket engine. In order to follow the reference trajectory, the tracking controller attempts to maneuver the chamber pressure from its initial value to its final value as quickly as possible.

Case 2: Varying reference ramp rate under fuzzy control. The fuzzy controller loop is activated in order to achieve a trade-off between performance and damage. The ramp rate of the pressure reference trajectory is allowed to change dynamically as a function of the current value of the damage rate in the O_2 and H_2 turbine blades as well as the current measure of performance following the procedure of Section 18.3.2.

360 LIFE-EXTENDING CONTROL OF MECHANICAL SYSTEMS

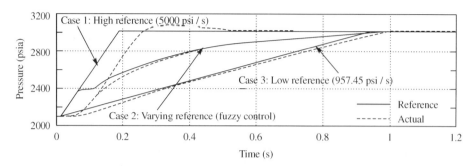

Figure 18.6 Transients of thrust chamber pressure under all three cases

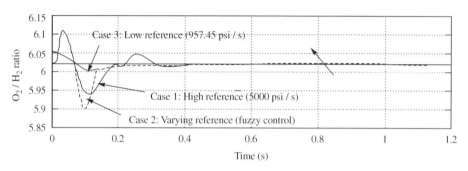

Figure 18.7 Transients of O_2/H_2 ratio under all three cases

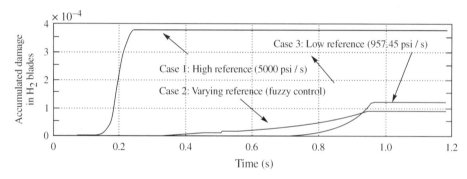

Figure 18.8 Accumulated damage in H_2 turbine blades under all three cases

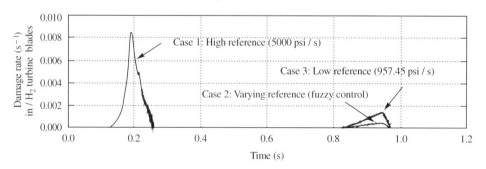

Figure 18.9 Damage rate in H_2 turbine blades under all three cases

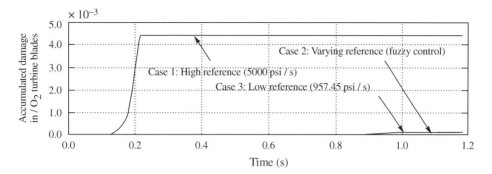

Figure 18.10 Accumulated damage in O_2 turbine blades under all three cases

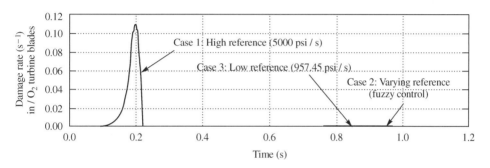

Figure 18.11 Damage rate in O_2 turbine blades under all three cases

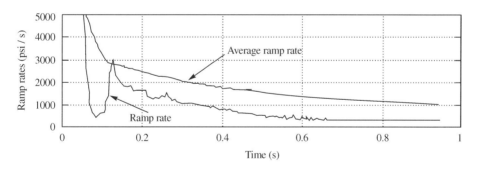

Figure 18.12 Ramp rate and average ramp rate under Fuzzy Control (case 2)

Case 3: Low constant reference ramp rate. This case is similar to Case 1 with the exception that the ramp rate is reduced to a constant value of 6.60 MPa/s (957.45 psi/s). This is the average ramp rate based on the time interval of about 0.94 s required for the pressure reference signal to reach 20.69 MPa (3000 psi) starting from 14.48 MPa (2100 psi) based on the results of the fuzzy controller simulation (Case 2).

Figure 18.6 shows that both the rise time and settling time are best for Case 1 and worst for Case 3. The 34.48 MPa/s (5000 psi/s) in Case 1 is the only one of the three cases that causes an overshoot in the pressure transients leading to a high damage rate. The pressure transients under fuzzy control in Case 2 are intermediate between the two extremes of fast and sluggish

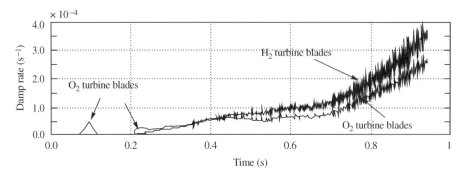

Figure 18.13 Damage rates in O_2 and H_2 turbine blades under fuzzy control (case 2)

responses of Case 1 and Case 3, respectively. For all three cases the O_2/H_2 ratio in Figure 18.7 is maintained within acceptable limits.

Figures 18.8–18.11 display the accumulated fatigue damage and damage rate of the H_2 and O_2 turbine blades for the three cases. For both blades, when the ramp rate is 34.48 MPa/s (5000 psi/s) in Case 1, the damage rate is extremely large for a brief period. This short burst of damage rate causes a relatively large accumulation of fatigue damage. The fatigue damage caused by the steep ramp rate in Case 1 is about 4.25 times more in the H_2 turbine blade and about 63.65 times more in the O_2 turbine blade than the respective damage for varying ramp rate under fuzzy control in Case 2. The damage accumulation in Case 3, with reduced ramp rate, is slightly higher than that in Case 2 although the pressure ramp rate is significantly more sluggish.

Figures 18.12 and 18.13 show how the fuzzy controller operates. Since the damage rate in both blades is very small from the starting time of 0 s to the instant of ~ 0.054 s, the fuzzy controller sets the pressure reference ramp rate to its maximum level of 34.48 MPa/s (5000 psi/s). At the instant of ~ 0.054 s the damage rate in the O_2 turbine blade begins to increase causing the fuzzy controller to abruptly decrease the reference ramp rate. This action prevents the large burst of damage rate that occurred in Case 1 under a constant ramp rate of 34.48 MPa/s (5000 psi/s). As the damage rate in the O_2 turbine blade begins to decrease at around 0.1 s, the fuzzy controller increases the ramp rate for the following reasons:

- the damage rate is decreasing; and
- the performance is now degraded as can be seen by the decrease in the average ramp rate plot in Figure 18.12.

Finally, as the pressure approaches its final value of 20.69 MPa (3000 psi), the damage rates of the two blades increase even though the fuzzy controller applies a small pressure ramp rate.

18.4 LIFE-EXTENDING CONTROL OF A POWER PLANT

A brief description of the power plant under consideration is given in Appendix B. The schematic diagram in Figure 18.14 shows a general architecture for the second configuration, LECS-2, that has three main functional modules where the discrete-time

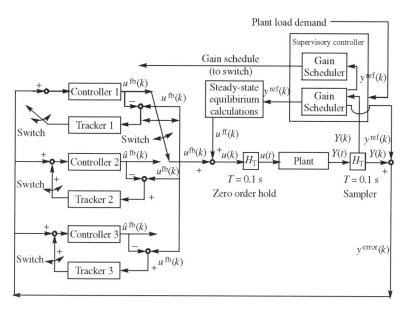

Figure 18.14 Schematic diagram LECS-2 for power plant

and continuous-time signals are denoted by k and t, respectively, in parentheses. It has an inner loop with the feedforward module and the linear feedback controllers. The supervisory controller constitutes the outer module. It consists of the gain scheduler and the fuzzy-logic-based controller. The gain scheduling of controllers is carried out based on the measured plant outputs $y^{\text{ref}}(k)$. In case of the power plant, it is the fourth element of $y^{\text{ref}}(k)$ that is the generated load in MW. Given a power plant operating strategy, the fuzzy-logic-based control module in the supervisory controller serves the role of generating $y^{\text{ref}}(k)$. The feedforward signal generated via equilibrium steady-state calculations, the same as in LECS-1. The robust feedback module is realized by a series of gain scheduled linear controllers. Only three controllers are depicted in Figure 18.14 although the number could vary depending on the plant and the performance requirements.

The sequence $u^{\text{ff}}(k)$ of the feedforward signal is updated every 1 s by the fuzzy controller, based on $y^{\text{ref}}(k)$, and is stored in the control computer a priori. The feedback control law $u^{\text{fb}}(k)$ is generated on a 0.1 s sampling time. At each sampling instant the feedforward and feedback signals are added together and are converted into a continuous-time signal using the zero-order hold (ZOH) logic, denoted by H_T in Figure 18.14. Since the feedforward sequence is based on a 1 s sampling time while the feedback control is based on a 0.1 s sampling time, each element in the feedforward sequence is applied for 10 consecutive sampling intervals. The four continuous-time plant outputs constitute $y(t)$ which are converted to $y(k)$ through a sampler. The error signal $y^{\text{error}}(k)$ is obtained by subtracting the reference signal $y^{\text{ref}}(k)$ from the plant output $y(k)$. The switch from continuous time to discrete and back allows us to use the powerful sample data technique for linear control synthesis.

The operating strategies for LECS-2 are load ramp-up and ramp-down. The control system is suited for any load variations at any power level. The first three elements, namely reference signals for throttle steams temperature (*THS*), hot reheat steam temperature (*THR*), and throttle steam pressure (*PHS*) of $y^{\text{ref}}(k)$ are functions of the fourth element electric power

(*JGN*). The main role of the supervisory controller is life extension without any significant loss of performance: thus it ensures stability, and supplements plant performance and tracker performance. Plant performance is the ability to meet any change in plant load demand in a timely fashion. The inputs to the supervisory module are plant load demand in MW and the actual plant outputs $y(k)$. Based on these inputs the fuzzy controller calculates a load ramp rate. This determines the $y^{\text{ref}}(k)$ for the next instance as shown by equation (18.2) for LECS-1. In the case of the power plant only the load (*JGN*) is ramped up, i.e. the fuzzy controller determines only the fourth output. The other three output references are functions of *JGN*. It will be shown, via simulation experiments, that at times when the feedback controllers fail, the supervisory controller can maintain robust stability.

Once the vector $\{y^{\text{ref}}(k)\}$ is completely determined, it can be used to generate feedforward input for the next sample. The linear feedback controller is a tracking controller similar to the one discussed in LECS-1. Owing to gain scheduling the role of these controllers is only to locally regulate the plant. At any instant one and only one linear controller is on-line and provides the feedback signal. The choice of the feedback controller is determined by the fourth element (*JGN*) of the vector $\{y^{\text{ref}}(k)\}$. In feedback controls terminology, this implies that the reference plant load is used as the scheduling variable. The controller in use in Figure 18.14 is controller 1. While a single specific controller is on-line, the trackers for the remaining two controllers, which are off-line, are functioning to ensure that the controllers are ready to switch smoothly under a sudden change in the plant load demand. The goal of the tracker is to ensure smooth dynamic switching between controllers. The design of the trackers is discussed in a subsequent section. As soon as the active controller goes off-line, its tracker is switched on. It is worth mentioning that the trackers are computationally very simple and can be run on-line even on a PC.

The next two subsections describe the synthesis of inner loop and the outer loop, specifically for the power plant model. The feedforward policy is created in a manner identical to the one for LECS-1 and is not discussed further. Since design techniques for linear feedback and the fuzzy controller are discussed in detail for SSME some of the details for the power plant are skipped to avoid repetetion.

18.4.1 Inner Loop Feedback and Gain Scheduling

Gain scheduling allows robust linear control of (continuous) non-linear plants over a wide operating range. It breaks the task of control synthesis into two steps (Rugh [19]). The first step is to synthesize a family of local linear controllers based on linearization of the non-linear plant at several different equilibrium points. The second step involves interpolation/ scheduling of the linear control laws at intermediate operating conditions. This scheduling algorithm is dependent on the operating conditions. Major issues in making gain scheduling decisions (Kallappa and Ray [10]) include the choice of the scheduling variable, constraints on the scheduling variable, the number of linear controllers and the algorithm for switching from one controller to another.

Shamma and Athans [20] have come up with rigorous proofs that guarantee stability of the gain scheduled non-linear systems. In effect this involves linearizing the plant at many points and finding the Lipschitz bounds on the A, B, C, and D matrices of the plant model. However, gain scheduling can still be implemented under the guidelines that the scheduling variables must vary slowly and capture the non-linearities of the plant dynamics. Slow variations of the scheduling variables ensure that high-frequency unmodeled dynamics of the plant are not excited. For the power plant control under consideration, a scheduling variable that effectively captures the plant non-linearities is the power plant load (MW). A slow variation in the plant load not only reduces the risk of instability but it also reduces the damage in most plant components.

A practical technique for gain scheduling of controllers for large-order systems is to generate a family of linear control laws at finitely many operating points by taking into account two factors:

1. robust stability and performance in the entire operating range for which the larger the number of controllers the better is the performance; and
2. impact of switching transients or interpolation of control signals on the plant performance.

Since a switching action may lead to an abrupt change in the closed-loop plant dynamics, the occurrence of such phenomena should be kept as small as possible. Based on the results of extensive simulation experiments of the power plant model under consideration, a set of three controllers is deemed to be most appropriate for gain scheduling over the entire operating range.

The controller scheduling strategy is chosen keeping in mind the large order of the controllers and the high non-linearity of the plant. The technique used here is called switched scheduling (Graebe and Ahlen [6]). Figure 18.15 shows this strategy. Let us assume, for the purpose of illustration, that both controllers are of the same order n although this assumption is not necessary. Let Controller 1 in Figure 18.15 be on-line, Controller 2 be off-line at the present instant and the feedback signal u^{fb} generated by Controller 1 manipulate the plant. At the instant of switching from one controller to the other, the feedback signal as well as their derivatives up to order $(n-1)$ should match according to the theory of observability (Chen [3]). Given that both controllers are observable, the switching problem can be recast as a tracking problem. As seen in Figure 18.15, the tracker generates the estimate $\hat{u}^{fb}$ to track the reference u^{fb} generated by Controller 1. There is an observer within the tracker which estimates the states of Controller 2 under the constraint that Controller 2 has the same input and output as Controller 1, i.e. y^{error} is the input and u^{fb} is the output of both controllers. Let the discrete-time state space representation of Controller 2 be

$$x(k+1) = A_{c2}x(k) + B_{c2}y^{error}(k), \tag{18.12}$$

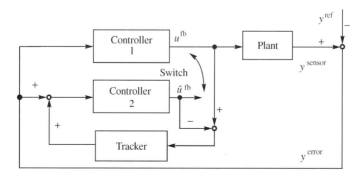

Figure 18.15 Controller switching strategy

and a state observer for this controller be

$$\hat{x}(k+1) = A_{c2}\hat{x}(k) + B_{c2}y^{\text{error}}(k) + L_{c2}\left(u^{\text{fb}}(k) - \hat{u}^{\text{fb}}(k)\right),$$
$$\hat{u}^{\text{fb}}(k) = C_{c2}\hat{x}(k) + D_{c2}y^{\text{error}}(k), \tag{18.13}$$

where L_{c2} is the observer gain matrix. Taking the z transform of equations (18.12) and (18.13) and a rearrangement reduces the transfer matrix of Controller 2 to

$$C_{c2}(z)(zI - A_{c2}(z))^{-1}B_{c2}(z) + D_{c2}(z) \tag{18.14}$$

and the transfer matrix of the tracker becomes

$$(C_{c2}(z)(zI - A_{c2}(z))^{-1}B_{c2}(z) + D_{c2}(z))^{-1}C(z)(zI - A_{c2}(z))^{-1}L_{c2}(z). \tag{18.15}$$

The only requirement for the above conversion is that the pair (A_{c2}, C_{c2}) be observable.

To summarize: the power plant has three gain scheduled linear feedback controllers with the reference plant load as the scheduling variable. Switched scheduling is implemented as shown in Figure 18.15.

18.4.1.1 Inner loop feedback controllers' design

It has been observed by extensive simulation studies (for the power plant under consideration) that three controllers, designed for linearized plant at 25%, 35% and 60% load, yield the best performance and damage mitigation over the entire operating range of 25%–100% full load. The plant dynamics are relatively less non-linear at higher load (e.g. between 50% and 100%) (Weng et al. [24]). Therefore, from a design perspective, one controller at 60% load is considered adequate to ensure good performance and damage mitigation for the load range of 45%–100%. However, for loads below 45%, a 5% change in load may lead to a much larger pole migration. Consequently, for loads below 50%, two controllers, one each at 35% and 25%, are designed.

The controllers are designed using the sampled data of Figure 18.2 and discussed in SSME. The (frequency-dependent) function W_{del} (plant uncertainty) and W_{dist} (external disturbance) remain the same for the three controllers. Input multiplicative modeling uncertainty is represented by

$$W_{\text{del}}(s) = 2\left(\frac{s + 0.05}{s + 1}\right), \tag{18.16}$$

which implies that the amount of plant uncertainty is being estimated as being 10% at low frequencies and 200% at high frequencies. The disturbance weighting function is chosen to be

$$W_{\text{dist}}(s) = \frac{0.1}{s + 0.1}, \tag{18.17}$$

which means that disturbances with frequency content of less than 0.1 rad/s are expected. However, the performance weights are chosen to be different for each of the three controllers and are based on several criteria as discussed below.

The physics of material degradation and operating experience lead to the observation that large oscillations of steam temperature and pressure are the major source of damage in power plant components, especially in the steam headers and steam generator tubes. Large oscillations in steam temperature may also cause high damage in the turbine blades but the pressure oscillations are relatively less damaging. The rationale is that the structural damage is caused primarily by creep flow and thermal stresses leading to cracks. Creep is an exponential function of temperature and rapid temperature oscillations cause high thermal stresses and stress oscillations. On the other hand, unlike an exponential function, mechanical stress cycling induced by pressure oscillations is governed by a relatively mild non-linear relationship. Therefore, pressure constraints are relaxed to enhance the quality of load-following performance. Since the dominant modes of thermal-hydraulic oscillations in a power plant are expected to be below 10 Hz (Weng et al. [24]), the amplitude of high-frequency oscillations (e.g. of the order of 10^2 Hz or more) of any output variables is likely to be insignificant. Therefore, a larger penalty is imposed on lower frequencies of each performance weighting function. However, due to high-frequency unmodeled dynamics, the risk of completely ignoring high-frequency oscillations is non-negligible because, rare as they might be, these incidents can cause instability leading to catastrophic failures or unscheduled plant shutdown. Based on the above observations, each performance weight is formulated as the sum of a low-pass filter and an all-pass filter.

The weights for the controller at 60% plant load are selected as follows:

$$W_{p1}(s) = 20 + \frac{100}{s+5} \quad \text{for THS}, \quad W_{p2}(s) = 20 + \frac{2}{s+0.1} \quad \text{for THR},$$

$$W_{p3}(s) = 10 + \frac{1}{s+0.1} \quad \text{for PHS}, \quad W_{p2}(s) = 20 + \frac{2}{s+0.1} \quad \text{for JGN}.$$

(18.18)

The performance weights for the controllers at 25% and 35% load impose a larger penalty on temperature oscillations because larger temperature variations are observed at lower load levels. This implies that the quality of dynamic performance is traded off for better damage mitigation and stability. The performance weights for the controllers at 25% and 35% load are selected as follows:

$$W_{p1}(s) = 30 + \frac{150}{s+5} \quad \text{for THS}, \quad W_{p2}(s) = 30 + \frac{3}{s+0.1} \quad \text{for THR},$$

$$W_{p3}(s) = 10 + \frac{1}{s+0.1} \quad \text{for PHS}, \quad W_{p2}(s) = 20 + \frac{2}{s+0.1} \quad \text{for JGN}.$$

(18.19)

In each case the generalized plant models (i.e. the augmentation of the linearized plant model with W_{del}, W_{dist} and W_p) have 47 states. The induced L_2 synthesis is performed through D-K iteration (Balas et al. [1]; Zhou et al. [27]). After Hankel model order reduction, the order of each controller is reduced to 26 states. The design of a tracker requires the state space representation (A, B, C, D) matrices and a state observer gain matrix, L, for each controller, whose existence is assured due to the observability of each controller.

18.4.2 Fuzzy Controller

In this application, the fuzzy control algorithm serves to achieve three interrelated goals:

1. *To reduce the damage rate in the critical components while satisfying the plant performance requirements.* Since slow dynamics of the tracking signal (i.e. the load-following rate) may lead to a loss of dynamic performance, the fuzzy control law is

designed to make a trade-off between plant performance and structural damage during these transients, while taking into consideration the required plant load.

2. *To maintain robust stability of the gain scheduled control system.* As stated earlier, slow variations in the scheduling variable (i.e. the plant load demand in MW) facilitates stability during gain scheduling. The fuzzy controller makes use of the available plant information on-line to limit the plant load variations to maintain stability without any significant loss of performance.

3. *To avoid abrupt dynamic changes in plant variables during controller switching.* Abrupt changes, even if they do not result in instability, may cause considerable damage to the structural material of critical plant components. The fuzzy controller enhances the smoothness of the switching mechanism and the tracker.

The inputs to the supervisory controller have to be quantities that can be measured by sensors and perceived by a human operator. Similarly, the outputs of the supervisory controller must be quantities that a human operator should be able to manipulate to achieve the goals. A thorough knowledge of the plant dynamics and damage dynamics is required to make these choices. Therefore, the critical plant states and outputs that affect stability and structural damage need to be identified as inputs. The patterns and behavior of the outputs that lead to appreciable damage and instability need to be identified, observed, and incorporated into the membership functions and rule bases. A unique feature of the output membership functions is their flexible nature. The output membership functions are not fixed but are a function of the rate of change of load demand, i.e. the required ramp rate. The non-fuzzy input to the fuzzy control system is the sensor data, $y(t)$, which is readily available. The non-fuzzy output of the fuzzy system is the load ramp rate and is the gain scheduling variable. This choice provides a convenient means for achieving the first goal. The remaining two goals can also be achieved through fuzzy logic via judicious choice of the membership functions. For example, if the goal is to achieve a smooth load increase from 30% to 60% at an average rate of 10% full load per minute, the fuzzy controller may decrease the ramp rate below 10% at certain points to maintain stability or reduce damage. On the other hand, if the sensor-based information indicates a low damage rate and stable operation, then the load ramp rate can be safely increased.

We have discussed, in earlier sections, the advantages of using the plant sensor outputs as fuzzy inputs and also the effects of the Main Steam Temperature (THS) and Hot Reheat Temperature (TRH) on the plant in terms of damage and their ability to reflect overall plant stability. In contrast, stability is not very sensitive to the other two plant outputs, the Main Steam Pressure (PHS) and the generated load (JGN). The extent to which PHS can be varied and its rate of variation is not as critical for stability as the temperatures THS and THR. Therefore, two temperature outputs, THS and THR, are used to derived the fuzzy controller while the plant load is maneuvered in terms of JGN.

The effects of temperatures can be critical in two ways. First, a rapid change in steam temperature may cause large damage to the plant components due to the induced thermal stresses. It can also result in the state vector moving out of the region of attraction of the nominal trajectory that the controllers are required to follow, thus resulting in instability. Since a negative change in temperature is almost as harmful as a positive change, magnitudes (i.e. absolute values) of the rate of change of the two temperatures are therefore used as non-fuzzy inputs to the fuzzy controller. Even a slow increase or decrease in temperature (i.e. without having very high rates of change) may lead to instability by gradually taking the controllers away from their region of attraction. To circumvent this problem, magnitudes of the two output temperature errors are also used as fuzzy inputs.

The output of the fuzzy controller is the magnitude of the load ramp rate, which is the derivative of the reference signal, $y^{\text{ref}}(4)$. During transient operations, such as ramp-up or ramp-down, the two temperatures *THS* and *THR* are major indicators of the damage accumulation rates. In order to obtain a better control of the damage-causing variables, slowing down the process dynamics is the most natural action of the supervisory controller. This implies a reduction in the load ramp rate. On the other hand, a good temperature performance can leave sufficient margins to increase the ramp rate, which a supervisor might choose to do. This justifies the choice of absolute value of the ramp rate as the fuzzy controller output. Depending on whether the current operation involves a power increment or reduction, a prescribed change in the load ramp rate is added to or subtracted from the current reference power, $y^{\text{ref}}(4)$, to update it for the next sampling period. As mentioned earlier, the output membership functions themselves are functions of the load ramp rate demand.

For the two temperature errors, the same universe of discourse and membership functions are used because the process variable, *THS* and *THR*, are functionally similar. The same argument holds for the rates of change of these two temperatures. A third membership function is required for the output. The universe of discourse is defined as follows:

- *Absolute Value of the Rate of Change of Temperature:*
  universe of discourse $X_1 = [0, \infty)$,
  membership functions $r = \{r_1, \ldots, r_{n1}\}$, where $r_i : X_1 \rightarrow [0, 1]$, $i = 1, \ldots, n1$.

- *Absolute Value of Temperature Error:*
  universe of discourse $X_2 = [0, \infty)$,
  membership functions $E = \{E_1, \ldots, E_{n2}\}$, where $E_i : X_2 \rightarrow [0, 1]$, $i = 1, \ldots, n2$.

- *Absolute Value of Ramp Rates:*
  universe of discourse $X_3 = [0, RR_{\max})$,
  membership functions $RR = \{RR_1, \ldots, RR_{n3}\}$, where $RR_i : X_3 \rightarrow [0, 1]]$, $i = 1, \ldots, n3$.

Each membership function set has a cardinality of five, i.e. $n1$, $n2$ and $n3$ are all equal to five. $RR_{\max}$ is a design variable implying a maximum allowable rate of load ramp-up or ramp-down and is set to 1.1 MW/s (i.e. ~ 12.5% full load per minute).

The membership functions are shown in Figures 18.16, 18.17 and 18.18. Unlike the membership functions of temperature rate and temperature error, the membership function of the load ramp rate in Figure 18.18 is not uniformly spaced. The spacings in Figure 18.18 are arrived at via trial and error over extensive simulation runs, similar to what a human operator would like to do to obtain an understanding of the physical process. The triangular shape (i.e. piecewise linearity) of each membership function is chosen for mathematical simplicity and produces sufficiently good results. The membership functions can be interpreted in a manner similar to the ones in LECS-1, e.g. r_1 = very low rate of change of temperature, r_2 = low rate of change of temperature, and so on. Similar labels can be assigned to temperature error and load ramp rate.

The output membership functions are themselves a function of the required ramp rate. This is obtained from the operation strategy, which in turn is obtained from the rate of change of the load (required ramp rate). If the required ramp rate is denoted by RR_{REQ}, the actual values

370 LIFE-EXTENDING CONTROL OF MECHANICAL SYSTEMS

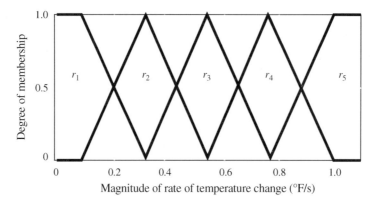

Figure 18.16 Membership functions for temperature rate of change

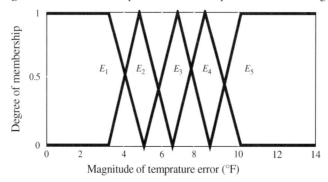

Figure 18.17 Membership functions for temperature error

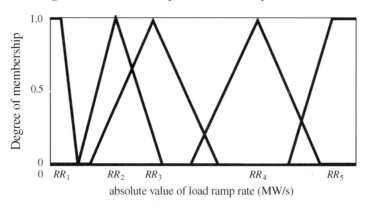

Figure 18.18 Membership function for load ramp rate. $RR_1 = 0.06\,RR_{REQ}$; $RR_2 = 0.3\,RR_{REQ}$; $RR_3 = 0.06\,RR_{REQ}$; $RR_4 = 0.9\,RR_{REQ}$; $RR_5 = 1.1s$

of the mean of each of the five output membership functions (Figure 18.18) are:

$RR_1 = 0.06\,RR_{REQ}$; $RR_2 = 0.3\,RR_{REQ}$; $RR_3 = 0.6\,RR_{REQ}$;
$RR_4 = 0.9\,RR_{REQ}$; $RR_5 = 1.1\,\text{MW/s}$.

These values are arrived at through trial and error and are based on human experience. They are specific to this power plant. These values can be modified very easily and on-line to achieve a different level of performance–damage trade-off. The highest value of 1.1 MW/s is fixed. This can be achieved only during near 'perfect' stability and very low damage rates. If the load demand changes quickly, then RR_4 might exceed 1.1 MW/s. If that happens the supervisor automatically reduces RR_{REQ} to 1.1 MW/s.

The membership functions are now combined into a set of fuzzy rules constituting a four-input–single-output fuzzy control system, with each input having a cardinality of five. This implies that there can be $5^4 (= 625)$ combinations of inputs and an If–Then rule is required for each combination. To simplify this situation, the fuzzy control system is partitioned into two parallel processing fuzzy systems, S1 and S2, as shown in Figure 18.19. The inputs to S1 are temperature rates and the output is the load ramp rate, while the inputs to S2 are the temperature errors and output is also the ramp rate. The junction '<' in Figure 18.19 represents an operation which picks the minimum of the two outputs, i.e. the slower ramp rate. Thus a conservative approach is adopted in order to simplify a large rule base possibly at the expense of performance. The advantage of this simplification is that, instead of 625 rules, two sets of 25 If–Then rules are now needed as listed in Table 18.2 and Table 18.3. For example, a rule *If r_1^{THS} and r_1^{THR} Then RR_5* represents: *If the rate of change of Main Steam*

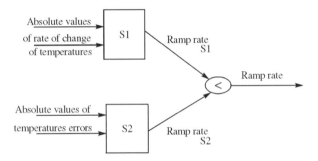

Figure 18.19 Parallel processing of the fuzzy control algorithm

Table 18.2 List of power plant input and output variables

No.	Input variables	Symbol	Unit	Output variables	Symbol	Unit
1	Governor valve area	AGV	–	Throttle steam temperature	THS	°F
2	Feedpump turbine valve area	APT	–	Hot reheat steam temperature	THR	°F
3	Fuel/air valve area	AFA	–	Throttle steam pressure	PHS	psi
4	Attemperator valve area	AAT	–	Electrical power	JGN	MW

Table 18.3 If–Then rules for temperature rate of change (fuzzy controller S1)

	r_1^{THS}	r_2^{THS}	r_3^{THS}	r_4^{THS}	r_5^{THS}
r_1^{THR}	RR_5	RR_4	RR_3	RR_2	RR_1
r_2^{THR}	RR_4	RR_4	RR_3	RR_2	RR_1
r_3^{THR}	RR_3	RR_3	RR_3	RR_2	RR_1
r_4^{THR}	RR_2	RR_2	RR_2	RR_2	RR_1
r_5^{THR}	RR_1	RR_1	RR_1	RR_1	RR_1

Temperature is very low and the rate of change of Hot Reheat Temperature is very low, then make the ramp rate very high.

The membership functions fuzzify the non-fuzzy inputs. The inference mechanism then determines the applicability of each rule to the present situation. The parameter λ_{ij} determines the applicability of each of the 25 rules to the present situation and takes a value in [0, 1] representing a measure of the amount the inputs satisfy the *if* part of the respective rule. The subscripts i and j represent the row and column of the rule matrices. For example, in Table 18.1, $\lambda_{ij} = \min\{r_i^{THS}, r_j^{THR}\}$ implies that λ_{ij} takes the minimum of the two values of the membership function involved in each rule.

The defuzzifier calculates one deterministic output in the form of the ramp rate. The output is calculated as a weighted average of the outcome of each rule ($RR_k, k = 1, 2, 3, 4, 5$) with the respective λ's as the weights. Since there are no probabilities associated with the fuzzy decision-making in the present controller, each outcome is concentrated on the geometric mean (i.e. center of gravity) of its membership function. The membership functions in Figure 18.18 are symmetric and the mean lies at the value with membership of one. Let the mean outcome of each rule be represented as $rr_{i,j}$, where $rr_{i,j}$ can take one of the five mean values depending on the outcome of the If–Then rule. Then, the final ramp rate is represented by

$$\text{ramp rate} = \sum_{i,j=1...5} \lambda_{ij} rr_{ij} / \sum_{i,j=1...5} \lambda_{ij}, \quad (18.20)$$

which is the weighted average of the geometric means of the output membership functions.

18.4.3 Results and Discussion

Simulation experiments are performed to demonstrate the performance, robustness and damage mitigating capabilities of LECS-2. This is accomplished by comparison of the plant dynamic performance under the following two feedback control configurations.

Case 1: Gain Scheduled without Fuzzy Logic ('gain sch.'). A combination of three gain scheduled controllers (without fuzzy logic) to operate over the entire range of plant operation.

Case 2: LECS-2/Gain Scheduled with Fuzzy Logic ('gain sch. with fuzzy'). A combination of three gain scheduled controllers along with a fuzzy logic to operate over the entire range of plant operation.

Each of these feedback control configurations is used in conjunction with the same feedforward policy. These two cases are compared based on output performance and structural damage. The plant performance requires generated plant load (JGN) to follow a predetermined trajectory. Each of the other three outputs, namely main steam temperature (THS), hot reheat temperature (THR) and main steam pressure (PHS) follow a trajectory based on the current plant load and is maintained within respective bounds. During these operations, damage accumulation in each of the main steam header, hot reheat header and superheater tubes is calculated using the damage models available from previous references. Simulation experiments are also performed to test the robustness of the control system under plant transients. Some of the plant parameters, time constants of valve dynamics, heat transfer coefficients, and turbine and pump efficiencies are perturbed

To test the closed-loop control system for nominal plant conditions a power ramp-up from 25% to 100% plant load is simulated. For the perturbed plant both power ramp-up and power ramp-down operations are performed. The recommended ramp rate is 10% per minute for both operations. This makes RR_{REQ} 0.875 MW/s. The desired operating conditions for the *THS*, *THR* and *PHS* at a given plant load (*JGN*) are a function of the *JGN*. The operating conditions are determined as the steady-state values of these outputs at the given plant load. At loads above 40% of the full power level, these set point values are maintained unchanged. The operating conditions for each load are as follows:

- 25% load − [*THS*, *THR*, *PHS*] = [935°F (501.7°C), 990°F (532.2°C), 2050 psi (14.13 MPa)];
- 30% load − [*THS*, *THR*, *PHS*] = [948°F (508.9°C), 998°F (536.7°C), 2285 psi (15.75 MPa)];
- 40% to 100% load − [*THS*, *THR*, *PHS*] = [950°F (510.0°C), 1000°F (537.8°C), 2415 psi (16.65 MPa)].

In between these loads the conditions are obtained through linear interpolation

Damage accumulation in each of the main steam header, hot reheat header and superheater tubes is measured. The steam temperatures and pressure, *THS*, *THR* and *PHS*, determine the stress conditions in the main steam and hot reheat steam headers as well as in the steam generator and reheater tubes and in the high pressure and intermediate pressure turbines. Each set of simulation experiments is performed by running LECS-2 ('gain sch. with fuzzy') first. This ensures a proper comparison of the performance and damage mitigation among the various control systems. The plots in the figures are marked with appropriate labels (e.g. 'gain sch.', and 'gain sch. with fuzzy' for LECS-2) to indicate different controller configurations under both ramp-up and ramp-down operations. The term 'Ref. Traj.' in the figures denotes 'Reference Trajectory'.

It takes 738 s for the ramp-up operation with an average ramp rate of 6.1% (of full load) as seen in Figure 18.20. The desired ramp rate was 10% per minute. Thus the fuzzy controller has modified the load ramp rate and reduced it in order to ensure stability and low damage rate. The plots in these figures show the respective initial steady-state loads held for the first 100 s, to demonstrate the absence of any initial (non-steady-state) transients. Similarly, the final steady states are held for an extended period of time to exhibit stability. In Figure 18.20 'gain sch. with fuzzy' outputs show excellent behavior for the steam temperature and pressure transients, *THS*, *THR* and *PHS*, that are directly responsible for damage reduction. The gain scheduled system has very large temperature oscillations, which can lead to catastrophic failures in the turbine casing. The load-following performance of both control systems is comparable. However, the responses of the other three outputs, *PHS*, *THS* and *THR*, are superior to those under the 'gain sch. with fuzzy' controller. The load-following performance can be improved by updating the fuzzy membership functions and allowing larger ramp rates in the membership functions of the plant outputs. For each case, the changes involve a trade-off, which is the designers' decision.

Figure 18.21 compares the damage under a ramp-up operation. The operation is preceded by 1000 s of steady state and followed by another 2000 s. This ensures that any delayed dynamics in damage will show up during steady-state operation. For each of the critical components, the 'gain sch. with fuzzy' controller yields better damage control. Maximum damage reduction takes place in the main steam header, because it is a thick pipe and is more prone to thermal stresses arising from larger temperature gradients across the wall. The hot

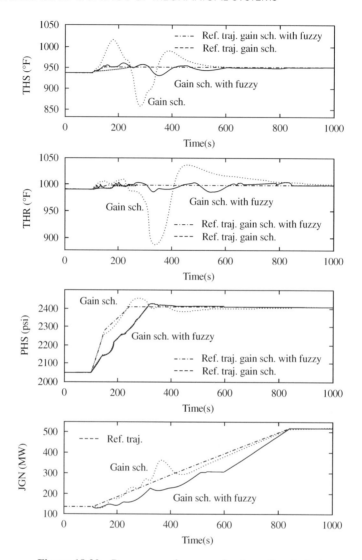

Figure 18.20 Ramp up performance for the national plant

reheat header and superheater tubes, on the other hand, are thinner pipes and damage is mainly due to the temperature and non-temperature gradients. For superheater and other steam generator tubes, the main cause of damage is the fireball size in the furnace, which is primarily responsible for the transfer of (radiant) thermal energy to the tubes. The fireball size is controlled by the air–fuel valve. Under nominal plant operations the feedforward control input to this valve is carefully designed to avoid any sudden change in fireball size and the feedback signal is responsible for fine-tuning only. However, a slight disturbance or perturbation can change the situation.

Simulation experiments are also conducted on the plant model with injected perturbations in order to test the robustness of the control system. The following perturbations are introduced:

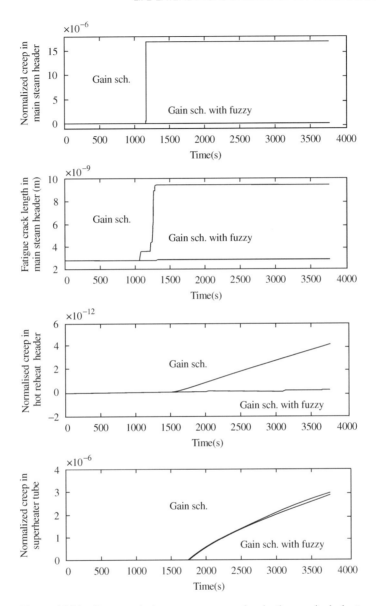

Figure 18.21 Damage during ramp-up operation in the nominal plant

- a 3% decrease in the efficiencies of all turbines and pumps due to structural degradation of rotating components;
- a 3% decrease in the heat transfer coefficients in the steam generator and reheater tubes resulting from possible scale formation on the inside wall;
- a 25% increase in the time constants of the governor, feedpump turbine and fuel–air valves due to possible degradation of the actuator components.

Figure 18.22 shows the ramp-down outputs for the perturbed plant under 'gain-sch.' and 'gain-sch. with fuzzy' controllers. A comparison of the two controllers in Figure 18.14

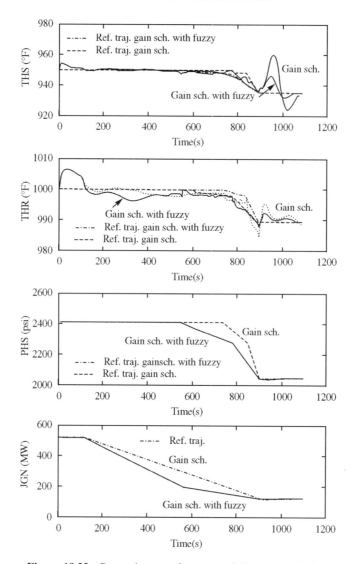

Figure 18.22 Ramp-down performance of the perturbed plant

shows an improvement in the steam temperatures, *THS* and *THR*, for 'gain sch. with fuzzy', especially at the end in terms of overshoot and oscillations. The response of the steam pressure, *PHS*, is also good for both cases. The generated power, *JGN*, exhibits the trade-offs. Without the fuzzy controller, the transient response of *JGN* is much better, while other variables, including damage in Figure 18.23, become worse. This also demonstrates the basic idea of the fuzzy controller. When the steam temperatures are well within specified limits, the fuzzy controller increases the ramp and when the temperatures begin to oscillate.

Figure 18.23 shows the damage for both controllers. Similar to the results in Figure 18.21, the damage is less for the 'gain sch. with fuzzy' system. But, unlike Figure 18.21, there is a marked improvement in damage control for superheater tubes. This is because, as mentioned in the previous section, damage mitigation is largely accomplished by the feedforward action

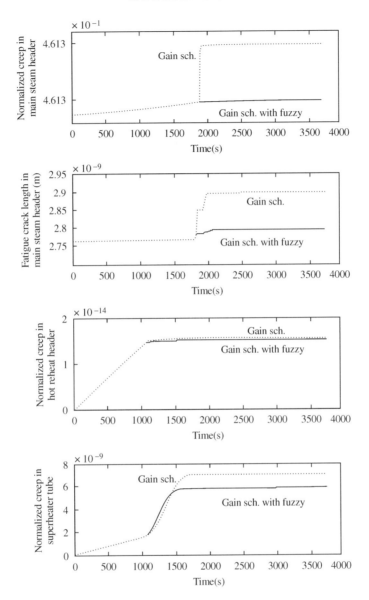

Figure 18.23 Damage during ramp-down in the perturbed plant

for the nominal plant. In contrast, for the perturbed plant, the feedforward action is no longer accurate and consequently the feedback action plays a relatively larger role. Thus, during ramp-down, for the perturbed plant, the 'gain sch. with fuzzy' system yields both better performance and damage control than the 'gain sch. with fuzzy' system yields both better performance and damage control than the 'gain sch.' system with a trade-off in load rate.

Figure 18.24 shows the ramp-up operation for the perturbed plant. While 'gain sch. with fuzzy' performs reasonably well, the control system becomes unstable without the fuzzy controller. The rationale for this observation is as follows. As the system starts to move away from the reference points, the fuzzy controller slows down the ramp rate and thereby the rate

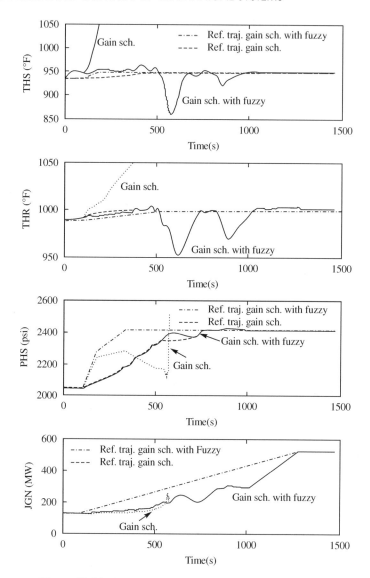

Figure 18.24 Ramp-up performance of the perturbed plant

Table 18.4 If–Then rules for temperature error (fuzzy controller S2)

	E_1^{THS}	E_2^{THS}	E_3^{THS}	E_4^{THS}	E_5^{THS}
E_1^{THR}	RR_5	RR_4	RR_3	RR_2	RR_1
E_2^{THR}	RR_4	RR_4	RR_3	RR_2	RR_1
E_3^{THR}	RR_3	RR_3	RR_3	RR_2	RR_1
E_4^{THR}	RR_2	RR_2	RR_2	RR_2	RR_1
E_5^{THR}	RR_1	RR_1	RR_1	RR_1	RR_1

of change of the plant load is reduced and stability is maintained. This is in accordance with the statement in Section 18.4.1 that slow variation of the gain scheduling variable, in this case the plant load, ensures stability. This observation clearly demonstrates the effectiveness of fuzzy logic in keeping the control system robust.

18.5 SUMMARY AND CONCLUSIONS

This chapter presents a procedure for the synthesis of fuzzy-logic-based life-extending control systems (LECS) over a wide operating range where the objective is to achieve a desired trade-off between performance and damage via feedback of on-line measurements of damage and damage rate. The proposed control system has a two-tier structure. Linear sampled-data controller(s) track a reference trajectory vector in the lower tier while a fuzzy-logic-based damage controller at the upper tier makes a trade-off between system performance and damage in critical plant components.

Two different configurations for the LECS architecture are proposed to enhance life extension capabilities, and increase stability robustness. The first configuration requires either on-line measurement(s) of actual structural damage or their estimation(s). The second configuration uses the plant process variables as indicators of structural damage and damage rate in the absence of on-line damage information. As such, the second configuration is ideally suited for large systems with many critical components, operating under extreme temperature and pressure conditions that make it difficult to measure structural damage and the large number of components results in on-line execution of too many damage models. Otherwise, these two configurations are structurally similar (e.g. both having a two-tier architecture). The inner tier/loop in both configurations uses linear feedback control of non-linear plants. The outer loop uses a fuzzy-logic-based controller for performance-damage trade-off. Syntheses of both systems require knowledge of the plant operations and the dynamics of both plant and structural damage. Robust wide-range controllers are designed for a rocket engine (similar to the Space Shuttle main engine) and a commercial-scale fossil-fuel power plant based on the first and second configurations, respectively.

The synthesis procedure, following the first configuration, is demonstrated by simulation experiments on the model of a reusable rocket engine (similar to the Space Shuttle main engine) where the fatigue crack damage in the turbine blades limits their service life. The performance variables of the rocket engine are expressed in terms of transients of main thrust chamber pressure and oxygen/hydrogen (O_2/H_2) mixture ratio. The results of simulation experiments for upthrust transients of the rocket engine operation show that the fuzzy controller is capable of regulating the performance/damage trade-off in the turbine blades. The fuzzy controller chooses the reference signal for the pressure ramp rate to be high at the beginning of the transients followed by a reduction in the ramp rate as the target pressure is approached. Since the fuzzy controller makes the decision to change the pressure reference signal as a function of on-line damage prediction, it can react to unexpected situations, such as damage caused by external disturbances that cannot be predicted a priori. From the perspective of the trade-off between performance and damage, the simulation results indicate that dynamically maneuvering the ramp rate of the chamber pressure reference signal under fuzzy control is superior to the application of a constant ramp rate throughout the maneuver.

The second control configuration in a commercial-scale fossil power plant leads to dynamically smooth switching between controllers under gain scheduling. It enhances robust stability of wide-range plant operations and simultaneously reduces structural damage. If the knowledge of the power plant and damage dynamics is available, the fuzzy controller is easy to design, implement and modify. It does not require any direct use of the non-linear plant model and damage models. The design of the fuzzy supervisor is based on a particular trade-off between load-following and damage rate. This level of trade-off can be altered by using a combination of the following strategies.

- Changing the mean values of the ramp rates (Figure 18.18): the mean values of the ramp rates are increased for better performance and they are reduced for improved stability and life extension. This is the easiest modification and can be executed on-line.
- Changing the input membership functions: the permissible errors and rates of change are reduced to maintain the damage rates low. They are increased for better performance.

Based on the results of simulation experiments on the power plant, it is apparent that there is virtually no trade-off in damage mitigation among the major critical components of the power plant under consideration. Reduction of damage in one component does not lead to increase in damage in other components. It also establishes the overall superiority of gain scheduling with fuzzy control, especially for power ramp-up operations.

The concept of life extending control is of significant importance to the operation of any plant where structural integrity is a critical issue. By including the on-line information of structural damage in the control scheme, not only the service life of the controlled process is extended, but also the mean time between major maintenance actions can be increased. For example, during a particular mission, if the damage rate in a plant component exceeds the expected level, it may be possible to modify the operation of the plant on-line so that the current mission can be completed with an acceptable amount of damage accumulation. The trade-off could be a (possibly) small reduction in the plant performance. Changing the operation of the plant in this situation may also prevent a potentially catastrophic situation caused by the failure of a critical plant component. The implementation of damage mitigating control has the potential of providing both economic benefits and enhancement of operational safety.

18.5.1 Control System Stability

Stability of the closed-loop system is the single most important requirement of any controller design. While a necessary and sufficient condition for stability is easily obtained from the eigenvalues of the A-matrix in finite-dimensional linear time-invariant systems, there is no such straight-forward condition for stability of non-linear and/or time-varying systems.

In the first configuration of the control architecture, fatigue damage models induce severe non-linearities, control systems containing a damage model in the feedback loop are non-linear, and, in some cases, time-varying as well. The inner control loop containing the discrete time tracking controller is a linear sampled-data system which has been shown to be robustly stable with respect to the uncertainty description shown in Figure 18.3. To avoid the possibility of any signal in the two-tier control system in Figure 18.1 becoming unbounded, the damage controller in the outer loop has been given limited authority in the sense that there are bounds on the ramp rates being generated by the fuzzy controller. In essence, the fuzzy controller does not have the ability to choose a chamber pressure reference signal that will cause the inner control loop to become unstable. However, this rate limit does not establish stability of the control system in the sense of Lyapunov. For example, there is no guarantee that phenomena like limit cycling of the reference signals will not occur. Analytical methods which may be successful in proving the stability and absence of limit cycles in damage-mitigating control systems include describing function and the absolute stability methods of Lur'e and Postnikov (Vidyasagar [23]). In the second configuration of the control architecture, the responsibility of stability is assigned to the outer loop fuzzy controller. Although stability is not guaranteed, the fuzzy controller actually adds robustness to the system as seen in Figure 18.24.

ACKNOWLEDGMENTS

The authors acknowledge the benefits of discussions with their colleagues at the Pennsylvania State University and acknowledge the support of National Science Foundation and NASA Lewis Research Center. The research work reported in this chapter is supported in part by National Science Foundation under Grant No. DMI-9424587, National Science Foundation under Grant No. CMS-9531835, and NASA Lewis Research Center under Grant No. NAG-3-1240.

APPENDIX A: BRIEF DESCRIPTION OF THE ROCKET ENGINE

The reusable rocket engine under consideration is similar to the Space Shuttle main engine (Ray and Dai [1995]). The propellants, namely liquid hydrogen fuel and liquid oxygen, are individually pressurized by separate closed cycle turbopumps. Pressurized cryogenic fuel and oxygen are pumped into two high-pressure preburners which feed the respective turbines with fuel-rich hot gas. The fuel and oxidizer turbopump speeds and hence the propellant flow into the main thrust are controlled by the respective preburner. The exhaust from each turbine is injected into the main combustion chamber where it burns with the oxidizer to make the most efficient use of the energy. The oxygen flow into each of the two preburners are independently controlled by the respective servo-valves while the valve position for oxygen flow into the main thrust chamber is held in a fixed position to derive the maximum possible power from the engine. The plant outputs of interest are the oxygen/hydrogen (O_2/H_2) mixture ratio and main thrust chamber pressure which are closely related to the rocket engine performance in terms of specific impulse, thrust-to-weight ratio, and combustion temperature.

A finite-dimensional state-space model of the rocket engine has been formulated via lumped parameter approximation of the partial differential equations describing mass, momentum, and energy conservation. The plant model is constructed by causal interconnection of the primary subsystem models such as main thrust chamber, preburners, turbopumps, valves, fuel and oxidizer supply header, and fixed nozzle regeneration cooling. The plant dynamic model consists of twenty state variables, two control inputs, and two measured variables (Ray and Dai [17]). The structural damage model of blades, made of a nickel-based superalloy, in each of the fuel and oxidizer turbines calculates the cyclic mechanical stresses at the root of a typical blade which is presumed to be a critical point in this study. The blade model for each of the two turbines is reported by Ray and Wu [15].

APPENDIX B: BRIEF DESCRIPTION OF THE POWER PLANT

The power plant dynamics under consideration is a fossil-fueled, generating unit having the rated capacity of 525 MWe. The plant dynamics have been represented by a 27th order nonlinear state-space model, which is described in detail by Weng *et al.* [24]. The following four valve commands are selected as control inputs (Table 18.1): high-pressure turbine governor valve area (AGVR); feedpump turbine control valve area (APTR); furnace fuel/air valve area (AFAR); and reheat spray attemperator valve area (AATR). The measured plant outputs are throttle steam temperature (*THS*), hot reheat steam temperature (*THR*), throttle steam pressure (*PHS*), and electric power (*JGN*). LECS-2 is used for life extension of three critical power plant components.

1. *Main Steam Header.* It is made of 2.25% chromium and 1% molybednum ferritic steel. It has an inner diameter of 4.5 in. (114.3 mm) and an outer diameter of 7.2 in. (182.9 mm). The main steam header is thermally insulated. Damage in the main steam header results from fatigue cracking and thickness reduction due to creep. Maximum growth occurs on the outer surface. All crack growth is monitored as growth in the radial direction, i.e. a crack starts on the outside and propagates inside. An initial value of the crack length is assumed. Normalized creep is calculated as the reduction in header thickness per unit original thickness and is designated 'Creep Thinning'.

2. *Hot Reheat Header.* It is made of 2.25% chromium and 1% molybednum ferritic steel. It has an inner diameter of 6.5 in. (152.4 mm) and an outer diameter of 7.0 in. (177.8 mm). It is also thermally insulated to reduce heat loss to the atmosphere. Damage in the hot reheat header is predominantly due to creep and is represented in a fashion identical to the creep damage in the main steam header.

3. *Superheater.* Each superheater tube is made of 2.25% chromium and 1% molybednum ferritic steel. It has an inner diameter of 1.0 in. (25.4 mm) and an outer diameter of 1.75 in. (44.45 mm). Damage in the superheater tubes is due to creep and is represented in a fashion identical to the creep damage in the main steam header.

The steam temperatures and pressure, *THS, THR* and *PHS*, determine the stress conditions in the main steam and hot reheat steam headers as well as in the steam generator and reheater tubes and in the high pressure and intermediate pressure turbines.

REFERENCES

1. Balas, G. J., J. C. Doyle, *et al.*, *μ-Analysis and Synthesis Toolbox*, The Math Works, Natick, 1993.
2. Bamieh, B. A. and J. B. Pearson, A general framework for linear periodic systems with applications to H_∞ sampled- data control, *IEEE Transactions on Automatic Control*, **37**(4), 1992, 418–435.
3. Chen, C.-T., *Linear System Theory and Design*, Harcourt Brace College Publishers, Fort Worth, 1984.
4. Dai, X. and A. Ray, Life prediction of the thrust chamber wall of a reusable rocket engine, *AIAA Journal of Propulsion and Power*, **11**(6), 1995, 1279–1287.
5. Dai X. and A. Ray, Damage-mitigating control of a reusable rocket engine: Parts I and II, *ASME Journal of Dynamic Systems, Measurement and Control*, **118**(3), 1996, 401–415.
6. Graebe, S. F. and A. L. B. Ahlen, Dynamic transfer among alternative controllers and its relation to antiwindup controller design, *IEEE Transactions on Control Systems Technology*, **4**(1), 1996, 92–99.

7. Holmes, M. S. and A. Ray, Fuzzy damage mitigating control of mechanical structures, *ASME Journal of Dynamic Systems, Measurement and Control*, **120**(2), 1998, 249–256.
8. Jamshidi, M., N. Vadiee and T. J. Ross (eds.), *Fuzzy Logic and Control*, 1st edn., Prentice-Hall, Engelwood Cliffs, 1993.
9. Kallappa, P., M. S. Holmes and A. Ray, Life-extending control of fossil fuel power plants, *Automatica*, **33**(6), 1997, 1101–1118.
10. Kallappa, P. and A. Ray, Fuzzy wide range control of power plants for life extension and performance enhancement, *Automatica*, **36**(1), 2000, 69–82.
11. Lele, M., A. Ray and P. Kallappa, Life extension of superheater tubes in fossil power plants, *ISA 96 POWID Conference*, Chicago, 1996.
12. Lorenzo, C. F. and W. C. Merrill, Life extending control: A concept paper, *Proceedings of the American Control Conference*, Boston, 1991, pp. 1080–1095.
13. Noll, T., E. Austin, S. Donley, G. Graham, T. Harris, I. Kaynes, B. Lee and J. Sparrow, Impact of active controls technology on structural integrity, *Proceedings of the Thirty-second AIAA/ASME/ASCE/AHS/ASC Structures, Structural Dynamics and Materials Conference*, Baltimore, 1991, pp. 1869–1878.
14. Pedrycz, W., *Fuzzy Control and Fuzzy Systems*, 2nd edn., John Wiley, New York, 1992.
15. Ray, A. and M.-K. Wu, *Damage-Mitigating Control of Space Propulsion Systems for High Performance and Extended Life*, NASA Contractor Report 19440 under Lewis Research Center Grant NAG 3-1240, 1994.
16. Ray, A. M.-K. Wu, M. Carpino and C. F. Lorenzo, Damage-Mitigating Control of Mechanical Systems: Parts I and II, *ASME Journal of Dynamic Systems, Measurement and Control*, **116**(3), 1994, 437–455.
17. Ray, A. and X. Dai, *Damage-Mitigating Control of a Reusable Rocket Engine*, NASA Contractor Report 4640 under Lewis Research Center Grant NAG 3-1240, 1995.
18. Ray, A. and S. Tangirala, A nonlinear stochastic model of fatigue crack dynamics, *Probabilistic Engineering Mechanics*, **12**(3), 1997, 33–40.
19. Rugh, W. J., 1991, Analytical framework for gain scheduling, *IEEE Control Systems Magazine*, **11**(1), 1991, 79–84.
20. Shamma, J. S. and M. Athans, Guaranteed properties of gain scheduled control for linear parameter-varying plants, *Automatica*, **27**(3), 1991, 559–564.
21. Sivashankar, N. and P. P. Khargonekar, Robust stability and performance analysis of sampled-data systems, *IEEE Transactions on Automatic Control*, **38**(1), 1993, 58–69.
22. Tangirala, S., M. Holmes, A. Ray and M. Carpino, Life-extending control of mechanical structures: Experimental Verification of the concept, *Automatica*, **34**(1), 1998, 3–14.
23. Vidyasagar, M., *Nonlinear Systems Analysis*, 2nd edn., Prentice-Hall, Englewood Cliffs, 1992.
24. Weng, C.-K., A. Ray, and X. Dai, Modelling of power plant dynamics and uncertainties for robust control synthesis, *Applied Mathematical Modelling*, **20**, 1996, 501–512.
25. Weng, C.-K. and A. Ray, Robust wide range control of steam-electric power plants, *IEEE Transactions on Control Systems Technology*, **5**(1), 1997, 74–88.
26. Yen, J., R. Langari and L. A. Zadeh, *Industrial Applications of Fuzzy Logic and Intelligent Systems*, IEEE Press, Piscataway, 1995.
27. Zhou, K., J. C. Doyle and K. Glover, *Robust and Optimal Control*, Prentice-Hall, Upper Saddle River, 1996.

Epilogue

As complexity rises, precise statements lose meaning and meaningful statements lose precision.

L.A. Zadeh

So far as the laws of mathematics refer to reality, they are not certain. And so far as they are certain, they do not refer to reality

A. Einstein

Noncontroversial is only that which is not of interest to us.

H. Laube

This volume touches on so many issues pertaining to the design and analysis of fuzzy control systems. The problems or examples provided in the chapters are of varied interests and degrees of difficulty. They are not necessarily the world's most interesting or difficult problems. The objective at the outset was never to be that absolutely global, if there is such an ambitious endeavor. They are nonetheless important problems in their respective domains. Moreover, the problems tackled illustrate how various approaches, fuzzy or otherwise may be applied in the modeling, analysis and synthesis of closed-loop systems that are controlled via fuzzy logic and fuzzy sets. Now, as a final point, the audience needs to be cautioned that the coverage is by no means exhaustive. Nevertheless, the essence should be clear. As control engineers and scientists, fuzzy control is not viewed as some sort of "voodoo" control methodology that would have to devise its own "off-the-wall" tools for analysis so that others may find it difficult to comprehend or find it fuzzy at best. And by so doing, one finds the impetus to claim completeness or even a basis for acceptability. Not even so for design. On the contrary, the chapters in this volume, where applicable, show that the same tools that exist for linear and nonlinear control systems design and analysis, conventional or modern, are out there to be used as long as the problems at hand can be prudently cast in the appropriate frameworks.

We have come a long way in automatic control: From the first application of the float regulator mechanism (Ktesibios, Greece 300-1 B.C.), the oil lamp (Phylon, 250 B.C.) that used this regulator, the temperature regulator (Drebbel, Holland, 1572), the pressure regulator (Papin, 1681), the flyball governor (Watt, 1769), the flyball stability analysis (Maxwell, 1868), stability from differential equations (Routh-Hurwitz, 1877), the mechanized assembly operation (Henry Ford, USA, 1913), feedback amplifiers (Bode, 1927), stability analysis (Nyquist, 1932), root locus (Evans, 1948) ... the state variable and optimal control (Kalman, 1960) ... the microprocessor (Hoff, 1969) and to robust control – H-infinity and Mu synthesis in the 1980's (Doyle, Khargonekar, Glover, Francis), to mention just a few. The emergence of fuzzy control systems since the late 1970's to the present day should be viewed only as yet another advance and enrichment, no more, no less, to the field of control engineering, and a major component in intelligent decision and control.

Shehu S. Farinwata
Dearborn, Michigan, USA
January 2000

Index

accumulated fatigue, 362
ad hoc analysis, 296
adaptive control, 284
adaptive FLC, 310
adaptive fuzzy control, 310
adaptive law, 288
adaptive learning, 48
admissible, 206
Advanced Technology Wing, 221,
algebraic loops, 127
approximate model, 204
approximation ability, 75
approximation error, 340
ARMA model, 323, 324
artifical neural network, 73
asymptotic stability, 95, 214
asymptotically stable, 118, 326
attraction basin, 146

backpropagation, 78
Barbalat's lemma, 207, 316
basis functions, 344
branch-and-bound, 32

causality, 145
cell partition, 203
cell-to-cell mapping, 228
cell-wise linearization, 165, 210
chaotic, 318
chattering, 314
chromosome encoding, 5
Class-K functions, 211
closed-loop stability, 241
closed-loop systems, 214
closeness of solutions, 339
clustered rulebase, 184
clustering, 76
co-dimension, 335
companion form, 295
compatibility, 29

condition number, 180
control horizon, 30
control law, 341
control objective, 183, 338
control performance, 256
control profile, 218, 219
convergence rate, 61, 79
convergence, 100
convex constraints, 31
convex hull, 126
coprime, 328
covering, 168, 208
cross sensitivity, 175, 189, 191

damage mitigation, 349
damage rates, 351
data mining, 166
data-driven, 23, 47
de facto stabilizer, 169
decay rate, 253, 258
decision process, 182
degree of fulfillment, 205, 230
degree of overlapping, 85
describing function, 145
destabilizing effects, 145
diffeomorphism, 210
diminishing variations, 186
Diophantine equations, 328
direct feed, 211
direct Lyapunov method, 103, 118
disk-and-pendulum, 342
dissipative map, 204
dissipative mapping, 208, 214
dissipative parameter, 129
dissipative, 122, 213
distance functions, 8
disturbance rejection, 253, 259
Duffing system, 317
dynamic fuzzy logic system, 284
dynamic performance, 349

dynamic system, 117

effective gains, 232
eigenvalue problem, 129
eigenvalues, 148, 177, 213, 216
eigenvector, 177
emerging technologies, 166
engine model, 182
engineered system, 197
equality constraint, 64
equilibrium fuzzy set, 106
error sensitivities, 170
estimation error, 250, 288
Euler-Lagrange systems, 114, 135, 137
excess if passivity, 119
exponential stability, 209
exponentially stable, 315
extension principle, 97, 99

feasible set, 206
feedback gains, 257, 263
feedback interconnection, 209
feedforward, 376
firing strengths, 116
fixed point, 167, 206
frequency response, 145
fuzziness, 21
fuzzy basis function, 311
fuzzy clustering, 25, 27, 35, 39
fuzzy clusters, 323,
fuzzy composition, 98
fuzzy control algorithm, 168
fuzzy control problem, 206
fuzzy controllers, 3
fuzzy coordinates, 210
fuzzy dynamic system, 95, 104
fuzzy model, 35, 47, 54, 60, 254
fuzzy multilayer perceptron, 74
fuzzy observer, 240, 244, 253, 262
fuzzy partitioning, 322
fuzzy partitions, 53
fuzzy regulator, 253, 255
fuzzy relation, 75, 97
fuzzy robustness, measure, 180, 181, 185
fuzzy rule base, 356
fuzzy rules, 74
fuzzy sensitivity, 189
fuzzy variable, 115
fuzzy weights, 116

gain function, 295

gain scheduling, 321, 351, 364
Gaussian function, 98
Gaussian, 51, 293
Gauss-Seidel, 64
generalized deflection command, 223
gershegorin theorem, 148
GEVP, 263
global asymptotic stability, 229
global gain, 6
global learning, 55, 56
global optimum, 57
global stability, 145, 213, 214, 216, 283
globally asymptotically stable, 218
globally bounded, 210
GMDH, 74
gustafson-Kessel clustering, 43

harmonic balance, 149, 150
high gain control, 336
HVAC system, 34
hybrid, 169
hyperstability, 114

identification error, 292
idle speed controller, 182
implicit states, 168
impulse response, 332
incremental control, 40
induction, 105
industrial furnace, 344
industrial processes, 338
inequality bounds, 165
information granularity, 3, 4, 11
information processing, 4
inner product, 340
input space 5, 16
input-output mapping, 233, 336
input-output stability, 118, 203
input-output, 52
integral sliding-mode, 310, 313
intelligent control, 166
interaction, 174
interior-point method, 268
internal stability, 150
interpolation, 321
intrinsic uncertainties, 113
invariance principle, 118
invariant rule, 219
invariant set, 95, 106
inverted pendulum, 249, 279

Jacobian matrix, 147

Kalman Filter, 244,
Kalman-Yakubovich, 212, 221

Lagrangian, 104
learning algorithm, 341
learning control, 314
learning epoch, 58
learning law, 314, 315
learning rule, 57, 59
least-squares, 27
levels of identification, 52
life-extending control system, 349
limit cycle, 145
limit cycling, 381
limit theorem, 160
limiting structures, 153
linear matrix inequality, 239, 243, 253
linear subsystems, 322
linguistic granule, 4, 13
linguistic information, 3, 49, 283
Lipschitz continuous, 141
LMI approach, 168
LMI problem, 233
LMI regions
local controller, 116
local data set, 57
local learning, 48, 63, 65
local mean squares, 56
Lukasiewicz fuzzy logic, 153, 157
Lyapunov equation, 311
Lyapunov function candidate, 296, 302, 305, 311, 314
Lyapunov function, 118, 239
Lyapunov inequalities, 242
Lyapunov stability, 95, 104, 203, 241, 255
Lyapunov synthesis, 287
Lyapunov-like lemma, 207

Mamdani controller, 131
Mamdani fuzzy controller, 153
manipulators, 335
matching conditions, 336
maximal firing level, 333
mean defuzzification, 140
mean Value theorem, 107
memoryless, 147
minimum eigenvalue, 303
missile autopilot, 214
model predictive control, 30

model-free estimator, 78
modeling error, 302
monic polynomial, 325
multi-level relay, 153, 161, 162
multimodel structure, 321

neural networks, 166
neuro-fuzzy, 54
nominal bound, 297
nominal dynamics, 315
nominal stability, 173
nominal system, 338
nominal values, 170, 185
non-active rules, 67
non-autonomous, 241
non-expansive, 213
non-linear control, 203
non-linear matrix inequality, 268
non-linear PID, 157
non-linear stability, 113
non-parametric, 333
normal form, 341
numeric manifestation, 5
numerical information, 283

observer gains, 246, 263
off-line learning, 48
open-loop stable, 211
optimal controller, 146, 197
optimum parameters, 181
optimum sensitivities, 178
output feedback, 320
output tracking, 335
overlapping coefficient, 87

parallel distributed compensation, 253, 255
parameter error vector, 315
parameter perturbation, 165, 172, 196
parameter variations, 88, 165
parameter vector, 288, 310
parametric uncertainties, 276, 278
parametric uncertainty, 336
Parseval theorem, 121
partial fuzzy model, 62
passivity theory, 203, 212
passivity, 113, 117, 126
PDC, 267, 268
performance assessment, 166
performance specifiers, 166
persistently exciting, 315, 341
pole placement, 274

polytope, 324
polytopic differential inclusions, 117, 128, 129
polytopic, 239, 322
positive real lemma, 212
positive real, 121
possibility measure, 18
power plant, 362, 382
prediction horizon, 23, 30
predictive control, 23
projection algorithm, 312
projections, 27
proportional-derivative control, 224
prototype control, 15

quadratic performance function, 261
qualitative information, 335
quasi-optimal, 90

radial basis function, 74, 77
Rayleigh constant, 213
Rayleigh quotient, 213
Rayleigh, 138
rectangular grid, 208
recursive equation, 66
recursive least squares, 63
redundancy, 27
regression matrix, 26
reinforced learning, 66
relative degree, 342
relaxed stability, 257
reliability, 349
reusable rocket engine, 359, 381
robust analysis, 149
robust linear control, 364
robust stability, 165, 355
robustness analysis, 168
robustness measure, 165
robustness set, 169
robustness, 4, 13
robust-optimal, 261
rocket engine, 352
Rolle's theorem, 106

safety, 349
schur stability, 324
schur-Cohn stability, 325
sensitivities, 165, 166, 172, 178, 186
sensitivity matrix, 173
separation principle, 262
separation property, 246
shadowed sets, 9

shift of origin, 207
shortage of passivity, 119
similarity measure, 28
similarity-driven, 24
singular points, 335
singular value decomposition, 63
singular values, 149, 165, 179, 180
singular, 179
sliding surface, 313
small changes, 173
small perturbations, 185
space Shuttle, 381
stability analysis, 227, 267
stability conditions, 324
stability convergence, 176, 204
stability criterion, 229, 269, 277
stability LMIs, 271
stability properties, 100, 291
stability regions, 217
stability, 113, 117, 145, 253, 239, 256, 380
state-feedback, 241
static system, 129
stochastic models, 97
structural degradation, 349
structural model, 352
structural parameters, 49
structure organizer, 82
sub-manifold, 335
sufficient condition, 242, 269
supervisory control, 287
supervisory controller, 368
surface deflection, 223
switching gain, 313

Takagi-Sugeno, 26, 113, 115, 126, 239, 240, 253, 254, 267, 268, 321
Takagi-Sugeno-Kang, 321
time-varying, 380
t-norms, 51, 78, 290
total sensitivity matrix, 175
tracking controller, 352
tracking error, 292, 339
training algorithm, 80
training increment, 297
training law, 291, 293
triangular co-norm, 96
triangular norm, 96
TSK compensators, 330, 327, 333
TSK model, 47, 322

uncertain blocks, 259

uncertain systems, 95
uncertainty, 3
unforced system, 207
uniform boundedness, 208
univariate, 77
universal approximators, 47, 133
unmodeled dynamics, 343, 365
update law, 316, 341

validity indices, 36

wind-tunnel model, 223

yaw axis, 214

zero-shifting, 120
zero-state detectable, 120